PROGRESS IN VIRUS RESEARCH

PROGRESS IN VIRUS RESEARCH

SABINE THEBAULT
Editor

Nova Biomedical Books
New York

For permission to use material from this book please contact us:
Telephone 631-231-7269; Fax 631-231-8175
Web Site: http://www.novapublishers.com

Library of Congress Cataloging-in-Publication Data
Available upon request

ISBN 1-59454-438-7

Published by Nova Science Publishers, Inc. ✣New York

Contents

Preface

The first part of the book deals with several aspects of different viruses : viral structure, function, replication and interplay between the virus and the host. Six viruses are used as examples, four RNA viruses (HTLV-I, HIV-1, MMTV and coxsackievirus B4) and two DNA viruses (EBV and KSHV). The second part of the book is devoted to the use of the knowledge on viruses to practical applications and also to the characterization of HIV inhibitors. Reviewing the results of research on different viruses is important since, although viruses possess vast degrees of complexity, they also share similar features. In addition, viruses are more and more used as models to solve molecular biology problems.

Human T-cell leukemia virus type I (HTLV-I) is a complex retrovirus first identified in 1980. It is estimated that between 15 and 20 million people are HTLV-I carriers world-wide. The primary routes of HTLV-I transmission are: (i) transmission from mother to child through maternal HTLV-I-infected T cells in breast milk, (ii) sexual transmission from male to female, male to male, and rarely female to male, (iii) transmission through the transfusion of blood containing infected T cells, and (iv) transmission among drug users through shared needles. The transmission of HTLV-I is quite similar to that of the human immunodeficiency virus (HIV) although HIV is more pathogenic since only 5% to 10% of HTLV-I-infected people develop pathologies. Prevalence rates for HTLV-I infection in the general population are greater than 1% in the Caribbean Basin, Central Africa, and South Japan. In most other areas in the world, HTLV-I infections are mainly found in immigrants from endemic areas and in intravenous injection users. As reported in chapter I, a high rate of infection was observed in the native Amerindian population in North America as well as South America. HTLV-I has also been found to be prevalent in certain population groups in the Middle East and India. Blood donors are routinely screened for HTLV-I in North America, several countries in Europe, Japan, and Taiwan.

Human T cell leukemia virus type I (HTLV-I) is associated with the development of adult T-cell leukemia (ATL) and the neurological disease tropical spastic paraparesis/ HTLV-I associated myelopathy (TSP/HAM). The major viral oncoprotein, Tax, is a transcriptional transactivator of viral and cellular genes. It is hypothesized that HTLV-I causes cancer by altering cellular gene expression and function leading to dysregulation of cellular processes including cell cycle progression and DNA repair. Chapter II will review the effects of Tax on cell cycle progression as a consequence of altering cellular gene expression and protein function. How these events may contribute to transformation of HTLV-I infected T cells and the subsequent development of ATL will also be discussed.

As with all viruses, human immunodeficiency virus type 1 (HIV-1) requires the use of cellular pathways and components in its replication cycle. Understanding these interactions between a virus and its host cell can assist in the discovery and development of antiviral treatments and strategies. Retroviruses produce infectious particles from a rather small number of components: three structural polyproteins and a homodimer of genomic RNA. These essential building blocks must all interact within the vast space of the host cell to form the organized structure of an infectious virion. While the current understanding of the retroviral proteins and genomic RNA functions have allowed for the creation of a reasonably detailed working model of how they assemble and their relative importance, the roles that cellular components and pathways play in retroviral assembly remain poorly understood. Several mechanistic questions remain unclear: How does the assembling virus find its genomic RNA in the cell? How is Gag targeted to the sites of assembly? What cellular components are required during assembly? Where does Gag assemble? How does the assembled virus release from the cell? What roles to do cellular proteins play in the virion? What potential antiviral therapies could interrupt these targets? Chapter III will examine the cellular side of these HIV-1 assembly questions and recent experimental advances that are beginning to unravel the secrets of the interactions between HIV-1 and cells. While clear answers to these assembly questions are not currently known, our partial understanding of these complex and required interactions between the assembling viral proteins and the host cell machinery has opened up exciting new avenues for HIV-1 assembly research as well as cell biology in general. Although this review is mostly restricted to HIV-1, lessons learned from this virus may apply to other retroviruses.

Recently, several reports have focused on the molecular aspects of immunodeficiency viral pathogenesis within myeloid antigen presenting cells such as the dendritic cells and macrophage. From the receptors that are hypothesised to attach immunodeficiency viruses in the mucosae during HIV transmission, mechanisms of viral harbouring and transfer and the use of the exosome synthesis machinery during APC infection. Chapter IV will address the concepts of viral attachment, infection and viral transfer in light of current observations of APC subsets. In relation to viral transfer, the authors will more specifically address the uptake compartments in DC, the internalised half-life of virus and finally the exit pathways for either "old" regurgitated virus versus "new," *de novo* produced virus within the DC.

An efficient host defense against viral pathogens operates at two levels: innate and adaptive immunity resulting in long-lasting protective immunity. Viruses, however, have developed mechanisms to evade both innate and adaptive immunity, and in some cases certain aspects of normal protective host responses have been redirected or modified to the advantage of these viruses. Susceptibility to retroviral pathogenesis is controlled by the genetic make-up of the host and can be studied in mice from different inbred backgrounds, because allelic differences enable cloning of the genes responsible for the phenotype and subsequent understanding of the mechanism of their action. Using mice from distinct genetic backgrounds we learned that mouse mammary tumor virus (MMTV) has developed strategies to both exploit and subvert the immune system. For example, MMTV activates profound T-cell proliferation necessary for virus propagation via virally-encoded superantigen (SAg). At the same time, MMTV can stimulate production of the immunosuppressive cytokine IL-10 by subverting innate immune signaling. This allows the virus to avoid the ongoing anti-virus immune response and to replicate without restraint. In addition to their unique ability to evade immune responses, retroviruses are also very efficient in causing tumors. It is well accepted that most retroviruses cause tumors by acting as insertional mutagens activating expression of cellular protoocogenes. The authors found, however, that MMTV activation of oncogenes is a necessary but insufficient step in mammary tumor induction. By using different viral variants and mice from distinct genetic backgrounds, it was discovered that virally-encoded genes must co-operate with oncogenes for efficient tumor induction in the MMTV model. Identification of novel pathways utilized by retroviruses to induce tumors will increase our understanding of the signal transduction pathways involved in tumorigenesis in general. Chapter V will focus on the usefulness of mouse strains with distinct virus-susceptible or virus-resistant phenotypes to gain understanding of the fundamental mechanisms of virus-host interactions.

Enteroviruses are small, naked viruses whose genome is a positive-strand RNA. Type 1 diabetes is an autoimmune pathology characterized by a loss of insulin due to destruction of insulin-secreting β-cells in pancreas Langerhans islets. Epidemiological studies on identical twin pairs pointed out that even though genetic factors are important factors in disease susceptibility, environmental factors, like virus infection, play also a role as descrined in chapter VI. Among viruses, coxsackieviruses B4 (CVB4) are suggested to be involved in disease emergence. Although the mechanism responsible for virus-induced diabetes is unknown, two mechanisms are thought to act at different levels of the immune system. Molecular mimicry is based on sequence homology between host glutamic acid decarboxylase 65 (GAD 65) and viral protein 2C (P2C). Virus can also act at the pancreas level with an islet direct infection followed by bystander activation. Recent research has demonstrated that CVB4 is able to infect thymic epithelial and lymphoid compartments of human thymus and to induce severe thymic dysfunction.

As reported in chapter VII, Epstein-Barr virus (EBV), a widespread human γ-herpes virus, is a well-established risk factor for Burkitt's lymphoma, Hodgkin's lymphoma, a subset of T cell lymphomas and various tumors of epithelial cell origin with nasopharyngeal carcinoma as the best studied example. Furthermore, EBV-associated lymphoproliferative disorders are a well-recognized major complication of immune suppression. These EBV-associated lymphoid and epithelial cell malignancies are characterized by peculiar geographic distributions and distinctive epidemiologic features. EBV establishes a life-long asymptomatic infection in immunocompetent hosts by colonizing the memory B-lymphocytes compartment and hijacking cellular signaling pathways that regulate antigen-dependent B-cell activation and differentiation. This multifaceted interaction with B cells is mediated by a set of EBV-encoded proteins that are differently expressed in normal and malignant cells. The quantification of circulating EBV DNA and EBV-epitope specific T cells are valuable tools in the diagnosis, clinical monitoring and prognosis of some EBV-associated tumors. Recent advances in elucidation of the mechanisms underlying EBV-induced growth transformation and immune evasion have been instrumental in guiding the design of novel approaches for the treatment EBV-associated malignancies.

Kaposi's sarcoma Associated Herpesvirus (KSHV) is the recently discovered human herpesvirus associated with various human malignancies including Body Cavity Based Lymphomas (BCBLs) or Primary Effusion Lymphomas (PELs) and Multicentric Castleman's Disease (MCDs). The viral DNA was detected in all forms of KS associated tumors and the sequence determination revealed that KSHV belongs to Gammherpesvirus sub-family of Herpesviridae and showed similarity to Herpesvirus saimiri and Rhesus Rhadinovirus. Chapter VIII reports that KSHV, like other Herpesviruses, establishes a latent infection and expresses a repertoire of latent genes. These latent genes have been shown to be important for the establishment of latent infection and tumorigenesis. Among the latent genes, Latency Associated Nuclear Antigen (LANA) is detected in all forms of KS-associated malignancies and is critical for episomal maintenance of the viral DNA by anchoring the viral genome to the host chromosome. Besides episomal maintenance, LANA has also been shown to modulate various cellular pathways. It down regulates the p53 and pRb tumor suppressor pathways involved in programmed cell death in response to cellular stress which includes viral infection. Additionally, other viral encoded genes are involved in driving cell proliferation and are critical components of the overall mechanism by which KSHV mediates cell transformation. Studies are underway to understand the complex signaling processes of cell cycle regulation and cell proliferation mediated by KSHV.

Retroviruses are used as leukemogenic agents because they can activate genes to be become transforming or inactivate tumor suppressors and, importantly, the cellular genes they alter can be easily identified as reported in chapter IX. During the retrovirus' life cycle, viral DNA integrates into the host DNA by a mechanism that is essentially random. If the mutagenic events contribute to an oncogenic phenotype, than cells carrying such insertions have a selective advantage for growth and/or survival, leading to clonal outgrowth and eventually leukemia. Retrovirus-induced leukemias usually have multiple integrated viruses (proviruses) with more than one contributing to the multistep disease process. Common mechanisms of oncogene activation include promoter insertion, enhancer insertion, and destabilization of mRNA. The retroviral sequences at the site of proviral insertion provide molecular tags to identify genetic loci. Although retroviral tagging has been carried out for nearly two decades, the methodology has recently expanded to include large-scale identification of cancer genes. The recent exploitation of this gene discovery method became possible with several new technical advances in PCR based strategies for cloning virus integration sites, sequencing, databases of sequences from the mouse genome, and informatics. Initial reports on retroviral tagging predominantly focused on loci involved in lymphomagenesis, but new ones include loci associated with myeloid leukemia. Retroviruses are currently being used to mutagenize the genome of mice that are already genetically engineered to be predisposed to cancer, for example transgenic mice and those with targeted deletions. Future experimentation in these cancer "sensitized" mice will surely provide a wealth of information regarding genes that cooperate in causing cancer and molecular cancer pathways.

Retroviral vectors are promising tools for gene transfer, in part because retroviral integration precisely links therapeutic sequences to host chromosomes. However, although providing stable transgene expression, vector integration may have deleterious effects, mainly due to insertional mutagenesis. In that sense, the sites of insertion of HIV-1 and MLV have been mainly mapped to RNA pol II highly transcribed genes, strengthening the potential danger of using these viral agents as vectors for gene therapy. Interestingly, other retrovirus-like elements are known to preserve the integrity of the host genome during their replication. This is the case of Ty3, a yeast gypsy-like retroelement, which integrates upstream of RNA pol III dependent promoter start sites without deleterious effects for host survival. Ty 5, another yeast retrotransposon, specifically inserts into regions of silent chromatin. Such site selection is driven by specific interactions occurring between the viral integration machinery and host proteins, allowing a balance between the fitness of the host and the ability of the retrotransposon to propagate and survive in the yeast genome. Understanding the precise interactions between cellular and viral partners involved in the integration process could certainly lead to the elaboration of safer therapeutic vectors, specifically targeting the corrective gene to chosen regions of the host genome allowing a reasonable level of transgene expression without affecting cellular gene transcription. The goal of Chapter X is to summarize our current knowledge on site specificity of integration of retrovirus-like elements, and the recent progress made in integration site selection.

Highly active antiretroviral therapy (HAART) based on combinations of drugs that target HIV enzymes has considerably reduced morbidity and mortality from HIV infection. However, despite a large number of antiretroviral drugs approved for the treatment of HIV infection, current drug regimens present challenges, urging for the development of new drugs. Candidate drugs with novel mechanisms of action are under investigation, such as the HIV entry inhibitors. This family of drugs targets the three steps of HIV entry: receptor binding (attachment), co-receptor binding and membrane fusion. Entry inhibitors currently in development are presented in chapter XI. Preliminary results are giving some insight on how these promising family of molecules may be used in future anti-HIV multi-therapies.

In: Progress in Virus Research
Editor: Sabine Thebault, pp. 1-38
ISBN 1-59454-438-7

Chapter I

Regulation of Viral and Cellular Transcription by the Human T-Cell Leukemia Virus Type I Tax Oncoprotein

Jean-Michel Mesnard *

Laboratoire Infections Rétrovirales et Signalisation Cellulairc, CNRS/UM I UMR 5121/IFR 122, Institut de Biologie, 4 Bd Henri IV, CS 89508, 34960 Cedex 2, Montpellier, France

Abstract

Human T-cell leukemia virus type I (HTLV-I) is a complex retrovirus first identified in 1980. It is estimated that between 15 and 20 million people are HTLV-I carriers worldwide. The primary routes of HTLV-I transmission are: (i) transmission from mother to child through maternal HTLV-I-infected T cells in breast milk, (ii) sexual transmission from male to female, male to male, and rarely female to male, (iii) transmission through the transfusion of blood containing infected T cells, and (iv) transmission among drug users through shared needles. The transmission of HTLV-I is quite similar to that of the human immunodeficiency virus (HIV) although HIV is more pathogenic since only 5% to 10% of HTLV-I-infected people develop pathologies. Prevalence rates for HTLV-I infection in the general population are greater than 1% in the Caribbean Basin, Central Africa, and South Japan. In most other areas in the world, HTLV-I infections are mainly found in immigrants from endemic areas and in intravenous injection users. A high rate of infection was observed in the native Amerindian population in North America as well as South America. HTLV-I has also been found to be prevalent in certain population groups in the Middle East and India. Blood donors are routinely screened for HTLV-I in North America, several countries in Europe, Japan, and Taiwan.

* Phone: (33) 4 67 60 86 60, Fax: (33) 4 67 60 44 20, E-mail: jean-michel.mesnard@univ-montp1.fr

Introduction

Human T-cell leukemia virus type I (HTLV-I) is a complex retrovirus first identified in 1980 (Hinuma et al., 1981; Poiesz et al., 1980). It is estimated that between 15 and 20 million people are HTLV-I carriers world-wide. The primary routes of HTLV-I transmission are (Höllsberg and Hafler, 1993): (i) transmission from mother to child through maternal HTLV-I-infected T cells in breast milk, (ii) sexual transmission from male to female, male to male, and rarely female to male, (iii) transmission through the transfusion of blood containing infected T cells, and (iv) transmission among drug users through shared needles. The transmission of HTLV-I is quite similar to that of the human immunodeficiency virus (HIV) although HIV is more pathogenic since only 5% to 10% of HTLV-I-infected people develop pathologies. Prevalence rates for HTLV-I infection in the general population are greater than 1% in the Caribbean Basin, Central Africa, and South Japan (Ferreira Jr et al., 1997). In most other areas in the world, HTLV-I infections are mainly found in immigrants from endemic areas and in intravenous injection users. A high rate of infection was observed in the native Amerindian population in North America as well as South America. HTLV-I has also been found to be prevalent in certain population groups in the Middle East and India. Blood donors are routinely screened for HTLV-I in North America, several countries in Europe, Japan, and Taiwan.

HTLV-I is etiologically linked to an aggressive and often fatal leukemia of $CD4^+$ T lymphocytes (Uchiyama et al., 1977) called adult T-cell leukemia (ATL). ATL develops after a prolonged latency period of up to 30-50 years postinfection (Murphy et al., 1989) and is a rapidly progressing disease with leukemic cell infiltration into various organs including skin, spleen, liver and lungs. ATL is characterized by a malignant proliferation of a mature T lymphocyte with a phenotype of $CD2^+$, $CD3^+$, $CD4^+$, $CD8^-$, and $HLA\text{-}DR^+$. In addition, ATL cells are characterized by high levels of expression of the IL-2 receptor α-chain (Deeper et al., 1984). Typically, cells with lobulated nuclei known as flower cells are seen. Clinically, there are several subtypes of ATL: smoldering, chronic, acute, and lymphoma type (Cann and Chen, 1996). About 30% of patients with clinical manifestations of HTLV-I infection have chronic/smoldering ATL. These forms of ATL are less aggressive than the acute stage of the disease. Patients in both the smoldering and chronic stages of ATL can progress into acute ATL within a period of months. Lymphadenopathy, hepatosplenomegaly, and skin lesions due to infiltration of leukemia cells characterize acute ATL. Patients manifest abnormalities in certain serum chemistries, including elevation of lactate dehydrogenase, hyperbilirubinemia, and hypercalcemia. Hypercalcemia is associated with the presence of lytic bone lesions detected on the skull and long bones. The median survival is approximately 6 months, despite chemotherapy.

HTLV-I infection is also associated with a neurological demyelinating disease (Gessain et al., 1985) known as tropical spastic paraparesis/HTLV-I-associated myelopathy (TSP/HAM). TSP/HAM has a shorter incubation period than ATL and can show an accelerated development within 1 month to 4 years after the transfusion of the infected blood product. TSP/HAM is characterized by a chronic progressive course with loss of pyramidal function, spastic paralysis, sphincter dysfunction, and sensory neuropathy, caused by demyelinization and atrophy of the spinal cord. The current hypothesises about the pathogenesis of TSP/HAM (Höllsberg, 1997) are: (i) the cytotoxic hypothesis for which cytotoxic $CD8^+$ T cells may recognize viral antigens on HTLV-I-infected central nervous system (CNS) cells causing a cellularly mediated cytotoxic demyelination, (ii) the autoimmune hypothesis for which HTLV-I-infected CNS-autoreactive $CD4^+$ T cells are activated by HTLV-I infection and migrate to the CNS where they encounter their antigen, (iii) the bystander hypothesis for which the presence of HTLV-I-infected $CD4^+$ T cells and their recognition by specific $CD8^+$ T cells in the CNS induce the secretion of cytokines toxic for the myelin.

Other diseases have been reported to be associated with HTLV-I-infection. Less frequently, the joints (HTLV-I arthropathy), the eyes (HTLV-I uveitis), the skin (infective dermatitis in children), the muscles (polymyositis), or the lungs (pulmonary infiltrative pneumonitis) can be affected.

Genomic Organization of HTLV-I

HTLV-I is classified in the HTLV-BLV group of the retrovirus family. Other group members include HTLV-II, bovine leukemia virus (BLV), and the simian T-cell leukemia viruses (STLV). All retroviruses share similarity in their genome organization with four genes encoding proteins necessary for the production of infectious virions, the order of these genes being invariably 5' *gag, pro, pol, env* 3' (Fig. 1). *gag* codes for internal structural proteins of the virion, *pro, pol, env* encoding the viral protease, the reverse transcriptase and the envelope glycoproteins of the virion, respectively. In addition, as a complex retrovirus, HTLV-I uses alternative spicing and internal initiation codons to produce several regulatory and accessory proteins located in the pX region of the viral genome (Fig. 1) between *env* and the 3' long terminal repeat (LTR). Alternative spliced forms of mRNA code for three accessory proteins, $p12^I$, $p13^{II}$, and $p30^{II}$ (Albrecht and Lairmore, 2002), and two regulatory proteins, Rex and Tax. Sequence analysis of $p12^I$ reveals two putative transmembrane regions and four putative proline-rich SH3-binding domains. This viral protein localizes in the endoplasmic reticulum and *cis*-Golgi compartment, and could be necessary to augment T-cell activation to facilitate the viral replication and productive infection (Ding et al., 2003). $p12^I$ would trigger early IL-2 expression by increasing the level of intracellular calcium and selectively activating NFAT (Kim et al., 2003).

On the other hand, less is known about the functions of the two other accessory proteins $p13^{II}$ and $p30^{II}$. The larger protein is encoded by a doubly spliced mRNA with an internal methionine codon in $p30^{II}$ that can be recognized to produce $p13^{II}$. $p13^{II}$ and $p30^{II}$ localize in the mitochondria and the nucleus, respectively (D'Agostino et al., 2002; Koralnik et al., 1993). Recently, it has been found that $p30^{II}$ is able to retain in the nucleus the *Tax* and *Rex* mRNAs (Nicot et al., 2004). Because Tax and Rex are positive regulators of viral gene expression, the inhibition of their expression by $p30^{II}$ could reduce virion production.

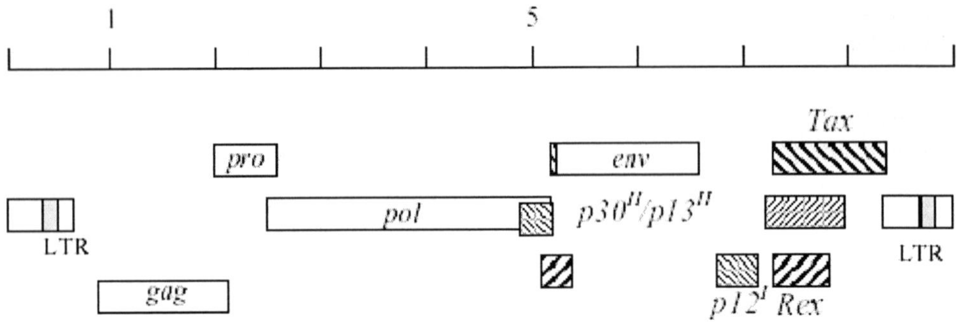

Figure 1. Genomic organization of HTLV-I DNA proviral. Position and relative sizes of the different open reading frames (ORFs) are shown as boxes. The ORFs present in the pX region are represented as differently striped boxes.

Rex is a 27-kDa protein that contains several domains critical for its function: (i) an arginine-rich N-terminal RNA binding domain that overlaps with a nuclear localization signal (NLS), (ii) a leucine rich activation domain encompassing the nuclear export signal (NES), and (iii) two regions flanking the NES important for Rex-Rex multimerization. Rex is required for efficient cytoplasmic expression of the unspliced and incompletely spliced viral transcripts coding for the viral structural and enzymatic proteins but suppresses expression of accessory and regulatory proteins (Hidaka et al., 1988). Indeed, Rex promotes the nucleocytoplasmic export and hence the translation of unspliced and singly spliced mRNAs that encode the *gag-pol* and *env* gene products. The functionality of Rex depends on its binding to the Rex response element, which correspond to *cis*-acting sequences located in the R regions of the viral RNAs (Askjaer and Kjems, 1998). However, it is not yet fully clarified how HTLV-I Rex mediates the accumulation of incompletely spliced mRNAs in the cytoplasm. The presence of the NES suggests that Rex could directly activate the transport of intron containing RNAs into the cytoplasm. However, Rex could also directly inhibit splicing, as suggested by the observation that Rex alters the balance between unspliced and spliced RNA in the nucleus as well as in the cytoplasm of T cells (Grone et al., 1996). In addition, it has been reported that Rex specifically prevents the unspliced RNA from degradation in the nucleus (Grone et al., 1996).

The Tax Protein is Oncogenic

HTLV-I involvement in ATL development was originally based on the finding of provirus integration in leukemic cell of the patients. Every ATL cell contained the proviral integrated in its chromosomal DNA and are monoclonally expanded, indicating that leukemic cells originated from a single infected T cell (Mortreux et al., 2003). However, HTLV-I has no preferential site of integration that could explain why a transformation process is initiated. In addition, the *cis*-acting effect of the LTR of integrated proviruses seems unlikely to be the mechanism for ATL. Some published results suggest that the viral Tax protein contributes to the development of ATL. Indirect evidence for the role of Tax in leukemogenesis comes from studies on malignant cells from ATL patients that have defective proviruses retaining the 3' end of the HTLV-I genome (Korber et al., 1991). From experiments on primary human T-lymphocyte immortalization by a recombinant herpesvirus vector carrying the HTLV-I 3' region (Grassmann et al., 1989), it has been shown that the *tax* gene is necessary and sufficient for cell immortalization (Grassmann et al., 1992). Tax can also transform established rodent fibroblasts (Smith and Greene, 1992; Tanaka et al., 1990; Yamaoka et al., 1996; Yamaoka et al., 1992). In addition, cotransfection of Tax with Ras can also induce transformation of primary rat embryo fibroblasts (Pozzatti et al., 1990).

Tax transgenic mice models were also developed to demonstrate the tumorigenic potential of Tax. Different enhancers and promoters were utilized to drive expression of Tax. In transgenic models utilizing the HTLV-I LTR, several transgenic mice developed neurofibromas composed primarily of fibroblastic cells (Hinrichs et al., 1987; Nerenberg et al., 1987; Ruddle et al., 1993). These mice also exhibited severe skeletal abnormalities characterized by high bone turnover. Tax mRNA and protein were highly expressed in bone but not in bone marrow or in any other tissues except salivary gland and neurofibromas. Within bone, Tax protein was detected in oeteoclasts and endosteal (Ruddle et al., 1993). In transgenic models utilizing the Thy1.2 promoter to target Tax expression specifically to thymocites, the mice showed expression in the thymus, as well as the brain, bone marrow, lung, and testes (Nerenberg et al., 1991). Mice developed fibrosarcoma on the tail, cars, and hind limbs. Always in an attempt to target Tax expression in the lymphoid cells, Benvenisty et al. (1992) developed a binary set of transgenic mice, one bearing the HTLV-I LTR positioned to drive the c-myc oncogene and another bearing a fusion transgene consisting of the immunoglobulin promoter/enhancer driving Tax. When these two transgenic lines were mated, bigenic offspring harboring both transgenes developed $CD4^+$ T-cell lymphomas and central nervous system tumors (Benvenisty et al., 1992). Finally, by using the human granzyme B promoter, Tax was also targeted to the mature T-lymphocyte compartment in transgenic mice that developed large granular lymphocytic leukemia (Grossman et al., 1995). Taken together, all these different transgenic mice models demonstrate that expression of Tax in the lymphocyte compartment is sufficient for the development of leukemia.

Tax *Trans*-Activates Viral Transcription in Association with CREB

Transcription of retroviruses is regulated by *cis* elements present in the 5' LTR composed of three regions, U3, R, and U5. Several cellular factors such as c-Jun (Jeang et al., 1991), p53 (Aoyama et al., 1992), Myb (Bosselut et al., 1992; Dasgupta et al., 1992), Elf-1 (Clark et al., 1993), YB-1 (Kashanchi et al., 1994), and Sp1 (Gegonne et al., 1993; Torgeman et al., 1999) have been suggested to bind to regulatory sequences in the HTLV-I LTR, but only a few have been shown to be actually involved in transcriptional regulation of the LTR *in vivo* (Lemasson et al., 2002). On the other hand, HTLV-I transcription is *trans*-activated by Tax *in vivo* (Cann et al., 1985; Felber et al., 1985; Fujisawa et al., 1985; Sodroski et al., 1985). The U3 region of the LTR contains important elements needed for Tax *trans*-activation corresponding to three repeats of 21-bp sequence (Tax-responsive element 1 or TxRE1)

(Brady et al., 1987; Fujisawa et al., 1986; Rosen et al., 1987; Shimotohno et al., 1986) and another element termed TxRE2 (Brady et al., 1987) that is localized between the promoter-central and promoter-proximal TxRE1s (Fig. 2). However, TxRE2 alone is totally inactive even in its multiple form and requires the 21-bp sequence to respond to Tax (Tanimura et al., 1998). TxRE2 contains binding sites for Ets1, Sp1, Myb, TIF-1, and Gli2 (Bosselut et al., 1992; Dan et al., 1999; Gegonne et al., 1993; Gitlin et al., 1993; Marriott et al., 1990). It has been proposed that Gli2 binds to TxRE2 and interacts with CREB bound to TxRE1, resulting in more recruitment of Tax onto U3 (Dan et al., 1999). However, this model was not confirmed by the results reported elsewhere since Gli2 was described as a repressor of Tax-mediated transcription (Smith et al., 2001). Another possibility is the presence of an Ets-responsive region (ERR1) in TxRE2 (Gegonne et al., 1993). It has been demonstrated that Tax and Ets1 increase HTLV-I transcription in a cooperative manner and that the functional interaction between Tax and Ets1 requires the presence of TxRE1 and ERR1 (Gitlin et al., 1993). Moreover, activity of ERR1 also depends on the integrity of an Sp1 binding site localized in TxRE2 (Gegonne et al., 1993) and Tax facilitates the interaction between Ets1 and Sp1, forming a ternary complex (Dittmer et al., 1997). Taken together, these results suggest that the Tax/Ets1 cooperative effect on HTLV-I transcription could be based on the ability of Tax, Ets1, and Sp1 to form a ternary complex on the viral LTR.

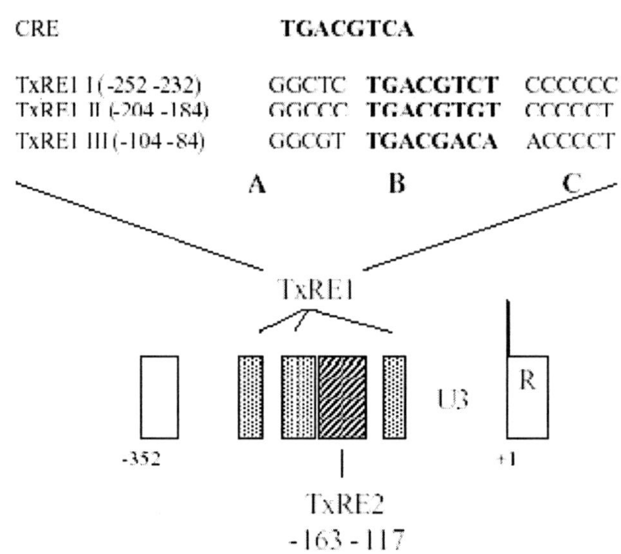

Figure 2. Schematic organization of the 5' LTR. The indicated positions are relative to the mRNA initiation site at the U3 boundary. The nucleotide sequences of the TxRE1s and the consensus sequence of CRE site are also shown.

Two direct repeats TxRE1s are sufficient to confer *trans*-activation of transcription by Tax (Fujisawa et al., 1986; Paskalis et al., 1986; Shimotohno et al., 1986), one copy of TxRE1 being weakly activated. Each TxRE1 can be subdivided into three motifs known as A, B, and C (Fig. 2). The central domains B correspond to imperfect cAMP response elements (CRE) that are able to interact *in vitro* with multiple members of the activating transcription factor/CRE-binding protein (ATF/CREB) family (Bantignies et al., 1996; Beimling and

Moelling, 1992; Franklin et al., 1993; Shnyreva and Munder, 1996; Tsujimoto et al., 1991; Yoshimura et al., 1990). The domains A and C can be recognized *in vitro* by cellular factors, such as AP-2, HEB-1, PRDII-BF1, and TAXREB107 (Montagne et al., 1990; Morita et al., 1993; Muchardt et al., 1992; Seeler et al., 1993).

To activate transcription of the HTLV-I genome, Tax does not bind specifically to DNA (Beimling and Moelling, 1989; Giam and Xu, 1989; Jeang et al., 1988) but interacts with ATF/CREB proteins that are able to interact with the B domain. The ATF/CREB proteins belong to the leucine zipper (ZIP) transcription factor family (Landschulz et al., 1988) that play an important role in regulating the expression of various cellular genes. The ZIP structure consists of a periodic repetition of leucine residues at every seventh position over a distance covering eight helical turns. The segments containing these periodic arrays of leucine residues seem to exist in an alpha-helical conformation. The leucine side chains extending from one alpha-helix interact with those from a similar alpha helix of a second polypeptide, facilitating dimerization. The structure formed by cooperation of these two regions forms a coiled coil (O'Shea, Rutkowski, and Kim, 1989). Thus, the ZIP motif allows the association of a single protein or potentially different proteins, explaining the heterotypic complex reported for some ZIP factors. Homo or heterodimerization is a prerequisite to DNA binding. A 30 amino acid region, rich in basic residues, immediately preceding the ZIP (bZIP domain) is required for DNA binding. Alteration of the register between the basic and ZIP regions by insertion or deletion of amino acids eliminates specific DNA binding, confirming the need for a continuous helical connection between the DNA-binding and dimerization motifs (Agre et al., 1989). The bZIP pattern is present in many transcription factors including the ATF/CREB proteins, the activator protein-1 (AP-1) transcription factors, the CCATT-box/enhancer binding protein (C/EBP), the yeast general control protein 4 (GCN4), the C-myc, L-myc and N-myc oncogenes, the octamer-binding transcription factor (Oct-2/OTF-2), the Maf transcription factors, and the Cap'n'Collar (CNC)-bZIP family.

Although several members of the ATF/CREB family have been described to interact with TxRE1 of the HTLV-I LTR, few of them are able to form a complex with both TxRE1 and Tax. For instance, ATF-2 binds to probes containing the 21-bp motifs (Yoshimura et al., 1990) but does not form a stable complex with Tax (Adya et al., 1994; Brauweiler et al., 1995; Maekawa et al., 1991; Yin et al., 1995a; Zhao and Giam, 1992). On the other hand, CREB or CREM forms a stable ternary complex with TxRE1 and Tax (Anderson and Dynan, 1994; Brauweiler et al., 1995; Kwok et al., 1994; Paca-Uccaralertkun et al., 1994; Suzuki et al., 1993; Yin and Gaynor, 1996). CREB was originally isolated from a human cDNA expression library with a double-stranded CRE probe (Hoeffler et al., 1988). Different isoforms of CREB are generated by alternative splicing, certain isoforms behaving as CREB inhibitors. Subsequent to the characterization of CREB, two other highly related gene products were characterized, ATF-1 and cAMP response element modulator (CREM). Like CREB, the CREM gene generates either activator or repressor forms of CREM by alternative splicing. CREB, ATF-1, and CREM bind to their DNA target sequence as homodimers. However, the presence of highly related dimerization domains (for instance, CREB and ATF-1 bZIPs are 91% identical) suggested that different members of the CREB/CREM/ATF-1 subgroup might be able to form heterodimers. Indeed, such heterodimers have been detected in a variety of cell lines, although the amount of heterodimers varies depending on the cell type. Moreover, it has been suggested that HTLV-I LTR *trans*-activation required the association of Tax and CREB/ATF-1 (Zhao and Giam, 1992). On the other hand, CREB,

CREM, and ATF-1 do not heterodimerize with many other bZIP factors, suggesting that they represent a distinct subgroup.

Tax Interacts with CBP

CREB and CREM possess an amino-terminal kinase-inducible domain (KID) flanked by glutamine-rich transcriptional activation domains. The KID contains a particular residue, serine 133 (Ser133), that is phosphorylated by a variety of protein kinases including protein kinase A (PKA), mitogen-acivated protein kinases (MAPKs), and Ca^{2+}/calmodulin-dependent protein kinases (CaMKs). After its phosphorylation, the protein is able to recruit the transcriptional coactivator CREB binding protein (CBP) and its homologue p300 (Chrivia et al., 1993), that are known to be histone acetyltransferases (HATs). The CBP/p300 recruitment to the promoter could induce local nucleosome modifications by histone acetylation and facilitate stable binding of components of the basal transcription machinery.

Many of the CREM isoforms, including CREMα and CREMΔ(C-G), are known to be inhibitors of transcription. By utilizing CREMΔ(C-G), that contains the conserved bZIP domain but not the KID, it has been demonstrated that the KID is not essential for viral promoter activation in the presence of Tax (Laurance et al., 1997). This result suggests that this activation is due to the ability of Tax to recruit the coactivator CBP to the viral promoter. Thus, Tax may provide a bridge between the HTLV-I promoter and the coactivator CBP through a mechanism that does not depend on phosphorylation. Effectively, Kwok et al. (1996) have demonstrated that Tax interacts directly with CBP. Tax recruits CBP through multiple interactions with both the amino- (C/H1 and KIX domains) and carboxy-terminal (CR2 domain) regions of the coactivator (Georges et al., 2002; Kwok et al., 1996; Lemasson and Nyborg, 2001). Two regions of Tax, respectively localized between amino acid residues 81 to 95 (Harrod et al., 1998) and 312 to 319 (Scoggin et al., 2001), participate in CBP binding (Fig. 3). Thus, to activate transcription of the HTLV-I genome, Tax interacts with CREB or CREM bound to the TxRE1s and then can recruit CBP. The recruitment of CBP to the HTLV-I promoter could be involved in the acetylation of the core histone tails (Gachon et al., 2002; Georges et al., 2002; Lu et al., 2002). Another acetyltransferase, the p300/CBP-associated factor p/CAF, has also been described to interact with Tax (Harrod et al., 2000; Jiang et al., 1999). However, the exact role of this acetyltransferase in Tax-mediated transcription still remains unknown. In addition, direct interactions between Tax and basal polymerase II factors such as the transcription factors TFIIA (Clemens et al., 1996) and TFIID through TATA-binding protein (TBP) (Caron et al., 1993) and TBP-associated factor $TAF_{II}28$ (Caron et al., 1997) have also been reported.

Tax Enhances the Binding Affinity of CREB for the Viral LTR

The recruitment of CBP to the TxRE1s is not the only molecular mechanism involved in stimulation of viral transcription by Tax. Indeed, several studies have clearly demonstrated that Tax stabilizes the binding of CREB and CREM to the TxRE1s (Brauweiler et al., 1995; Yin and Gaynor, 1996; Zhao and Giam, 1992). Tax enhances the binding of CREB and CREM only to the TxRE1s but not to the cellular CRE motif (Brauweiler et al., 1995; Gachon et al., 2000; Paca-Uccaralertkun et al., 1994; Yin and Gaynor, 1996) suggesting that the GC-rich sequences flanking the CRE-like motif of the TxRE1s (Fig. 2) are required for the effect of Tax. Mutagenesis experiments confirm this hypothesis since mutations in these sequences (domains A and C) reduce HTLV-I transcription (Paca-Uccaralertkun et al., 1994; Seeler et al., 1993). The role of the GC-rich sequences has been studied by footprinting analyses and it has been demonstrated that Tax extends the protection of CREB from the TxRE1s into the GC-rich flanking sequences (Kimzey and Dynan, 1998; Lenzmeier et al., 1998; Lundblad et al., 1998). In addition, cross-linking studies show that Tax itself makes contact with the DNA minor groove (Kimzey and Dynan, 1998; Lenzmeier et al., 1998; Lundblad et al., 1998) at two symmetric positions 14-nucleotide apart on either side of the TxRE1 (Kimzey and Dynan, 1998). Amino acids in Tax that contact DNA are localized between residues 89 and 110 (Kimzey and Dynan, 1999). Thus, Tax stimulates the binding of CREB and CREM to the TxRE1s by enhancing protein-DNA interactions.

CREB-2 Activates Viral Transcription in Cooperation with Tax

Different observations suggest that CREB and CREM are not the only CRE-binding proteins that can interact with Tax *in vivo*. Others and we have isolated by the two-hybrid approach another bZIP protein, CREB-2, that interacts with Tax to enhance HTLV-I transcription (Gachon et al., 1998; Reddy et al., 1997). Moreover, CREB-2 binds to the TxRE1s and this binding is enhanced by Tax (Gachon et al., 2000). CREB-2 has first been isolated from a Jurkat T-cell cDNA library by using the human T-cell receptor α enhancer CRE motif (Karpinski et al., 1992) or the HTLV-I TxRE1 sequence (Tsujimoto et al., 1991) as probes. CREB-2, also called ATF-4 or TAXREB 67 (Hai et al., 1989; Tsujimoto et al., 1991), represents with the mouse mATF-4, mTR67, C/ATF, and the Aplysia ApCREB-2 a subgroup of the ATF/CREB family. Although CREB-2 was originally described as a repressor of CRE-dependent transcription (Karpinski et al., 1992), it contains a constitutive activation domain of transcription and can directly interact with CBP/p300 (Gachon et al., 2000; Liang and Hai, 1997).

Unlike CREB, CREB-2 forms *in vivo* heterodimers with other bZIP transcription factors including C/EBPβ and C/EBPγ (Podust, Krezel, and Kim, 2001; Vallejo et al., 1993; Vinson, Hai, and Boyd, 1993), Nrf1 and Nrf2 (He et al., 2001; Murphy and Kolsto, 2000), and CHOP (Gachon et al., 2001). Such heterodimers are unable to *trans*-activate viral transcription in the presence of Tax (Gachon et al., 2001; Gaudray et al., 2002; Hivin et al., 2004). Actually, only CREB-2 homodimers are able to enhance HTLV-I transcription and Tax obviously promotes the homodimerization of CREB-2 (Gaudray et al., 2002). Indeed, it has been demonstrated that Tax could stimulate the DNA binding activity of bZIP factors by enhancing their homodimerization (Wagner and Green, 1993). Moreover, Tax also forms homodimers (Basbous et al., 2003b; Tie et al., 1996), which contribute to optimal transcriptional activity from the HTLV-I promoter (Jin and Jeang, 1997; Tie et al., 1996). Although Tax dimerization is dictated more by secondary structure than by primary structure (Basbous et al., 2003b; Chun et al., 2000), it has been possible to delimit a central region (Fig. 3) involved in Tax dimerization that encompasses the residues 127 to 228 (Basbous et al., 2003b; Tie et al., 1996). This region can be divided into three subdomains of dimerization (Basbous et al., 2003b): DD1 (residues 127 to 146), DD2 (residues 181 to 194), and DD3 (residues 213 to 228). Mechanistically, Tax self-association could guide the homodimerization of CREB and CREB-2. In this model, a Tax monomer could interact with a CREB or CREB-2 monomer and then Tax dimerization would bring two bZIP domains into close proximity to dimerize (Jin and Jeang, 1997; Tie et al., 1996).

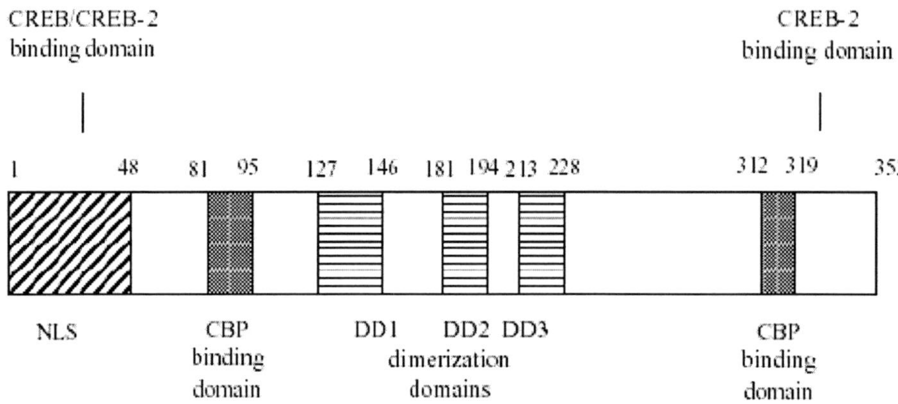

Figure 3. Molecular organization of Tax. The numbers above the diagram correspond to the position of the different domains of Tax.

Tax differently interacts with CREB and CREB-2. For instance, CREB is unable to interact with Tax in yeast two-hybrid assay (Bantignies et al., 1996; Shnyreva and Munder, 1996). It has been proposed that the masking of the Tax N-terminus by the GAL4 DNA binding domain is probably responsible for the inability of Tax to interact with CREB. Indeed, the N-terminal cysteine-rich domain of Tax (Fig. 3), that corresponds to a zinc finger motif (Semmes and Jeang, 1992), has been shown to be required for interaction with CREB (Adya and Giam, 1995; Goren et al., 1995; Yin et al., 1995a). On the other hand, the isolation of CREB-2 by yeast two-hybrid approach using Tax as a bait (Gachon et al., 1998; Reddy et al., 1997) confirms that CREB-2 interacts with Tax in a different way than that of CREB. Indeed, the N-terminal cysteine-rich domain of Tax is also necessary for binding to CREB-2 but not sufficient by itself; the 46 C-terminal amino acids of Tax, encompassing its acidic domain, are involved in this interaction. Moreover, in the case of CREB, it has been demonstrated that the amino acid sequence of the basic domain and its ZIP were required for interactions with Tax (Yin et al., 1995b). On the other hand, the basic subdomain of the CREB-2 bZIP is not necessary for its association with Tax. Indeed, whereas its ZIP structure is always indispensable, the basic domain of CREB-2 required for the interaction with Tax corresponds to a short basic C-terminal domain located downstream of the ZIP (Gachon et al., 2000). However, it is noteworthy that for both cellular proteins the ZIP structure, that serves as a dimerization interface, is absolutely necessary. This observation confirms the involvement of Tax in bZIP protein dimerization (Wagner and Green, 1993).

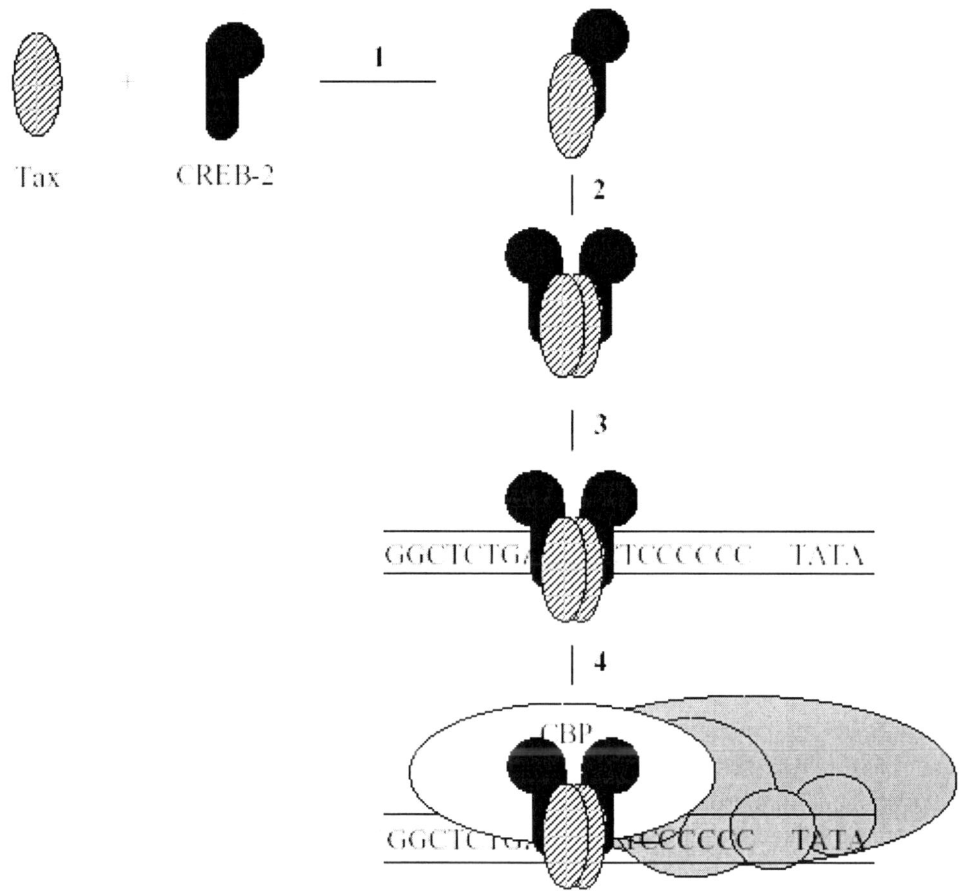

Figure 4. Model for Tax-mediated transcription of HTLV-I. (1) A Tax monomer interacts with a CREB-2 monomer, (2) Tax self-association guides the homodimerization of CREB-2, (3) the CREB-2 homodimer binds to the CRE motif of the TxRE1 and allows the fixation of Tax on the viral promoter, (4) Tax stabilizes the TxRE1-bound complex and recruits CBP, which induces local nucleosome modifications by histone acetylation and facilitates stable binding of components of the basal transcription machinery.

The finding that both CREB and CREB-2 form a complex with Tax and the TxRE1s raises the question as to why two different members of the ATF/CREB family are involved in regulation of HTLV-I transcription. One explanation is that CREB and CREB-2 could regulate transcription in different stages of the cell. Thus, two different factors provide alternative mechanisms to ensure that stimulation of HTLV-I transcription occurs. Another possibility is that CREB and CREB-2 could have distinct roles in HTLV-I transcription. Indeed, CREB stimulates transcription of HTLV-I reporter genes in the presence of Tax in transfected F9 teratocarcinoma cells (Giebler et al., 1997; Kwok et al., 1996) but represses Tax-mediated transcription in T-cell lines (Bodor et al., 1995; Gachon et al., 2000). Moreover, some members of the CREM subfamily have also been shown to be inhibitors of Tax-mediated HTLV-I expression in Jurkat, NIH3T3, and peripheral blood mononuclear cells

(Bodor et al., 1996; Bodor et al., 1995; Newbound et al., 2000). On the other hand, in the presence of Tax, CREB-2 activates viral transcription in transfected T-cell lines (Gachon et al., 1998; Reddy et al., 1997). It is difficult to explain such a difference between CREB and CREB-2 in T-cell lines. CREB has already been described as an inhibitor of multiple viral activators such as HSV VP16, adenovirus E1A, and HIV-1 Tat (Lemaigre et al., 1993). Balance between the production of CREB and CREB-2 could be involved in regulation of HTLV-I transcription in the infected T cells. Concerning this hypothesis, it is worth noting that CREB-2 expression is stimulated in HTLV-I-infected cells (Gachon et al., 2000; Harhaj et al., 1999).

In conclusion and as shown in Fig. 4, Tax can act at different levels to stimulate HTLV-I transcription. At first, Tax interacts with CREB or CREB-2 and then enhances their DNA-binding activity for TxRE1 by stabilizing bZIP dimer formation and by contacting nucleotides flanking TxRE1. Then, Tax recruits CBP/p300 to the HTLV-I promoter. The recruitment of CBP/p300 could induce local nucleosome modifications by histone acetylation and facilitate stable binding of components of the basal transcription machinery allowing the stimulation of viral transcription.

Down-Regulation of HTLV-I Transcription and Viral Latency

Results from many studies indicate that the majority of HTLV-I-infected cells in peripheral blood mononuclear cells (PBMC) carry a provirus that is transcriptionally silent. Low levels of cells expressing HTLV-I transcripts were detected in PBMCs by reverse transcriptase PCR in asymptomatic carriers and TSP/HAM patients (Furukawa et al., 1995; Gessain et al., 1990). Furthermore, fresh ATL cells do not express viral antigens until they are cultured in the presence of fetal calf serum, when in some cases ATL cells become positive for viral antigens or viral transcripts. However, the possibility that viral transcripts are derived from contaminating untransformed HTLV-I-infected cells is not ruled out. Thus, it appears that most HTLV-I-infected cells *in vivo* harbor a provirus that is transcriptionally silent and that latent infection may be the norm for HTLV-I *in vivo*.

Different levels of viral transcription can be targeted in order to establish HTLV-I latency. Thus, hypermethylation of the 5' LTR is an important mechanism by which HTLV-I gene expression is repressed (Fig. 5). Different analysis of LTR methylation showed that ATL cells and HTLV-I-infected cells in the PBMCs of asymptotic carriers have integrated proviral DNA carrying hypermethylated 5' LTRs versus hypomethylation of 3' LTR (Koiwa et al., 2002). These results raise interesting questions concerning the mechanisms underlying differential methylation of the two LTRs and also the possible transcriptional activity of the 3' LTR. Effectively, we have recently characterized a novel HTLV-I protein (Gaudray et al., 2002) synthesized from the minus-strand RNA (Fig. 6A) that is transcribed by a functional promoter present in the antisens strand of the 3' terminus of the proviral genome containing the 3' LTR (Larocca et al., 1989). As this protein contains a bZIP domain in its C-terminal region, we have named it HBZ for HTLV-I bZIP factor.

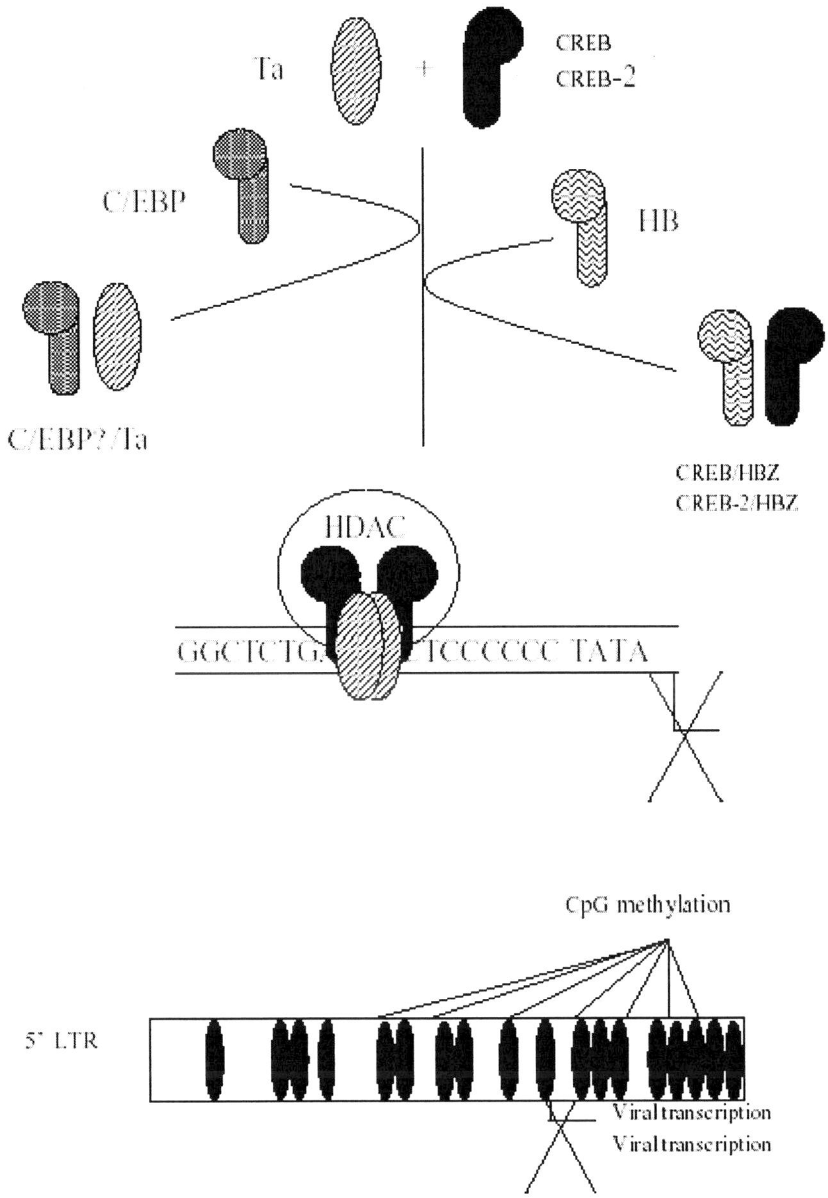

Figure 5. HTLV-I transcription can be down-regulated or inhibited at different levels. Firstly, C/EBPβ and HBZ can block the binding of an active complex on the viral promoter by interacting with Tax and CREB or CREB-2, respectively. Secondly, HDAC1 competes with CBP in binding with Tax. Lastly, hypermethylation of the 5' LTR represses HTLV-I gene expression.

HBZ is a 31-kDa protein that resembles to a prototypical bZIP transcription factor, with an N-terminal transcriptional activation domain, a central domain involved in its nuclear transport, and a C-terminal bZIP domain (Fig. 6B). HBZ interacts with cellular bZIP factors including CREB, CREB-2, c-Jun, JunB and JunD (Basbous et al., 2003a; Gaudray et al., 2002; Thébault et al., 2004). Moreover, HBZ acts as a repressor of HTLV-I transcription by forming heterodimers with CREB or CREB-2 that are no longer able to bind to the TxRE1s (Fig. 5). Consequently, Tax cannot be positioned on the viral promoter and then is unable to *trans*-activate viral transcription. In addition, HBZ is also capable of inhibiting Tax-independent transcription by interacting with c-Jun, that is involved in basal HTLV-I transcription.

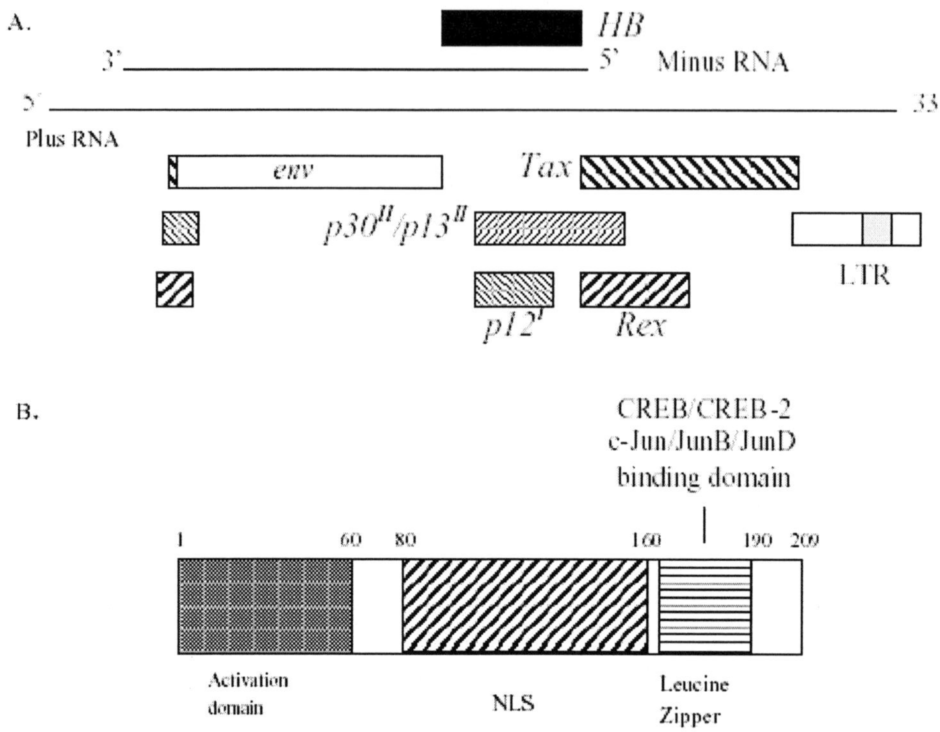

Figure 6. The complementary strand of the HTLV-I RNA genome codes for HBZ. (A) Genomic organization of the pX region of the HTLV-I provirus. The HBZ gene encoded by the minus strand RNA is represented by a black box. (B) Molecular organization of HBZ. The numbers above the diagram correspond to the position of the different domains of HBZ.

Cellular proteins, such as C/EBPβ and histone deacetylase1 (HDAC1), that interact with Tax, can also down-regulate HTLV-I transcription. The C/EBP family encompasses factors with structural homologies, which suggest an evolutionary history of genetic duplications with subsequent pressure to diversify. The prototypic C/EBP factor contains a domain responsible for transcriptional activation located in the N-terminal region and a C-terminus bZIP domain. The different C/EBP family members homo- and heterodimerize through their ZIP and bind to the C/EBP site, GATTGCGCAATC (Lekstrom-Himes and Xanthopoulos, 1998). They also interact with transcription factors of other bZIP protein families forming heterodimers that bind to chimerical DNA sites (Vinson et al., 1993). The C/EBP factors are involved in regulation of many genes including several cytokine and liver-specific genes, many viral genomes, and the immunoglobulin heavy chain and kappa loci. Among the C/EBP protein family members, C/EBPβ, also called NF-IL6, LAP, or CRP2, can physically interact with Tax (Hivin et al., 2004; Tsukada et al., 1997). We have recently demonstrated that Tax is no more able to *trans*-activate viral transcription is the presence of C/EBPβ (Hivin et al., 2004). We have shown that Tax is no longer able to activate the transcription from the TxRE1s because the interaction between C/EBPβ and Tax blocks the formation of an active complex on the viral promoter (Fig. 5). The incapacity of Tax and C/EBPβ to cooperate together to *trans*-activate viral transcription is probably due to the mode of interaction between the two proteins. As already mentioned in this review, Tax stimulates the DNA-binding activity of CREB and CREB-2 by interacting with their ZIP domain. However, the bZIP domain of C/EBPβ is not required for the interaction between C/EBPβ and Tax. Our observations suggest that C/EBPβ may help infected T cells to escape the cytotoxic-T-lymphocyte (CTL) response by inducing a low level of Tax production in infected T cells, as already proposed with another cellular bZIP factor, ICER (inducible cAMP early repressor) (Newbound et al., 2000). ICER is a potent transcriptional *trans*-repressor without activation domain that suppresses Tax-dependent transcription (Bodor et al., 1996; Newbound et al., 2000). Indeed, the binding of ICER to TxRE1 is enhanced by Tax and blocks viral transcription in activated PBMC (Newbound et al., 2000).

Acetylation of nucleosomal histones by histone acetyltransferases generally stimulates transcription, whereas deacetylation of nucleosomes by HDACs is correlated with transcriptional repression. Indeed, HDACs repress transcription by removing the acetyl group from histone tails resulting in tighter packaging of nucleosomes and reduced access of transcription factors to regulatory regions of genes. Since the discovery of histone deacetylase RPD3 in *Saccharomyces cerevisiae*, numerous HDACs have been identified in mammalian cells, and are grouped into three classes based on sequence homologies (Gray and Eksrtom, 2001). Members of the class I include RPD3 and are nuclear proteins that are generally small in size (40-55 kDa) and ubiquitously expressed. In contrast, the class II histone deacetylases, that closely resemble the yeast HDA1 protein, consist of larger proteins ranging from 100 to 130 kDa. Their expression patterns tend to be tissue-specific. Finally, the yeast SIR2 protein and its homologues form the third class. These enzymes contain nicotinamide adenine dinucleotide (NAD+)-dependent activities that are insensitive to the histone deacetylase inhibitor trichostatin A. It has been demonstrated that the class I-member HDAC1 can interact with Tax and that HDAC1 represses the *trans*-activation function of Tax (Ego et al., 2002). Thus, HDAC1 could compete with CBP in binding with Tax to control negatively viral transcription (Fig. 5).

Tax is well known to be responsible for a strong CTL response to HTLV-I-infected cells. In asymptotic human carriers of HTLV-I and in patients with TSP /HAM, the CTL response is mainly directed against Tax (Jacobson et al., 1990). Moreover, CTLs specific to Gag, Pol, and Env have also been detected (Parker et al., 1992). Therefore viral expression has to be tightly regulated, otherwise the infected cells would be eliminated by the host immune system and then unable to survive. It is clear that HTLV-I-infected T cell has developed different strategies to escape the CTL response in order to maintain a state of persistent infection. The fact that the fraction of infected cell expressing HTLV-I proteins is considerably lower than cells containing provirus confirm that repression of HTLV-I transcription is one of these strategies.

Tax Repression of P53 is Mediated Through Competition for CBP

The transcription factor p53 is a tumor suppressor that plays important roles in cell cycle progression. p53 is activated in response to DNA damage and induction of p53 results in either cellular arrest in G_1 (mediated by the cyclin-dependent kinase inhibitor $p21^{WAF1/CIP1}$, for instance) or apoptosis (mediated by the tumor suppressor Bax, for example) (Evan and Littlewood, 1998; Gottlieb and Oren, 1996; Ko and Prives, 1996). Inactivation of p53 function allows the growth of cells that accumulate a number of mutations in various genes such as oncogenes, tumor suppressor genes, and related genes. The accumulation of DNA damage in HTLV-I transformed cells has been associated with Tax expression suggesting the inhibition of p53 function by Tax (Kao et al., 2000). Concerning the mechanism involves in the p53 inactivation, it has been first proposed that Tax could repress p53 gene transcription (Uittenbogaard et al., 1995) as already demonstrated for the genes encoding β-polymerase and $p56^{lck}$ kinase (Jeang et al., 1990; Lemasson et al., 1997; Uittenbogaard et al., 1995). However, elevated expressions of p53 found in several HTLV-I-transformed T-lymphocyte lines (Gartenhaus and Wang, 1995; Reid et al., 1993; Yamato et al., 1993) and Tax-immortalized T cells (Akagi et al., 1997) do not support this hypothesis. On the other hand, data obtained from transfection experiments using a p53-responsive reporter plasmid demonstrate that Tax alone is capable of suppressing the p53 transcriptional activity (Akagi et al., 1997; Cereseto et al., 1996; Mulloy et al., 1998; Pise-Masison et al., 1998a). Some viral oncoproteins, such as adenovirus E1B 55K, SV40 large T antigen, and papillomavirus E6 protein inactivate p53 by direct binding but it is not the case for HTLV-I since Tax is unable to interact with p53 (Yamato et al., 1993). Thus, Tax appears to compromise p53 function via an indirect mechanism.

Several reports have shown that the repression occurs as a consequence of competition between Tax and p53 for CBP since both proteins utilize the cellular cofactor as a mediator of transcriptional activation (Ariumi et al., 2000; Suzuki, Uchida-Toita, and Yoshida, 1999). Indeed, Tax mutated in its CBP/p300 binding site (Harrod et al., 1998) is no more capable of repressing p53 activity (Ariumi et al., 2000; Van Orden et al., 1999a). Since CBP/p300 has been described to be present at limiting level in the cell, binding of CBP/p300 to Tax could down-regulate p53-mediated transcription by reducing the level of CBP/p300 on the p53-responsive promoters (Ariumi et al., 2000; Suzuki, Uchida-Toita, and Yoshida, 1999; Van Orden et al., 1999a). This model has also been proposed to explain the functional impairment of p73 and p51, the p53-related factors, and the bHLH (helix-loop-helix) transcription factors by Tax (Kaida et al., 2000; Suzuki, Uchida-Toita, and Yoshida, 1999). Tax could also attenuate p53 activity by abrogating the interaction between p53 and the KIX domain of CBP (Fig. 7). Indeed, it has been demonstrated that the activating domain of p53 participates in KIX binding and that this interaction is critical for p53 activity (Van Orden et al., 1999a). Thus, Tax, that has been shown to bind this region, could abrogate the association between p53 and CBP/p300 by masking the KIX domain. Such a mechanism could also be involved in repression by Tax of other cellular transcription factors such as c-Myb, c-jun, and MyoD (Colgin and Nyborg, 1998; Riou et al., 2000; Van Orden et al., 1999b).

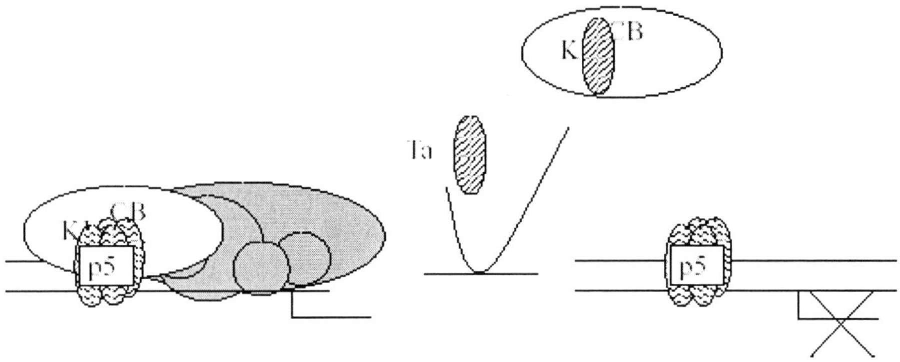

Figure 7. Model of p53 repression by Tax. By competiting with p53 for CBP binding, Tax destabilizes this complex and inhibits p53-dependent transcription.

Alternatively, other studies have indicated that Tax repression of p53 may occur through alteration in the phosphorylation state of p53 by post-transcriptional modification (Pise-Masison et al., 2000; Pise-Masison et al., 1998b). Pise-Masison et al. have suggested that p53 would be inactivated by Tax in T cells by phosphorylation of p53 on serines 15 and 392 (Pise-Masison et al., 2000; Pise-Masison et al., 2001). This hyperphosphorylated form of p53 would be unable to interact *in vivo* with TBP and Mdm-2 that is a cellular protein involved in p53 degradation (Pise-Masison et al., 1998b). It is noteworthy that the half-life of p53 is actually increased in HTLV-I-transformed T lymphocytes (Reid et al., 1993) compared with the rapid turnover observed in normal cells. However, the mechanism by which Tax could activate the phosphorylation of p53 and the kinase involved in this modification remain still unknown. Moreover, these observations are based on the use of the Tax mutants M22, that fails to activate NF-κB pathway (Smith and Greene, 1990), and V89A, that is unable to interact with CBP (Harrod et al., 1998). By using these mutants, Pise-Masison et al. showed that Tax activation of NF-κB was critical for Tax-mediated p53 inhibition and that there was no correlation between CBP binding and Tax-mediated p53 inhibition (Pise-Masison et al., 2000). Moreover, in Jurkat cells, expression of exogenous p300 fails to rescue p53 activity in the presence of Tax. Taken together, these results argue against the model that suggests that the interaction of Tax with the coactivator CBP is responsible for sequestering the coactivator from p53 and thus attenuating its transcriptional function. To explain this difference, they suggest that Tax utilizes distinct mechanisms to inhibit p53 function that are cell type-dependent. Whereas the CREB/ATF pathway would be important for p53 inhibition in HeLa, H1299, and Saos-2 cells, the NF-κB pathway would be involved in p53 inhibition in Jurkat cells and mouse embryo fibroblasts (Pise-Masison et al., 2001). However, other results do not favor such an explanation. Indeed, Van et al. (2001) found that $CREB^+$ $NF\kappa B^-$ but not $CREB^-$ $NFkB^+$ Tax mutants repressed p53 activity in HeLa, Soas-2, and Jurkat lymphocytes. It is difficult to know the exact reasons for this difference.

Tax Activates NF-κB Pathway

The NF-κB proteins are critical for activating the expression of cellular genes that are involved in the control of the immune and inflammatory response and in protecting cells from apoptosis in response to a variety of stress stimuli. The prototypical NF-κB complex corresponds to a heterodimer of the p50 (NF-κB1) and p65 (RelA) members of the NF-κB/Rel family of transcription factors. Other factors in this family include RelB, c-Rel, and p52 (NF-κB2). NF-κB is sequestered in the cytoplasm of most cells, where it is bound to a family of inhibitory proteins, including IκBα, IκBβ, and the related proteins p105 (NF-κB1) and p100 (NF-κB2) (Baldwin Jr, 1996). Cellular stimulation with a variety of agents (TCR/CD28 ligands, mitogens, cytokines) increases the activity of IκB kinases IKKα and IKKβ, resulting in enhanced phosphorylation of the IκB proteins (Karin, 1999). IKKβ is the critical kinase required for cytokine-induced phosphorylation of the IκB proteins, leading to their ubiquitination and subsequent degradation by the proteasome (Mercurio et al., 1997; Woronicz et al., 1997; Zandi et al., 1997). In contrast, IKKα is not required for IκB degradation, but is involved in an alternative pathway of NF-κB activation that leads to the processing of p100 to p52 (Senftleben et al., 2001). In addition to IKKα and IKKβ, an additional factor known as IKKγ/NEMO is also critical for the activity of the IKK complex (Li et al., 1999b; Mercurio et al., 1999; Rothwarf et al., 1998). IKKγ/NEMO interacts with IKKα and IKKβ and is a component of the high molecular mass IKK complex that migrates between 600 and 900 kDa following gel filtration chromatography. Although IKKγ/NEMO itself does not have kinase activity, it is essential for NF-κB activation likely via a scaffold function that is required for IKK activity. Additional proteins have been found to interact with the IKK complex including MEK kinase (MEKK1), NF-kB-inducing kinase (NIK), receptor-interacting protein (RIP), and IKK complex associated protein. IKKα and IKKβ share significant homology and contain three identical structural domains, namely a protein kinase domain at their N termini, a ZIP and a HLH motif at their C termini. IKKα and IKKβ can form homo- and heterodimers through their ZIP. IKKγ/NEMO does not contain a catalytic domain and is composed of three large α-helical regions: an N-terminal domain involved in its interactions with IKKβ, a coiled-coil domain, which mediates its oligomerization critical in stimulating IKK activity, and a C-terminal ZIP involved in the recruitment of upstream factors such as RIP.

Tax is a potent inducer of NF-κB activity. Tax induces translocation of NF-κB from the cytoplasm into the nucleus, apparently by liberating NF-κB from several distinct cytoplasmic complexes (Kanno et al., 1994; Sun et al., 1994). In particular, degradation of IκB-α is enhanced in the presence of Tax, suggesting that some NF-κB can be released by rapid turnover of this inhibitor (Lacoste et al., 1995). The question arises how Tax induces degradation of the inhibitory proteins. Based on observed binding of Tax to various IκB and NF-κB proteins (Kanno et al., 1994; Munoz et al., 1994; Suzuki et al., 1995), it has been proposed that Tax effects dissociation of the inhibitor and then favors nuclear translocation of NF-κB. Moreover, Tax could also anchor IκB proteins to the 26S proteasome (Petropoulos and Hiscott, 1998; Rousset et al., 1996). However, in contrast to these results, Sun et al. (1994) demonstrated that the association of Tax and p100 leads to the cytoplasmic sequestration of Tax and an inhibition of Tax-mediated activation rather than participating in the induction of nuclear transcription factor by promoting cleavage of p100 to p52. Moreover, different observations suggest that Tax may have an indirect effect, possibly by affecting physiologic signal transduction pathways, which then leads to activation of NF-κB. Indirect support for the latter model comes from the observation that Tax-expressing cells have detectable amounts of phosphorylated IκB-α (Kanno et al., 1995; Sun et al., 1994) and that the IKK complex is the primary cellular target of HTLV-I. Indeed, the IKK complex is constitutively activated in HTLV-I-infected cells, but not in Tax-deficient Jurkat T cells (Chu et al., 1998; Geleziunas et al., 1998; Li et al., 1999a; Uhlik et al., 1998). In addition, Tax is stably associated with the IKK complex in HTLV-I-infected T cells and in Tax-expressing cells (Chu et al., 1998; Li et al., 1999a). At last, several studies indicate that Tax-mediated IKK activation is due in part to a direct interaction between Tax and IKK-γ (Chu et al., 1999; Harhaj and Sun, 1999; Jin et al., 1999).

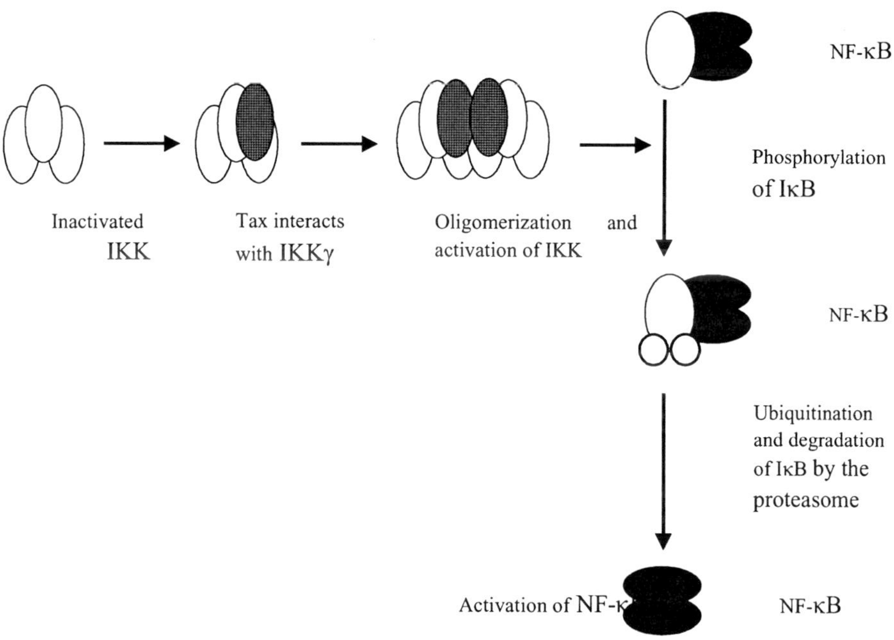

Figure 8. Proposed mechanism of NF-κB activation by Tax. Tax interacts with the IKK complex by binding to IKKγ. Tax self-association guides IKK oligomerization, which triggers IKK activation. Phosphorylation of IκB proteins leads to their ubiquitination and then their degradation by the proteasome.

A potential model that could explain how the IKK complex is stimulated by Tax is activation by oligomerization. In support of this model, after TNF stimulation, the IKK complex undergoes a structural change that leads to a significant increase in its molecular mass, as observed using gel filtration. Thus, prior to TNF stimulation, the IKK components exist as a ~650-kDa heteromeric complex of IKKα, IKKβ, and IKKγ. Following TNF stimulation, RIP bound to the TNF receptor TNF-R1 recruits this IKK complex through interaction with IKKγ and induces its oligomerization. This oligomerization results in activation of the IκB kinases. The activated, high molecular mass ~1.3 MDa IKK complex is then released from the TNF receptor and is able to phosphorylate the IκB proteins (Poyet et al., 2000). The oligomerization of IKKγ is mediated through a C-terminal domain comprising residues 251-419, whereas its N-terminal half interacts with IKKα and IKKβ. Tax has also been shown to enhance IKKγ oligomerization by interacting with a central domain of IKKγ comprising residues 201-250 (Huang et al., 2002). Thus, as already described for ATF/CREB proteins, Tax self-association could guide the oligomerization of the IKK complex. In this model, a Tax monomer could interact with inactivated IKK complex via its binding to an IKKγ monomer and then Tax dimerization would bring them into close proximity to oligomerize (Fig. 8). In support of this model, it is worth to note that the Tax mutant M22, that does not activate NF-kB pathway, is unable to dimerize (Basbous et al., 2003b; Tie et al., 1996).

HTLV-I Activates AP-1 Pathway

The AP-1 transcription factors play an important role in regulating the expression of various genes involved in the control of cell proliferation, death, and differentiation (Shaulian and Karin, 2002). Various AP-1 components, including c-Jun, are essential for embryonal development and can induce oncogenic transformation upon chronic activation in avian and mammalian cells (Mechta-Grigoriou et al., 2001; Van Dam and Castellazzi, 2001; Vogt, 2001). The AP-1 transcription factor consists of heterodimers between the Jun (c-Jun, JunB, and JunD) and Fos (c-Fos, FosB, Fra-1, and Fra-2) family members through their ZIP domains. Jun proteins form both homodimers and heterodimers with Fos proteins, but Fos proteins are unable to homodimerize and require heterodimerization to bind to DNA. Dimers formed by Fos and Jun bind with the highest affinity to the asymmetric AP-1 site TGA(C/G)TCA but they also bind to the CRE sequence with only slightly lower affinity (Chinenov and Kerppola, 2001; Hai et al., 1988; Ryseck and Bravo, 1991). The relative binding affinities of the AP-1 transcription factors depend on the specific DNA sequence, on the promoter context, and on the dimer combinations (Halazonetis et al., 1988; Kerppola and Curran, 1991; Ryseck and Bravo, 1991). For instance, c-Jun can bind to DNA either as homodimers or as heterodimers but binding is stronger for c-Jun/c-Fos heterodimers, resulting in a most potent transcriptional activation (Halazonetis et al., 1988). Moreover, c-Fos and FosB contain transcriptional activation domains whereas Fra-1, Fra-2 and the alternative splicing variants of FosB, FosB-2 and ΔFosB, do not. For c-Jun, dimer combinations with weaker transactivating capacities include cJun/JunB, c-Jun/ΔFosB, and c-Jun/Fra-2 (Deng and Karin, 1993; Nakabeppu and Nathans, 1991; Suzuki et al., 1991). The AP-1 complexes are not limited to Jun and Fos dimers, since Jun and Fos proteins have been shown to dimerize with other bZIP proteins, including members of the ATF/CREB family (ATF-2, ATFa, ATF-3, and ATF-4), and the Maf transcription factors (c-Maf, MafA, MafB, MafG/F/K and Nrl), generating complexes with altered binding specificities.

Previous studies have also shown that T-cell lines transformed by HTLV-I express high levels of AP-1 activity (Iwai et al., 2001; Mori et al., 2000) with increased levels of mRNAs encoding the AP-1 proteins c-Jun, JunB, JunD, c-Fos, and Fra-1 (Fujii et al., 1991; Hooper et al., 1991), Tax inducing expression of genes coding for c-Fos (Alexandre and Verrier, 1991; Fujii et al., 1988; Nagata et al., 1989), Fra-1 (Fujii et al., 1991; Tsuchiya et al., 1993), c-Jun (Fujii et al., 1991), and JunD (Fujii et al., 1991). In addition, AP-1 sites have been characterized as responsive elements for Tax in different cellular genes including *fra-1* (Tsuchiya et al., 1993), *IL-2* (Curtiss et al., 1996), *IL-5* (Yamagata et al., 1997), *IL-8* (Mori et al., 1998), *TR3/nur77* (Liu et al., 1999), and *TIMP-1* (Uchijima et al., 1994). It has been suggested that Tax could be involved in the AP-1 regulation at a posttranscriptional level (Iwai et al., 2001; Mori et al., 2000) but the exact mechanism of activation remains poorly understood. On the other hand, it is obvious that Tax is not the only player for constitutive activation of AP-1 in HTLV-I-infected cells since the leukemic cells of patients with ATL display constitutive AP-1 DNA binding activity involving JunD (Mori et al., 2000) but do not express significant levels of Tax.

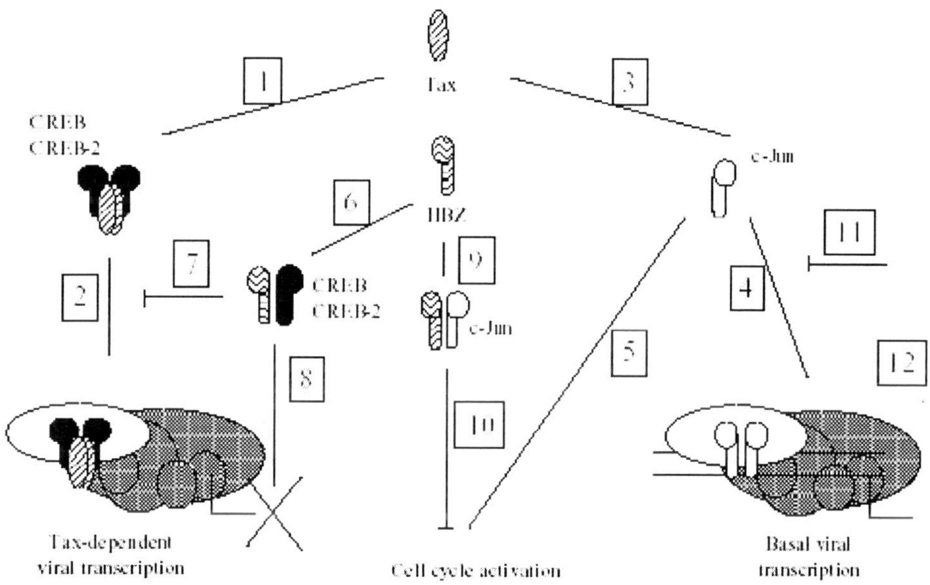

Figure 9. Schematic representation of the effects of Tax and HBZ on viral and cellular transcription. By interacting with CREB or CREB-2, Tax is capable of binding to the HTLV-I promoter [1]. Then, Tax recruits CBP, which induces viral transcription [2]. By stimulating c-Jun activity [3], not only Tax can favor the basal viral transcription (4) but also activates cell cycle of the infected T lymphocyte [5]. On the other hand, HBZ has an opposite effect on viral transcription and cell cycle progression. Indeed, by interacting with CREB or CREB-2 [6], HBZ inhibits the recruitment of Tax on the HTLV-I promoter [7] and then down-regulates Tax-dependent viral transcription [8]. HBZ can also interact with c-Jun [9], which attenuates cell cycle activation of the infected lymphocyte (10), and which blocks the fixation of c-Jun on the viral promoter [11] and for this reason the basal viral transcription [12].

Another viral product, HBZ, is able to modulate Jun protein activity via protein-protein interactions (Basbous et al., 2003a; Thébault et al., 2004). Whereas HBZ inhibits c-Jun activity (Basbous et al., 2003a), this viral factor is capable of activating JunB and JunD (Thébault et al., 2004). However, c-Jun and JunB or JunD display opposite effects on cell cycle progression. While c-Jun is a strong transcriptional activator of cyclin D1, JunD is a weak positive activator and JunB even an inhibitor of this promoter. Increasing the abundance of JunB or JunD relative to c-Jun decreases cyclin D1 transcription. The temporal changes in AP-1 composition and activity during the M-G1 transition may constitute an interesting way to control cell cycle progression (Mechta-Grigoriou et al., 2001). Overexpression of c-Jun or JunD exerts antagonistic effects on fibroblasts. c-Jun promotes cell proliferation and cooperates with activated *ras* to transform cells, whereas JunD overexpression slows cell growth and antagonizes transformation by *ras*. From these obsrvations, it is tempting to speculate that HBZ could control cell cycle progression. Tax has been described to accelerate cell cycle progression (Lemoine and Marriott, 2001; Low et al., 1997; Schmitt et al., 1998). This effect is mediated at a point between entry into the cell cycle at G1 phase and the G1/S phase transition (Lemoine and Marriott, 2001). On the other hand, HBZ could inhibit infected cell proliferation by negatively controlling Tax expression and by modulating AP-1 activity (Fig. 9).

Conclusion

It is clear that the main function of Tax is to take control of the infected cell by acting at the transcriptional level. By interacting with CREB and CREB-2, Tax *trans*-activates viral transcription. By stimulating NF-κB (via interaction with IKKγ) and AP-1 pathways, Tax favors the proliferation of the infected cell. This proliferation increases the proviral load and allows HTLV-I to disseminate through body compartment such as the CNS in the case of TSP/HAM. However, the T cells activated by the HTLV-I infection represent a pool of proliferating cells that are not oncogenic themselves, but which provide a population from which a malignant clone may subsequently arise. This accounts for the long latent period between infection and manifestation of a tumor in HTLV-I disease. Thus T-cell activation by Tax and oncogenesis are separate but related processes. Tax may initiate a cascade of events leading toward transformation that at later stages may not require expression of Tax itself. Indeed, even when detected in ATL cells, Tax expression is low and occurs in only a small percent of cells (Takeda et al., 2004).

Among these events, the loss of genomic stability is an important step in the transformation process. Tax negatively interferes with some DNA repair functions of the host cell. Tax represses transcription from the human DNA polymerase β promoter (Jeang et al., 1990), an enzyme involved in base excision repair. In the same way, Tax inhibits nucleotide excision repair by *trans*-activating the promoter of the proliferating cell nuclear antigen (PCNA) (Ressler et al., 1997). PCNA is a required cofactor of DNA Pol δ, an enzyme involved in both DNA replication and DNA repair. By increasing the PCNA-to-Pol δ ratio, Tax may deregulate Pol δ DNA repair activity (Kao and Marriott, 1999). Furthermore, Tax functionally inhibits the human mitotic checkpoint MAD1 (Jin et al., 1998). In the presence of Tax, MAD1 is mislocalized from the nucleus to the cytoplasm (Kasai et al., 2002). Loss of MAD1 function leads to a loss of the mitotic spindle assembly checkpoint and to chromosomal aberrations. Thus, by inhibiting p53 and MAD1 functions, Tax favors the accumulation of DNA mutations and chromosomal aberrations in the infected cell and can play a role in the initiation of leukemogenesis.

However, Tax can also be a brake for cellular proliferation of the infected cell. Indeed, Tax-expressing cells are rejected by the host immune response. Therefore, during leukemogenesis, there may be a stage at which balance of competition develops between the advantages and disadvantages of Tax for cell proliferation. At such a stage, it is obvious that proliferated cells that elicit weaker CTL responses are selected during the development of ATL.

Aknowledgements

This work was supported by institutional grants from the Centre National de la Recherche Scientifique (CNRS) and the Université Montpellier I (UM I), and a grant from the Association pour la Recherche sur le Cancer (ARC n° 3366).

References

Adya, N., and Giam, C.-Z. (1995). Distinct regions in human T-cell lymphotropic virus type 1 Tax mediate interactions with activator protein CREB and basal transcription factors. *J. Virol* 69, 1834-1841.

Adya, N., Zhao, L.-J., Huang, W., Boros, I., and Giam, C.-Z. (1994). Expansion of CREB's DNA recognition specificity by Tax results from interaction with Ala-Ala-Arg at positions 282-284 near the conserved DNA-binding domain of CREB. *Proc. Natl. Acad. Sci. USA* 91, 5642-5646.

Agre, P., Johnson, P. F., and McKnight, S. L. (1989) Cognate DNA binding specificity retained after leucine zipper exchange between GCN4 and C/EBP. *Science* 246, 922-926.

Akagi, T., Ono, H., Tsuchida, N., and Shimotohno, K. (1997). Aberrant expression and function of p53 in T-cells immortalized by HTLV-I Tax1. *FEBS Lett.* 406, 263-266.

Albrecht, B., and Lairmore, M. D. (2002). Critical role of human T-lymphotropic virus type 1 accessory proteins in viral replication and pathogenesis. *Microbiol. Mol. Biol. Rev.* 66, 396-406.

Alexandre, C., and Verrier, B. (1991). Four regulatory elements in the human c-*fos* promoter mediate transactivation by HTLV-1 Tax protein. *Oncogene* 6, 543-551.

Anderson, M. G., and Dynan, W. S. (1994). Quantitative studies of the effect of HTLV-I Tax protein on CREB protein-DNA binding. *Nucleic Acids Res.* 22, 3194-3201.

Aoyama, N., Nagase, T., Sawasaki, T., Mizuguchi, G., Nakagoshi, H., Fujisawa, J.-I., Yoshida, M., and Ishii, S. (1992). Overlap of the p53-responsive element and cAMP-responsive element in the enhancer of human T-cell leukemia virus type I. *Proc. Natl. Acad. Sci. USA* 89, 5403-5407.

Ariumi, Y., Kaida, A., Lin, J.-Y., Hirota, M., Masui, O., Yamaoka, S., Taya, Y., and Shimotohno, K. (2000). HTLV-1 Tax oncoprotein represses the p53-mediated *trans*-activation function through coactivator CBP sequestration. *Oncogene* 19, 1491-1499.

Askjaer, P., and Kjems, J. (1998). Mapping of multiple RNA binding sites of human T-cell lymphotropic vitus type I Rex protein within 5'- and 3'-Rex response elements. *J. Biol. Chem.* 273, 11463-11471.

Baldwin Jr, A. S. (1996). The NF-κB and IκB proteins: New discoveries and insights. *Annu. Rev. Immunol.* 14, 649-683.

Bantignies, F., Rousset, R., Desbois, C., and Jalinot, P. (1996). Genetic characterization of transactivation of the human T-cell leukemia virus type 1 promoter: binding of Tax to Tax-responsive element 1 is mediated by the cyclic AMP-responsive members of the CREB/ATF family of transcription factors. *Mol. Cell. Biol.* 16, 2174-2182.

Basbous, J., Arpin, C., Gaudray, G., Piechaczyk, M., Devaux, C., and Mesnard, J. M. (2003a). HBZ factor of HTLV-I dimerizes with transcription factors JunB and c-Jun and modulates their transcriptional activity. *J. Biol. Chem.* 278, 43620-43627.

Basbous, J., Bazarbachi, A., Granier, C., Devaux, C., and Mesnard, J. M. (2003b). The central region of human T-cell leukemia virus type 1 Tax protein contains distinct domains involved in subunit dimerization. *J. Virol.* 77, 13028-13035.

Beimling, P., and Moelling, K. (1989). Isolation and characterization of the tax protein of HTLV-I. *Oncogene* 4, 511-516.

Beimling, P., and Moelling, K. (1992). Direct interaction of CREB protein with 21 bp Tax-response elements of HTLV-I LTR. *Oncogene* 7, 257-262.

Benvenisty, N., Ornitz, D. M., Bennett, G. L., Sahagan, B. G., Kuo, A., Cardiff, R. D., and Leder, P. (1992). Brain tumours and lymphomas in transgenic mice that carry HTLV-I LTR/c-myc and Ig/tax genes. *Oncogene* 7, 2399-2405.

Bodor, J., Spetz, A.-L., Strominger, J. L., and Habener, J. F. (1996). cAMP inducibility of transcriptional repressor ICER in developing and mature human T lymphocytes. *Proc. Natl. Acad. Sci. USA* 93, 3536-3541.

Bodor, J., Walker, W., Flemington, E., Spetz, A.-L., and Habener, J. F. (1995). Modulation of Tax and PKA-mediated expression of HTLV-I promoter via cAMP response element binding and modulator proteins CREB and CREM. *FEBS Letters* 377, 413-418.

Bosselut, R., Lim, F., Romond, P.-C., Frampton, J., Brady, J., and Ghysdael, J. (1992). Myb protein binds to multiple sites in the human T cell lymphotropic virus type 1 long terminal repeat and transactivtes LTR-mediated expression. *Virology* 186, 764-769.

Brady, J., Jeang, K. T., Duvall, J., and Khoury, G. (1987). Identification of p40x-responsive regulatory sequences within human T-cell leukemia virus type I long terminal repeat. *J. Virol.* 61, 2175-2181.

Brauweiler, A., Garl, P., Franklin, A. A., Giebler, H. A., and Nyborg, J. K. (1995). A molecular mechanism for human T-cell leukemia virus latency and Tax transactivation. *J. Biol. Chem.* 270, 12814-12822.

Cann, A. J., and Chen, I. S. Y. (1996). Human T-cell leukemia virus type I and II. *Fields Virology, Third Edition* Chapter 59, 1849-1880.

Cann, A. J., Rosenblatt, J. D., Wachsman, W., Shah, N. P., and Chen, I. S. (1985). Identification of the gene responsible for human T-cell leukemia virus transcriptional regulation. *Nature* 318, 571-574.

Caron, C., Mengus, G., Dubrowskaya, V., Roisin, A., Davidson, I., and Jalinot, P. (1997). Human $TAF_{II}28$ interacts with the human T cell leukemia virus type I Tax transactivator and promotes its transcriptional activity. *Proc. Natl. Acad. Sci. USA* 94, 3662-3667.

Caron, C., Rousset, R., Béraud, C., Moncollin, V., Egly, J.-M., and Jalinot, P. (1993). Functional and biochemical interaction of the HTLV-I Tax1 transactivator with TBP. *EMBO J.* 12, 4269-4278.

Cereseto, A., Diella, F., Mulloy, J. C., Cara, A., Michieli, P., Grassmann, R., Franchini, G., and Klotman, M. E. (1996). p53 functional impairment and high $p21^{waf1/cip1}$ expression in human T-cell lymphotropic/leukemia virus type I-transformed T cells. *Blood* 88, 1551-1560.

Chinenov, Y., and Kerppola, T. K. (2001). Close encounters of many kinds: Fos-Jun interactions that mediate transcription regulatory specificity. *Oncogene* 20, 2438-2452.

Chrivia, J. C., Kwok, R. P. S., Lamb, N., Hagiwara, M., Montminy, M. R., and Goodman, R. H. (1993). Phosphorylated CREB binds specifically to the nuclear protein CBP. *Nature* 376, 855-859.

Chu, Z.-L., DiDonato, J. A., Hawiger, J., and Ballard, D. W. (1998). The Tax oncoprotein of human T-cell leukemia virus type 1 associates with and persistently activates IκB kinases containing IKKα and IKKβ. *J. Biol. Chem.* 273, 15891-15894.

Chu, Z.-L., Shin, Y.-A., Yang, J.-M., DiDonato, J. A., and Ballard, D. W. (1999). IKKγ mediates the interaction of cellular IκB kinases with the Tax transforming protein of human T-cell leukemia virus type 1. *J. Biol. Chem.* 274, 15297-15300.

Chun, A., Zhou, Y., Wong, C.-M., Kung, H.-K., Jeang, K.-T., and Jin, D.-Y. (2000). Coiled-coil motif as a structural basis for the interaction of HTLV type I Tax with cellular cofactors. *AIDS Res. Hum. Retrov.* 16, 1689-1694.

Clark, N. M., Smith, M. J., Hilfinger, J. M., and Markovitz, D. M. (1993). Activation of the human T-cell leukemia virus type I enhancer is mediated by binding sites for Elf-1 and the pets factor. *J. Virol.* 67, 5522-5528.

Clemens, K. E., Piras, G., Radonovich, M. F., Choi, K. S., Duvall, J. F., DeJong, J., Roeder, R., and Brady, J. N. (1996). Interaction of the human T-cell lymphotropic virus type 1 Tax transactivator with transcription factor IIA. *Mol. Cell. Biol.* 16, 4656-4664.

Colgin, M. A., and Nyborg, J. K. (1998). The human T-cell leukemia virus type I oncoprotein Tax inhibits the transcriptional activity of c-Myb through competition for the CREB binding protein. *J. Virol.* 72, 9396-9399.

Curtiss, V. E., Smilde, R., and McGuire, K. L. (1996). Requirements for interleukin 2 promoter transactivation by the Tax protein of human T-cell leukemia virus type 1. *Mol. Cell. Biol.* 16, 3567-3575.

D'Agostino, D. M., Ranzato, L., Arrigoni, G., Cavallari, I., Belleudi, F., Torrisi, M. R., Silic-Benussi, M., Ferro, T., Petronilli, V., Marin, O., Chieco-Bianchi, L., Bernardi, P., and Ciminale, V. (2002). Mitochondrial alterations induced by the $p13^{II}$ protein of human T-cell leukemia virus type I. *J. Biol. Chem.* 277, 34424-34433.

Dan, S., Tanimura, A., and Yoshida, M. (1999). Interaction of Gli2 with CREB protein on DNA elements in the long terminal repeat of human T-cell leukemia virus type 1 is responsible for transcriptional activation by Tax protein. *J. Virol* 73, 3258-3263.

Dasgupta, P., Reddy, C. D., Saikumar, P., and Reddy, E. P. (1992). The cellular proto-oncogene product Myb acts as transcriptional activator of the long terminal repeat of human T-lymphotropic virus type I. *J. Virol.* 66, 270-276.

Deeper, J. M., Leonard, W. J., Kronke, M., Waldmann, T. A., and Greene, W. C. (1984). Augmented T-cell growth receptor expression in HTLV-I-infected human leukemic T cells. *J. Immunol.* 133-140.

Deng, T., and Karin, M. (1993). JunB differs from c-Jun in its DNA-binding and dimerization domains, and represses c-Jun by formation of inactive heterodimers. *Genes Dev.* 7, 479-490.

Ding, W., Kim, S.-J., Nair, A. M., Michael, B., Boris-Lawrie, K., Tripp, A., Feuer, G., and Lairmore, M. D. (2003). Human T-cell lymphotropic Virus type 1 $p12^{I}$ enhances interleukin-2 production during T-cell activation. *J. Virol.* 77, 11027-11039.

Dittmer, J., Pise-Masison, C. A., Clemens, K. E., Choi, K. S., and Brady, J. N. (1997). Interaction of human T-cell leukemia virus type I Tax, Ets1, and Sp1 in transactivation of the PTHrP P2 promoter. *J. Biol. Chem.* 272, 4953-4958.

Ego, T., Ariumi, Y., and Shimotohno, K. (2002). The interaction of HTLV-1 Tax with HDAC1 negatively regulates the viral gene expression. *Oncogene* 21, 7241-7246.

Evan, G., and Littlewood, T. (1998). A matter of life and cell death. *Science* 281, 1317-1326.

Felber, B. K., Paskalis, H., Kleinman-Ewing, C., Wong-Staal, F., and Pavlakis, G. N. (1985). The pX protein of HTLV-I is a transcriptional activator of its long terminal repeats. *Science* 229, 675-679.

Ferreira Jr, O. C., Planelles, V., and Rosenblatt, J. D. (1997). Human T-cell leukemia viruses: epidemiology, biology, and pathogenesis. *Blood reviews* 11, 91-104.

Franklin, A. A., Kubik, M. F., Uittenbogaard, M. N., Brauweiler, A., Utaisincharoen, P., Matthews, M. A., Dynan, W. S., Hoeffler, J. P., and Nyborg, J. K. (1993). Transactivation by the human T-cell leukemia virus Tax protein is mediated through enhanced binding of activating transcription factor-2 (ATF-2) ATF-2 response and cAMP element-binding protein (CREB). *J. Biol. Chem.* 268, 21225-21231.

Fujii, M., Niki, T., Mori, T., Matsuda, T., Matsui, M., Nomura, N., and Seiki, M. (1991). HTLV-1 Tax induces expression of various immediate early serum responsive genes. *Oncogene* 6, 1023-1029.

Fujii, M., Sassone-Corsi, P., and Verma, I. M. (1988). c-fos promoter trans-activation by the tax1 protein of human T-cell leukemia virus type I. *Proc. Natl. Acad. Sci. USA* 85, 8526-8530.

Fujisawa, J. I., Seiki, M., Kiyokawa, T., and Yoshida, M. (1985). Functional activation of the long terminal repeat of human T-cell leukemia virus type I by a trans-acting factor. *Proc. Natl. Acad. Sci. USA* 82, 2277-81.

Fujisawa, J. I., Seiki, M., Sato, M., and Yoshida, M. (1986). A transcriptional enhancer sequence of HTLV-I is responsible for *trans*-activation mediated by p40x of HTLV-I. *EMBO J.* 5, 713-718.

Furukawa, Y., Osame, M., Kubota, R., Tara, M., and Yoshida, M. (1995). Human T-cell leukemia virus type-1 (HTLV-1) Tax is expressed at the same level in infected cells of HTLV-1-associated myelopathy or tropical spastic paraparesis patients as in asymptotic carriers but at a lower level in adult T-cell leukemia cells. *Blood* 85, 1865-1870.

Gachon, F., Devaux, C., and Mesnard, J. M. (2002). Activation of HTLV-I transcription in the presence of Tax is independent of the acetylation of CREB-2 (ATF-4). *Virology* 299, 271-278.

Gachon, F., Gaudray, G., Thebault, S., Basbous, J., Koffi, A. J., Devaux, C., and Mesnard, J. M. (2001). The cAMP response element binding protein-2 (CREB-2) can interact with the C/EBP-homologous protein (CHOP). *FEBS lett.* 502, 57-62.

Gachon, F., Péléraux, A., Thébault, S., Dick, J., Lemasson, I., Devaux, C., and Mesnard, J. M. (1998). CREB-2, a cellular CRE-dependent transcription repressor, functions in association with Tax as an activator of the human T-cell leukemia virus type 1 promoter. *J. Virol.* 72, 8332-8337.

Gachon, F., Thebault, S., Peleraux, A., Devaux, C., and Mesnard, J. M. (2000). Molecular interactions involved in the transactivation of the human T-cell leukemia virus type 1 promoter mediated by Tax and CREB-2 (ATF-4). *Mol. Cell. Biol.* 20, 3470-3481.

Gartenhaus, R. B., and Wang, P. (1995). Functional inactivation of wild-type p53 protein correlates with loss of IL-2 dependence in HTLV-I transformed human T lymphocytes. *Leukemia* 9, 2082-2086.

Gaudray, G., Gachon, F., Basbous, J., Biard-Piechaczyk, M., Devaux, C., and Mesnard, J. M. (2002). The complementary strand of HTLV-1 RNA genome encodes a bZIP transcription factor that down-regulates the viral transcription. *J. Virol.* 76, 12813-12822.

Gegonne, A., Bosselut, R., Bailly, R. A., and Ghysdaal, J. (1993). Synergistic activation of the HTLV-I LTR ets-responsive region by transcription factor ets1 and Sp1. *EMBO J.* 2, 1169-1178.

Geleziunas, R., Ferrel, S., Lin, X., Mu, Y., Cunningham, E. T., Grant, M., Connely, M. A., Hambor, J. E., Marcu, K. B., and Greene, W. C. (1998). Human T-cell leukemia virus type 1 Tax induction of NF-κB involves activation of the IκB kinase α (IKKα) and IKKβ cellular kinases. *Mol. Cell. Biol.* 18, 5157-5165.

Georges, S. A., Kraus, W. L., Luger, K., Nyborg, J. K., and Laybourn, P. J. (2002). p300-mediated Tax transactivation from recombinant chromatin: histone tail deletion mimics coactivator function. *Mol. Cell. Biol.* 22, 127-137.

Gessain, A., Saal, F., Gout, O., Daniel, M.-T., Flandrin, G., de The, G., Peries, J., and Sigaux, F. (1990). High human T-cell lymphotropic virus type I proviral DNA load with polyclonal integration in peripheral blood mononyclear cells of French West Indian, Guianese, and African patients with tropical spastic paraparesis. *Blood* 75, 428-433.

Gessain, A., Vernant, J. C., Maurs, L., Gout, O., Calander, A., and De Thé, G. (1985). Antibodies to human T-lymphotropic virus type 1 in patients with tropical spastic paraparesis. *Lancet* 2, 407-410.

Giam, C.-Z., and Xu, Y.-L. (1989). HTLV-I *Tax* gene product activates transcription via pre-existing cellular factors and cAMP responsive element. *J. Biol. Chem.* 264, 15236-15241.

Giebler, H. A., Loring, J. E., Van Orden, K., Colgin, M. A., Garrus, J. E., Escudero, K. W., Brauweiler, A., and Nyborg, J. K. (1997). Anchoring of CREB binding protein to the human T-cell leukemia virus type I promoter: a molecular mechanism of Tax transactivation. *Mol. Cell. Biol.* 17, 5156-5164.

Gitlin, S. D., Dittmer, J., Shin, R. C., and Brady, J. N. (1993). Transcriptional activation of the human T-lymphotropic virus type I long treminal repeat by functional interaction of Tax and Ets1. *J. Virol.* 67, 7307-7316.

Goren, I., Semmes, O. J., Jeang, K.-T., and Moelling, K. (1995). The amino terminus of Tax is required for interaction with the cyclic AMP response element binding protein. *J. Virol.* 69, 5806-5811.

Gottlieb, T. M., and Oren, M. (1996). p53 in growth control and neoplasia. *Biochem. Biophys. Acta* 1287, 77-102.

Grassmann, R., Berchtold, S., Radant, I., Alt, M., Fleckenstein, B., Sodroski, J. G., Haseltine, W. A., and Ramstedt, U. (1992). Role of human T-cell leukemia virus type I X-region proteins in immortalization of primary human lymphocytes in culture. *J. Virol.* 66, 4570-4575.

Grassmann, R., Dengler, C., Muller-Fleckenstein, I., McGuire, K., Dokhler, M. C., Sodroski, J. G., and Haseltine, W. A. (1989). Transformation to continuous growth of primary human T lymphocytes by human T-cell leukemia virus type I X-region genes transduced by a herpesvirus saimiri vector. *Proc. Natl. Acad. Sci. USA* 86, 3351-3355.

Gray, S. G., and Eksrtom, T. J. (2001). The human histone deacetylase family. *Exp. Cell. Res.* 262, 75-83.

Grone, M., Koch, C., and Grassmann, R. (1996). The HTLV-1 Rex protein induces nuclear accumulation of unspliced viral RNA by avoiding intron excision and degradation. *Virology* 218, 316-325.

Grossman, W. J., Kimata, J. T., Wong, F.-H., Zutter, M., Ley, T. J., and Ratner, L. (1995). Development of leukemia in mice transgenic for the *tax* gene of human T-cell leukemia virus type I. *Proc. Natl. Acad. Sci. USA* 92, 1057-1061.

Hai, T., Liu, F., Coukos, W. J., and Green, M. R. (1989). Transcription factor ATF cDNA clones: an extensive family of leucine zipper proteins able to selectively form DNA-binding heterodimers. *Genes Dev.* 3, 2083-2090.

Hai, T. W., Liu, F., Allegretto, E. A., Karin, M., and Green, M. R. (1988). A family of immunologically related transcription factors that includes multiple forms of ATF and AP-1. *Genes Dev.* 2, 1216-1226.

Halazonetis, T. D., Georgopoulos, K., Greenberg, M. E., and Leder, P. (1988). c-Jun dimerizes with itself and with c-Fos, forming complexes of different DNA binding affinities. *Cell* 55, 917-924.

Harhaj, E. W., Good, L., Xiao, G., and Sun, S.-C. (1999). Gene expression profiles in HTLV-I-immortalized T cells: Deregulated expression of genes involved in apoptosis regulation. *Oncogene* 18, 1341-1349.

Harhaj, E. W., and Sun, S.-C. (1999). IKKγ serves as a docking subunit of the IκB kinase (IKK) and mediates interaction of IKK with the human T-cell leukemia virus Tax protein. *J. Biol. Chem.* 274, 22911-22914.

Harrod, R., Kuo, Y.-L., Tang, Y., Yao, Y., Vassilev, A., Nakatani, Y., and Giam, C.-Z. (2000). p300 and p300/cAMP-responsive element-binding protein associated factor interact with human T-cell lymphotropic virus type-1 Tax in a multi-histone acetyltransferase/activator-enhancer complex. *J. Biol. Chem.* 275, 11852-11857.

Harrod, R., Tang, Y., Nicot, C., Lu, H. S., Vassilev, A., Nakatani, Y., and Giam, C.-Z. (1998). An exposed KID-like domain in human T-cell lymphotropic virus type 1 Tax is responsible for the recruitment of coactivators CBP/p300. *Mol. Cell. Biol.* 18, 5052-5061.

He, C. H., Gong, P., Hu, B., Stewart, D., Choi, M. E., Choi, A. M. K., and Alam, J. (2001). Identification of activating transcription factor 4 (ATF4) as an Nrf2-interacting protein. *J. Biol. Chem.* 276, 20858-20865.

Hidaka, M., Inoue, J., Yoshida, M., and Seiki, M. (1988). Post-transcriptional regulator (rex) of HTLV-1 initiates expression of viral structural proteins but suppresses expression of regulatory proteins. *EMBO J.* 7, 519-523.

Hinrichs, S. H., Nerenberg, M., Reynolds, R. K., Khoury, G., and Jay, G. (1987). A transgenic mouse model for human neurofibromatosis. *Science* 237, 1340-1343.

Hinuma, Y., Nagata, K., Hanaoka, M., Nakai, M., Matsumoto, T., Kinoshita, K. I., Shirakawa, S., and Miyoshi, I. (1981). Adult T-cell leukemia: antigen in an ATL cell line and detection of antibodies to the antigen in human sera. *Proc. Natl. Acad. Sci. USA* 78, 6476-6480.

Hivin, P., Gaudray, G., Devaux, C., and Mesnard, J. M. (2004). Interaction between C/EBPβ and Tax down-regulates human T-cell leukemia virus type I transcription. *Virology* 318, 556-565.

Hoeffler, J. P., Meyer, T. E., Yun, Y., Jameson, J. L., and Habener, J. F. (1988). Cyclic AMP-responsive DNA-binding protein: structure based on a cloned placental cDNA. *Science* 242, 1430-1433.

Höllsberg, P. (1997). Pathogenesis of chronic progressive myelopathy associated with human T-cell lymphotropic virus type I. *Acta Neurol. Scand.* 169, 86-93.

Höllsberg, P., and Hafler, D. A. (1993). Seminars in medicine of the Beth Israel Hospital, Boston. Pathogenesis of diseases induced by human lymphotropic virus type I infection. *N. Engl. J. Med.* 328, 1173-1182.

Hooper, W. C., Rudolph, D. L., Lairmore, M. D., and Lal, R. B. (1991). Constitutive expression of c-Jun and Jun-B in cell lines infected with human T-lymphotropic virus types I and II. *Biochem. Biophys. Res. Commun.* 181, 976-980.

Huang, G.-J., Zhang, Z.-Q., and Jin, D.-Y. (2002). Stimulation of IKK-γ oligomerization by the human T-cell leukemia virus oncoprotein Tax. *FEBS lett.* 531, 494-498.

Iwai, K., Mori, N., Oie, M., Yamamoto, N., and Fujii, M. (2001). Human T-cell leukemia virus type 1 Tax protein activates transcription through AP-1 site by inducing DNA binding activity in T cells. *Virology* 279, 38-46.

Jacobson, S., Shida, H., McFarlin, D. E., Fauci, A. S., and Koenig, S. (1990). Circulating CD8+ cytotoxic T lymphocytes specific to HTLV-I pX in patients with HTLV-I associated neurological disease. *Nature* 348, 245-248.

Jeang, K.-T., Chiu, R., Santos, E., and Kim, S.-J. (1991). Induction of the HTLV-I LTR by Jun occurs through the Tax-responsive 21-bp elements. *Virology* 181, 218-227.

Jeang, K.-T., Widen, S. G., Semmes IV, O. J., and Wilson, S. H. (1990). HTLV-I trans-activator protein, Tax, is a transrepressor of the human β–polymerase gene. *Science* 247, 1082-1084.

Jeang, K. T., Boros, I., Brady, J., Radanovich, M., and Khoury, G. (1988). Characterization of cellular factors that interact with the human T-cell leukemia virus type I p40x-responsive 21-base-pair sequence. *J. Virol.* 62, 4499-4509.

Jiang, H., Lu, H., Schiltz, R. L., Pise-Masison, C. A., Ogryzko, V. V., Nakatani, Y., and Brady, J., N. (1999). PCAF interacts with Tax and stimulates Tax transactivation in a histone acetyltransferase-independent manner. *Mol. Cell. Biol.* 19, 8136-8145.

Jin, D.-J., Giordano, V., Kibler, K. V., Nakano, H., and Jeang, K.-T. (1999). Role of adapter function in oncoprotein-mediated activation of NF-κB. *J. Biol. Chem.* 17402-17405.

Jin, D.-Y., and Jeang, K.-T. (1997). HTLV-I Tax self-association in optimal *trans*-activation function. *Nucleic Acids Res.* 25, 379-387.

Jin, D.-Y., Spencer, F., and Jeang, K.-T. (1998). Human T cell leukemia virus type 1 oncoprotein Tax targets the human mitotic checkpoint protein MAD1. *Cell* 93, 81-91.

Kaida, a., Ariumi, Y., Ueda, Y., Lin, J.-Y., Hijikata, M., Ikawa, S., and Shimotohno, K. (2000). Functional impairment of p73 and p51, the p53-related proteins, by the human T-cell leukemia virus type 1 Tax oncoprotein. *Oncogene* 19, 827-830.

Kanno, T., Brown, K., and Siebenlist, U. (1995). Evidence in support of a role for human T-cell leukemia virus type I Tax in activating NF-κB via stimulation of signaling pathway. *J. Biol. Chem.* 270, 11745-11748.

Kanno, T., Franzoso, G., and Siebenlist, U. (1994). Human T-cell leukemia virus type I Tax-protein-mediated activation of NF-κB from p100 (NF-κB)-inhibited cytoplasmic reservoirs. *Proc. Natl. Acad. Sci. USA* 91, 12634-12638.

Kao, S.-Y., and Marriott, S. J. (1999). Disruption of nucleotide excision repair by the human T-cell leukemia virus type 1 Tax protein. *J. Virol.* 73, 4299-4304.

Kao, S. Y., Lemoine, F. J., and Marriott, S. J. (2000). Suppression of DNA repair by HTLV-I Tax is rescued by a functional p53 signaling pathway. *J. Biol. Chem.* 275, 35926-35931.

Karin, M. (1999). How NF-κB is activated: The role of the IkappaB kinase (IKK) complex. *Oncogene* 18, 6867-6874.

Karpinski, B. A., Morle, G. D., Huggenvik, J., Uhler, M. D., and Leiden, J. M. (1992). Molecular cloning of human CREB-2: an ATF/CREB transcription factor that can negatively regulate transcription from the cAMP response element. *Proc. Natl. Acad. Sci. USA* 89, 4820-4824.

Kasai, T., Iwanaga, Y., Iha, H., and Jeang, K.-T. (2002). Prevalent loss of mitotic spindle checkpoint in adult T-cell leukemia confers resistance to microtubule inhibitors. *J. Biol. Chem.* 277, 5187-5193.

Kashanchi, F., Duvall, J. F., Dittmer, J., Mireskandari, A., Reid, R. L., Gitlin, S. D., and Brady, J. N. (1994). Involvement of transcription factor YB-1 in human T-cell lymphotropic virus type I basal gene expression. *J. Virol.* 68, 561-565.

Kerppola, T. K., and Curran, T. (1991). Fos-jun heterodimers and jun homodimers bend DNA in opposite orientations: implication for transcription factor cooperativity. *Cell* 66, 317-326.

Kim, S.-J., Ding, W., Albrecht, B., Green, P. L., and Lairmore, M. D. (2003). A conserved calcineurin-binding motif in human T lymphotropoic virus type 1 p12^{I} functions to modulate nuclear factor of activated T cell activation. *J. Biol. Chem.* 278, 15550-15557.

Kimzey, A. L., and Dynan, W. S. (1998). Specific regions of contact between human T-cell leukemia virus type I Tax protein and DNA identified by photocross-linking. *J. Biol. Chem.* 273, 13768-13775.

Kimzey, A. L., and Dynan, W. S. (1999). Identification of a human T-cell leukemia virus type I Tax peptide in contact with DNA. *J. Biol. Chem.* 274, 34226-34232.

Ko, L. J., and Prives, C. (1996). p53: puzzle and paradigm. *Genes Dev.* 10, 1054-1072.

Koiwa, T., Hamano-Usami, A., Ishida, T., Okayama, A., Yamaguchi, K., Kamahira, S., and Watanabe, T. (2002). 5'-long terminal repeat-selective CpG methylation of latent human T-cell leukemia virus type I provirus in vitro and in vivo. *J. Virol.* 76, 9389-9397.

Koralnik, I. J., Fullen, J., and Franchini, G. (1993). The p12^{I}, p13^{II}, and p30^{II} proteins encoded by human T-cell leukemia/lymphotropic virus type I open reading frames I and II are localized in three different cellular compartments. *J. Virol.* 67, 2360-2366.

Korber, B., Okayama, A., Donnelly, R., Tachibana, N., and Essex, M. (1991). Polymerase chain reaction analysis of defective human T-cell leukemia virus type I proviral genomes in leukemic cells of patients with adult T-cell leukemia. *J. Virol* 65, 5471-5476.

Kwok, R. P. S., Laurance, M. E., Lundblad, J. R., Goldman, P. S., Shih, H.-M., Connor, L. M., Marriott, S. J., and Goodman, R. H. (1996). Control of cAMP-regulated enhancers by the viral transactivator Tax through CREB and the coactivator CBP. *Nature* 370, 223-226.

Kwok, R. P. S., Lundblad, J. R., Chrivia, J. C., Richards, J. P., Bächinger, H. P., Brennan, R. G., Roberts, S. G. E., Green, M. R., and Goodman, R. H. (1994). Nuclear protein CBP is a coactivator for the transcription factor CREB. *Nature* 370, 223-226.

Lacoste, J., Petropoulos, L., Pepin, N., and Hiscott, J. (1995). Constitutive phosphorylation and turnover of IkBa in human T-cell leukemie virus type I-infected and Tax-expressing T cells. *J. Virol.* 69, 564-569.

Landschulz, W. H., Johnson, P. F., and McKnight, S. L. (1988). The leucine zipper: a hypothetical structure common to a new class of DNA binding proteins. *Science* 240, 1759-1764.

Larocca, D., Chao, L. A., Seto, M. H., and Brunck, T. K. (1989). Human T-cell leukemia virus minus strand transcription in infected T-cells. *Bioch. Biophys. Res. Comm.* 163, 1006-1013.

Laurance, M. E., Kwok, R. P. S., Huang, M. S., Richards, J. P., Lundblad, J. R., and Goodman, R. H. (1997). Differential activation of viral and cellular promoters by human T-cell lymphotropic virus-1 Tax and cAMP-responsive element modulator isoforms. *J. Biol. Chem.* 272, 2646-2651.

Lekstrom-Himes, J., and Xanthopoulos, K. G. (1998). Biological role of the CCAAT/enhancer-binding protein family of transcription factors. *J. Biol. Chem.* 273, 28545-28548.

Lemaigre, F. P., Ace, C. I., and Green, M. R. (1993). The cAMP response element binding protein, CREB, is a potent inhibitor of diverse transcriptional activators. *Nucleic Acids Res.* 21, 2907-2911.

Lemasson, I., and Nyborg, J. K. (2001). Human T-cell leukemia virus type I Tax repression of p73β is mediated through competition for the C/H1 domain of CBP. *J. Biol. Chem.* 276, 15720-15727.

Lemasson, I., Polakowski, N. J., Laybourn, P. J., and Nyborg, J. K. (2002). Transcription factor binding and histone modifications on the integrated proviral promoter in HTLV-I-infected T cells. *J. Biol. Chem.* 277, 49459-49465.

Lemasson, I., Robert-Hebmann, V., Hamaia, S., Duc Dodon, M., Gazzolo, L., and Devaux, C. (1997). Transrepression of *lck* gene expression by human T-cell leukemia virus type I-encoded $p40^{tax}$. *J. Virol.* 71, 1975-1983.

Lemoine, F. J., and Marriott, S. J. (2001). Accelerated G_1 phase progression induced by the human T cell leukemia virus type I (HTLV-I) Tax oncoprotein. *Oncogene* 276, 31851-31857.

Lenzmeier, B. A., Giebler, H. A., and Nyborg, J. K. (1998). Human T-cell leukemia virus type 1 Tax requires direct access to DNA for recruitment of CREB binding protein to the viral promoter. *Mol. Cell. Biol.* 18, 721-731.

Li, X. H., Murphy, K. M., Palka, K. T., Surabhi, R. M., and Gaynor, R. B. (1999a). The human T-cell leukemia vitus type-1 Tax protein regulates the activity of the IκB kinase complex. *J. Biol. Chem.* 274, 34417-34424.

Li, Y., Kang, J., Friedman, J., Tarassishin, L., Ye, J., Kovalenko, A., Wallach, D., and Horwitz, M. S. (1999b). Identification of a cell protein(FIP-3) as a modulator of NF-κB activity and as a target of an adenovirus inhibitor of tumor necrosis factor α-induced apoptosis. *Proc. Natl. Acad. Sci. U. S. A.*(96), 1042-1047.

Liang, G., and Hai, T. (1997). Characterization of human activating transcription factor 4, a transcriptional activator that interacts with multiple domains of cAMP-responsive element-binding protein (CREB)-binding protein (CBP). *J. Biol. Chem.* 272, 24088-24095.

Liu, X., Chen, X., Zachar, V., Chang, C., and Ebbesen, P. (1999). Transcriptional activation of human TR3/nur77 gene expression by human T-lymphotropic virus type I Tax protein through two AP-1-like elements. *J. Gen. Virol.* 80, 3073-3081.

Low, K. G., Dorner, L. F., Fernando, D. B., Grossman, J., Jeang, K., and Comb, M. J. (1997). Human T-cell leukemia virus type I Tax releases cell cycle arrest induced by p16^{INK4A}. *J. Virol.* 71, 1956-1962.

Lu, H., Pise-Masison, C. A., Fletcher, T. M., Schiltz, R. L., Nagaich, A. K., Radonovich, M., Hager, G., Cole, P. A., and Brady, J. N. (2002). Acetylation of nucleosomal histones by p300 facilitates transcription from Tax-responsive human T-cell leukemia virus type 1 chromatin template. *Mol. Cell. Biol.* 22, 4450-4462.

Lundblad, J. R., Kwok, R. P. S., Laurance, M. E., Huang, M. S., Richards, J. P., Brennan, R. G., and Goodman, R. H. (1998). The human T-cell leukemia virus-1 transcriptional activator Tax enhances cAMP-responsive element-binding protein (CREB) binding activity through interactions with the DNA minor groove. *J. Biol. Chem.* 273, 19251-19259.

Maekawa, T., Matsuda, S., Fujisawa, J.-I., Yoshida, M., and Ishii, S. (1991). Cyclic AMP response element-binding protein, CRE-BP1, mediates the E1A-induced but not the Tax-induced *trans*-activation. *Oncogene* 6, 627-632.

Marriott, S. J., Lindholm, P. F., Brown, K. M., Gitlin, S. D., Duvall, J. F., Radonovich, M. F., and Brady, J. N. (1990). A 36-kilodalton cellular transcription factor mediates an indirect interaction of human T-cell leukemia/lymphoma virus type I TAX with a responsive element in the viral long terminal repeat. *Mol. Cell. Biol.* 10, 4192-4201.

Mechta-Grigoriou, F., Ferald, D., and Yaniv, M. (2001). The mammalian Jun proteins: redundancy and specificity. *Oncogene* 20, 2378-2389.

Mercurio, F., Murray, B. W., Shevchenko, A., Bennett, B. L., Young, D. B., Li, J. W., Pascual, G., Motiwala, A., Zhu, H., Mann, M., and Manning, A. M. M. C. B., 1526-1538. (1999). IkappaB kinase (IKK)-associated protein 1, a common component of the heterogeneous IKK complex. *Mol. Cell. Biol.* 19, 1526-1538.

Mercurio, F., Zhu, H., Murray, B. W., Shevchenko, A., Bennett, B. L., Li, J., Young, D. B., Barbosa, M., Mann, M., and Manning, A. M. S., 860-866. (1997). IKK-1 and IKK-2: cytokine-activated IκB kinases essential for NF-κB activation. *Science* 278, 860-866.

Montagne, J., Béraud, C., Crenon, I., Lombert-Platet, G., Gazzolo, L., Sergeant, A., and Jalinot, P. (1990). Tax1 induction of the HTLV-I 21 bp enhancer requires cooperation between two cellular DNA-binding proteins. *EMBO J.* 9, 957-964.

Mori, N., Fujii, M., Iwai, K., Ikeda, S., Yamasaki, Y., Hata, T., Yamada, Y., Tanaka, Y., Tomonaga, M., and Yamamoto, N. (2000). Constitutive activation of transcription factor AP-1 in primary adult T-cell leukemia cells. *Blood* 95, 3915-3921.

Mori, N., Mukaida, N., Ballard, D. W., Matsushima, K., and Yamamoto, N. (1998). Human T-cell leukemia virus type I Tax transactivates human interleukin 8 gene through acting concurrently on AP-1 and nuclear factor-kappaB-like sites. *Cancer Res.* 58, 3993-4000.

Morita, T., Sato, T., Nyunoya, H., Tsujimoto, A., Takahara, J., Irino, S., and Shimotohno, K. (1993). Isolation of a cDNA clone encoding DNA-binding protein (TAXREB107) that binds specifically to domain C of the *tax*-responsive enhancer element in the long terminal repeat of human T-cell leukemia virus type I. *AIDS Res. Hum. Retrov.* 9, 115-121.

Mortreux, F., Gabet, A. S., and Wattel, E. (2003). Molecular and cellular aspects of HTLV-1 associated leukemogenesis in vivo. *Leukemia* 17, 26-38.

Muchardt, C., Seeler, J.-S., Nirula, A., Gong, S., and Gaynor, R. (1992). Transcription factor AP-2 activates gene expression of HTLV-I. *EMBO J.* 11, 2573-2581.

Mulloy, J. C., Kislyakova, T., Cereseto, A., Casareto, L., LoMonico, A., Fullen, J., Lorenzi, M. V., Cara, A., Nicot, C., Giam, C.-Z., and Franchini, G. (1998). Human T-cell lymphotropic/leukemia virus type 1 Tax abrogates p53-induced cell cycle arrest and apoptosis through its CREB/ATF functional domain. *J. Virol.* 72, 8852-8860.

Munoz, E., Courtois, G., Veschambre, P., Jalinot, P., and Israël, A. (1994). Tax induces nuclear translocation of NF-κB through dissociation of cytoplasmic complexes containing p105 or p100 but does not induce degradation of IκBα/MAD3. *J. Virology* 68, 8035-8044.

Murphy, E. L., Hanchard, B., Figueroa, J. P., Gibbs, W. N., Lofters, W. S., Campbell, M., Goedert, J. J., Blattner, W. A., and 15;43(2):250-3, F. (1989). Modelling the risk of adult T-cell leukemia/lymphoma in persons infected with human T-lymphotropic virus type I. *Int. J. Cancer* 43, 250-253.

Murphy, P., and Kolsto, A. (2000). Expression of the bZIP transcription factor TCF11 and its potential dimerization partners during development. *Mech. Dev.* 97, 141-148.

Nagata, K., Ohtani, K., Nakamura, M., and Sugamura, K. (1989). Activation of endogenous c-*fos* proto-oncogene expression by human T-cell leukemia virus type I-encoded $p40^{tax}$ protein in the human T-cell line, Jurkat. *J. Virol.* 63, 3220-3226.

Nakabeppu, Y., and Nathans, D. (1991). A naturally occuring truncated form of FosB that inhibits Fos/Jun transcriptional activity. *Cell* 64, 751-759.

Nerenberg, M., Hinrichs, S. H., Reynolds, R. K., Khoury, G., and Jay, G. (1987). The *tat* gene of human T-lymphotropic virus type 1 induces mesenchymal tumors in transgenic mice. *Science* 237, 1324-1329.

Nerenberg, M. I., Minor, T., Price, J., Ernst, D. N., Shinohara, T., and Schwarz, H. (1991). Transgenic thymocytes are refractory to transformation by the human T-cell leukemia virus type I tax gene. *J. Virol.* 65, 3349-3353.

Newbound, G. C., Q'Rourke, J. P., Collins, N. D., Andrews, J. M., DeWille, J., and Lairmore, M. D. (2000). Repression of Tax-mediated human T-lymphotropic virus type 1 transcription by inducible cAMP early repressor (ICER) protein in peripheral blood mononuclear cells. *J. Medical Virol.* 62, 286-292.

Nicot, C., Dundr, M., Johnson, J. M., Fullen, J. R., Alonzo, N., Fukumoto, R., Princler, G. L., Derse, D., Misteli, T., and Franchini, G. (2004). HTLV-1-encoded p30(II) is a post-transcriptional negative regulator of viral replication. *Nature Med.*

O'Shea, E. K., Rutkowski, R., and Kim, P. S. (1989). Evidence that the leucine zipper is a coiled coil. *Science* 243, 538-542.

Paca-Uccaralertkun, S., Zhao, L.-J., Adya, N., Cross, J. V., Cullen, B. R., Boros, I. M., and Giam, C.-Z. (1994). In vitro selection of DNA elements highly responsive to the human T-cell lymphotropic virus type I transcriptional activator. *Mol. Cell. Biol.* 14, 456-462.

Parker, C. E., Daenke, S., Nightingale, S., and Bangham, C. R. (1992). Activated HTLV-I-specific cytotoxic T-lymphocytes are found in healthy seropositives as well as in patients with tropical spastic paraparesis. *Virology* 188, 628-636.

Paskalis, H., Felber, B. K., and Pavlakis, G. N. (1986). *Cis*-acting sequences responsible for the transcriptional activation of human T-cell leukemia virus type I constitute a conditional enhancer. *Proc. Natl. Acad. Sci. USA* 83, 6558-6562.

Petropoulos, L., and Hiscott, J. (1998). Association between HTLV-1 Tax and IκBα is dependent on the IkBa phosphorylation state. *Virology* 252, 189-199.

Pise-Masison, C. A., Choi, K.-S., Radonovich, M., Dittmer, J., Kim, S.-J., and Brady, J. N. (1998a). Inhibition of p53 transactivation function by the human T-cell lymphotropic virus type 1 Tax protein. *J. Virol.* 72, 1165-1170.

Pise-Masison, C. A., Mahieux, R., Jiang, H., Ashcroft, M., Rodonovich, M., Duvall, J., Guillerm, C., and Brady, J. N. (2000). Inactivation of p53 by human T-cell lymphotropic virus type 1 Tax requires activation of the NF-κB pathway and is dependent on p53 phosphorylation. *Mol. Cell. Biol.* 20, 3377-3386.

Pise-Masison, C. A., Mahieux, R., Rodonovich, M., Jiang, H., and Brady, J. N. (2001). Human T-lymphotropic virus type I Tax protein utilizes distinct pathways for p53 inhibition that are cell-type-dependent. *J. Biol. Chem.* 276, 200-205.

Pise-Masison, C. A., Radonovich, M., Sakaguchi, K., Appella, E., and Brady, J. N. (1998b). Phosphorylation of p53: a novel pathway for p53 inactivation in human T-cell lymphotropic virus type 1-transformed cells. *J. Virol.* 72, 6348-6355.

Podust, L. M., Krezel, A. M., and Kim, Y. (2001). Crystal structure of the CCAAT box/enhancer-binding protein β activating transcription factor-4 basic leucine zipper heterodimer in the absence of DNA. *J. Biol. Chem.* 276, 505-513.

Poiesz, B. J., Ruscetti, F. W., Gazdar, A. F., Bunn, P. A., Minna, J. D., and Gallo, R. C. (1980). Detection and isolation of type C retrovirus particles from fresh and cultured lymphocytes of a patient with cutaneous T-cell lymphoma. *Proc. Natl. Acad. Sci. USA* 77, 7415-7419.

Poyet, J.-L., Srinivasula, S. M., Lin, J.-H., Fernandes-Alnemri, T., Yamaoka, S., Tsichlis, P. N., and Alnemri, E. S. (2000). Activation of the IκB kinases by RIP via IKKγ/NEMO-mediated oligomerization. *J. Biol. Chem.* 275, 37966-37977.

Pozzatti, R., Vogel, J., and Jay, G. (1990). The human T-cell leukemia virus type I *tax* gene can cooperate with the *ras* oncogene to induce neoplastic transformation of cells. *Mol. Cell. Biol.* 10, 413-417.

Reddy, T. R., Tang, H., Li, X., and Wong-Staal, F. (1997). Functional interaction of the HTLV-1 transactivator Tax with activating transcription factor-4 (ATF4). *Oncogene* 14, 2785-2792.

Reid, R. L., Lindholm, P. F., Mireskandari, A., Dittmer, J., and Brady, J. N. (1993). Stabilization of wild-type p53 in human T-lymphocytes transformed by HTLV-I. *Oncogene* 8, 3029-3036.

Ressler, S., Morris, G. F., and Marriott, S. J. (1997). Human T-cell leukemia virus type 1 Tax transactivates the human proliferating cell nuclear antigen promoter. *J. Virol.* 71, 1181-1190.

Riou, P., Bex, F., and Gazzolo, L. (2000). The human T-cell leukemia/lymphotropic virus type 1 Tax protein represses MyoD-dependent transcription by inhibiting MyoD-binding to the KIX domain of p300. *J. Biol. Chem.* 275, 10551-10560.

Rosen, C. A., Park, R., Sodroski, J. G., and Haseltine, W. A. (1987). Multiple sequence elements are required for regulation of human T-cell leukemia virus gene expression. *Proc. Natl. Acad. Sci. USA* 84, 4919-4923.

Rothwarf, D. M., Zandi, E., Natoli, G., and Karin, M. (1998). IKK-gamma is an essential regulatory subunit of the IkappaB kinase complex. *Nature* 395, 297-300.

Rousset, R., Desbois, C., Bantignies, F., and Jalinot, P. (1996). Effects on NF-kB1/p105 processing of the interaction between the HTLV-1 transactivator Tax and the proteasome. *Nature* 381, 328-331.

Ruddle, N. H., Li, C. B., Horne, W. C., Santiago, P., Troiano, N., Jay, G., Horowitz, M., and Baron, R. (1993). Mice transgenic for HTLV-I LTR-tax exhibit tax expression in bone, skeletal alterations, and high bone turnover. *Virology* 197, 196-204.

Ryseck, R.-P., and Bravo, R. (1991). c-JUN, JUN B, and JUN D differ in their binding affinities to AP-1 and CRE consensus sequences: effect of FOS proteins. *Oncogene* 6, 533-542.

Schmitt, I., Rosin, O., Rohwer, P., Gossen, M., and Grassmann, R. (1998). Stimulation of cyclin-dependent kinase activity and G_1- to S-phase transition in human lymphocytes by the human T-cell leukemia/lymphotropic virus type 1 Tax protein. *J. Virol.* 72, 633-640.

Scoggin, K. E. S., Ulloa, A., and Nyborg, J. K. (2001). The oncoprotein Tax binds the SRC-1-interacting doamin of CBP/p300 to mediate transcriptional activation. *Mol. Cell. Biol.* 21, 5520-5530.

Seeler, J.-S., Muchardt, C., Podar, M., and Gaynor, R. B. (1993). Regulatory elements involved in Tax-mediated transactivation of the HTLV-I LTR. *Virology* 196, 442-450.

Semmes, O. J., and Jeang, K.-T. (1992). HTLV-I Tax is a zinc-binding protein: Role of zinc in Tax structure and function. *Virology* 188, 754-764.

Senftleben, U., Cao, Y., Xiao, G., Greten, F. R., Krahn, G., Bonizzi, G., Chen, Y., Hu, Y., Fong, A., Sun, S. C., and Karin, M. (2001). Activation by IKKα of a second, evolutionary conserved, NF-κB signaling pathway. *Science* 293, 1495-1499.

Shaulian, E., and Karin, M. (2002). AP-1 as a regulator of cell life and death. *Nature Cell Biol.* 4, 131-136.

Shimotohno, K., Takano, M., Teruuchi, T., and Miwa, M. (1986). Requirement of multiple copies of a 21 nucleotide sequence in the U3 region of human T-cell leukemia virus type I and type II long terminal repeats for *trans*-acting activation of transcription. *Proc. Natl. Acad. Sci. USA* 83, 8112-8116.

Shnyreva, M., and Munder, T. (1996). The oncoprotein Tax of the human T-cell leukemia virus type 1 activates transcription via interaction with cellular ATF-1/CREB factors in *Saccharomyces cerevisiae*. *J. Virol.* 70, 7478-7484.

Smith, M. J., Gitlin, S. D., Browning, C. M., Lane, B. R., Clark, N. M., Shah, N., Rainier, S., and Markovitz, D. M. (2001). GLI-2 modulates retroviral gene expression. *J. Virol.* 75, 2301-2313.

Smith, M. R., and Greene, W. C. (1990). Identification of HTLV-I *tax trans*-activator mutants exhibiting novel transcriptional phenotypes. *Genes Dev.* 4, 1875-1885.

Smith, M. R., and Greene, W. C. (1992). Characterization of a novel nuclear localization signal in the HTLV-I Tax transactivator protein. *Virology* 187, 316-320.

Sodroski, J. G., Patarca, R., Rosen, C. A., Wong-Staal, F., and Haseltine, W. A. (1985). Location of the trans-activating region on the genome of human T-lymphotrophic virus type III. *Science* 229, 74-75.

Sun, S. C., Elwood, J., Beraud, C., and Green, M. R. (1994). Human T-cell leukemia virus type I Tax activation of NK-κB/Rel involves phosphorylation and degradation of IκBα and RelA (p65)-mediated induction of the c-*rel* gene. *Mol. Cell. Biol.* 14, 7377-7384.

Suzuki, T., Fujisawa, J. I., Toita, M., and Yoshida, M. (1993). The *trans*-activator Tax of human T-cell leukemia virus type I (HTLV-I) interacts with cAMP-responsive element (CRE) binding and CRE modulator proteins that bind to the 21-base-pair enhancer of HTLV-I. *Proc. Natl. Acad. Sci. USA* 90, 610-614.

Suzuki, T., Hirai, H., Murakami, T., and Yoshida, M. (1995). Tax protein of HTLV-1 destabilizes the complexes of NF-κB and IκB-a and induces nuclera translocation of NF-κB for transcriptional activation. *Oncogene* 10, 1199-1207.

Suzuki, T., Okuno, H., Yoshida, T., Endo, T., Nishina, H., and Iba, H. (1991). Difference in transcriptional regulatory function between c-Fos and Fra-2. *Nucleic Acids Res.* 19, 5537-5542.

Suzuki, T., Uchida-Toita, M., and Yoshida, M. (1999). Tax protein of HTLV-1 inhibits CBP/p300-mediated transcription by interfering with recruitment of CBP/p300 onto DNA element of E-box or p53 binding site. *Oncogene* 18, 4137-4143.

Takeda, S., Maeda, M., Morikawa, S., Taniguchi, Y., Yasunaga, J., Nosaka, K., Tanaka, Y., and Matsuoka, M. (2004). Genetic and epigenetic inactivation of tax gene in adult T-cell leukemia cells. *Int. J. Cancer* 20, 559-567.

Tanaka, A., Takahashi, C., Yamaoka, S., Nosaka, T., Maki, M., and Hatanaka, M. (1990). Oncogenic transformation by the Tax gene of human T-cell leukemia virus type I in vitro. *Proc. Natl. Acad. Sci. USA* 87, 1071-1075.

Tanimura, A., Dan, S., and Yoshida, M. (1998). Cloning of novel isoforms of the human Gli2 oncogene and their activities to enhance Tax-dependent transcription of the human T-cell leukemia virus type 1 genome. *J. Virol.* 72, 3958-3964.

Thébault, S., Basbous, J., Hivin, P., Devaux, C., and Mesnard, J. M. (2004). HBZ interacts with JunD and stimulates its transcriptional activity. *FEBS lett.* 562, 165-170.

Tie, T., Adya, N., Greene, W. C., and Giam, C.-Z. (1996). Interaction of the human T-lymphotropic virus type 1 Tax dimer with CREB and the viral 21-base-pair repeat. *J. Virol.* 70, 8368-8374.

Torgeman, A., Mor-Vaknin, N., and Aboud, M. (1999). Sp1 is involved in a protein kinase C-independent activation of human T-cell leukemia virus type I long terminal repeat by 12-O-tetradecanoylphorbol-13-acetate. *Virology* 254, 279-287.

Tsuchiya, H., Fujii, M., Niki, T., Tokuhara, M., Matsui, M., and Seiki, M. (1993). Human T-cell leukemia virus type 1 Tax activates transcription of the human fra-1 gene through multiple cis elements responsive to transmembrane signals. *J. Virol.* 67, 7001-7007.

Tsujimoto, A., Nyunoya, H., Morita, T., Sato, T., and Shimotohno, K. (1991). Isolation of cDNAs for DNA-binding proteins which specifically bind to a *tax*-responsive enhancer element in the long terminal repeat of human T-cell leukemia virus type I. *J. Virol.* 65, 1420-1426.

Tsukada, J., Misago, M., Serino, Y., Ogawa, R., Murakami, S., Nakanishi, M., Tonai, S., Kominato, Y., Moritono, I., Auron, P. E., and Eto, S. (1997). Human T-cell leukemia virus type I Tax transactivates the promoter of human prointerleukin-1β gene through association with two transcription factors, nuclear factor-interleukin-6 and Spi-1. *Blood* 90, 3142-3153.

Uchijima, M., Sato, H., Fujii, M., and Seiki, M. (1994). Tax proteins of human T-cell leukemia type 1 and 2 induce expression of the gene encoding erythroid-potentiating activity (tissue inhibitor of metalloproteinases-1, TIMP-1). *J. Biol. Chem.* 269, 14946-14950.

Uchiyama, T., Yodoi, J., Sagawa, K., Takatsuki, K., and Uchino, H. (1977). Adult T-cell leukemia: clinical and hematologic features of 16 cases. *Blood* 50, 481-492.

Uhlik, M., Good, L., Xiao, G., Harhaj, E. W., Zandi, E., Karin, M., and Sun, S.-C. (1998). NF-κB-inducing kinase and IκB kinase participate in human T-cell leukemia virus type I Tax-mediated NF-κB activation. *J. Biol. Chem.* 273, 21132-21136.

Uittenbogaard, M. N., Giebler, H. A., Reisman, D., and Nyborg, J. K. (1995). Transcriptional repression of *p53* by human T-cell leukemia virus type I Tax protein. *J. Biol. Chem.* 270, 28503-28506.

Vallejo, M., Ron, D., Miller, C. P., and Habener, J. F. (1993). C/ATF, a member of the activating transcription factor family of DNA-binding proteins, dimerizes with CAAT/enhancer-binding proteins and directs their binding to cAMP response elements. *Proc. Natl. Acad. Sci. USA* 90, 4679-4683.

Van Dam, H., and Castellazzi, M. (2001). Distinct roles of Jun:Fos and Jun:ATF dimers in oncogenesis. *Oncogene* 20, 2453-2464.

Van Orden, K., Giebler, H. A., Lemasson, I., Gonzales, M., and Nyborg, J. K. (1999a). Binding of p53 to the KIX domain of CREB binding protein. *J. Biol. Chem.* 274, 26321-26328.

Van Orden, K., Jian-ping, Y., Ulloa, A., and Nyborg, J. K. (1999b). Binding of the human T-cell leukemia virus Tax protein to the coactivator CBP interferes with CBP-mediated transcriptional control. *Oncogene* 18, 3766-3772.

Van, P. L., Kyong-Wook, Y., Jin, D.-Y., Dapolito, G., Kurimasa, A., and Jeang, K.-T. (2001). Genetic evidence of a role for ATM in functional interaction between human T-cell leukemia virus type 1 Tax and p53. *J. Virol.* 75, 396-407.

Vinson, C. R., Hai, T., and Boyd, M. (1993). Dimerization specificity of the leucine zipper-containing bZIP motif on DNA binding: prediction and rational design. *Genes Dev.* 7, 1047-1058.

Vogt, P. K. (2001). Jun, the oncoprotein. *Oncogene* 20, 2365-2377.

Wagner, S., and Green, M. R. (1993). HTLV-I Tax protein stimulation of DNA binding of bZIP proteins by enhancing dimerization. *Science* 262, 395-399.

Woronicz, J. D., Gao, X., Cao, Z., Rothe, M., Goeddel, D. V., and Rao, A. (1997). IκB kinase-β: NF-κB activation and complex formation with IκB kinase-α and NIK. *Science* 278, 866-869.

Yamagata, T., Mitani, K., Ueno, H., Kanda, Y., Yazaki, Y., and Hirai, H. (1997). Triple synergism of human T-lymphotropic virus type 1-encoded Tax, GATA-binding protein, and AP-1 is required for constitutive expression of the interleukin-5 gene in adult T-cell leukemia cells. *Mol. Cell. Biol.* 17, 4272-4281.

Yamaoka, S., Inoue, H., Sakurai, M., Sugiyama, T., Hazama, M., Yamada, T., and Hatanaka, M. (1996). Constitutive activation of NF-κB is essential for transformation of rat fibroblasts by the human T-cell leukemia virus type I Tax protein. *EMBO J.* 15, 873-887.

Yamaoka, S., Tobe, T., and Hatanaka, M. (1992). Tax protein of T-cell leukemia virus type I is required for maintenance of the transformed phenotype. *Oncogene* 7, 433-437.

Yamato, K., Oka, T., Hiroi, M., Iwahara, Y., Sugito, S., Tsuchida, N., and Miyoshi, I. (1993). Aberrant expression of the p53 tumor suppressor gene in adult T-cell leukemia and HTLV-I-infected cells. *Jpn. J. Cancer Res.* 84, 4-8.

Yin, M.-J., and Gaynor, R. B. (1996). HTLV-I 21-bp repeat sequences facilitate stable association between Tax and CREB to increase CREB binding affinity. *J. Mol. Biol.* 264, 20-31.

Yin, M.-J., Paulssen, E. J., Seeler, J.-S., and Gaynor, R. B. (1995a). Protein domains involved in both *in vivo* and *in vitro* interaction between human T-cell leukemia virus type I Tax and CREB. *J. Virol.* 69, 3420-3432.

Yin, M. J., Paulssen, E., Seeler, J., and Gaynor, R. B. (1995b). Chimeric proteins composed of Jun and CREB define domains required for interection with the human T-cell leukemia virus type 1 Tax protein. *J. Virol.* 69, 6209-6218.

Yoshimura, T., Fujisawa, J.-I., and Yoshida, M. (1990). Multiple cDNA clones encoding nuclear proteins that bind to the *tax*-dependent enhancer of HTLV-1: all contain a leucine zipper structure and basic amino acid domain. *EMBO J.* 9, 2537-2542.

Zandi, E., Rothwarf, D. M., Delhase, M., Hayakawa, M., and Karin, M. (1997). The IκB kinase complex (IKK) contains two kinases subunits, IKKα and IKKβ, necessary for IκB phosphorylation and NK-κB activation. *Cell* 91, 243-252.

Zhao, L. J., and Giam, C.-Z. (1992). Human T-cell lymphotropic virus type I (HTLV-I) transcriptional activator, Tax, enhances CREB binding to HTLV-I 21-base-pair repeats by protein-protein interaction. *Proc. Natl. Acad. Sci. USA* 89, 7070-7074.

In: Progress in Virus Research
Editor: Sabine Thebault, pp. 39-63

ISBN 1-59454-438-7

Chapter II

De-Regulation of the Cell Cycle by Human T Cell Leukemia Virus Type I (HTLV-I) Tax: Mechanisms and Effects

Julie C. Watt and Susan J. Marriott*
Department of Molecular Virology and Microbiology
Baylor College of Medicine; One Baylor Plaza; Houston, TX 77030

Abstract

Human T cell leukemia virus type I (HTLV-I) is associated with the development of adult T-cell leukemia (ATL) and the neurological disease tropical spastic paraparesis/ HTLV-I associated myelopathy (TSP/HAM). The major viral oncoprotein, Tax, is a transcriptional transactivator of viral and cellular genes. It is hypothesized that HTLV-I causes cancer by altering cellular gene expression and function leading to dysregulation of cellular processes including cell cycle progression and DNA repair. This chapter will review the effects of Tax on cell cycle progression as a consequence of altering cellular gene expression and protein function. How these events may contribute to transformation of HTLV-I infected T cells and the subsequent development of ATL will also be discussed.

Human T Cell Leukemia Virus Type I (HTLV-I)

Human T cell leukemia virus type I (HTLV-I) [reviewed in 80] was the first retrovirus to be associated with human disease. It was first found to be associated with adult T cell leukemia (ATL) [reviewed in], which is a clonal malignancy of infected mature CD4+ helper T cells [88;145;185;190]. HTLV-I infection was later found to be associated with tropical spastic paraparesis/HTLV-I associated myelopathy (TSP/HAM), a chronic demyelinating

* Corresponding author: Susan J. Marriot, phone: (713) 798-4440, fax: (713) 798-3490 susanm@bcm.tmc.edu

disease affecting the spinal cord and brain [50; reviewed in .11;12] Infection with HTLV-I is endemic in parts of Japan, South America, Africa, and the Caribbean [17;41;66;67;155;162]. However, the majority of infected individuals never develop ATL or TSP/HAM. In the case of ATL, the cumulative lifetime risk of an HTLV-I carrier developing ATL is 4% and the disease does not develop until 20-30 years after initial infection [126]. Because of the long clinical latency and low percentage of individuals who develop ATL, it is hypothesized that T cell transformation induced by HTLV-I occurs after a series of cellular alterations and/or mutations. A statistical analysis of the relationship between age and disease onset has predicted that five independent genetic events are required for the development of ATL [139].

The HTLV-I genome encodes several proteins that regulate viral infection and contribute to oncogenesis [reviewed in 5;47], but the viral transcriptional transactivator protein, Tax, has been shown to be the major viral oncoprotein responsible for cellular transformation [51]. Tax modulates the activity of transcription factors such as CREB, NF-κB, and SRF resulting in alterations in cellular gene expression. Tax also interacts with specific cellular proteins causing changes in their function. Tax-mediated transformation is thought to be a consequence of its effects on gene expression and protein function [189]. However, transformed cells isolated from ATL patients express little or no Tax [48;91], suggesting that Tax functions early in infection to disrupt cellular processes and initiate a series of events that, over a period of decades, can lead to the development of cancer. This chapter will discuss the effects of Tax on the expression of genes and function of gene products involved in cell cycle regulation. How these effects of Tax lead to cell cycle de-regulation, including enhanced cellular proliferation, loss of cell cycle regulatory mechanisms, loss of controls that ensure accurate DNA replication, and disruption of normal apoptotic responses will also be discussed. All of these alterations in cell cycle control increase the likelihood of introducing mistakes during DNA replication that promote cellular transformation.

Each cell in an organism has a distinct function and undergoes cell division only when appropriate for that cell type. Some cell types, such as epithelial cells, divide frequently, and others, such as neurons in a mature individual, almost never divide. Multiple levels of control, both internal and external to the cell, ensure that proliferation only occurs when conditions are favorable for accurate completion of cell division. If a cell's ability to control proliferation is disrupted or lost, DNA replication and cell division can occur at inappropriate times. Such a cell may be destroyed by apoptotic mechanisms, but if these mechanisms fail, the cell may continue to proliferate and cellular transformation can result. This chapter will discuss the effects of Tax on both internal and external mechanisms that control cell proliferation.

Tax Disrupts Cellular Responses to Exogenous Growth Factors

Interleukin 2 (IL-2)

HTLV-I infects mature CD4+ T lymphocytes, and these cells utilize external control mechanisms to ensure that proliferation of T cells is restricted to those cells specific for the antigen presented during an adaptive immune response. Antigen presentation to T lymphocytes, in combination with proper co-stimulatory signals, causes T cells to produce interleukin 2 (IL-2). Resting T lymphocytes express low levels of the IL-2 receptor (IL-2R) on their surface and require high levels of IL-2 to stimulate progression through the cell cycle. The binding of IL-2 to IL-2R on the T cell surface triggers increased production of IL-2 as well as IL-2Rα, a subunit of the IL-2 receptor that increases the affinity of the receptor for IL-2. Cells expressing the high affinity IL-2R are more sensitive to IL-2 and continue to proliferate if the proper co-stimulatory signal is present [69].

HTLV-I induces infected T cells to proliferate in the absence of proper antigenic stimulation and co-stimulatory signals [4;157]. In the early stages following HTLV-I infection *in vitro*, IL-2 is required to induce proliferation of the infected cells. However, at later stages of infection, the proliferation of HTLV-I infected cells becomes IL-2 independent [191]. A similar situation appears to occur *in vivo*; however, when and how the switch to IL-2 independence occurs has not been clearly determined. Some ATL cells isolated from patients proliferate in response to IL-2, and some do not [6;7;85;113;172]. It has been suggested that IL-2 dependent growth occurs in the early, pre-leukemic stages of infection and that IL-2 independent T cell proliferation occurs during late stages of ATL [20]. Other exogenous factors, such as IL-4, may also contribute to the development of ATL by stimulating T cell proliferation *in vivo* [20;176].

Tax has several effects on the IL-2 pathway that may promote IL-2 independent cell growth. *In vitro*, Tax has been shown to transactivate both the IL-2 and IL-2Rα promoters through an NF-κB-like DNA element, leading to increased secretion of IL-2 and expression of the high affinity IL-2R on the surface of infected cells [10;34;75;102;115;149;161]. However, while ATL cells do express significantly elevated levels of IL-2R and IL-2Rα *in vivo*, they produce low or undetectable amounts of IL-2 [71;92;117;175;181]. It is possible that other cell types present *in vivo* secrete IL-2 and help to stimulate heterologous proliferation of ATL cells that are IL-2 dependent. Alternatively, because of their increased IL-2R expression, infected cells may be more sensitive to IL-2 and proliferate in response to lower levels of IL-2 than would be required to stimulate the proliferation of uninfected cells. It is also possible that ATL cells do not require IL-2 at all, due to alterations in their signaling pathways.

IL-2 independent, HTLV-I infected T cells and Tax expressing cell lines have a constitutively activated JAK-STAT pathway, which is involved in IL-2R signal transduction, and these cells no longer require IL-2 to activate the pathway and proliferate. JAK-STAT activity is usually regulated by SHP1, a phosphatase that dephosphorylates and inactivates JAK proteins [27;193]. SHP1 expression is reduced in HTLV-I infected cells at both the mRNA and protein levels, resulting in constitutive phosphorylation of the JAK and STAT proteins [119;183]. Loss of SHP1 may therefore account for the constitutive activation of the JAK-STAT pathway in HTLV-I infected cells. Consistent with these findings, HTLV-I infected T cells that still require IL-2 to proliferate do not have a constitutively activated JAK-STAT pathway [119;193]. Additionally, leukemic cells from ATL patients that have a constitutively active JAK-STAT pathway proliferate more rapidly than cells from patients without a constitutively activated pathway [170;193]. Whether constitutive activation of the JAK-STAT pathway in ATL cells correlates with IL-2 independence *in vivo* has not been determined.

The effects of Tax on the IL-2 and IL-2R genes and the JAK-STAT pathway likely contribute to the inappropriate proliferation of HTLV-I infected T cells. However, it is unclear how these events contribute to cellular transformation *in vivo*. It has been reported that healthy HTLV-I carriers produce little IL-2R while ATL patients secrete high levels of soluble IL-2R [124;187]. This suggests that the upregulation of IL-2R is likely to influence transformation and the development of disease late in infection, rather than during initial stages of infection.

Transforming Growth Factor β (TGF-β)

In addition to promoting cell proliferation in the absence of growth stimulators such as IL-2, Tax can also free cells from the negative effects of growth suppressors such as TGF-β. ATL cells secrete an increased amount of TGF-β due to Tax transactivation of the TGF-β promoter [89;137]. It was somewhat surprising, however, to find that HTLV-I infected T cells are resistant to growth inhibition by TGF-β [70]. This resistance is due to several different effects of Tax on the TGF-β pathway. First, Tax represses the TGF-β signaling pathway by binding and sequestering the transcription coactivators CBP and p300, making them unavailable to interact with and activate Smad proteins that are required to transcribe TGF-β responsive genes [97;122]. Tax also interacts directly with some of the Smad proteins, preventing their interaction with any remaining free CBP/p300 and reinforcing the block in transcription of TGF-β responsive genes [97]. In addition, Tax induces the activity of JNK (c-Jun N-terminal kinase) through unknown mechanisms, and active JNK then phosphorylates c-Jun [183]. Phophorylated c-Jun can bind to Smad3, impairing its ability to bind DNA and activate the transcription of TGF-β responsive genes [9]. Finally, inhibition of the TGF-β responsive protein $p15^{INK4b}$ appears to be important in the process of Tax transformation. Approximately 19% of ATL patients have deletions in the $p15^{INK4b}$ gene [62;64]. In addition, Tax interferes with the function of $p15^{INK4b}$ by associating with the protein and preventing it from inhibiting the kinase activity of cdk4 [57;168].

So, although Tax enhances TGF-β production, infected cells are resistant to TGF-β-mediated growth inhibition because Tax inhibits the activation of TGF-β responsive genes and interferes with the function of growth inhibiting proteins induced by TGF-β. The fact that Tax blocks TGF-β-mediated growth inhibition by numerous different mechanisms suggests that this is a critical function of the protein. However, if blocking TGF-β growth inhibition is so important, then why does Tax also upregulate TGF-β expression? Tax transactivates the TGF-β promoter through an AP-1 like transcription element [89]. Tax activates gene expression through several transcription pathways including AP-1 [76], suggesting that TGF-β upregulation is a consequence of global upregulation of the AP-1 pathway by Tax and is not directly beneficial to the virus. Because the presence of TGF-β is detrimental to the virus, it has developed numerous mechanisms to counteract the effects of TGF-β.

The ability of Tax to promote cell proliferation in the absence of growth factors, such as IL-2, and to block the activity of growth inhibitors, such as TGF-β, provides an environment where cell proliferation is favored. As a result, infected T lymphocytes are more likely to proliferate. In addition, Tax alters the internal cell cycle machinery to favor cell cycle progression and cellular proliferation.

Tax Dysregulates Cell Cycle Machinery

Overview of the Cell Cycle

In addition to releasing HTLV-I infected T cells from IL-2 dependence and TGF-β growth suppression, Tax also disrupts the cell cycle machinery that regulates DNA replication and cell division [reviewed in 47;49;63;118;131;132;188;189]. The cell cycle can be divided into five phases: G_0, G_1, S, G_2, and M (Figure 1). G_0 is a quiescent phase, in which the cell is not actively cycling. During G_1 phase, the cell grows and produces factors needed for DNA replication, which occurs during S phase. During G_2, the cell continues to grow and prepares to undergo mitosis. Finally, during M (mitosis) phase, the cell segregates its chromosomes and divides into two daughter cells. Mistakes during any of these phases can have catastrophic effects; however, mistakes during S phase are particularly detrimental as they can lead to DNA replication errors that result in fixation of mutations in the genome. The cell contains multiple, overlapping pathways that function during G_1 and the G_1/S phase transition to ensure that DNA replication is not initiated if conditions are not optimal for its successful completion. Some of the regulatory pathways that block cell cycle progression during unfavorable conditions will be discussed in a later section.

Progression through, and transition between, each phase of the cell cycle is controlled by phosphorylation of specific target proteins that regulate cell cycle progression by specific cyclin/cdk (cyclin dependent kinase) complexes (Figure 1). Cdk levels remain relatively constant throughout cell cycle, and cyclin levels fluctuate depending on the phase of cell cycle [73]. Thus cyclin/cdk kinase activity is determined by cyclin levels not cdk levels. Cells express numerous cdk inhibitors that can alter cyclin/cdk activity in response to various stimuli. The function of some specific cdk inhibitors will be discussed later.

As mentioned above, replication of DNA under suboptimal conditions can have catastrophic consequences for the cell including mutation and possibly cell death. Therefore, the onset of DNA replication is one of the most tightly controlled processes within a cell. DNA replication absolutely requires active E2F transcription factor to transcribe numerous genes that are needed in S phase for DNA replication [reviewed in 1;165]. The retinoblastoma protein (Rb) regulates E2F activity. Hypophosphorylated Rb binds to and inactivates E2F. Cyclin D/cdk4 or cdk6, or cyclin E/cdk2 complexes can hyperphosphorylate Rb during G_1 phase or at the G_1/S transition, respectively, causing Rb to dissociate from E2F. Active unbound E2F is then free to transcribe the genes necessary for S phase (Figure 1).

Although the Rb gene is mutated in many types of cancers, it is only mutated in about 5% of ATL patients [64]. However, the function of Rb is disrupted in leukemic cells from ATL patients because Tax interferes with the phosphorylation state of Rb and increases the levels of active E2F protein, resulting in constitutive activation of E2F responsive genes. In Tax expressing cells and ATL cells in culture, Tax stimulates cyclin/cdk activity (discussed below), resulting in hyperphosphorylation of Rb and activation of E2F. Tax also enhances transcription of the E2F gene itself, resulting in elevated levels of E2F protein [77;99;128;133;138]. It is thought that the combined effects of hyperphosphorylated Rb, which cannot bind and inactivate E2F, and increased levels of E2F protein contribute to the accelerated entry into S phase that is observed in Tax expressing cells [101;107].

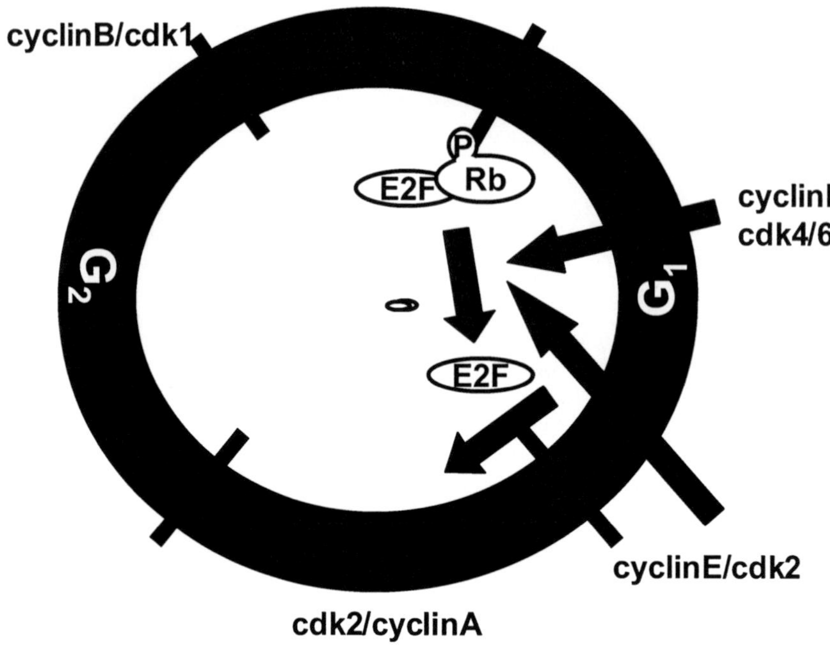

Figure 1. Cell cycle regulation. The phases of cell cycle are shown above (G_0 is not shown). Progression through and between phases is controlled by cyclin/cdk partners. During G_1 phase, cyclinD/cdk4/6 and cyclinE/cdk2 phophorylate Rb, causing it to dissociate from E2F. Free E2F is then active and can transcribe the genes needed for DNA replication during S phase and cell cycle progression.

Tax also affects the levels and activity of several cyclin/cdk complexes that regulate G_1 progression and S phase entry. These include genes encoding the G_1 cyclins, D1 and D2, which are transactivated by Tax, and the cyclin D3 gene, which is repressed by Tax [2;72;77;87;121;154]. Tax interacts with cdk4 and enhances its binding to cyclin D2 resulting in an earlier onset of cdk kinase activity and an earlier accumulation of hyperphosphorylated Rb in Tax expressing cells than in cells not expressing Tax [55;56;104;133;156]. Tax also interacts with cdk6 and increases its kinase activity, but whether this binding stabilizes its interaction with a particular cyclin partner has not been reported [56;156]. Tax has been shown to increase the kinase activity of cdk2, but it has not been reported to interact with cdk2 or to stabilize the interaction of cdk2 with a cyclin partner [77]. The combined effects of Tax on cdk4/6/2 activity result in Rb phosphorylation, and promote G_1 progression and G_1/S transition. However, the ability of Tax to repress cyclin D3 expression is an exception whose effects on cell cycle progression are not yet understood.

Cell Cycle Arrest

The basic cell cycle machinery has been described above, and we have discussed how Tax accelerates cell cycle progression through G_1 and entry into S phase. However, conditions are not always favorable for cell cycle progression due to various circumstances including, but not limited to, DNA damage. The presence of DNA damage while DNA replication is underway greatly increases the likelihood of introducing mutations into the genome. Because mutations are generally detrimental to the cell, multiple mechanisms exist to arrest the cell cycle if conditions are not favorable for faithful DNA replication. These mechanisms include cdk inhibitors, which block cyclin/cdk kinase activity, and cell cycle checkpoints, which can induce cell cycle arrest during various phases of the cell cycle including G_1 phase, the G_1/S transition, S phase, G_2 phase, and early mitosis. Of these, the effects of Tax on cdk inhibitors, the G_1/S checkpoint, and the mitotic spindle checkpoint (in early mitosis) will be discussed. Inactivation of checkpoints together with acceleration of cell cycle progression through G_1 and entry into S phase creates an environment where mutations arise more frequently, a phenomenon that has been referred to as the mutator phenotype, and transformation is more likely to occur.

1. p16^{INK4a} and other members of the INK4 family of G_1 cdk inhibitors. Cdk inhibitors are proteins that associate with cdks and prevent them from interacting with their cyclin partners, thus blocking cdk kinase activity. Multiple cdk inhibitors target different cyclin/cdk complexes. The cdk inhibitor p16^{INK4a} interacts with cdk4, preventing it from forming an active complex with cyclin D [158]. This prevents cyclin D/cdk4 from phosphorylating Rb. As a result, E2F remains bound to Rb and is unable to transcribe the genes needed for S phase, thereby preventing progression through G_1 and into S phase. This mechanism is disabled in Tax-expressing cells because Tax binds to p16^{INK4a} and prevents its association with cdk4. In this manner, Tax promotes formation of the active cyclin D/cdk4 complex which is consistent with stimulation of G_1 progression [111;167]. It is also interesting to note that deletions in the p16^{INK4a} gene occur in about 27% of ATL patients [62;64], which could reinforce the ability of Tax to inhibit p16^{INK4a} function. However, the inactivation of p16^{INK4a} may not completely explain elevated cdk4 kinase activity because in p16^{INK4a} null cells, Tax is still able to increase cdk4 activity, perhaps by increasing cyclin D protein levels as discussed earlier [133].

Tax also disrupts the levels and/or functions of other cdk inhibitors including p15^{INK4b}, p18^{INK4c}, and p19^{INK4d}. As mentioned in the section on TGF-β, Tax interferes with the cdk inhibitory function of p15^{INK4b} and about 19% of ATL patients have deletions in the p15^{INK4b} gene [62;64;168]. Tax has also been shown to repress transcription of p18^{INK4c} resulting in reduced expression of p18^{INK4c} protein [168]. Tax expressing cells also have lower levels of p19^{INK4d} protein, which may result from both transcriptional and post-transcriptional effects of Tax [77]. By disrupting cdk inhibitor protein levels and function, Tax interferes with normal cellular responses to DNA damage and promotes active G_1 cyclin/cdk complexes resulting in G_1 to S phase cell cycle progression.

2. G_1/S checkpoint. The transition from G_1 phase to S phase is a critical point in the cell cycle that requires active cyclin E/cdk2 [93]. At this point, the cell must decide whether or not DNA replication is appropriate. The pathway that monitors this transition and arrests the cell cycle, if necessary, has been examined thoroughly in recent years. Events such as DNA damage trigger the activation of either ATM or ATR kinase. ATM then phosphorylates and activates a second kinase, Chk2, while ATR phosphorylates and activates Chk1. Activated Chk1 and Chk2 then phosphorylate Cdc25A, marking it for ubiquitination and degradation. Cdc25A is a phosphatase that dephosphorylates cdk2 and thus activates the cyclin E/cdk2 complex. Since the activity of the cyclin E/cdk2 complex is necessary for G_1 to S phase progression, degradation of Cdc25A prevents its activation, thus initiating the G_1/S checkpoint (Figure 2). The portion of the pathway discussed so far, which initiates the DNA damage induced G_1/S cell cycle checkpoint, is triggered within the first hour after DNA damage and lasts for several hours [22;33;45;53;68;79;94;98;114;152;159;163;182].

Recently, Tax was shown to co-localize and co-immunoprecipitate with Chk2, and increased levels of Chk2 were observed in Tax-expressing cells, but the mechanism for increased Chk2 expression is unknown [58]. Potential effects of the interaction between Tax and Chk2 on the G_1/S checkpoint have not been fully characterized. To date however, this is the only report of a Tax function that might influence the initiation of this checkpoint.

A second phase of the G_1/S checkpoint maintains cell cycle arrest. The maintenance phase requires gene transcription and is not fully activated until several hours after DNA damage. The maintenance phase remains active for a length of time that is determined by the extent of DNA damage and other cellular factors. In the maintenance phase, ATM and/or ATR phosphorylate and activate p53, which then transcriptionally upregulates p21. Increased levels of the p21 protein associate with cyclin E/cdk2 complexes and inactivate them (Figure 2) [38;112; complete G_1/S checkpoint reviewed in 13;14].

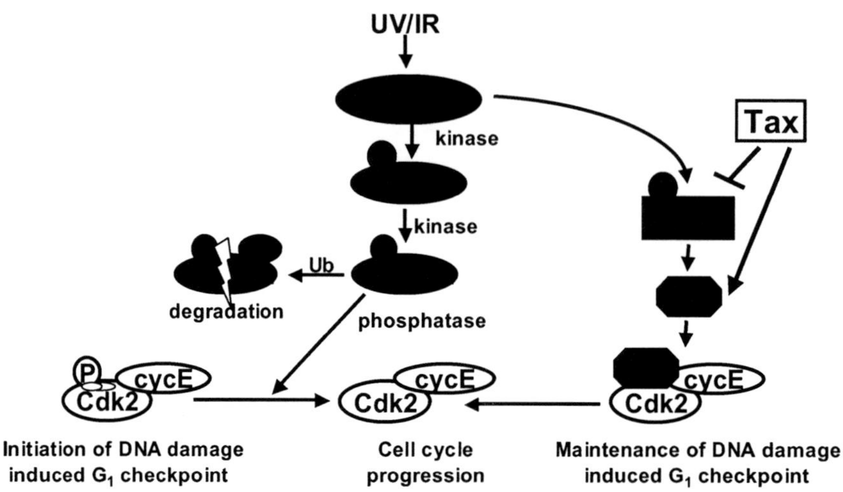

Figure 2. DNA damage induced G_1/S checkpoint. DNA damage activates the kinase ATR/ATM. In the initiation phase of the checkpoint, ATR/ATM activates a second kinase Chk1/2, triggering degradation of the phosphatase cdc25A, which can activate cyclinE/cdk2. During the maintenance phase, ATR/ATM activates p53, which then upregulates p21 expression. p21 then inactivates cyclinE/cdk2. Tax inactivates p53 and transcriptionally upregulates p21 expression. The downstream effects of this are not fully understood, but Tax does disrupt the DNA damage induced G_1/S checkpoint.

The maintenance pathway was discovered before the initiation pathway and is better characterized. Tax is known to have effects on several components of this checkpoint including inactivation of p53 and transcriptional activation of p21 (both of these effects will be thoroughly discussed in the next two sections). These activities are consistent with the ability of Tax to disrupt the G_1/S checkpoint [59].

a. p53. p53 is one of the most important regulators of the cell cycle, playing a central role in cell cycle arrest, apoptosis, and numerous other cellular functions. p53 functions as a transcription factor that, when active, recruits the coactivators CBP and p300 resulting in altered expression of numerous genes that regulate cellular functions [reviewed in 150]. The regulatory functions of p53 play a critical role in tumorigenesis as demonstrated by p53 knockout mice, which develop numerous tumors and have a shorter life expectancy than wildtype mice [39].

The p53 gene is mutated in about one fourth of ATL patients [127;134;153]. However, numerous studies using ATL cells from patients and infected cells in culture containing wildtype p53 have shown that Tax stabilizes p53 leading to increased protein levels, while inactivating the transactivating functions of p53 [3;23;141;146;171;186]. Thus, even though p53 levels are elevated in Tax-expressing cells, the protein is not functional. Several studies have investigated the mechanism by which Tax inactivates p53, but the results remain controversial.

Since Tax competitively binds the transcription cofactors CBP and p300, which are required for p53 mediated transactivation, it has been proposed that Tax sequesters CBP/p300, making them unavailable to bind p53 [8;108;125;169;178;179]. Alternatively, it has been proposed that Tax inactivates p53 by altering its phosphorylation. In the presence of Tax, the p53 residues serine-15 and serine-392 are constitutively phosphorylated and phosphorylation of these residues has been associated with inactivation of p53 [142;144]. In an attempt to address both of these mechanisms, Pise-Masison et al. [143] showed that p53 was inactivated by phosphorylation in T lymphocyte cell lines while in certain other cell lines p53 was inactivated through CBP/p300 sequestration. They suggested that sequestration of CBP/p300 can inactivate p53 in cell lines where CBP/p300 levels are low, when CBP/p300 levels are higher, sequestration can no longer inactivate p53. They further demonstrated that T lymphocytes have high levels of CBP/p300, and therefore they propose that phosphorylation is the relevant mechanism by which Tax inactivates p53 in T lymphocytes *in vivo*. The ability of Tax to inactivate p53 is likely a major mechanism by which Tax disrupts the G_1/S checkpoint.

b. p21. Originally, p21 was characterized as a cdk inhibitor, which could bind cdk4, cdk6 and cdk2 to inactivate cyclin D/cdk4 (or cdk6) or cyclin E/cdk2 complexes [60;61]. p21 gene expression is largely controlled by p53. Activation of p53 in response to a stimulus such as DNA damage allows p53 to activate p21 gene expression and p21 subsequently binds and inactivates cyclin/cdk complexes to arrest the cell cycle during G_1 or at the G_1/S checkpoint (Figure 2) [40;42]. Later, when the stimulus that activated p53 is resolved or removed, p21 is degraded, alleviating the inhibition of cyclin/cdk complexes and allowing the cell cycle to resume [18]. However, it was soon discovered that the function of p21 is not so straightforward.

The level of p21 protein in relation to cyclin D/cdk4 or cdk6 has now been shown to be critical for its function. Low amounts of p21 facilitate the interaction of cyclin D with cdk4 or cdk6 and do not inhibit the cdk activity of the complex. Higher amounts of p21 function as a cdk inhibitor of cyclin D/ckd4/6 as originally described [95]. To date, p21 has not been shown to similarly facilitate cyclin E binding to cdk2 and has only been shown, at higher concentrations, to inhibit activity of the assembled cyclin E/cdk2 complex.

The ability of HTLV-I Tax to inactivate p53 suggests that Tax expressing cells would express very little p21 protein. However, this is not observed. Rather, Tax has been shown to transcriptionally transactivate the p21 promoter leading to increased levels of p21 protein [2;23;30;35;36;77]. It is therefore important to determine the effect of increased p21 levels on Tax-expressing cells. Since some cell cycle checkpoints are defective in Tax-expressing cells [59], it appears that the p21 in these cells does not inhibit cdk function. The increased levels of cdk4, cdk6, and some of the cyclin D proteins in Tax expressing cells [2;55;56;72;77;121;133;154;156] might recruit p21 to facilitate formation of the cyclin/cdk complex without inhibiting its function. However, careful studies to determine the relative levels of p21, cyclin D and cdk4/6 in Tax-expressing cells have not been reported. Alternatively, Tax may inactivate the function of p21, as it does with p53, but further studies are needed to address this question.

p21 also can bind to PCNA (proliferating cell nuclear antigen), a DNA replication processivity factor, inhibiting its function in DNA replication and some types of DNA repair [46]. Recently it was reported that the use of low doses of ultraviolet (UV) light to induce DNA damage resulted in degradation, rather than upregulation of p21 [16]. These authors suggested that p21 degradation would release functional PCNA that is required for nucleotide excision repair (NER) of the DNA damage. These cells should arrest for only a short period of time because the initiation phase, but not the maintenance phase, of the G_1/S checkpoint would be activated. At higher doses of UV irradiation, p21 expression is upregulated rather than being degraded [16]. This would result in activation of the initiation and maintenance phases of the G_1/S checkpoint leading to a longer cell cycle arrest and providing time for the repair of more extensive damage. Since Tax can transcriptionally transactivate the PCNA promoter leading to increased amounts of PCNA protein [147], it is possible that some of the excess p21 present in Tax-expressing cells is bound to the excess PCNA and is thus unavailable for binding and inhibition of cyclin/cdk complexes. The ability of Tax to disrupt NER may be due in part to the sequestration of free PCNA by p21 in Tax expressing cells [59;83;100].

3. Mitotic spindle checkpoint. A critical event in mitosis is the separation of sister chromatids by the mitotic spindle, a process that is regulated by the mitotic spindle checkpoint. Chromatid separation is achieved by degradation of cohesion molecules holding the chromatids together by the protease separase (separin) [52;65;177;180]. The mitotic spindle checkpoint regulates the activity of the ubiquitin ligase anaphase promoting complex (APC). The APC activates separase by mediating the degradation of securin, which normally binds to and inactivates separase [31;164;194]. Inactivation of separase also occurs following the binding of cyclin B to cdk1, activating the kinase which subsequently promotes inhibitory phosphorylation of separase [32;44;164]. The APC mediates cyclin B degradation, resulting in inactivation of cdk1, again favoring separase activation [90;166; reviewed in 140;192]. A recent report demonstrating that Tax expressing cells contain reduced levels of cyclin B, possibly caused by premature degradation, suggests one mechanism by which Tax might disable the mitotic spindle checkpoint (Figure 3) [109].

Misaligned chromosomes and spindle damage activate the mitotic spindle checkpoint [reviewed in 103]. After several initial signal transduction events, HsMAD1 (homo sapiens mitotic arrest-defective 1) homodimerizes and then heterodimerizes with HsMAD2 resulting in inhibition of APC activity [78;105;106]. HsMAD1 was first identified as a protein that interacts with Tax. This interaction was subsequently shown to prevent the homodimerization of HsMAD1 and in this manner, disrupts the mitotic spindle checkpoint (Figure 3) [74;78;84]. However, because it was later found that the characterized HsMAD1 gene sequence contained a mutation yielding a different sized protein than HsMAD1, there is some uncertainty about the exact effects of Tax on HsMAD1 [21]. Regardless of the mechanism involved, disruption of the mitotic spindle checkpoint by Tax could lead to improper segregation of chromosomes and the aneuploidy that is frequently observed in ATL cells [160].

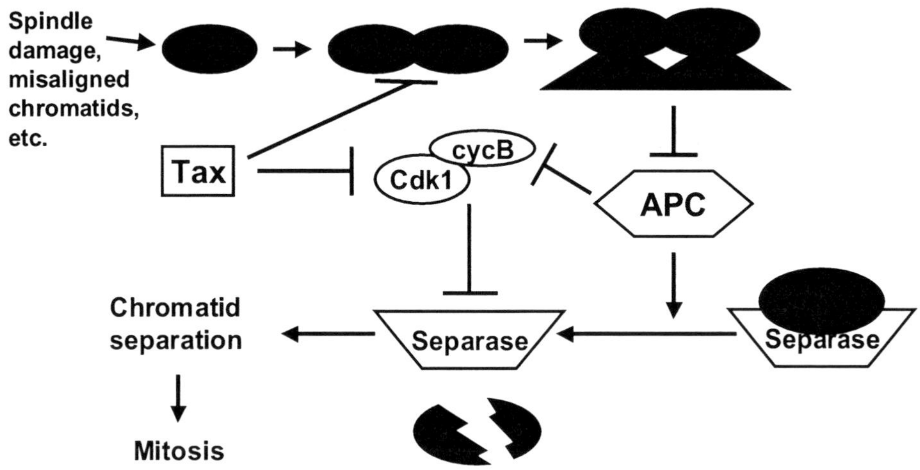

Figure 3. Mitotic spindle checkpoint. APC activates separase by triggering the degradation of securin and cyclin B. If spindle damage, misaligned chromosomes, or other stimuli trigger the mitotic spindle checkpoint, MAD1 homodimerizes, then heterodimerizes with MAD2 to inhibit APC. Tax disrupts the mitotic spindle checkpoint by associating with MAD1 and preventing its homodimerization resulting in constitutive activation of APC.

Apoptosis

Cell cycle regulation is intricately linked to programmed cell death by apoptosis. When the cell cycle checkpoints discussed above are activated, the cell either arrests and attempts to repair the damage, or undergoes apoptosis. Numerous studies have tried to delineate the relationship between HTLV-I Tax and apoptosis, yet it remains unclear how Tax affects apoptosis. The current literature on this subject will be discussed in this section.

There is general agreement in the literature that transient Tax expression leads to apoptosis in T lymphocytes [25;26;29;110] and in tumor-derived cells from Tax transgenic mice [54]. It is not yet clear whether Tax actually induces apoptosis, or rather, interferes with mechanisms that block apoptosis in response to other stimuli. Several studies have found that Tax enhances FasL (Fas ligand) expression through activation of the NF-AT transcription factor pathway, suggesting a mechanism by which Tax might directly induce apoptosis [25;26;28;148]. However, other mechanisms are likely also involved.

In cells that stably express Tax, apoptosis is sometimes increased and sometimes decreased [135;151]. There are several mechanisms by which Tax might decrease apoptosis. Tax transactivates the promoter of Bcl-xL through NF-κB transcription elements. Bcl-xL is an anti-apoptotic factor, whose expression promotes cell survival over cell death [120;136;173]. Tax has also been shown to repress the Bax promoter, through a putative helix-loop-helix transcription factor binding site, leading to decreased levels of Bax protein [19]. Since Bax binds and inhibits the anti-apoptotic factor Bcl-2, its overexpression could also promote cell survival rather than apoptosis.

Activation of the NF-κB transcription pathway can have both pro- and anti-apoptotic functions. Tax is well known to constitutively activate the NF-κB pathway [96], which could have downstream effects on apoptosis. Some reports have shown that constitutive activation of the NF-κB pathway in Tax expressing cells protects cells from apoptosis [43;86;123;129;130], whereas others have shown that it promotes apoptosis [110;135]. Differences in cell type and/or experimental approaches used in these studies may explain the variation in apoptotic outcomes, and more studies will be necessary to determine the precise role of NF-κB activation in Tax-related apoptosis.

DNA damage and serum starvation have been shown to induce apoptosis in Tax expressing cells [37;81;82;184]. Interestingly, Cereseto et al. found that IL-2 dependent HTLV-I infected T cells are more likely to undergo apoptosis after serum starvation and IL-2 deprivation than IL-2 independent HTLV-I infected T cells [24], suggesting that IL-2 independent infected cells may be resistant to apoptosis. In the process of becoming IL-2 independent, infected T cells may undergo changes in gene expression or protein function, which can alter cellular stress responses that normally would induce apoptosis.

Tax expression dramatically alters numerous cellular functions as discussed previously. After initial Tax expression, these cellular effects probably induce apoptosis in many cells; however, some cells may be resistant to apoptosis and survive. The differences between cells that survive initial Tax expression and those that undergo apoptosis have not been defined but it is likely that newly infected cells *in vivo* often undergo apoptosis. The long-term effects of Tax expression on apoptosis in the surviving cells may be indirect and depend on cell type and other factors. As mentioned, the transcription of FasL, Bcl-xL, and Bax is altered by Tax disruption of general transcription pathways like NF-κB. These altered transcription pathways affect numerous cellular functions in addition to apoptosis. It is possible that the overall effect of Tax on apoptosis depends on how Tax affects each transcription factor pathway in a particular experimental system. This could explain why there is disagreement in the literature regarding the long-term effects of Tax expression on apoptosis.

Concluding Remarks

HTLV-I Tax invokes several mechanisms that allow cells to escape from normal cell cycle control. Tax frees T lymphocytes from IL-2 dependence and TGF-β growth inhibition. Tax also accelerates cell cycle progression and inactivates several cell cycle checkpoints, including the G_1/S checkpoint and mitotic spindle checkpoint, allowing infected cells to continue growing and dividing during unfavorable circumstances. Tax largely exerts these effects by transcriptional transactivation of cellular gene expression and inactivation of protein function. As a result of these activities, infected cells are more likely to replicate their DNA at inappropriate times, leading to replication errors and replication of damaged DNA. These events can introduce permanent mutations into the cellular genome. Accumulation of mutations is commonly thought to promote cellular transformation which, in the case of HTLV-I, is manifested as adult T cell leukemia.

Given the plethora of cellular changes discussed in this chapter that are associated with Tax expression, it is important to ask what benefit these activities might have for the virus? HTLV-I infects T cells and integrates into the cellular genome where it is replicated whenever the cellular DNA is replicated. Since T cells are typically non-dividing, proliferation of infected cells benefits the virus by creating a larger pool of infected cells. The ability of Tax to stimulate cell proliferation is clearly an important step in achieving this goal. Many of the effects of Tax are consistent with this hypothesis. Tax has numerous and diverse functions that are mediated by changes in transcription or alterations in protein function. Tax alters general cellular transcription pathways, resulting in increased expression of several factors that promote cell cycle progression such as cyclin D1 and cyclin D2. Tax also disrupts the function of other proteins that cause cell cycle arrest such as p53 and p16^{INK4a}. Other effects of Tax such as disruption of DNA repair and possible induction of apoptosis are likely to be indirect consequences of enhanced cellular proliferation caused by Tax. Transformation and ATL are also probably unintended byproducts of virus infection given that the cumulative lifetime risk of developing ATL is less than 4% and it takes decades to develop disease. Additionally, ATL is usually fatal and does not enhance viral spread to new individuals, so the disease is not directly beneficial to the virus.

Although cell proliferation may be the ultimate goal of the virus, the indirect increase in mutation frequency observed in Tax-expressing cells may also be beneficial. It has been proposed that evolution is favored in an environment on the "edge of chaos". In terms of mutation frequency, this means that a rigid genome that tolerates no mutations does not allow sampling of new genetic information and evolution. Conversely, a very high mutation frequency (chaos) does not allow sufficient genomic stability for survival. Therefore, it is optimal to exist in an environment at the edge of chaos, where genetic changes are frequent, but not lethal. We propose that the HTLV-I Tax protein establishes such an environment, providing an evolutionary advantage to the virus. As a consequence, the cell also undergoes mutations, which can ultimately lead to the development of cancer.

Acknowledgements

We acknowledge research support from United States Public Service Grants CA-77371 and CA-55684 from National Cancer Institute, National Institutes of Health, awarded to S.J. Marriott. We thank Edward J. Parulski for helpful discussions. J.C. Watt is supported in part by National Institutes of Health Training Grant CA-09197.

References

[1] Adams PD. Regulation of the retinoblastoma tumor suppressor protein by cyclin/cdks. *Biochim Biophys Acta* 2001; 1471(3):M123-M133.

[2] Akagi T, Ono H, Shimotohno K. Expression of cell-cycle regulatory genes in HTLV-I infected T- cell lines: Possible involvement of Tax1 in the altered expression of cyclin D2, p18^{Ink4} and p21$^{Waf1/Cip1/Sdi1}$. *Oncogene* 1996; 12(8):1645-1652.

[3] Akagi T, Ono H, Tsuchida N, Shimotohno K. Aberrant expression and function of p53 in T-cells immortalized by HTLV-I Tax1. *FEBS Lett* 1997; 406(3):263-266.

[4] Akagi T, Shimotohno K. Proliferative response of Tax1-transduced primary human T cells to anti-CD3 antibody stimulation by an interleukin-2 independent pathway. *J Virol* 1993; 67:1211-1217.

[5] Albrecht B, Lairmore MD. Critical role of human T-lymphotropic virus type 1 accessory proteins in viral replication and pathogenesis. *Microbiol Mol Biol Rev* 2002; 66(3):396-406.

[6] Arima N, Daitoku Y, Ohgaki S, Fukumori J, Tanaka H, Yamamoto Y et al. Autocrine growth of interleukin 2-producing leukemic cells in a patient with adult T cell leukemia. *Blood* 1986; 68(3):779-782.

[7] Arima N, Daitoku Y, Yamamoto Y, Fujimoto K, Ohgaki S, Kojima K et al. Heterogeneity in response to interleukin 2 and interleukin 2-producing ability of adult T cell leukemic cells. *J Immunol* 1987; 138(9):3069-3074.

[8] Ariumi Y, Yamaoka S, Lin JY, Hirota M, Masui O, Taya Y et al. HTLV-1 tax oncoprotein represses the p53-mediated trans-activation function through coactivator CBP sequestration. *Oncogene* 2000; 19(12):1491-1499.

[9] Arnulf B, Villemain A, Nicot C, Mordelet E, Charneau P, Kersual J et al. Human T-cell lymphotropic virus oncoprotein Tax represses TGF-beta 1 signaling in human T cells via c-Jun activation: a potential mechanism of HTLV-I leukemogenesis. *Blood* 2002; 100(12):4129-4138.

[10] Ballard DW, Bohnlein E, Lowenthal JW, Wano Y, Franza BR, Greene WC. HTLV-1 Tax induces cellular proteins that activate the kB element in the IL-2 receptor gene. *Science* 1988; 241:1652-1655.

[11] Barmak K, Harhaj E, Grant C, Alefantis T, Wigdahl B. Human T cell leukemia virus type I-induced disease: pathways to cancer and neurodegeneration. *Virology* 2003; 308(1):1-12.

[12] Barmak K, Harhaj EW, Wigdahl B. Mediators of central nervous system damage during the progression of human T-cell leukemia type I-associated myelopathy/tropical spastic paraparesis. *J Neurovirol* 2003; 9(5):522-529.

[13] Bartek J, Lukas J. Mammalian G1- and S-phase checkpoints in response to DNA damage. *Curr Opin Cell Biol* 2001; 13(6):738-747.

[14] Bartek J, Lukas J. Pathways governing G1/S transition and their response to DNA damage. *FEBS Lett* 2001; 490(3):117-122.

[15] Bazarbachi A, Hermine O. Treatment of adult T-cell leukaemia/lymphoma: current strategy and future perspectives. *Virus Res* 2001; 78(1-2):79-92.

[16] Bendjennat M, Boulaire J, Jascur T, Brickner H, Barbier V, Sarasin A et al. UV irradiation triggers ubiquitin-dependent degradation of p21(WAF1) to promote DNA repair. *Cell* 2003; 114(5):599-610.

[17] Blattner WA, Kalyanaraman VS, Robert-Guroff M, Lister TA, Galton DA, Sarin PS et al. The human type-C retrovirus, HTLV, in Blacks from the Caribbean region, and relationship to adult T-cell leukemia/lymphoma. *Int J Cancer* 1982; 30(3):257-264.

[18] Bloom J, Amador V, Bartolini F, DeMartino G, Pagano M. Proteasome-mediated degradation of p21 via N-terminal ubiquitinylation. Cell 2003; 115(1):71-82.

[19] Brauweiler A, Garrus JE, Reed JC, Nyborg JK. Repression of Bax gene expression by the HTLV-I Tax protein: implications for suppression of apoptosis in virally infected cells. *Virology* 1997; 231:135-140.

[20] Burton JD, Bamford RN, Peters C, Grant AJ, Kurys G, Goldman CK et al. A lymphokine, provisionally designated interleukin T and produced by a human adult T-cell leukemia line, stimulates T-cell proliferation and the induction of lymphokine-activated killer cells. *Proc Natl Acad Sci USA* 1994; 91(11):4935-4939.

[21] Campbell MS, Chan GK, Yen TJ. Mitotic checkpoint proteins HsMAD1 and HsMAD2 are associated with nuclear pore complexes in interphase. *J Cell Sci* 2001; 114(Pt 5):953-963.

[22] Capasso H, Palermo C, Wan S, Rao H, John UP, O'Connell MJ et al. Phosphorylation activates Chk1 and is required for checkpoint-mediated cell cycle arrest. *J Cell Sci* 2002; 115(Pt 23):4555-4564.

[23] Cereseto A, Diella F, Mulloy JC, Cara A, Michieli P, Grassmann R et al. p53 functional impairment and high p21waf1/cip1 expression in human T-cell lymphotropic/leukemia virus type I-transformed T cells. *Blood* 1996; 88:1551-1560.

[24] Cereseto A, Kislyakova T, Washington PR, Nicot C, Franchini G. Differential response to genotoxic stress in immortalized or transformed human T-lymphotropic virus type I-infected T-cells. *J Gen Virol* 1999; 80 (Pt 7):1575-1581.

[25] Chaudhry S, Freebern WJ, Smith JL, Butscher WG, Haggerty CM, Gardner K. Cross-regulation of T cell growth factor expression by p53 and the Tax oncogene. *J Immunol* 2002; 169(12):6767-6778.

[26] Chen XL, Zachar V, Zdravkovic M, Guo M, Ebbesen P, Liu XD. Role of the Fas/Fas ligand pathway in apoptotic cell death induced by the human T cell lymphotropic virus type I Tax transactivator. *J Gen Virol* 1997; 78(12):3277-3285.

[27] Cheng J, Zhang D, Zhou C, Marasco WA. Down-regulation of SHP1 and up-regulation of negative regulators of JAK/STAT signaling in HTLV-1 transformed cell lines and freshly transformed human peripheral blood CD4(+) T-Cells. *Leuk Res* 2004; 28(1):71-82.

[28] Chlichlia K, Busslinger M, Peter ME, Walczak H, Krammer PH, Schirrmacher V et al. ICE-proteases mediate HTLV-1 tax function: T-cell activation and apoptosis. *Oncogene* 1997; 14:2265-2272.

[29] Chlichlia K, Moldenhauer G, Daniel PT, Busslinger M, Gazzolo L, Schirrmacher V et al. Immediate effects of reversible HTLV-I *tax* function: T-cell activation and apoptosis. *Oncogene* 1995; 10:269-277.

[30] Chowdhury IH, Farhadi A, Wang XF, Robb ML, Birx DL, Kim JH. Human T-cell leukemia virus type 1 Tax activates cyclin-dependent kinase inhibitor p21/Waf1/Cip1 expression through a p53-independent mechanism: Inhibition of cdk2. *Int J Cancer* 2003; 107(4):603-611.

[31] Ciosk R, Zachariae W, Michaelis C, Shevchenko A, Mann M, Nasmyth K. An ESP1/PDS1 complex regulates loss of sister chromatid cohesion at the metaphase to anaphase transition in yeast. *Cell* 1998; 93(6):1067-1076.

[32] Clute P, Pines J. Temporal and spatial control of cyclin B1 destruction in metaphase. *Nat Cell Biol* 1999; 1(2):82-87.

[33] Costanzo V, Robertson K, Ying CY, Kim E, Avvedimento E, Gottesman M et al. Reconstitution of an ATM-dependent checkpoint that inhibits chromosomal DNA replication following DNA damage. *Mol Cell* 2000; 6(3):649-659.

[34] Cross SL, Feinberg MB, Wolf JB, Holbrook NJ, Wong-Staal F, Leonard WJ. Regulation of the human interleukin-2 receptor α-chain promoter: Activation of a non-functional promoter by the transactivator gene of HTLV-I. *Cell* 1987; 49:47-56.

[35] De La Fuente C, Deng L, Santiago F, Arce L, Wang L, Kashanchi F. Gene expression array of HTLV type 1-infected T cells: Up-regulation of transcription factors and cell cycle genes. *AIDS Res Hum Retroviruses* 2000; 16(16):1695-1700.

[36] De La Fuente C, Santiago F, Chong SY, Deng L, Mayhood T, Fu P et al. Overexpression of p21(wafl) in human T-cell lymphotropic virus type 1-infected cells and its association with cyclin A/cdk2. *J Virol* 2000; 74(16):7270-7283.

[37] De La Fuente C, Wang L, Wang D, Deng L, Wu K, Li H et al. Paradoxical effects of a stress signal on pro- and anti-apoptotic machinery in HTLV-1 Tax expressing cells. *Mol Cell Biochem* 2003; 245(1-2):99-113.

[38] Di Leonardo A, Linke SP, Clarkin K, Wahl GM. DNA damage triggers a prolonged p53-dependent G1 arrest and long-term induction of Cip1 in normal human fibroblasts. *Genes Dev* 1994; 8(21):2540-2551.

[39] Donehower LA, Slagle BL, McArthur MJ, Butel JS, Bradley A. Mice deficient for p53 are developmentally normal but susceptible to spontaneous tumours. Nature 1992; 356:215-221.

[40] Dulic V, Kaufmann WK, Wilson SJ, Tlsty TD, Lees E, Harper JW et al. p53-dependent inhibition of cyclin-dependent kinase activities in human fibroblasts during radiation-induced G1 arrest. *Cell* 1994; 76(6):1013-1023.

[41] Edlich RF, Hill LG, Williams FM. Global epidemic of human T-cell lymphotrophic virus type-I (HTLV-I): an update. *J Long Term Eff Med Implants* 2003; 13(2):127-140.

[42] El-Deiry WS, Tokino T, Velculescu VE, Levy DB, Parsons R, Trent JM et al. WAF1, a potential mediator of p53 tumor suppression. *Cell* 1993; 75:817-825.

[43] El Sabban ME, Nasr R, Dbaibo G, Hermine O, Abboushi N, Quignon F et al. Arsenic-interferon-alpha-triggered apoptosis in HTLV-I transformed cells is associated with tax down-regulation and reversal of NF-kappa B activation. *Blood* 2000; 96(8):2849-2855.

[44] Evans T, Rosenthal ET, Youngblom J, Distel D, Hunt T. Cyclin: a protein specified by maternal mRNA in sea urchin eggs that is destroyed at each cleavage division. *Cell* 1983; 33(2):389-396.

[45] Falck J, Mailand N, Syljuasen RG, Bartek J, Lukas J. The ATM-Chk2-Cdc25A checkpoint pathway guards against radioresistant DNA synthesis. *Nature* 2001; 410(6830):842-847.

[46] Flores-Rozas H, Kelman Z, Dean FB, Pan Z-Q, Harper JW, Elledge SJ et al. Cdk-interacting protein 1 directly binds with proliferating cell nuclear antigen and inhibits DNA replication catalyzed by the DNA polymerase δ holoenzyme. *Proc Natl Acad Sci* USA 1994; 91:8655-8659.

[47] Franchini G, Nicot C, Johnson JM. Seizing of T cells by human T-cell leukemia/lymphoma virus type 1. *Adv Cancer Res* 2003; 89:69-132.

[48] Franchini G, Wong-Staal F, Gallo RC. Human T-cell leukemia virus (HTLV-I) transcripts in fresh and cultured cells of patients with adult T-cell leukemia. *Proc Natl Acad Sci* USA 1984; 81:6207-6211.

[49] Gatza ML, Watt JC, Marriott SJ. Cellular transformation by the HTLV-I Tax protein, a jack-of-all-trades. *Oncogene* 2003; 22(33):5141-5149.

[50] Gessain A, Barin F, Vernant JC, Gout O, Maurs L, Calander A et al. Antibodies to human T-lymphotropic virus type I in patients with tropical spastic paraparesis. *Lancet* 1985; 2:407-409.

[51] Grassmann R, Dengler C, Muller-Fleckenstein I, McGuire K, Dokhelar MC, Sodroski JG et al. Transformation to continuous growth of primary human T lymphocytes by human T cell leukemia virus type I X-region genes transduced by a herpesvirus saimiri vector. *Proc Natl Acad Sci* USA 1989; 86:3551-3355.

[52] Guacci V, Koshland D, Strunnikov A. A direct link between sister chromatid cohesion and chromosome condensation revealed through the analysis of MCD1 in S. cerevisiae. *Cell* 1997; 91(1):47-57.

[53] Guo Z, Kumagai A, Wang SX, Dunphy WG. Requirement for Atr in phosphorylation of Chk1 and cell cycle regulation in response to DNA replication blocks and UV-damaged DNA in Xenopus egg extracts. *Genes Dev* 2000; 14(21):2745-2756.

[54] Hall AP, Irvine J, Blyth K, Cameron ER, Onions DE, Campbell ME. Tumors derived from HTLV-1 tax transgenic mice are characterized by enhanced levels of apoptosis and oncogene expression. *J Rheumatol* 1998; 186:209-124.

[55] Haller K, Ruckes T, Schmitt I, Saul D, Derow E, Grassmann R. Tax-dependent stimulation of G1 phase-specific cyclin-dependent kinases and increased expression of signal transduction genes characterize HTLV type 1-transformed T cells. *AIDS Res Hum Retroviruses* 2000; 16(16):1683-1688.

[56] Haller K, Wu Y, Derow E, Schmitt I, Jeang KT, Grassmann R. Physical interaction of human T-cell leukemia virus type 1 tax with cyclin-dependent kinase 4 stimulates the phosphorylation of retinoblastoma protein. *Mol Cell Biol* 2002; 22(10):3327-3338.

[57] Hannon GJ, Beach D. p15INK4B is a potential effector of TGF-beta-induced cell cycle arrest. *Nature* 1994; 371(6494):257-261.

[58] Haoudi A, Daniels RC, Wong E, Kupfer G, Semmes OJ. Human T-cell leukemia virus-I tax oncoprotein functionally targets a subnuclear complex involved in cellular DNA damage-response. *J Biol Chem* 2003; 278(39):37736-37744.

[59] Haoudi A, Semmes OJ. The HTLV-1 tax oncoprotein attenuates DNA damage induced G1 arrest and enhances apoptosis in p53 null cells. *Virology* 2003; 305(2):229-239.

[60] Harper JW, Adami GR, Wei N, Keyomarsi K, Elledge SJ. The p21 Cdk-interacting protein Cip1 is a potent inhibitor of G1 cyclin-dependent kinases. *Cell* 1993; 75(4):805-816.

[61] Harper JW, Elledge SJ, Keyomarsi K, Dynlacht B, Tsai LH, Zhang P et al. Inhibition of cyclin-dependent kinases by p21. *Mol Biol Cell* 1995; 6(4):387-400.

[62] Hatta Y, Hirama T, Miller CW, Yamada Y, Tomonaga M, Koeffler HP. Homozygous deletions of the p15 (MTS2) and p16 (CDKN2/MTS1) genes in adult T-cell leukemia. *Blood* 1995; 85:2699-2704.

[63] Hatta Y, Koeffler HP. Role of tumor suppressor genes in the development of adult T cell leukemia/lymphoma (ATLL). *Leukemia* 2002; 16(6):1069-1085.

[64] Hatta Y, Yamada Y, Tomonaga M, Koeffler HP. Extensive analysis of the retinoblastoma gene in adult T cell leukemia/lymphoma (ATL). *Leukemia* 1997; 11:984-989.

[65] Hauf S, Waizenegger IC, Peters JM. Cohesin cleavage by separase required for anaphase and cytokinesis in human cells. *Science* 2001; 293(5533):1320-1323.

[66] Hinuma Y, Komoda T, Hosa T, Ondo T, Ohakura M, Akenaka T et al. Antibodies to adult T-cell leukemia-virus-associated antigen (ATLA) in sera from patients with ATL and controls in Japan: a nation-wide sero-epidemiologic study. *Int J Cancer* 1982; 29:631-635.

[67] Hinuma Y, Nagata K, Misoka M, Nakai T, Matsumoto T, Kiroshita K et al. Adult T-cell leukemia: Antigen in ATL cell line and detection of antibodies to the antigen in human sera. *Proc Natl Acad Sci* USA 1981; 78:6476-6480.

[68] Hoffmann I, Draetta G, Karsenti E. Activation of the phosphatase activity of human cdc25A by a cdk2-cyclin E dependent phosphorylation at the G1/S transition. *EMBO J* 1994; 13(18):4302-4310.

[69] Hollsberg P. Mechanisms of T-cell activation by human T-cell lymphotropic virus type I. *Microbiol Mol Biol Rev* 1999; 63(2):308-333.

[70] Hollsberg P, Ausubel LJ, Hafler DA. Human T cell lymphotropic virus type I-induced T cell activation. Resistance to TGF-beta 1-induced suppression. *J Immunol* 1994; 153(2):566-573.

[71] Hori T, Uchiyama T, Umadome H, Tamori S, Tsudo M, Araki K et al. Dissociation of interleukin-2-mediated cell proliferation and interleukin-2 receptor upregulation in adult T-cell leukemia cells. *Leuk Res* 1986; 10(12):1447-1453.

[72] Huang Y, Ohtani K, Iwanaga R, Matsumura Y, Nakamura M. Direct trans-activation of the human cyclin D2 gene by the oncogene product Tax of human T-cell leukemia virus type I. *Oncogene* 2001; 20(9):1094 -102.

[73] Hunter T, Pines J. Cyclins and cancer. II: Cyclin D and CDK inhibitors come of age. *Cell* 1994; 79(4):573-582.

[74] Iha H, Kasai T, Kibler KV, Iwanaga Y, Tsurugi K, Jeang KT. Pleiotropic effects of HTLV type 1 tax protein on cellular metabolism: mitotic checkpoint abrogation and NF-κB activation. *AIDS Res Hum Retroviruses* 2000; 16:1633-1638.

[75] Inoue J, Seiki M, Taniguchi T, Tsuru S, Yoshida M. Induction of interleukin 2 receptor gene expression by $p40^x$ encoded by human T-cell leukemia virus type I. *EMBO J* 1986; 5:2883-2888.

[76] Iwai K, Mori N, Oie M, Yamamoto N, Fujii M. Human T-cell leukemia virus type 1 tax protein activates transcription through AP-1 site by inducing DNA binding activity in T cells. *Virology* 2001; 279(1):38-46.

[77] Iwanaga R, Ohtani K, Hayashi T, Nakamura M. Molecular mechanism of cell cycle progression induced by the oncogene product Tax of human T-cell leukemia virus type I. *Oncogene* 2001; 20(17):2055-2067.

[78] Jin DY, Spencer F, Jeang KT. Human T cell leukemia virus type 1 oncoprotein tax targets the human mitotic checkpoint protein MAD1. *Cell* 1998; 93(1):81-91.

[79] Jinno S, Suto K, Nagata A, Igarashi M, Kanaoka Y, Nojima H et al. Cdc25A is a novel phosphatase functioning early in the cell cycle. EMBO J 1994; 13(7):1549-1556.

[80] Johnson JM, Harrod R, Franchini G. Molecular biology and pathogenesis of the human T-cell leukaemia/lymphotropic virus Type-1 (HTLV-1). *Int J Exp Pathol* 2001; 82(3):135-147.

[81] Kao S-Y, Lemoine FJ, Marriott SJ. HTLV-1 Tax protein sensitizes cells to apoptotic cell death induced by DNA damaging agents. *Oncogene* 2000; 19:2240-2248.

[82] Kao S-Y, Lemoine FJ, Marriott SJ. p53-independent induction of apoptosis by the human T cell leukemia virus type I Tax protein following UV irradiation. *Virology* 2001; 291:292-298.

[83] Kao S-Y, Marriott SJ. Disruption of nucleotide excision repair by the human T-cell leukemia virus type 1 Tax protein. *J Virol* 1999; 73(5):4299-4304.

[84] Kasai T, Iwanaga Y, Iha H, Jeang KT. Prevalent loss of mitotic spindle checkpoint in adult T-cell leukemia confers resistance to microtubule inhibitors. *J Biol Chem* 2002; 277(7):5187-5193.

[85] Katoh T, Harada T, Morikawa S, Wakutani T. IL-2- and IL-2-R- independent proliferation of T-cell lines from adult T-cell leukemia/lymphoma patients. *Int J Cancer* 1986; 38(2):265-274.

[86] Kawakami A, Nakashima T, Sakai H, Urayama S, Yamasaki S, Hida A et al. Inhibition of caspase cascade by HTLV-I tax through induction of NF-kappaB nuclear translocation. *Blood* 1999; 94(11):3847-3854.

[87] Kibler KV, Jeang KT. CREB/ATF-dependent repression of cyclin a by human T-cell leukemia virus type 1 Tax protein. *J Virol* 2001; 75(5):2161-2173.

[88] Kikuchi M, Mitsui T, Takeshita M, Okamura H, Naitoh H, Eimoto T. Virus associated adult T-cell leukemia (ATL) in Japan: clinical, histological and immunological studies. *Hematol Oncol* 1986; 4(1):67-81.

[89] Kim SJ, Kehrl JH, Burton J, Tendler CL, Jeang KT, Danielpour D et al. Transactivation of the transforming growth factor beta 1 (TGF-beta1) gene by human T lymphotropic virus type 1 tax: A potential mechanism for the increased expression of TGF-beta1 in adult T cell leukemia. *J Exp Med* 1990; 172:121-129.

[90] King RW, Peters JM, Tugendreich S, Rolfe M, Hieter P, Kirschner MW. A 20S complex containing CDC27 and CDC16 catalyzes the mitosis-specific conjugation of ubiquitin to cyclin B. *Cell* 1995; 81(2):279-288.

[91] Kinoshita T, Shimoyama M, Tobianai K, Ito M, Ito S, Ikeda S et al. Detection of mRNA for the tax1/rex1 gene of human T-cell leukemia virus type I in fresh peripheral blood mononuclear cells of adult T-cell leukemia patients and viral carriers by using the polymerase cain reaction. *Proc Natl Acad Sci* USA 1989; 86:5620-5624.

[92] Kodaka T, Uchiyama T, Umadome H, Uchino H. Expression of cytokine mRNA in leukemic cells from adult T cell leukemia patients. *Jpn J Cancer Res* 1989; 80(6):531-536.

[93] Koff A, Giordano A, Desai D, Yamashita K, Harper JW, Elledge S et al. Formation and activation of a cyclin E-cdk2 complex during the G1 phase of the human cell cycle. *Science* 1992; 257(5077):1689-1694.

[94] Kumagai A, Guo Z, Emami KH, Wang SX, Dunphy WG. The Xenopus Chk1 protein kinase mediates a caffeine-sensitive pathway of checkpoint control in cell-free extracts. *J Cell Biol* 1998; 142(6):1559-1569.

[95] LaBaer J, Garrett MD, Stevenson LF, Slingerland JM, Sandhu C, Chou HS et al. New functional activities for the p21 family of CDK inhibitors. *Genes Dev* 1997; 11(7):847-862.

[96] Lacoste J, Cohen L, Hiscott J. NF-kappa B activity in T cells stably expressing the Tax protein of human T cell lymphotropic virus type I. *Virology* 1991; 184(2):553-562.

[97] Lee DK, Kim BC, Brady JN, Jeang KT, Kim SJ. Human T-cell Lymphotropic Virus Type 1 Tax Inhibits Transforming Growth Factor-beta Signaling by Blocking the Association of Smad Proteins with Smad-binding Element. *J Biol Chem* 2002; 277(37):33766-33775.

[98] Lee H, Larner JM, Hamlin JL. A p53-independent damage-sensing mechanism that functions as a checkpoint at the G1/S transition in Chinese hamster ovary cells. *Proc Natl Acad Sci* USA 1997; 94(2):526-531.

[99] Lemasson I, Thébault S, Sardet C, Devaux C, Mesnard JM. Activation of E2F-mediated transcription by human T-cell leukemia virus type I tax protein in a $p16^{INK4A}$-negative T-cell line. *J Biol Chem* 1998; 273(36):23598-23604.

[100] Lemoine FJ, Kao S-Y, Marriott SJ. Suppression of DNA Repair by HTLV-I Tax correlates with Tax transactivation of PCNA gene expression. *AIDS Res Hum Retroviruses* 2000; 16:1623-1627.

[101] Lemoine FJ, Marriott SJ. Accelerated G_1 phase progression induced by the human T cell leukemia virus type I (HTLV-I) Tax oncoprotein. *J Biol Chem* 2001; 276:31851-31857.

[102] Leung KY, Nabel GJ. HTLV-1 transactivator induces interleukin-2 receptor expression through an NF-kB-like factor. *Nature* 1988; 333:776-778.

[103] Lew DJ, Burke DJ. The spindle assembly and spindle position checkpoints. *Annu Rev Genet* 2003; 37:251-282.

[104] Li J, Li H, Tsai MD. Direct binding of the N-terminus of HTLV-1 tax oncoprotein to cyclin-dependent kinase 4 is a dominant path to stimulate the kinase activity. *Biochemistry* 2003; 42(22):6921-6928.

[105] Li R, Murray AW. Feedback control of mitosis in budding yeast. *Cell* 1991; 66:519-531.

[106] Li Y, Benezra R. Identification of a human mitotic checkpoint gene: hsMAD2. *Science* 1996; 274(5285):246-248.

[107] Liang MH, Geisbert T, Yao Y, Hinrichs SH, Giam CZ. Human T-lymphotropic virus type I oncoprotein tax promotes S-phase entry but blocks mitosis. *J Virol* 2002; 76:4022-4033.

[108] Lill NL, Grossman SR, Ginsberg D, DeCaprio J, Livingston DM. Binding and modulation of p53 by p300/CBP coactivators. *Nature* 1997; 387(6635):823-827.

[109] Liu B, Liang MH, Kuo YL, Liao W, Boros I, Kleinberger T et al. Human T-lymphotropic virus type 1 oncoprotein tax promotes unscheduled degradation of Pds1p/securin and Clb2p/cyclin B1 and causes chromosomal instability. *Mol Cell Biol* 2003; 23(15):5269-5281.

[110] Los M, Khazaie K, Schulze-Osthoff K, Baeuerle PA, Schirrmacher V, Chlichlia K. Human T cell leukemia virus-I (HTLV-I) Tax-mediated apoptosis in activated T cells requires an enhanced intracellular prooxidant state. *J Immunol* 1998; 161(6):3050-3055.

[111] Low KG, Dorner LF, Fernando DB, Grossman J, Jeang KT, Comb MJ. Human T-cell leukemia virus type 1 tax releases cell cycle arrest induced by $p16^{INK4a}$. *J Virol* 1997; 71(3):1956-1962.

[112] Lu X, Lane DP. Differential induction of transcriptionally active p53 following UV or ionizing radiation: defects in chromosome instability syndromes? *Cell* 1993; 75(4):765-778.

[113] Maeda M, Arima N, Daitoku Y, Kashihara M, Okamoto H, Uchiyama T et al. Evidence for the interleukin-2 dependent expansion of leukemic cells in adult T cell leukemia. *Blood* 1987; 70(5):1407-1411.

[114] Mailand N, Falck J, Lukas C, Syljuasen RG, Welker M, Bartek J et al. Rapid destruction of human Cdc25A in response to DNA damage. *Science* 2000; 288:1425-1429.

[115] Maruyama M, Shibuya H, Harada H, Hatakeyama M, Seiki M, Fujita T et al. Evidence for abberrant activation of the interleukin-2 autocrine loop by HTLV-I encoded $p40^x$ and T3/Ti complex triggering. *Cell* 1987; 48:343-350.

[116] Matsuoka M. Human T-cell leukemia virus type I and adult T-cell leukemia. *Oncogene* 2003; 22(33):5131-5140.

[117] Matsushita K, Arima N, Fujiyoshi T, Daitoku Y, Hidaka S, Ohtsubo H et al. Interleukin-2-mediated growth of leukemic cells in lymph nodes of patients with adult T-cell leukemia/lymphoma. *Leuk Res* 1996; 20(2):135-141.
[118] Mesnard JM, Devaux C. Multiple control levels of cell proliferation by human T-cell leukemia virus type 1 tax protein. *Virology* 1999; 257(2):277-284.
[119] Migone TS, Lin J-X, Cereseto A, Mulloy JC, O'Shea JJ, Franchini G et al. Constitutively activated Jak-STAT pathway in T cells transformed with HTLV-I. *Science* 1995; 269:79-81.
[120] Mori N, Fujii M, Cheng G, Ikeda S, Yamasaki Y, Yamada Y et al. Human T-cell leukemia virus type I tax protein induces the expression of anti-apoptotic gene Bcl-xL in human T-cells through nuclear factor-kappaB and c-AMP responsive element binding protein pathways. *Virus Genes* 2001; 22(3):279 -87.
[121] Mori N, Fujii M, Hinz M, Nakayama K, Yamada Y, Ikeda S et al. Activation of cyclin D1 and D2 promoters by human T-cell leukemia virus type I tax protein is associated with IL-2-independent growth of T cells. *Int J Cancer* 2002; 99(3):378-385.
[122] Mori N, Morishita M, Tsukazaki T, Giam CZ, Kumatori A, Tanaka Y et al. Human T-cell leukemia virus type I oncoprotein Tax represses Smad-dependent transforming growth factor beta signaling through interaction with CREB-binding protein/p300. *Blood* 2001; 97(7):2137-2144.
[123] Mori N, Yamada Y, Ikeda S, Yamasaki Y, Tsukasaki K, Tanaka Y et al. Bay 11-7082 inhibits transcription factor NF-kappaB and induces apoptosis of HTLV-I-infected T-cell lines and primary adult T-cell leukemia cells. *Blood* 2002; 100(5):1828-1834.
[124] Motoi T, Uchiyama T, Uchino H, Ueda R, Araki K. Serum soluble interleukin-2 receptor levels in patients with adult T-cell leukemia and human T-cell leukemia/lymphoma virus type-I seropositive healthy carriers. *Jpn J Cancer Res* 1988; 79(5):593-599.
[125] Mulloy JC, Kislyakova T, Cereseto A, Casareto L, LoMonico A, Fullen J et al. Human T-cell lymphotropic/leukemia virus type 1 Tax abrogates p53-induced cell cycle arrest and apoptosis through its CREB/ATF functional domain. *J Virol* 1998; 72(11):8852-8860.
[126] Murphy EL, Hanchard B, Figueroa JP, Gibbs WN, Lofters WS, Campbell M et al. Modelling the risk of adult T-cell leukemia/lymphoma in persons infected with human T-lymphotropic virus type 1. *Int J Cancer* 1989; 43:250-253.
[127] Nagai H, Kinoshita T, Imamura J, Murakami Y, Hayashi K, Mukai K et al. Genetic alteration of p53 in some patients with adult T-cell leukemia. *Jpn J Can Res* 1991; 82:1421-1427.
[128] Nakayama K, Yamada Y, Koji T, Hayashi T, Tomonaga M, Kamihira S. Expression and phosphorylation status of retinoblastoma protein in adult T-cell leukemia/lymphoma. *Leuk Res* 2000; 24(4):299-305.
[129] Nasr R, Rosenwald A, El Sabban ME, Arnulf B, Zalloua P, Lepelletier Y et al. Arsenic/interferon specifically reverses 2 distinct gene networks critical for the survival of HTLV-1-infected leukemic cells. *Blood* 2003; 101(11):4576-4582.
[130] Nawata H, Maeda Y, Sumimoto Y, Miyatake J, Kanamaru A. A mechanism of apoptosis induced by all-trans retinoic acid on adult T-cell leukemia cells: a possible involvement of the Tax/NF-kappaB signaling pathway. Leuk Res 2001; 25(4):323-331.

[131] Neuveut C, Jeang KT. HTLV-I Tax and cell cycle progression. *Prog Cell Cycle Res* 2000; 4:157-162.
[132] Neuveut C, Jeang KT. Cell cycle dysregulation by HTLV-I: role of the tax oncoprotein. *Front Biosci* 2002; 7:d157-63.
[133] Neuveut C, Low KG, Maldarelli F, Schmitt I, Majone F, Grassmann R et al. Human T-cell leukemia virus type 1 Tax and cell cycle progression: Role of cyclin D-cdk and p110Rb. *Mol Cell Biol* 1998; 18(6):3620-3632.
[134] Newcomb EW. P53 gene mutations in lymphoid diseases and their possible relevance to drug resistance. *Leuk Lymphoma* 1995; 17(3-4):211-221.
[135] Nicot C, Harrod R. Distinct p300-responsive mechanisms promote caspase-dependent apoptosis by human T-cell lymphotropic virus type 1 Tax protein. *Mol Cell Biol* 2000; 20(22):8580-8589.
[136] Nicot C, Mahieux R, Takemoto S, Franchini G. Bcl-$X_{(L)}$ is up-regulated by HTLV-I and HTLV-II in vitro and in ex vivo ATLL samples. *Blood* 2000; 96:275-281.
[137] Niitsu Y, Urushizaki Y, Koshida Y, Terui K, Mahara K, Kohgo Y et al. Expression of TGF-beta gene in adult T cell leukemia. *Blood* 1988; 71(1):263-266.
[138] Ohtani K, Iwanaga R, Arai M, Huang Y, Matsumura Y, Nakamura M. Cell type-specific E2F activation and cell cycle progression induced by the oncogene product Tax of human T-cell leukemia virus type I. *J Biol Chem* 2000; 275(15):11154-11163.
[139] Okamoto T, Ohno Y, Tsugane S, Watanabe S, Shimoyama M, Tajima K et al. Multi-step carcinogenesis model for adult T-cell leukemia. *Jpn J Can Res* 1989; 80:191-195.
[140] Peters JM. The anaphase-promoting complex: proteolysis in mitosis and beyond. *Mol Cell* 2002; 9(5):931-943.
[141] Pise-Masison CA, Choi KS, Radonovich M, Dittmer J, Kim SJ, Brady JN. Inhibition of p53 transactivation function by the human T-Cell lymphotropic virus type 1 Tax protein. *J Virol* 1998; 72(2):1165-1170.
[142] Pise-Masison CA, Mahieux R, Jiang H, Ashcroft M, Radonovich M, Duvall J et al. Inactivation of p53 by human T-cell lymphotropic virus type 1 Tax requires activation of the NF-kB pathway and is dependent on p53 phosphorylation. *Mol Cell Biol* 2000; 20:3377-3386.
[143] Pise-Masison CA, Mahieux R, Radonovich M, Jiang H, Brady JN. Human T-lymphotropic virus type I Tax protein utilizes distinct pathways for p53 inhibition that are cell type-dependent. *J Biol Chem* 2001; 276(1):200-205.
[144] Pise-Masison CA, Radonovich M, Sakaguchi K, Appella E, Brady JN. Phosphorylation of p53: a novel pathway for p53 inactivation in human T-cell lymphotropic virus type 1-transformed cells. *J Virol* 1998; 72:6348-6355.
[145] Poiesz BJ, Ruscetti FW, Gadzar AF, Bunn PA, Minna JD, Gallo RC. Detection and isolation of type C retrovirus particles from fresh and cultured lymphocytes of a patient tropism of human T-cell leukemia virus type I. *Proc Natl Acad Sci* USA 1980; 77:7415-7419.
[146] Reid RL, Lindholm PF, Mireskandari A, Dittmer J, Brady JN. Stabilization of wild-type p53 in human T-lymphocytes transformed by HTLV-I. *Oncogene* 1993; 8:3029-3036.
[147] Ressler S, Morris GF, Marriott SJ. Human T-cell leukemia virus type 1 Tax transactivates the human proliferating cell nuclear antigen promoter. *J Virol* 1997; 71(2):1181-1190.

[148] Rivera I, Harhaj EW, Sun SC. Involvement of NF-AT in type I human T-cell leukemia virus tax- mediated Fas ligand promoter transactivation. *J Biol Chem* 1998; 273(35):22382-22388.

[149] Ruben S, Poteat H, Tan T, Kawakami K, Roeder R, Haseltine W et al. Cellular transcription factors and regulation of IL-2 receptor gene expression by HTLV-I *tax* gene product. *Science* 1988; 241:89-92.

[150] Ryan KM, Phillips AC, Vousden KH. Regulation and function of the p53 tumor suppressor protein. *Curr Opin Cell Biol* 2001; 13(3):332-337.

[151] Saggioro D, Barp S, Chieco-Bianchi L. Block of a mitochondrial-mediated apoptotic pathway in Tax-expressing murine fibroblasts. Exp Cell Res 2001; 269(2):245-255.

[152] Saha P, Eichbaum Q, Silberman ED, Mayer BJ, Dutta A. p21CIP1 and Cdc25A: competition between an inhibitor and an activator of cyclin-dependent kinases. *Mol Cell Biol* 1997; 17(8):4338-4345.

[153] Sakashita A, Hattori T, Miller CW, Suzushima H, Asou N, Takatsuki K et al. Mutations of the p53 gene in adult T-cell leukemia. *Blood* 1992; 79:477-480.

[154] Santiago F, Clark E, Chong S, Molina CA, Mozafari F, Mahieux R et al. Transcriptional up-regulation of the cyclin D2 gene and acquisition of new cyclin-dependent partners in human T-cell leukemia virus type 1-infected cells. *J Virol* 1999; 73:9917-9927.

[155] Saxinger W, Blattner WA, Levine PH, Clark J, Biggar R, Hoh M et al. Human T-cell leukemia virus (HTLV-I) antibodies in Africa. *Science* 1984; 225(4669):1473-1476.

[156] Schmitt I, Rosin O, Rohwer P, Gossen M, Grassmann R. Stimulation of cyclin-dependent kinase activity and G_1- to S- Phase transition in human lymphocytes by the human T-cell leukemia/lymphotropic virus type 1 Tax protein. *J Virol* 1998; 72(1):633-640.

[157] Scholz C, Freeman GJ, Greenfield EA, Hafler DA, Hollsberg P. Activation of human T cell lymphotropic virus type I-infected T cells is independent of B7 costimulation. *J Immunol* 1996; 157(7):2932-2938.

[158] Serrano M, Hannon GJ, Beach D. A new regulatory motif in cell-cycle control causing specific inhibition of cyclin D/CDK4. *Nature* 1993; 366(6456):704-707.

[159] Sexl V, Diehl JA, Sherr CJ, Ashmun R, Beach D, Roussel MF. A rate limiting function of cdc25A for S phase entry inversely correlates with tyrosine dephosphorylation of Cdk2. *Oncogene* 1999; 18(3):573-582.

[160] Shimoyama M. Diagnostic criteria and classification of clinical subtypes of adult T-cell leukemia/lymphoma. A report from the Lymphoma Study Group (1984-87). *Br J Haematol* 1991; 79:428-437.

[161] Siekevitz M, Feinberg MB, Holbrook N, Wong-Staal F, Greene WC. Activation of interleukin 2 and interleukin 2 receptor (Tac) promoter expression by the trans-activator (tat) gene product of human T-cell leukemia virus, type I. *Proc Natl Acad Sci* USA 1987; 84:5389-5393.

[162] Slattery JP, Franchini G, Gessain A. Genomic evolution, patterns of global dissemination, and interspecies transmission of human and simian T-cell leukemia/lymphotropic viruses. *Genome Res* 1999; 9(6):525-540.

[163] Sorensen CS, Syljuasen RG, Falck J, Schroeder T, Ronnstrand L, Khanna KK et al. Chk1 regulates the S phase checkpoint by coupling the physiological turnover and

ionizing radiation-induced accelerated proteolysis of Cdc25A. *Cancer Cell* 2003; 3(3):247-258.

[164] Stemmann O, Zou H, Gerber SA, Gygi SP, Kirschner MW. Dual inhibition of sister chromatid separation at metaphase. *Cell* 2001; 107(6):715-726.

[165] Stevens C, La Thangue NB. E2F and cell cycle control: a double-edged sword. *Arch Biochem Biophys* 2003; 412(2):157-169.

[166] Sudakin V, Ganoth D, Dahan A, Heller H, Hershko J, Luca FC et al. The cyclosome, a large complex containing cyclin-selective ubiquitin ligase activity, targets cyclins for destruction at the end of mitosis. *Mol Biol Cell* 1995; 6(2):185-197.

[167] Suzuki T, Kitao S, Matsushime H, Yoshida M. HTLV-1 Tax protein interacts with cyclin-dependent kinase inhibitor $p16^{INK4A}$ and counteracts its inhibitory activity towards CDK4. *EMBO J* 1996; 15(7):1607-1614.

[168] Suzuki T, Narita T, Uchida-Toita M, Yoshida M. Down-regulation of the INK4 family of cyclin-dependent kinase inhibitors by tax protein of HTLV-1 through two distinct mechanisms. *Virology* 1999; 259:384-391.

[169] Suzuki T, Uchida-Toita M, Yoshida M. Tax protein of HTLV-1 inhibits CBP/p300-mediated transcription by interfering with recruitment of CBP/p300 onto DNA element of E-box or p53 binding site. *Oncogene* 1999; 18(28):4137-4143.

[170] Takemoto S, Mulloy JC, Cereseto A, Migone TS, Patel BK, Matsuoka M et al. Proliferation of adult T cell leukemia/lymphoma cells is associated with the constitutive activation of JAK/STAT proteins. *Proc Natl Acad Sci* USA 1997; 94(25):13897-13902.

[171] Takemoto S, Trovato R, Cereseto A, Nicot C, Kislyakova T, Casareto L et al. p53 stabilization and functional impairment in the absence of genetic mutation or the alteration of the p14(ARF)-MDM2 loop in ex vivo and cultured adult T-cell leukemia/lymphoma cells. *Blood* 2000; 95(12):3939-3944.

[172] Tendler CL, Greenberg SJ, Blattner WA, Manns A, Murphy E, Fleisher T et al. Transactivation of interleukin 2 and its receptor induces immune activation in human T-cell lymphotropic virus type I-associated myelopathy: pathogenic implications and a rationale for immunotherapy. *Proc Natl Acad Sci* USA 1990; 87:5218-5222.

[173] Tsukahara T, Kannagi M, Ohashi T, Kato H, Arai M, Nunez G et al. Induction of Bcl-xL expression by human T-cell leukemia virus type 1Tax through NF-kB in apoptosis-resistant T-cell transfectants with Tax. *J Virol* 1999; 73:7981-7987.

[174] Uchiyama T. Human T cell leukemia virus type I (HTLV-I) and human diseases. *Annu Rev Immunol* 1997; 15:15-37.

[175] Uchiyama T, Hori T, Tsudo M, Wano Y, Umadome H, Tamori S et al. Interleukin-2 receptor (Tac antigen) expressed on adult T cell leukemia cells. *J Clin Invest* 1985; 76:446-453.

[176] Uchiyama T, Kamia M, Kodaka T, Tamori S, Fukuhara S, Amakawa R et al. Leukemic cells from some adult T-cell leukemia patients proliferate in response to interleukin-4. *Blood* 1988; 72:1182-1186.

[177] Uhlmann F, Wernic D, Poupart MA, Koonin EV, Nasmyth K. Cleavage of cohesin by the CD clan protease separin triggers anaphase in yeast. *Cell* 2000; 103(3):375-386.

[178] Van Orden K, Yan JP, Ulloa A, Nyborg JK. Binding of the human T-cell leukemia virus Tax protein to the coactivator CBP interferes with CBP-mediated transcriptional control. *Oncogene* 1999; 18(25):3766-3772.

[179] Van PL, Yim KW, Jin DY, Dapolito G, Kurimasa A, Jeang KT. Genetic evidence of a role for ATM in functional interaction between human T-cell leukemia virus type 1 Tax and p53. *J Virol* 2001; 75:396-407.

[180] Waizenegger IC, Hauf S, Meinke A, Peters JM. Two distinct pathways remove mammalian cohesin from chromosome arms in prophase and from centromeres in anaphase. *Cell* 2000; 103(3):399-410.

[181] Waldmann TA. The interleukin-2 receptor on normal and malignant lymphocytes. *Adv Exp Med Biol* 1987; 213:129-137.

[182] Xiao Z, Chen Z, Gunasekera AH, Sowin TJ, Rosenberg SH, Fesik S et al. Chk1 mediates S and G2 arrests through Cdc25A degradation in response to DNA-damaging agents. *J Biol Chem* 2003; 278(24):21767-21773.

[183] Xu X, Kang SH, Heidenreich O, Okerholm M, O'Shea JJ, Nerenberg MI. Constitutive activation of different Jak tyrosine kinases in human T cell leukemia virus type 1 (HTLV-1) tax protein or virus- transformed cells. *J Clin Invest* 1995; 96(3):1548-1555.

[184] Yamada T, Yamaoka S, Goto T, Nakai M, Tsujimoto Y, Hatanaka M. The human T-cell leukemia virus type I Tax protein induces apoptosis which is blocked by the Bcl-2 protein. *J Virol* 1994; 68:3374-3379.

[185] Yamada Y. Phenotypic and functional analysis of leukemic cells from 16 patients with adult T-cell leukemia/lymphoma. *Blood* 1983; 61(1):192-199.

[186] Yamato K, Oka T, Hiroi M, Iwahara Y, Sugito S, Tsuchida N et al. Aberrant expression of the p53 tumor suppressor gene in adult T-cell leukemia and HTLV-1-infected cells. *Jap J Cancer Res* 1993; 84:4-8.

[187] Yasuda N, Lai PK, Ip SH, Kung PC, Hinuma Y, Matsuoka M et al. Soluble interleukin 2 receptors in sera of Japanese patients with adult T cell leukemia mark activity of disease. *Blood* 1988; 71(4):1021-1026.

[188] Yoshida M. Molecular biology of HTLV-I: Recent progress. J Acquir Immune Defic *Syndr Hum Retrovirol* 1996; 13:S63-S68.

[189] Yoshida M. Multiple viral strategies of HTLV-1 for dysregulation of cell growth control. *Annu Rev Immunol* 2001; 19:475-96.

[190] Yoshida M, Miyoshi I, Hinuma Y. Isolation and characterization of retrovirus from cell lines of human adult T-cell leukemia and its importance in the disease. *Proc Natl Acad Sci* USA 1982; 79:2031-2035.

[191] Yssel H, de Waal MR, Duc D, Blanchard D, Gazzolo L, de Vries JE et al. Human T cell leukemia/lymphoma virus type I infection of a CD4+ proliferative/cytotoxic T cell clone progresses in at least two distinct phases based on changes in function and phenotype of the infected cells. *J Immunol* 1989; 142(7):2279-2289.

[192] Yu H. Regulation of APC-Cdc20 by the spindle checkpoint. *Curr Opin Cell Biol* 2002; 14(6):706-714.

[193] Zhang QA, Lee B, Korecka M, Li G, Weyland C, Eck S et al. Differences in phosphorylation of the IL-2R associated JAK/STAT proteins between HTLV-I (+), IL-2-independent and IL-2-dependent cell lines and uncultured leukemic cells from patients with adult T-cell lymphoma/leukemia. *Leuk Res* 1999; 23(4):373-384.

[194] Zou H, McGarry TJ, Bernal T, Kirschner MW. Identification of a vertebrate sister-chromatid separation inhibitor involved in transformation and tumorigenesis. *Science* 1999; 285(5426):418-422.

In: Progress in Virus Research
Editor: Sabine Thebault, pp. 65-100
ISBN 1-59454-438-7

Chapter III

Advances in Understanding the Roles of the Cell in HIV-1 Assembly

David E. Ott [*]
Basic Research Program, SAIC-Frederick, Inc.,
National Cancer Institute at Frederick,
Frederick, Maryland 21702-1201

Abstract

As with all viruses, human immunodeficiency virus type 1 (HIV-1) requires the use of cellular pathways and components in its replication cycle. Understanding these interactions between a virus and its host cell can assist in the discovery and development of antiviral treatments and strategies. Retroviruses produce infectious particles from a rather small number of components: three structural polyproteins and a homodimer of genomic RNA. These essential building blocks must all interact within the vast space of the host cell to form the organized structure of an infectious virion. While our current understanding of the retroviral proteins and genomic RNA functions have allowed for the creation of a reasonably detailed working model of how they assemble and their relative importance, the roles that cellular components and pathways play in retroviral assembly remain poorly understood. Several mechanistic questions remain unclear: How does the assembling virus find its genomic RNA in the cell? How is Gag targeted to the sites of assembly? What cellular components are required during assembly? Where does Gag assemble? How does the assembled virus release from the cell? What roles to do cellular proteins play in the virion? What potential antiviral therapies could interrupt these targets? This review will examine the cellular side of these HIV-1 assembly questions and recent experimental advances that are beginning to unravel the secrets of the interactions between HIV-1 and cells. While clear answers to these assembly questions are not currently known, our partial understanding of these complex and required interactions between the assembling viral proteins and the host cell machinery has opened up exciting new avenues for HIV-1 assembly research as well as cell biology in general.

[*] Corresponding author: David E. Ott, Telephone:(301) 846-5723, Fax:(301) 846-5588, Email: ott@ncifcrf.gov

Although this review is mostly restricted to HIV-1, lessons learned from this virus may apply to other retroviruses.

Introduction

Viruses require a host cell to provide them with many of the processes needed for replication. Therefore, by definition they are not living things; rather, they exist as simpler-than-life forms that steal from and subvert the host cell. This dependence allows viruses to be relatively small with a limited number of viral proteins, thus presenting fewer antigens for the host's immune system to detect. The use of cell systems makes it quite difficult to stop viral replication and disease since, similar to cancer, most of the essential systems that could be attacked by a therapeutic are those of the cell itself. Thus, viruses present few virus-specific targets, since the viral machinery has commandeered cellular pathways for its own purposes.

Different viruses rely on the host to different extents: viruses with larger genomes, usually DNA viruses, can express many nonessential viral proteins and enzymes in addition to their essential enzymes and structural proteins. These nonessential proteins often modulate the various systems of the host (e.g., the immune system, host gene expression, or a biosynthetic pathway) or duplicate existing cellular activities for a specific viral functions. On the other hand, RNA viruses, including retroviruses, have relatively small genomes, due in part to the inherent instability of RNA and infidelity of RNA replication that precludes larger genomes. Thus, these smaller viruses have to rely on the cell to a larger extent for their replication as a trade-off for a compact genome. RNA viruses do have an advantage over DNA viruses in that they can mutate more easily than DNA viruses, due to the error-prone nature of RNA replication[1]. The ability of RNA viruses to rapidly mutate presents a severe problem for clinically useful antiviral therapies and for vaccines since mutants are readily generated that escape some immune responses and antiviral therapies[2].

An RNA genome can be replicated in several ways, mostly by employing an RNA-dependent RNA polymerase. Instead of this direct approach, retroviruses replicate their genome by reversing the normal transcription process of biology, changing their RNA genome to double-stranded DNA copy, and inserting this complimentary DNA (cDNA) genome into the genome of the host, becoming a set of genes. To do this, retroviruses possess their "trademark" reverse transcriptase that copies their 9.3 kb RNA genome into DNA and an integrase enzyme that places this cDNA genome into the host genome. The integrated viral DNA, a provirus, functions essentially as a complex genetic locus, expressing one RNA that is differentially spliced to express multiple transcripts that produce viral proteins[3].

The ability to integrate into cellular DNA is the key to the great success of retroviruses, as evidenced by the many known mammalian retroviruses discovered to date[3], as well as those in fish[4] and reptiles[4]. Retrovirus-like elements are even found in yeast[5]. Retroviruses may be the consummate intracellular parasites since, unlike other viruses, they have crossed over from infecting somatic cells to integrating into the host's germline, thus being transmitted vertically (i.e., genetically) as well as horizontally[3]. Indeed, approximately 1% of the human genome is made up of endogenous retroviral sequences[6]. This is a relatively large portion of the genome, as only approximately 3% of the human genome is used for protein-encoding genes[6]. These human endogenous retroviral sequences, HERVs, no longer produce extracellular viruses and mostly exist as unexpressed or weakly expressed defective viruses in the human genome. Some of these sequences have been implicated in autoimmunity and cancer[6]. Interestingly, retrovirus-like endogenous sequences have been found in many vertebrate genomes[7].

While retroviruses are routinely used in the laboratory as vectors for introducing genes and DNA constructs into cells as well as human gene therapy vehicles[8], the worldwide focus on retroviruses is due to the HIV-induced acquired immune deficiency syndrome (AIDS) pandemic[9]. For most people, HIV-1 infection eventually causes a severe depletion in $CD4^+$ T cells that ultimately results in the collapse of the immune system by several mechanisms[10]. This outcome occurs despite a considerable host immune response elicited by HIV-1 infection. Part of the intractable nature of AIDS is the ability of HIV-1 to integrate into the genome of the host, thereby becoming a permanent part of the cell. Ironically, the primary property that provides a helpful tool to the molecular biologist and the gene therapist is one of the biggest obstacles for antiviral treatment. While other acutely replicating viruses must present themselves to the host immune system, retroviruses can essentially stay in the cell for its life, in some cases lying dormant for long periods of time. Since some cells do not actively produce virus, there appears to be a latent, reactivatible reservoir of infected cells that escape immune surveillance and anti-viral treatments[11].

All replication-competent retroviruses express three genes, *gag*, *pol*, and *env*, which produce the essential structural proteins of the virus. While "simple" retroviruses (e.g., murine or avian retroviruses) only contain these absolutely required genes, HIV-1 is much more complex, expressing at least 9 genes. Although all of the 6 additional genes are required for normal replication and disease induction in the host[12-15], these additional genes have varying importance for the expression and replication of HIV in cell culture. Two of these additional proteins, Tat and Rev, coordinate the expression of the HIV-1 genome to a degree that they are also essential for viral replication. Tat increases viral RNA expression by activating RNA Pol II transcript elongation, thus causing the viral proteins to be produced at relatively high levels[16]. Rev regulates viral RNA splicing, preventing the multiple splicing of the viral RNA that produces Tat and Rev mRNA. In this way, Rev allows the export of full-length viral RNA from the nucleus for use as both genomic RNA and mRNA for the production of Gag and Gag-Pol from unspliced and Env from singly spliced viral mRNA[17].

Other nonessential proteins, sometimes referred to as accessory proteins, appear to be important for HIV-1 replication *in vivo*, but not necessarily *in vitro*[18]. Nef somehow modulates the cellular signaling of infected cells, producing virions with increased infectivity. Vpr appears to play an important role in virus expression and causes G2 arrest of infected cells[19]. Vpu increases the efficiency of virus release from cells[20.] Vif induces the degradation of a cellular protein, Apobec3g, and also blocks its incorporation into virions which prevents it from damaging the viral genome with its cytidine deaminase activity [21-24](see Cellular Proteins section below). While Vpr is incorporated into virions at approximately 10% of the molar Gag ratio by binding to the p6 domain of Gag [25,26], Nef and Vif have been shown to be present inside HIV-1 in small amounts. Vif appears to be incorporated by an interaction with the viral RNA [27] while the mechanism for Nef incorporation is not clear[14]

Unlike the expression and accessory proteins, structural proteins are expressed as polyproteins (Gag, Pol, and Env) that are then proteolytically cleaved into their mature forms (Fig.1). Each polyprotein contributes a separate function for the virion. The Gag polyprotein and its mature cleavage products make up most of the mass of the virion[28] (approximately

90% of the protein in the virus). It is the only protein that is required for the production of particles; thus, Gag contains all the functions required for particle assembly. The mature Gag proteins are arranged in a matrix (MA, or p17)-capsid (CA, or p24)-spacer peptide 1 (SP1, or p2)-nucleocapsid (NC or p7)-spacer peptide 2 (SP2 or p1)-p6 configuration within the Gag polyprotein[14]. This polyprotein has a strong propensity to assemble into multimer. In this way, Gag assembling at the plasma membrane is the driving force for virion production, strongly associating with other Gag proteins to produce a viral particle. The expression, transport, and assembly of functionally linked proteins as polyproteins make for a very efficient assembly mechanism, as all the mature proteins are incorporated by just two types of protein interactions, rather than each one individually. In this way, the correct stoichiometry (1-to-1) can be easily maintained.

However, after virion formation, this tight association is no longer needed in the virus particle; in fact, this close association makes infection processes, reverse transcription and integration, impossible. So, in a theme common to many viruses[29], the polyprotein needs to be converted into its mature proteins that function in the infectious process. For HIV-1, Gag processing by the viral protease starts during assembly and budding and is completed after release of the virion from the cell[30,31]. In this way, proteolytic processing switches Gag from a structure that assembles to one that carries out post-entry infection steps. Each of the mature cleavage products appears to have a function in the Gag precursor: the myristylated matrix domain, MA, directs the precursors to the plasma membrane; the capsid domain, CA, forms the dense immature structure and mature core of the virion; the nucleocapsid domain, NC, specifically binds the viral RNA and brings it into the virion; and the C-terminal p6 domain is required for efficient budding of HIV-1 from the plasma membrane[14,32]. Present together within Gag, all these proteins work in concert to effect particle assembly. However, the role of these proteins in infection is somewhat less understood. Mutations in CA can prevent the infectivity of virus, providing evidence for a direct function of CA in the infection process[33,34]. In contrast, NC has been shown to be an important participant in the reverse transcription and integration steps[35-39].

Pol contains all the viral enzymes required for replication. Like Gag, the protease, reverse transcriptase, and integrase proteins are expressed as a polyprotein encoded by the *pol* gene (Fig. 1). Pol is expressed by a -1 frameshift during Gag translation that occurs just after the NC region in Gag in about 1 in 20 products, switching the translating machinery from the Gag reading frame to the overlapping Pol frame, producing a Gag-Pol fusion polyprotein. The protein is incorporated into the virion by interactions of the Gag portion of the Gag-Pol fusion with the assembling Gag proteins. The protease within Gag-Pol is the enzyme which cleaves both Gag and Gag-Pol proteins into their components. The other enzymes play roles in the infectivity rather than assembly process: the reverse transcriptase enzyme copies the genomic RNA into double-stranded DNA that is then integrated into the host genome by integrase.

The Env precursor contains the proteins required for viral binding to and entry into the host cell[40]. Unlike the viral protease processing of Gag and Gag-Pol, this polyprotein is cleaved in the Golgi compartment by a cellular protease[41], to produce two mature glycoproteins, the surface and transmembrane proteins (Fig. 1). These two proteins associate together to form the envelope protein complex on the surface of the plasma membrane that is incorporated into the viral envelope during budding. The surface glycoprotein, SU protein (or $gp120^{SU}$), is responsible for the specific virus binding to the HIV-1 receptor molecule, CD4, on the host cell[14]. It is tethered to the viral lipid envelope by noncovalent interactions with the transmembrane protein, TM (gp41[TM]). Binding to the CD4 receptor produces a conformational change in SU that allows it to interact with one of several chemokine receptors that can serve as a co-receptor molecule for HIV-1[40]. This second binding event induces another conformational change in SU that exposes the fusion domain of the TM protein. Viral penetration into the cell is then mediated by this region of TM, which causes a fusion event between the host and viral membranes.

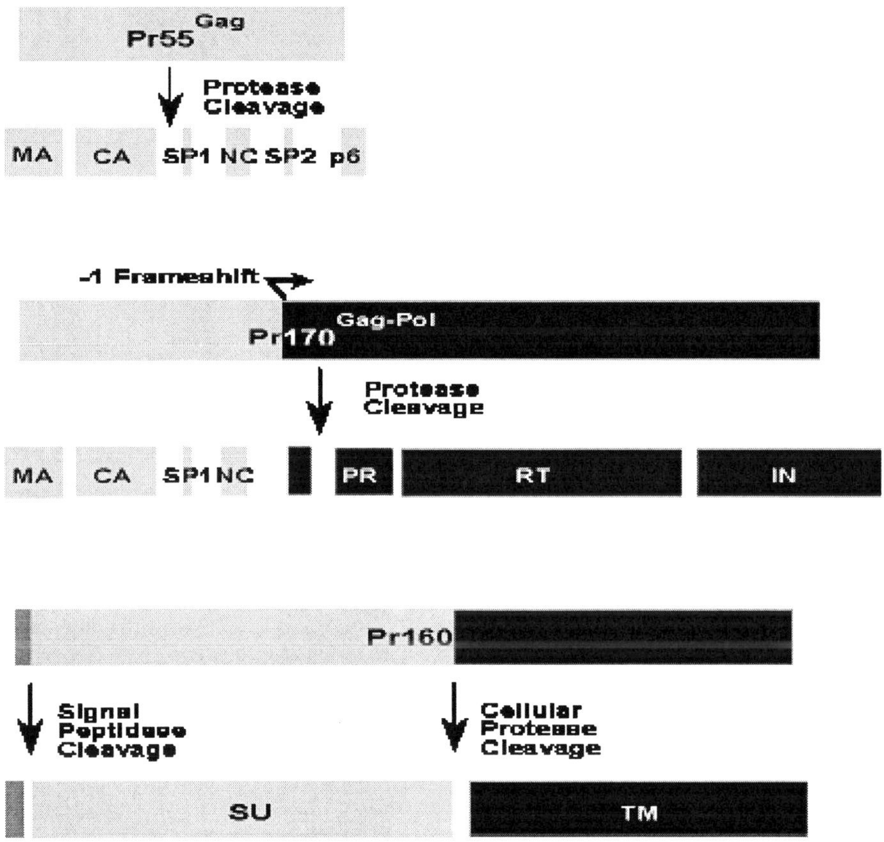

Figure 1, Gag and Gag-Pol organization and processing. Gag regions are shaded gray while Pol regions are in black and Env in open boxes.

The first visible step of HIV-1 assembly occurs on the cytoplasmic side of cellular membranes. In T cells and epithelial cell lines used as HIV-1 model systems, assembly appears to occur at the plasma membrane[3]. However, in monocytes and macrophages, HIV-1 buds into intracellular vacuoles and there is much less virus budding from the plasma memebrane[42]. Based on electron microscopic studies, assembling Gag structures in T cells are first seen as bar-like, electron dense structures associated with the inner leaflet plasma membrane that likely consists of Gag, Gag-Pol and the genomic RNA[43,44] (Fig. 2). As more Gag polyproteins assemble, this structure grows and forms a spherical shell that distends the plasma membrane. This structure results in the formation of a bud that is tethered to the cell surface by a thin stalk of membrane, which eventually pinches off from the cell, releasing the virion into the extracellular space. By budding at the cell surface, the virion acquires a lipid envelope, derived from the host plasma membrane that contains the Env protein complex. In addition to viral proteins, other cellular proteins are also incorporated, in some cases selectively by mechanisms and for purposes that are not completely clear[45.]

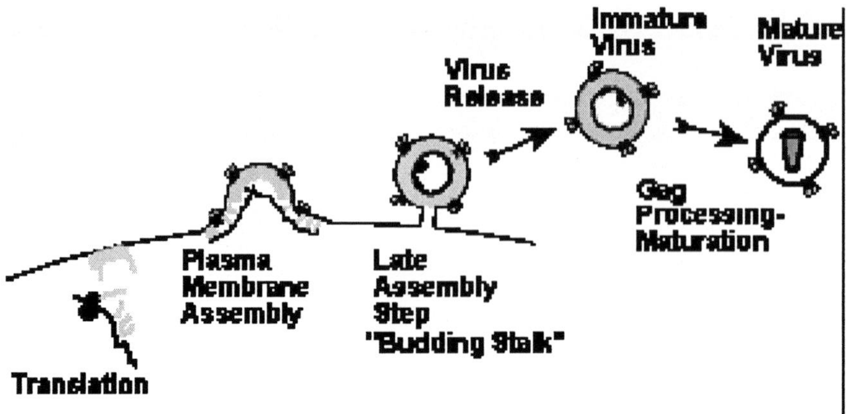

Figure 2. HIV budding summary. RNA is denoted by a black line while Gag is in gray.

Assembly Questions

Great strides have been made in understanding the role of HIV-1 proteins in virus assembly. However, the role of the cell is certainly less understood. The following questions highlight some recent advances and some basic principles in the cellular contributions to viral assembly.

- How does the assembling virus find its genomic RNA in the cell?
- How is Gag targeted to the sites of assembly?
- What cellular components are required during assembly?
- Where does Gag assemble?
- How does the assembled virus release from the cell?
- What roles to do cellular proteins play in the virion?
- What potential antiviral therapies could interrupt these targets?

How Does the Assembling Virus Find its Genomic RNA in the Cell?

The incorporation of the HIV-1 genomic RNA into the virions is crucial for the virus. But a significant challenge to this process is the rather large volume of the cell and the complexity of proteins and nucleic acids inside compared to the virus and its components. For example, HIV-1 genomic RNA, the viral full-length (unspliced) mRNA, is less than 1% of the total mRNA in an infected cell [3,46,47], yet it is almost exclusively packaged into virions, even in a gross excess of cellular RNAs. Thus, it is inconceivable that viral proteins and RNA simply meet by diffusion. So how does Gag pick out its genomic RNA over all these other RNAs?

The most widely recognized mechanism for specific packaging is the recognition of the psi site in the 5-prime untranslated region of the genomic RNA by Gag, specifically the NC domain in Gag[3,46,48]. Psi contains a high degree of secondary structure that is required for this interaction, and experimental and structural studies suggest that psi is recognized by Gag through a combination of these specific sequences and secondary structures [49 and references therein]. While psi has been well defined in murine leukemia virus, MuLV, the sequences responsible for packaging in HIV-1 are less well characterized[3]. Furthermore, binding assays show that NC binds psi-containing RNAs only 10-fold higher than those that do not[50,51] and deletion of some of the psi secondary structure leads to a 10-fold reduction of genomic RNA in particles, yet a several log decrease in infectivity[52,53], in contrast, HIV-1 without NC binds essentially no genomic RNA[45]. So there appears to be more to RNA recognition than a simple psi sequence-NC interaction.

Like all retroviruses, the HIV-1 unspliced RNA can act as both a messenger RNA, that produces both Gag and Gag-Pol polyproteins, as well as a genomic RNA that is packaged into the virion[3]. So could a full-length RNA perform both roles, or are there two separate populations of RNAs that function exclusively as either one or the other? Surprisingly, the answer appears to be that the different types of retroviruses employ either of these mechanisms[54,55]. For HIV-1, it appears, based on experiments with transcription inhibitors[54] and translation inhibitors[56], that full-length RNA can act as both mRNA and genomic RNA. These findings are quite different than those for MuLV indicating, based on analogous experiments[54,57,58], that this virus apparently has separate genomic RNA and mRNA pools used for assembly.

The dual role of HIV-1 RNA as both genome and translated mRNA opens up the possibility that translation and genome packaging could be linked. Early evidence for this came from the finding that some cellular proteins involved in translation are also incorporated into HIV-1 virions[59]. Since NC has a strong nonspecific affinity for RNA, it is possible that, as translation of Gag proceeds, the NC region of the newly formed Gag could bind to the rather large (7 kb) non-translated region 3-prime of the Gag coding sequence (that encodes *pol* and *env*), packaging the mRNA for use as genomic RNA *in cis* (Fig. 3). In this way, full-length mRNA can be packaged by the same Gag it expresses. This mechanism contrasts with a *trans* packaging scheme where the viral RNA and the proteins must be transported to the same location where the viral RNA is recognized and packaged (Fig. 3). Recent experiments have shown that an RNA that produces functional Gag is packaged with a 12-fold preference over an RNA that does not produce Gag, entirely consistent with a *cis* packaging mechanism[60]. Therefore, it appears that packaging and translation are linked. Despite this association, translation itself clearly is not required for HIV-1 genome packaging, as has been proven by biochemical studies[56] and by the ability of HIV-based vectors to be packaged[61]. *Cis* encapsidation of viral RNA is important for HIV-2 genome packaging[62] and required for hepatitis B virus genome packaging[63,64].

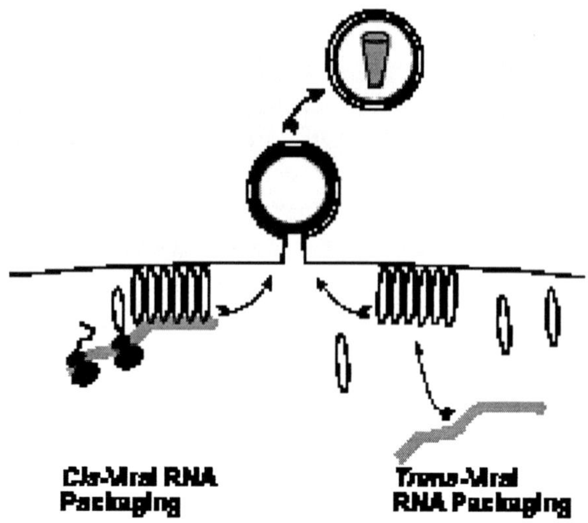

Figure 3. *Cis* versus *trans* RNA packaging models. RNA (grey line) is recognized by Gag and packaged in either a translation-linked mode, *cis* packaging or in a bimolecular reaction, *trans* packaging.

Since hepadnaviruses and retroviruses can use the same RNA as a message and as a genome (for the former, RNA is a precursor that is reverse-transcribed after packaging into the DNA genome found in the virus), this might be a general mechanism to promote efficient RNA recognition during assembly. The linkage of translation, packaging, and assembly puts these internal structural components in close proximity where they can interact efficiently without having to locate each other in the vast complexity of the cell. This is important, as viral RNA is a small portion of the total mRNA population and coordinating production of Gag with genome packaging and assembly would solve the "1%" problem. Therefore, *cis* packaging could work in concert with psi-mediated RNA recognition, bringing another level to the remarkable specificity RNA recognition by HIV-1.

How is Gag Targeted to the Sites of Assembly?

Since HIV-1 assembly takes place near the plasma membrane, the linkage of translation, assembly and packaging together, as outlined above, suggests that HIV-1 translation occurs near the plasma membrane. So instead of translating Gag in the cell interior and actively transporting Gag through the cell, viral RNA could be localized by the cell to the cortical regions of the cell, where it is translated near the plasma membrane rather than being transported through the cytoplasm. This is consistent with several pieces of data. The initial steps of the assembly process visible by electron microscopy are at the plasma membrane[43]. As noted above, Gag is rapidly associated with membranes after synthesis[65]. In addition, fluorescence microscopy using green fluorescence protein-labeled Gag observes Gag as punctuate loci at the cell surface with little evidence for diffuse internal staining that would be expected for general cytoplasmic expression[66-68]. Also, Staufen, a protein that can function to localize mRNAs in the cell, has been found in HIV-1 virions[69]. Cellular RNAs are often found in RNA granules that contain Staufen, tRNA synthetases, and elongation factor-1α[70] that appear to be part of the RNA transport and translation system. Supporting this linkage,

HIV-1 RNA has sequences that signal its association with RNA transport and RNA granules[71].

RNA transport and spatial translation as a method for protein localization have been described for several proteins[72]and are mediated by a zip-code sequence, usually located in the 3′ untranslated region of the RNA[73]. In this way, the cell can move one RNA molecule to make many proteins at the correct location, rather than move many proteins, providing an economy of effort within the cell. Electron microscopic *in situ* hybridization experiments on HIV-1-infected lymphocytes found viral RNA localized to the plasma membrane, in many instances without observable Gag structures[74]. These RNAs could be translating Gag just under the surface. .Also, MuLV full-length RNA appears to be actively transported to the cell surface by on late endosomes [75]. Interestingly, HIV-1 assembly has been linked to the endocytic pathway too (see below). Perhaps HIV-1 has exploited this cellular RNA localization mechanism for its own expression, assembly and genome encapsidation. While this is an interesting proposal, much remains to be further demonstrated for this model to be proven.

Alternatively, Gag could be transported to the cell surface by cellular motors that carry proteins along the cytoskeleton. IIIV-1 can associate with a microtubule-motor protein KIF4[76], consistent with an active transport system along the tubulin system. Others have found that the integrity of the cytoskeleton is important for virus budding. Gag associates with the actin cytoskeleton, though not the microtubule network[77]. Also, experiments found that interrupting the actin cytoskeleton and myosin, an actin motor, function decreased HIV-1 production[78]. However, there does not appear to be clear support for an established cellular system to transport HIV-1 Gag from intracellular places to membranes. Prolonged transport of Gag from intracellular sites to the assembly site is not consistent with either the finding that Gag associates with membranes remarkably soon after synthesis[65] or the imaging data[66-68]. However, Gag could associate with an internal membrane structure that is then transported to the cell surface (see below).

What Cellular Components are Required During Assembly?

Lipid Rafts?

As examined above, HIV-1 Gag assembly is closely linked with membrane-binding. One model proposes that HIV-1 assembles in contact with and with the assistance of lipid rafts[79]. Several types of lipids are present in the plasma membrane, and some of the bulkier lipids can exist in an ordered state. Bulky lipids, mostly cholesterol and sphingolipid, appear to be present in ordered patches or lipid rafts on the outer leaflet cholesterol, though cholesterol probably forms these structures on the inner leaflet, too[80-82]. Lipid rafts have been implicated in many processes, especially membrane signaling and receptor organization[80]. While a large body of research implicates lipid rafts in many cellular signaling and receptor organization processes, their size, stability and even functional existence are controversial[81,82]. Currently, it appears that lipid rafts are fairly small, approximately 70 nm, and relatively transient, forming and disassociating on the order of milliseconds lifetimes. There are even doubts about the interpretation of two of the most used methods for studying rafts, detergent insolubility and extraction of cholesterol with chelating compounds. The concern is that these treatments and dynamic nature of membranes and rafts could produce artifactual interactions that make proteins partition into insoluble lipid-bound fractions. So while ordered lipids certainly exist, exactly what they are and how best to biochemically study them are not clear.

Despite these caveats, there is a good body of evidence for cholesterol (and presumably raft) involvement in viral budding. Initially, HIV-1 was found to contain two GPI-linked proteins that are normally partitioned into lipid rafts, implying that rafts were incorporated into HIV particles[83]. Gag also was found to be localized near lipid rafts by microscopy and was biochemically-associated with lipid rafts in detergent studies[83]. This initial study was followed up by other groups that found Gag associated with insoluble lipid fractions[84,85]. Cholesterol depletion of virus-producing cells does reduce the production of virions[85,86]. Lipid raft association appears after membrane binding[85], suggesting that raft binding is not required for Gag-Gag association. This is consistent with *in vitro* assembly experiments that demonstrated that Gag can form virus-like particles with only RNA and without membranes [87 and references therein], although inositol phosphates appear to be required for proper assembly *in vitro*[88]. Additional experiments detected the presence of large Gag complexes associated with lipids that were referred to as lipid barges[86]. However, another report found that detergent insolubility was not raft-associated; rather, it was a property of the large Gag complexes themselves [89]. So for assembling HIV-1, the question remains: is the association of the assembling virus and rafts truly physiologically important?

Regardless of whether HIV-1 requires rafts for assembly, cholesterol and possibly rafts are incorporated into HIV-1 particles. HIV-1 particles are sensitive to cholesterol chelating agents, suggesting that the viral structural integrity and function, fusion and entry, are dependent on virion-associated lipid rafts[85,90-92]. One potential pitfall of depletion experiments is that removing a large constituent of the viral membrane might cause unexpected effects (e.g., perturbing the viral membrane by simply shrinking it) that complicate the interpretation of the results. Experiments that exchanged cholesterol for analogs that are unable to form lipid rafts still found a severe infectivity defect supporting the involvement of the rafts themselves and not simply the removal of lipid from the membrane[88]. Despite the controversy, lipid rafts and their impact on HIV-1 assembly remain an important research direction that could provide antiviral strategies. Cholesterol chelating-compounds, such as beta-cyclodextrins, could be used as inexpensive topical microbicides[90,91]. Also current drugs that inhibit cholesterol (namely statins) could be used to lower cholesterol levels, potentially inhibiting the replication of HIV-1[85].

RNA?

In addition to lipids, another important cellular structural component is RNA. HIV Gag requires RNA for assembly *in vitro*[87,93]. Membrane-stripped retroviral cores are sensitive to RNAses, demonstrating that RNA is a structural component of virions[94]. It is important to note that genomic RNA is not required for assembly of particles, just infectious particles. In the absence of viral genomic RNA, HIV-1 and other retroviruses incorporate cellular RNAs[3]. Thus, there appears to be no strict specificity for the incorporation of RNA, yet

RNA is an important cellular factor for viral assembly. From several convincing experiments, it has been proposed that RNA not only provides a structural element to the virion, but also serves as a scaffold upon which Gag assembles [87,94,95]. However, some experiments have suggested that, rather than a requirement for assembly, RNA stabilizes HIV-1 virions and those that are deficient in RNA readily form but are unstable and fall apart shortly after release[96,97].

Where Does Gag Assemble?

One way Gag can be localized to the site of assembly is by binding a cellular membrane. Gag associates with membranes by the interaction of two different structures in the N-terminal region of Gag, within MA. The very N-terminus of MA is co-translationally myristylated, a modification that assists in membrane targeting and assembly[98-101]. However, this modification is does not completely provide for membrane association as, when myristylation is blocked by mutation or chemical blocking and HIV-1 no longer buds from the cell, Gag binds to interior membranes rather than to the plasma membrane[102,103]. Thus myristylation is not strictly required for Gag membrane association, but is needed for plasma membrane targeting. Even though the mechanism for this is not known, compounds that interfere with myristylation have been proposed as antiviral therapeutics[101]. Gag also contains several positively charged residues in the N-terminus of MA[103,104]. They are in the correct configuration in the structure of the matrix to properly bind the negatively charged surface of lipid bilayers[105]. Mutation of these resides interferes with Gag assembly while MA mutants with changes at other positions cause Gag to be even more tightly associated with the plasma membrane[106]. Other Gag domains are also involved in membrane binding though indirectly. Sequences within CA and NC have been found to be important in membrane binding, probably because of their role in proper Gag-Gag association[107,108]. Since the binding of an individual Gag to the membrane by myristate and positive charge interactions is relatively weak, aggregation of Gag should stabilize membrane binding. This is experimentally seen by the rapid membrane association upon oligomerization of Gag soon after synthesis[65].

Based on early experiments, Gag expression was localized to free cytoplasmic ribosomes, as opposed to the membrane-bound ribosomes. While this provides a major distinction between two translation systems, thereby eliminating the secretory system, this does not determine where virus is expressed, as free ribosomes could be in nearly any cytosolic location in the cell. Even though Gag is produced on free ribosomes, it is clear that HIV Gag is associated with membranes and that this occurs soon after synthesis[65]. So Gag and Gag-Pol are probably produced somewhat near the assembling membrane, consistent with the *cis* packaging model above (Fig. 3,4). This conclusion is supported by several observations. Based on imaging studies in epithelial cell lines used to model infection, Gag mostly appears to be present predominantly near the plasma membrane, with little present in specific cytoplasmic regions[67,68,106]. Similar results were obtained in infected T cells[109]. These results are consistent with Gag being produced near membranes.

Instead of the prototypical retroviral plasma membrane budding, HIV-1 buds into intracellular structures in macrophages[42] (Fig. 4). Unfortunately, the distribution of Gag in these cells has been studied less than those of other systems, though Gag recently has been found to be localized primarily to these internal virus-laden structures in macrophages[110,111], supporting a linkage of translation localization and assembly even at these internal sites. Immunostaining and immunoprecipitations with various cellular markers have identified the virus-containing vacuoles as late endosomes, potentially multivesicular body (MVB)-like structures that are part of the endocytic pathway. The endocytosis system is used extensively by both macrophages and T cells. It consists of a rather intricate vesicular pathway that constantly removes proteins from the cell surface, brings them into the cell, as well as sends both external and intracellular proteins to the lysosome [112,113]. MVBs play an important role in this endocytic trafficking. They are formed when many small protein-laden vesicles bud into the lumen of the MVB and begin to fill this larger vesicular structure with the small particles, earning the MVB its descriptive name. This structure can then carry the vesicles into lysosomes for degradation or to the surface[113], where the luminal vesicles can be released from the cell as exosomes[79,113,114] or reabsorbed in the limiting membrane (that of the MVB), which is then fused to the plasma membrane[115]. Since MVBs can fuse with the plasma membrane and release their contents, they likely provide the mechanism for HIV-1 budding from macrophages (Fig. 4).

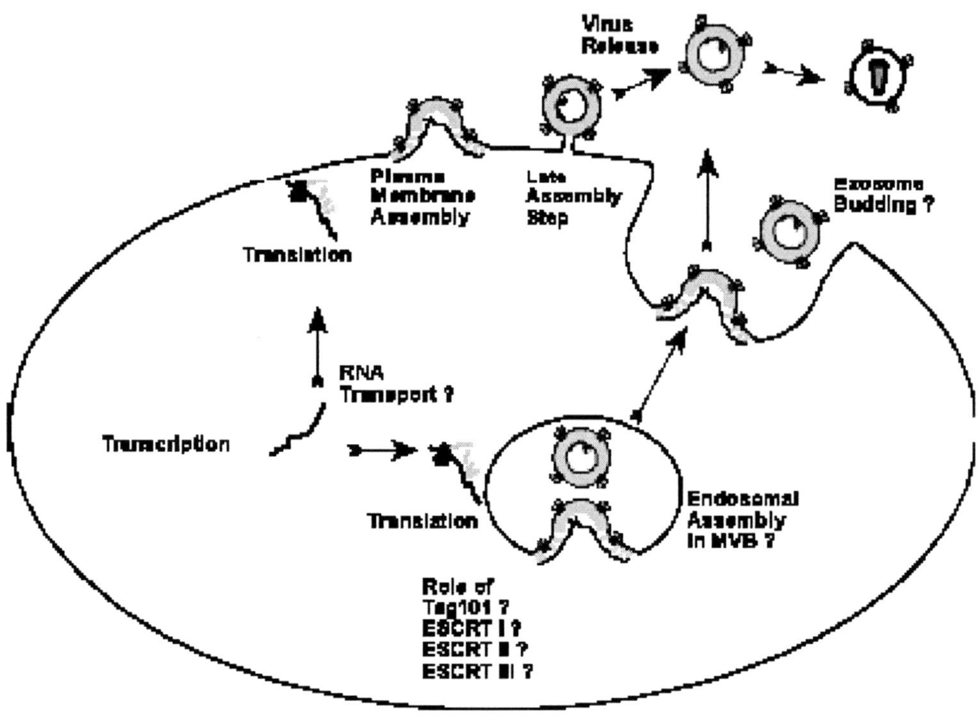

Figure 4. Assembly and budding pathways for HIV-1. Plasma membrane and endosome/MVB assembly, budding, and release is presented. Pertinent question are highlighted.

Despite the visible budding of HIV-1 from the plasma membrane, there is good reason to consider an involvement of the endosome system for HIV-1 budding in cells other than macrophages. As reviewed below (see “How Are Particles Released?”), protein components of the endosome-MVB pathway have been found to be an important part of the HIV-1 budding machinery[116]. While this could be an instance of Gag recruiting and redirecting enodcytic machinery and complexes from their normal location to the plasma membrane, it is also possible that Gag could begin to assemble with these complexes in the endocytic compartments in the same way that behaves in macrophages. However, rather than bud into endosomal vesicles, HIV-1 associated with endosomal membranes could then be transported to the plasma membrane though the endosome trafficking pathway. Once near the surface, fusion of the endosomal membrane with the plasma membrane would allow for viral budding of the Gag from endosomal patches that were fused to the plasma membrane[117]. This is consistent with the classical budding forms observed by electron microscopy[43,44] (Fig. 2). Support for this proposal is found in that some assembling and budding HIV-1 from non-macrophage cell lines associates with endocytic markers and structures[117-120]. These results have led several groups to propose that HIV-1 uses a part of the endocytic pathway for virus budding[117-119,121,122.]

In macrophages, HIV-1 budding seems to mimic that of exosomes, budding into internal bodies that are then released at the plasma membrane. Exosomes themselves appear to be a mechanism for intercellular communication being released from the cell into the extra cellular space, much in the same way the HIV-1 buds from macrophages[114,123]. These particles and particles like them (microvesicles) even can contaminate density-purified virus preparations, since they have similar size and shape characteristics as retroviruses[124,125]. This similarity between HIV-1 budding and exosome release has led to the proposal that HIV-1 has mimicked the exosome pathway, not only for budding and release in T cells, but also for uptake and entry into the cell[121]. However, this hypothesis still remains to be conclusively proven.

How is the Particle Released from the Cell?

The assembly of HIV-1 into a viral bud on the cell surface appears to be driven by primarily by Gag-Gag interactions with little active cellular assistance other than the RNA and lipid factors mentioned above[3]. In contrast, the release of budding virus from the cell requires extensive assistance from the cell[15,126]. The pinching off of this budding structure from the cell surface is not a simple process, as membranes must be broken and resealed to separate the plasma membrane from the viral membrane. This is not a spontaneous process, because membranes themselves are relatively very stable structures in an aqueous environment. For the cell, vesicle budding and fusion are used in both intracellular and extracellular transport processes[112]. The secretory and endocytic systems are two prominent systems that use complex protein systems to sort proteins into, contain them within, and release them from various structures[112]. The primary feature of Gag that appears to be required for this release process is the late assembly (L) domain (reviewed in 126,127]. These are rather short sequences (centered around a core of 4 amino acids) that are required for efficient particle release; mutations in these sequences either cause the accumulation of the virions on the surface as budding stalks that are arrested at a pinching-off step or cause virions to be released in large, connected chains of virions[128]. In both cases, there is a severe defect in virus replication, representing an inability of the virus to efficiently bud from the cell. The HIV-1 L domain core is a PTAP (proline-threonine-alanine-proline) sequence that is present in the C terminal p6 region of Gag. Other retroviruses use other core

L domain sequences, PPPY (3-prolines and a tyrosine) and YPDL (tyrosine-proline-aspartic acid-leucine). Generally, L domains can be functionally exchanged between different viruses[129,130].

It is clear that L domains interact with cellular proteins and that these likely provide the necessary function for virus budding. For HIV-1, the recent progress made in this field has been staggering. The primary breakthrough began with the finding that a curious protein, Tsg101, bound to the PTAP HIV-1 L domain[127,131,132]. Tsg101 was originally isolated as a putative tumor suppressor gene during random homologous knockout screening of mice[133]. This protein has played a role in many functions in the cell, transcription factor degradation, sorting proteins into the MVB/endocytosis/lysosome pathway, and is an essential gene for cells[134-136]. Tsg101 has an interesting structure[127,137,138]: it is very similar to that of the E2-ubiquitin-conjugating enzyme family that normally function in the ubiquitination pathway. An active role for ubiquitination of Gag has been implicated in virus budding[139] (see "Role of Ubiquitin/Proteasome System" below). However, while this protein can bind ubiquitin, it can not ubiquitinate proteins as a conjugating enzyme, due to the presence of a tyrosine instead of cysteine at the enzyme active site. Interestingly, mutation of this tyrosine abolishes binding of Tsg101 to Gag[131]. Of the vital roles that Tsg101 can play, the most attractive with respect to virus budding is its regulatory role in sorting proteins into the MVB budding pathway[134,140,141]. The tantalizing feature of this Tsg101 function is that the budding of vesicles into the lumen of MVB is topologically similar to virus budding: it is an outward budding from the cytoplasm into the luminal space of the MVB or extracellular space of the cell (Fig. 2,4). Perhaps in this way, retroviruses have commandeered this budding mechanism to solve their release problem.

A stunning amount of data links virus budding with the Tsg101 machinery. Depletion of Tsg101 causes a defect in virus budding that is analogous to an L domain mutant phenotype[141]. Interestingly, overexpression of Tsg101 can also interrupt virus budding, suggesting that the balance of a whole system involving Tsg101 must be maintained[128.] Expression of a fragment of Tsg101 that contains only the UEV (ubiquitin E2 variant) PTAP-binding region and not the portion that interacts with the sorting pathway specifically arrests HIV budding but not that of other retroviruses that use PPPY L domains[142]. However, the expression of a C-terminal fragment of Tsg101 which binds and interrupts the endosomal sorting machinery blocks both HIV-1 and PPPY-containing retroviruses[142]. This indicates that while the endosomal pathway is involved in retroviral budding, Tsg101 seems to be a specific interaction partner for HIV-1 and potentially for other PTAP-containing viruses. Microscopy shows that Tsg101 can also be recruited by Gag to the plasma membrane, and genetic studies have shown that HIV-1 L domain mutants can be complemented by restoring Tsg101 binding[143]. Currently, several of the proteins in this pathway have been tied to HIV-1 budding[122,141,144]. One of these is the ESCRT I complex, the first step in the MVB/endosomal sorting process[145-147]. Tsg101 is a part of the ESCRT I complex and thus might act as a link for HIV-1 release into the MVB budding machinery[122,141,144,148]. Other downstream proteins that are part of this process have also been shown to be important: an additional sequence in the C-terminal portion of p6 has been discovered that binds to AIP1/ALIX protein, which is also binds the ESCRT III complex[122,149,150]. ESCRT III performs the last step in the MVB sorting pathway, effecting the budding of the vesicles into the MVB lumen. Similar to Tsg101, expression of some mutant forms AIP1 protein can decrease the budding of HIV-1[149]. Thus, it seems that two separate p6 sequences within Gag might interact with components of the beginning and last components in the MVB sorting pathway to bring about virus release from the cell. This is consistent with the recent finding that HIV-1 requires more sequences than simply the PTAP L domain signal for efficient release[151].

While much progress has been made on the mechanism of HIV-1 release, questions about the role of Tsg101 and ESCRT proteins still remain. It is possible that perturbing Tsg101 and the MVB system causes overall cellular defects that decrease HIV budding by a global disruption of the cell rather than altering the specific pathway used by the virus. In this case, the action of Tsg101 would be indirect. Although this has not been widely examined, it seems unlikely, since Tsg101 directly interacts with HIV-1, and the finding that some Tsg101 mutants specifically reduce HIV budding while other mutants affect other retroviruses, too [152]. Taken together, these observations provide strong evidence for direct HIV interaction with the MVB/endosome system. The larger question is, does HIV-1 recruit this system to the plasma membrane for virus release similar to the proposed function of the Hrs protein[144] or does it interact with the endosomal system in the interior of the cell during assembly and follow the natural exosome release pathway [121]?

What is the Role of the Ubiquitin/Proteasome Pathway in Virus Budding?

The monoubiqitination of cell surface proteins causes them to be internalized, sorted into the MVB system by Tsg101 and the ESCRT pathway, and then delivered to the lysosome for degradation[145-147]. Thus, monoubiquitination serves as a tag for this pathway. Interestingly, a small amount of HIV-1 Gag (approximately 2%) is monoubiquitinated on p6, the same region that contains the L domain[153]. Ubiquitin is an 8.7 kDa protein that is attached to the ε-amino group of lysines on proteins to mark them for several cellular processes [reviewed in 154,155]. The ubiquitination pathway consists of a cascade that begins with the E1 ubiquitin-activating enzyme transferring ubiquitin to one of many E2 ubiquitin-conjugating enzymes, which either alone or in concert with one of the many E3 ubiquitin-protein ligase enzymes, attaches ubiquitin to the target protein[154,155]. In addition to attachment to other proteins, internal lysines of ubiquitin can also be ubiquitinated themselves (Fig. 5), which allows for the construction of polymeric ubiquitin chains on proteins. The most common function for ubiquitin is the marking of a target protein with a polymeric chain of 4 or more molecules linked on ubiquitin lysine 48. This polyubiquitination targets a protein for degradation by proteasomes[156-158]. However, monoubiquitination, the attachment of a single ubiquitin, sends a different signal; it appears to serve as a signal for recognition of proteins by the endocytosis pathway, histone transcription-regulation pathway, and possibly other unidentified functions[159] (Fig. 5).

The presence of ubiquitin-Gag conjugates in virions demonstrates that Gag interacts with ubiquitinating enzymes during budding[153,160]. Interestingly, the region in HIV-1 Gag that is ubiquitinated, p6, also contains the viral L domain, and ubiquitination is dependent on the presence of an L domain in Gag[161]. Other retroviruses are ubiquitinated in Gag near their respective L domains[153,162,163], suggesting a common linkage between late assembly and the monoubiquitination of Gag. So could monoubiquitnation of Gag assist in its recognition by Tsg101 and subsequent sorting into the MVB machinery and ultimately budding? Despite all of these compelling correlations, the ubiquitination of p6 is not required for HIV-1 replication, as mutating the lysines in this region of Gag produced no observable effect on the replication of these mutants[160] and deleting these sites has a similar wild-type phenotype[128]. This is not consistent with a model that requires ubiquitination of Gag for recognition by Tsg101; apparently, the interaction of Tsg101 with PTAP is sufficient for ESCRT I binding and endosome pathway entry. It is possible that lysines in other regions of Gag might be ubiquitinated, although these forms have not been unambiguously detected in the virions[160]. However, ubiquitinated Gag could be difficult to detect because of potentially small amounts of conjugates present and the rapidly reversible nature of ubiquitination itself, being subject to removal by ubiquitin hydrolases[155]. In fact, elongation factor-1α, which is also incorporated into HIV-1 particles, is reported to have a ubiquitin hydrolase activity[164,165]. Therefore, other regions of Gag could be transiently ubiquitinated during assembly but not be detectable in virions. Currently, a functional role for Gag ubiquitination is not clear[139].

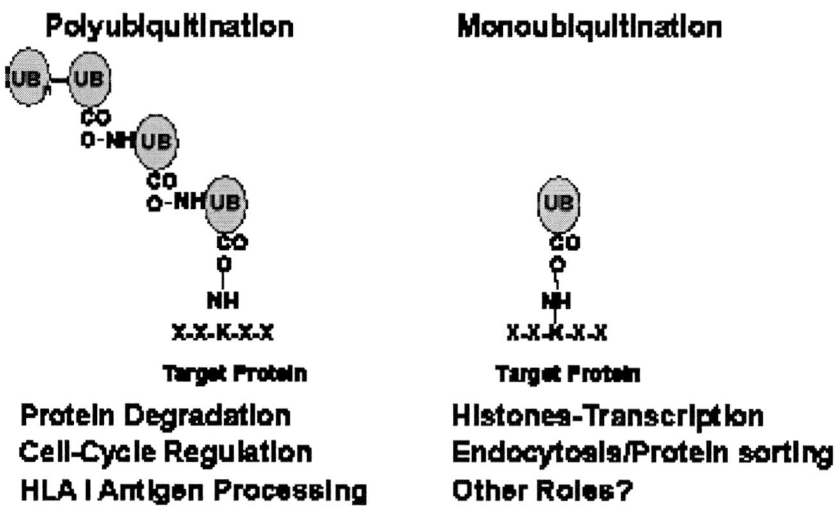

Figure 5. Ubiquitination and its roles.

Since the monoubiquitination of Gag does not seem to be required for virus budding, why is it present and why are Gag proteins from many viruses ubiquitinated near their L domain? One strong possibility is that this modification in the p6 region of Gag might simply be due to a ubiquitinating activity this is associated with the same machinery that recognizes the L domain and releases the virus. In this way, lysines in p6 could be ubiquitinated by an enzyme that normally recognizes another protein. This can happen because, while the recognition of a target protein by a ubiquitinating enzyme is specific, the addition of ubiquitin to the protein is not: a protein-protein interaction results in a ubiquitination on another part of the protein or on another protein in the complex. Therefore, ubiquitin simply could be a bystander rather than an active member in this interaction. While it is intriguing to imagine that the ubiquitination of Gag could play a role in Tsg101 function for HIV-1 budding, the apparent nonessential nature of ubiquitination and the failure to find any other ubiquitination sites in Gag argue against this possibility.

Another link between ubiquitin and HIV-1 budding is the finding that proteasome function is required for HIV-1 budding[166]. HIV-1 budding and Gag processing by the viral protease are both decreased 4-fold in the presence of proteasome inhibitors. Similar to the L domain mutants and Tsg101 depletion, HIV-1 budding is arrested in the budded stalk form (Fig. 6). One important complication to these inhibitor experiments is that proteasomes are required for many cell functions[155-157]. Therefore, inhibitor experiments must be done with great care and rapidly to avoid the inevitable toxic effects of the cells. Also, similar to the caveats on the interpretation of the Tsg101 experiments, proteasome inhibitors can have wide-ranging effects on the cell: thus, a block in virus release may not be due to a direct effect on the virus. This dependence of viral budding on proteasomes has also been found for a diverse group of retroviruses: HIV-1 and -2[166], simian immunodeficiency virus[161], Rous sarcoma virus (RSV)[167], MuLV[163], Mason-Pfizer monkey virus[163], and human T cell leukemia virus type 1[168]. Proteasome inhibitors can also decrease the budding of the vesicular stomatitis virus matrix protein[169], suggesting that this is a common feature of enveloped virus budding. Despite this widespread interruption of virus budding, equine infectious anemia virus, EIAV,[162,170] and mouse mammary tumor virus, MMTV,[163] are two retroviruses whose budding is not reduced by proteasome inhibitors. Influenza virus budding is also sensitive to proteasome inhibitors[171]. These findings show that while the requirement for proteasomes is widespread, it is not universal; retrovirus budding can occur without active proteasomes. It has been proposed that the insensitivity of EIAV budding to proteasome inhibitors might be due to a ubiquitin-like motif (NVKEKDQ) in the $p9^{Gag}$ region of EAIV Gag that could function to promote virus release in the presence of these inhibitors[170]. However, this result remains to be thoroughly tested and still does not explain the MMTV data as its Gag does not contain a ubiquitin-like sequence[163].

While the basis for this differential in proteasome requirement among retroviruses is not clear, one correlation is that sensitive viruses use PTAP or PPPY-based L domains, while those using other types of L domains are insensitive to the effects of proteasome inhibitors[163]. Considering that L domains interact with cellular proteins to allow budding, these differences in proteasome requirement may be a reflection of the L domain specific pathway Gag takes for budding. Progress made in identifying L domain binding partners (Tsg101 and ESCRT) and the effect the proteasome pathway might have on them might open the way to a better understanding of how virus budding is accomplished.

Why do many retroviruses need active proteasomes for efficient budding? While the answer is not clear[139], experiments with RSV found that budding of its Gag can be made resistant to the proteasome inhibitors by fusing ubiquitin to the C-terminus of this avian Gag[167]. This approach was used previously to demonstrate the role of monoubiquitination in targeting the endocytosis of surface proteins[172]. Based on these experiments with avian viruses, a model has been proposed that posits that ubiquitin and the ubiquitination of Gag is a vital part of the retroviral budding machinery: the ubiquitination of Gag is recognized by cellular proteins that then mediate virus budding[167]. Proteasome inhibitors can interfere with the ubiquitination system by preventing the recycling of ubiquitin from degraded polyubiquitinated proteins, depleting the amount of ubiquitin in the cell (Fig. 6). In turn, this could starve the ubiquitination system and prevent the ubiquitination of Gag. Thus, if ubiquitin is required for efficient budding, then a reduction in ubiquitinated Gag would be followed by a reduction in viral budding[167]. While it is attractive to integrate the RSV-ubiquitin fusion result with the HIV-1 Gag-Tsg101 observations, RSV uses a PPPY L domain that does not interact with Tsg101 [126,127,173], so it is unlikely that ubiquitin recognition by Tsg101 plays are role in PPPY L domain-mediated budding. Also RSV Gag has not been found to be ubiquitinated[174].

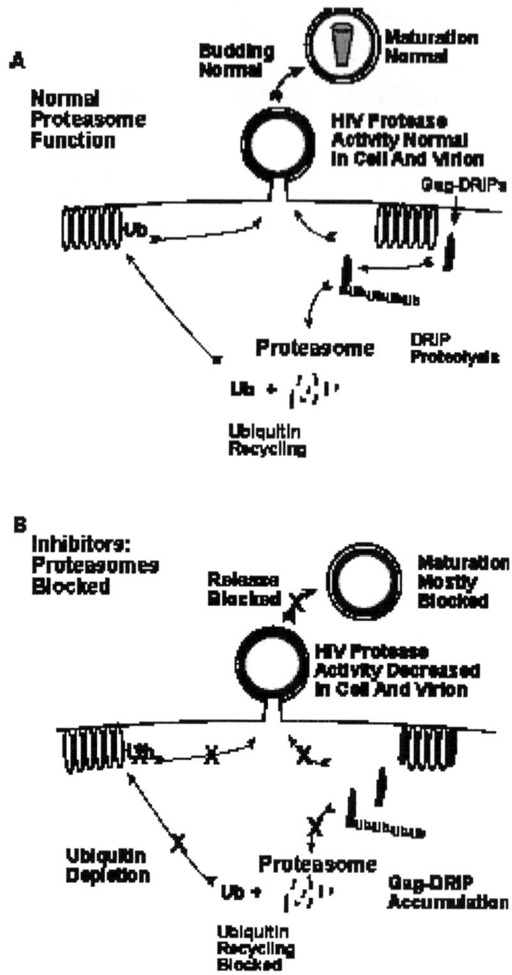

Figure 6. Proteasome function and proteasome inhibitor models. A. normal proteasome function with ubiquitin recycling. B. proteasome models to explain the dependence of HIV-1 budding on proteasome activity. On the left -hand side, a model that proposes that the direct ubiquitination of Gag is required for release is presented (ubiquitin depletion). On the right-hand side, the model for the involvement of Gag DRiPs is outlined (Gag-DRiP accumulation).

While this ubiquitin-budding observation seemingly holds true for avian retroviruses, this model has not been demonstrated for HIV-1 nor does it account for the insensitivity of EIAV and MMTV to proteasome inhibitors[162,170]. One of the major assumptions of this model is that proteasome inhibitors rapidly deplete ubiquitin levels (within 1-2 Hr), thereby depriving the viral assembly machine of an important factor. While rapid ubiquitin depletion was observed in the RSV system[167], it has not been observed in experiments with other retroviruses[163]. Over long periods of treatment (36 h), ubiquitin does decline 6-fold [166]. However, during the relatively short time-frame (8 h kinetics) of proteasome inhibitor experiments on viral budding, ubiquitin levels do not decrease, even though virus budding was clearly reduced[163]. Therefore, a simple decrease in free ubiquitin is not the simple cause of the reduced virus budding for HIV-1 and other viruses. Despite the relative stability of free ubiquitin levels, a marker for ubiquitination, the monoubiquitination of histones, was clearly inhibited after 1 to 2 hours of proteasome inhibitor treatment in all of the cell lines used in this study, suggesting that the ubiquitination of Gag would also be affected[163]. However, the inhibitors still decrease HIV-1 processing earlier than any observable effect on the cell (unpublished data), indicating that they act before the inhibition of ubiquitination. This, combined with the failure to observe a direct requirement for ubiquitin in HIV-1 budding, suggests that, unlike avian viruses, HIV-1 does not require ubiquitin for budding.

An alternative mechanism for the proteasome requirement for viral budding is centered around the accumulation of Gag proteins as defective ribosomal products (DRiPs) upon proteasome inhibition[166]. DRiPs are misfolded or improperly modified proteins that result from errors during or immediately after protein synthesis[175,176]. Approximately one-third of the many proteins produced in the cell, especially Gag, become DRiPS[177]. These dead-end proteins are normally removed from the cytoplasm by polyubiquitination and subsequent degradation by the 26S proteasome. It is important to understand that DRiPs are the non-functional proteins themselves and not necessarily polyubiquitinated. This large percentage of defective proteins is startling and, at first, seems to be inefficient.

However, one important function of the DRiP/proteasome pathway is that it provides for a constant and timely sampling of the cellular proteins by the major histocompatibility complex I system for immune surveillance[175,176,178] to recognize the presence of viral and cancer antigens in cells by the cell-mediated immune system. In the presence of inactivating levels of proteasome inhibitors DRiPs accumulate rapidly[166,175,176]. In the case of a highly cooperative process such as virion assembly, defective Gag proteins could interfere with the Gag-Gag interactions during particle formation (Fig. 6).

Since Gag-DRiPs are normally polyubiquitinated co-translationally and degraded soon after translation, they would never interfere with the assembly process. However, proteasome inactivation increases the levels of these Gag DRiPs that, in turn, might act in a dominant negative fashion to interfere with assembly and decrease virus budding. The Gag DRiP model is supported by the relatively rapid and co-translational effect that proteasome inhibitors have on virus budding even after 30 minutes of treatment (unpublished data), well before ubiquitin pools are reduced.

Given the potential importance of the proteasome protein degradation pathway in viral budding, it is important to consider the lysosomal degradation pathway and virus budding. One of the many roles of Tsg101 that could be interrupted by its depletion is its sorting of proteins into the MVB pathway for degradation by the lysosome[135,136]. Therefore, a potential disruption of cellular degradation systems is a common aspect of both proteasome inhibitor and Tsg101 experiments. Both lysosomes and proteasomes can degrade proteins that have lost their function. Perhaps lysosomes also remove defective Gag or another protein that interferes with budding. Since the effects of Tsg101 depletion and proteasome inhibition on virus budding are very similar, perhaps part of this commonality is the accumulation of defective Gag.

The potential importance of the proper folding of Gag for virus assembly highlighted above predicts that Gag might interact with protein folding chaperones. Protein-folding chaperones are found in HIV-1 virions: Hsp70, Hsp60, and Hsc70 are present in virions, though in small amounts (less than 5% of Gag level)[179]. Also, Hsp27, Hsp70, and Hsp78 interact with Gag inside the cell[180]. These proteins have been found to interact with other viruses (other than retroviruses) and in some cases functionally interact, assisting with viral functions. Curiously, in addition to conventional chaperones, elongation factor-1α is incorporated into virions. One of the many activities of this versatile protein is a protein-folding activity that is chaperone-like[181,182]. Also, ubiquitin, when fused to the C-terminus of another protein might have some chaperone-like role[183]. Perhaps the proteasome inhibitor resistance results obtained by fusing ubiquitin with RSV-Gag is due to a ubiquitin-mediated chaperone activity that rescues Gag-DRiPs. Another association of Gag assembly with a cellular protein is the finding that HP68, a protein that binds RNase L, associates with HIV Gag assembly complexes and expression of mutant forms of this protein in Gag-producing cells reduce viral budding[184]. While no functional data yet exists for the importance of chaperones in the HIV-1 life cycle, the presence of several proteins inside the virus that mediate folding and their interaction with Gag inside the cell make it an attractive speculation that correct Gag folding and chaperones are an important part of the assembly process.

Roles of Cellular Proteins in the Virion?

HIV-1 virions contain many cellular proteins, both on the surface of and inside the virion during the plasma membrane assembly and release stage[45,185,186]. Three basic mechanisms for the incorporation of host proteins into HIV-1 particles are: the protein was simply present at the site of assembly, thus is not specifically required; the protein was interacting with a viral protein, the genomic RNA or plasma membrane and was swept up into the virion during budding by this specific association; or the virus purposefully incorporated it to perform a function for the virus. While from the viral perspective, only the last mechanism is important, for the researcher, any of these potential mechanisms could provide important clues for viral assembly. A case in point is the incorporation of raft-associated GPI-linked proteins that are presumably acquired nonspecifically by viral budding through lipid rafts. This implicated rafts as an HIV-1 assembly partner as reviewed above. Thus, for any of these scenarios, it is important to study the cellular proteins in the virion, as they provide clues about the assembly site, the proteins required for assembly, and those required for roles in replication.

Cyclophilin A

Of the many proteins found in virions, few have been found to have important roles in the viral replication cycle. Cyclophilin A is found inside virions and is incorporated by specific interaction in the CA region of the Gag polyprotein, and its incorporation is required for efficient infection[45,187-189]. Virions without cyclophilin A are blocked in early reverse transcription[190]. The cellular role of this protein is primarily as a chaperone, using its peptidyl-prolyl *cis-trans* isomerase activity to facilitate bond rotation around prolines in polypeptides. One proposed role for this protein is that this isomerase activity is needed to unfold the mature viral core structure, allowing for early infection events inside the cell[191].

However, cyclophilin A does not have a clear destabilizing effect on the mature Gag core, as expected[192]. It is possible that cyclophilin A could also act during viral assembly to ensure the proper Gag conformation for infection. So the role of cyclophilin A in HIV is not yet settled[45]. Virion production is not affected by the lack of cyclphilin A activity; thus, this differs from the aberrant protein-folding defect of DRiPs. Given the specific folding activity of this protein and the effect removing its activity has on the virus, it is unlikely that this chaperone resolves folding dead-ends that would prevent Gag from assembling.

HLA Class II

HLA Class II-DR (HLA-II) is found on the surface of HIV-1 produced from HLA-II$^+$ cells (reviewed in 45,186,193). While the mechanism for HLA-II incorporation remains unknown, experiments with molecularly engineered HIV-1 mutants in lymphoid cells demonstrated that efficient HLA-II incorporation onto HIV-1 required the packaging of Env, suggesting an active mechanism for incorporation[194]. The presence of HLA-II on HIV-1 can provide modest increases in virion infectivity[195]. Another potential role for HLA-II is suppressing antiviral immune responses, thereby facilitating immune escape[196]. If HLA-II on virions can functionally bind the T cell receptor on CD4$^+$ T cells in conjunction with HIV-1 Env-CD4 binding, then these interactions could mimic part of the normal stimulation signal sent between antigen-presenting cells and T cells. However, HIV-1 does not incorporate CD80, a protein on antigen-presenting cells that binds CD28 on the T cell. This binding normally sends a required co-stimulatory signal that is required for effective T cell stimulation[197]. Receptor engagement in the absence of co-stimulation can result in T cell anergy (a loss of antigen responsiveness) or apoptosis (programmed cell death). Therefore, it is possible that HLA-II on the surface of virions could send just a single signal that downregulates T cell function, causing immunodeficiency without T cell infection[196]. This mechanism is consistent with the observation that, in infected individuals, most apoptotic death occurs in cells that are not productively infected, i.e., "bystanders"[198]. Supporting this hypothesis, virus-associated HLA-II can present superantigen to T cells, indicating that HLA-II on the virion surface retains its functional activity[199]. Also, non-infectious HIV-1 containing conformationally and functionally intact Env and HLA-II protein complexes caused apoptosis in T cells, while similar virions without HLA-II did not[200]. These results support a role for HLA-II on HIV-1 as a potential contributor to virus-mediated immunodeficiency. Thus, HIV-1 might specifically acquire HLA-II to disable the CD4$^+$ compartment, a classic mechanism of AIDS. Since viruses incorporate HLA-II from infected cells, the HLA-II on the virus might carry viral peptides that could target antiviral T cells for anergy or apoptosis.

Apobec3g

Unlike the two examples above, Apobec3g is a cellular protein whose absence in virions is important. This protein is a member of a family of cytidine deaminases that are thought to function mostly as RNA editing enzymes by removing the amino group from the cytosine bases in RNA, causing them to be read by the ribosome as uracil and thus changing the coding information of the RNA[201]. While members of this family are involved in many different cellular processes, Apobec3g appears to be the long sought-after answer to one of HIV-1's puzzling riddles: why does the Vif protein need to be present at the time of assembly for the released virions to be able to undergo the next round of replication in primary cells as well as in many model T cell lines[202]. This restriction can not be overcome by expression of Vif in the target cells; Vif must be present during assembly. The riddle was answered starting with the finding that the presence of a cellular protein Apobec3g (CEM15) in the virus-producing cell was responsible for the restriction[22]. This protein appears to damage the viral RNA, causing severe errors in the viral genome, making it unable to express functional virus[202-206]. Vif can bind to Apobec3g and prevent it from being incorporated into virions, thus the requirement for Vif [21,207]. More importantly, Vif also has been found to suppress the translation of Apobec3g and cause it to be ubiquitinated and degraded by the ubiquitin-proteasome pathway thereby reducing its levels in virus-producing cells [22,207-210]. Together, these finding explain the Vif requirement. So unlike the incorporation of cyclophilin A and HLA-II, it is what is not incorporated (Apobec3g) that is important in this case.

Integrase interacting factor-1 (INI-1) was identified as a cellular protein that bound to integrase *in vitro* and is incorporated into virions[211]. Interestingly, the presence of a truncated version of INI-1 inhibits virus assembly in a dominant negative manner, presumably by interfering with Gag and Gag-Pol interactions. This is surprising, as Gag itself is sufficient for assembly, although some integrase mutants can inhibit assembly. Together, these data suggest an important interplay between Gag and Gag-Pol, one that could be blocked by dominant negative mutants of INI1 or small molecule inhibitors for an antiviral therapy.

What Potential Antiviral Therapies Could Interrupt these Targets?

Several antiviral approaches either have been proposed or are currently being tested that target some of the cellular interactions with assembling HIV-1 (summarized in Table 1). While these proposals are promising, they have yet to make the transition from a concept to a practical treatment. One benefit in targeting virus/cell interactions is that it should be harder for HIV-1 to develop resistance to therapies that interfere with critical viral interactions with a cellular protein. Viral resistance often requires that the residues in a viral protein to be mutated to avoid the therapeutic yet maintain the interaction with other partners. For interactions between viral proteins, a drug-resistant mutation in one partner could be readily compensated by another mutation in its partner to maintain the required interaction. This has been observed in reverse transcriptase and especially in protease inhibitor escape mutants where other viral proteins changed to accommodate the changes in activity or specificity of the mutant enzymes[3]. Since there is no selection pressure on the cell to accommodate the mutated HIV-1 proteins, the ability of the virus to mutate around a therapeutic based on a cellular target. The difficulty in this approach is that the host needs these cellular systems for its own viability. Thus, any effective antiviral approach needs to focus on targeting the specific virus/cellular contact, possibly by small molecule inhibitors.

Table 1.Proposed anti-HIV-1 therapies using cellular targets

Therapy	Targeted Cellular Feature
Cyclosporin A	Cyclophilin A[191
Proteasome inhibitors	Ubiquitin-virus interaction\protein folding[166]
INI1-mimicking small molecules	Interrupt Gag interaction with Gag-Pol[211]
Statins: decrease cholesterol	Lipid rafts and HIV-1 assembly[85]
Topical detergents	Interrupt lipid rafts in virions[91]
Tsg101-blocking small molecules	L domain-mediated virus budding[142]
HLA II-linked toxin	HLA II on the surface of virions[213]
Vif-blocking small molecules	Increase Apobec3g incorporation[21,23,24,204,205,210]

Final Comments

Though this review, I have provided only partial answers and point to future directions in the search for answers some assembly questions. Clearly, we do not have definitive answers, but we do have a much better understanding and insight into the cellular aspects of virus assembly. I have posed only a few questions about cellular contributions to virus assembly; many others remain, e.g., Env targeting and incorporation. While viral proteins and cellular proteins have traditionally been studied separately by the virologist, advances in cell biology and in the tools available now make it more feasible to study these two aspects together. Viral protein translation, RNA packaging, virus assembly, and virus release are usually studied individually and not as a linked process within the cell as the virus naturally uses. One future direction of assembly research is to consider these processes as a continuous pathway where all the components act together rather than a compartmentalized set of steps. While this is a daunting task, the benefits of looking at virus assembly as a whole are potentially great.

The emphasis on HIV-1 research has spilled over into immunology and vaccine research, providing for new insights in these disciplines. Now that this large field is starting to overlap cell biology, it seems very likely that more will be learned about cellular systems themselves through HIV-1 research. In fact, this is already happening; HIV-1 research has lead to new insights into the function of the Hrs protein[212]. Finally, the AIDS epidemic demands that better methods of preventing infection and managing the virus in infected people be developed. These goals can only be assisted by a better understanding of HIV-1 assembly in the context of the cell.

Acknowledgments

This project has been funded in whole or in part with Federal funds from the National Cancer Institute, National Institutes of Health, under Contract No. NO1-CO-12400. The content of this publication does not necessarily reflect the views or policies of the Department of Health and Human Services, nor does mention of trade names, commercial products, or organization imply endorsement by the U.S. Government.

References

[1] Rouzine IM, Rodrigo A, Coffin JM. Transition between stochastic evolution and deterministic evolution in the presence of selection: general theory and application to virology. Microbiol. *Mol. Biol. Rev.* 2001;65(1):151-185.

[2] Coffin JM. HIV population dynamics in vivo: implications for genetic variation, pathogenesis and therapy. *Science* 1995;267:483-489.

[3] Coffin J, Hughes S, Varmus H. Retroviruses. 1997.

[4] Holzschu D, Lapierre LA, Lairmore MD. Comparative pathogenesis of epsilonretroviruses. *J. Virol.* 2003;77(23):12385-12391.

[5] Boeke JD, Devine SE. Yeast retrotransposons: finding a nice quiet neighborhood. *Cell* 1998;93(7):1087-1089.

[6] Nelson PN, Carnegie PR, Martin J, Davari Ejtehadi H, Hooley P, Roden D, Rowland-Jones S, Warren P, Astley J, Murray PG. Demystified . . . Human endogenous retroviruses. *Mol. Pathol.* 2003;56(1):11-18.

[7] Herniou E, Martin J, Miller K, Cook J, Wilkinson M, Tristem M. Retroviral diversity and distribution in vertebrates. *J. Virol.* 1998;72(7):5955-5966.

[8] Anderson WF. Gene therapy. *Sci Am* 1995;273(3):124-128.

[9] UNAIDS. AIDS epidemic update 2003. Volume 2003: UNAIDS; 2003.

[10] Shearer GM. HIV-induced immunopathogenesis. *Immunity* 1998;9(5):587-593.

[11] Blankson JN, Persaud D, Siliciano RF. The challenge of viral reservoirs in HIV-1 infection. *Annu. Rev. Med.* 2002;53:557-593.

[12] Lang SM, Desrosiers RC. Relative importance and functional role of genetic elements targeted for a live attenuated AIDS vaccine. *Antibiot. Chemother.* 1996;48:92-99.

[13] Desrosiers RC. HIV with multiple gene deletions as a live attenuated vaccine for AIDS. AIDS Res. *Hum. Retroviruses* 1992;8(3):411-421.

[14] Coffin J, Hughes S, Varmus H. Retroviruses. Plainview, NY: Cold Spring Harbor Press; 1997.

[15] Freed EO. HIV-1 replication. Somat. *Cell Mol. Genet.* 2001;26(1-6):13-33.

[16] Karn J. Tackling Tat. *J. Mol. Biol.* 1999;293(2):235-254.

[17] Pollard VW, Malim MH. The HIV-1 Rev protein. *Annu. Rev. Microbiol.* 1998;52:491-532.

[18] Steffens CM, Hope TJ. Recent advances in the understanding of HIV accessory protein function. *AIDS* 2001;15 Suppl 5:S21-26.

[19] Sherman MP, De Noronha CM, Williams SA, Greene WC. Insights into the biology of HIV-1 viral protein R. *DNA Cell. Biol.* 2002;21(9):679-688.

[20] Bour S, Strebel K. The HIV-1 Vpu protein: a multifunctional enhancer of viral particle release. Microbes. *Infect.* 2003;5(11):1029-1039.

[21] Mariani R, Chen D, Schrofelbauer B, Navarro F, Konig R, Bollman B, Munk C, Nymark-McMahon H, Landau NR. Species-specific exclusion of APOBEC3G from HIV-1 virions by Vif. *Cell* 2003;114(1):21-31.

[22] Sheehy AM, Gaddis NC, Choi JD, Malim MH. Isolation of a human gene that inhibits HIV-1 infection and is suppressed by the viral Vif protein. *Nature* 2002;418(6898):646-650.

[23] Sheehy AM, Gaddis NC, Malim MH. The antiretroviral enzyme APOBEC3G is degraded by the proteasome in response to HIV-1 Vif. *Nat. Med.* 2003;9(11):1404-1407.

[24] Yu X, Yu Y, Liu B, Luo K, Kong W, Mao P, Yu XF. Induction of APOBEC3G ubiquitination and degradation by an HIV-1 Vif-Cul5-SCF complex. *Science* 2003;302(5647):1056-1060.

[25] Kondo E, Mammano F, Cohen EA, Gottlinger HG. The p6gag domain of human immunodeficiency virus type 1 is sufficient for the incorporation of Vpr into heterologous viral particles. *J. Virol.* 1995;69(5):2759-2764.

[26] Muller B, Tessmer U, Schubert U, Krausslich HG. Human immunodeficiency virus type 1 Vpr protein is incorporated into the virion in significantly smaller amounts than gag and is phosphorylated in infected cells. *J. Virol.* 2000;74(20):9727-9731.

[27] Khan MA, Aberham C, Kao S, Akari H, Gorelick R, Bour S, Strebel K. Human immunodeficiency virus type 1 Vif protein is packaged into the nucleoprotein complex through an interaction with viral genomic RNA. *J. Virol.* 2001;75(16):7252-7265.

[28] Vogt VM, Simon MN. Mass determination of Rous sarcoma virus virions by scanning transmission clectron microscopy. *J. Virol.* 1999;73(8):7050-7055.

[29] Dougherty WG, Semler BL. Expression of virus-encoded proteinases: functional and structural similarities with cellular enzymes. *Microbiol. Rev.* 1993;57(4):781-822.

[30] Kaplan AH, Manchester M, Swanstrom R. The activity of the protease of human immunodeficiency virus type 1 is initiated at the membrane of infected cells before the release of viral proteins and is required for release to occur with maximum efficiency. *J. Virol.* 1994;68:6782-6786.

[31] Kaplan AH, Swanstrom R. HIV-1 gag proteins are processed in two cellular compartments. Proc. *Natl. Acad. Sci.* (USA) 1991;88:4528-4532.

[32] Freed EO. HIV-1 gag proteins: diverse functions in the virus life cycle. *Virology* 1998;251(1):1-15.

[33] Forshey BM, von Schwedler U, Sundquist WI, Aiken C. Formation of a human immunodeficiency virus type 1 core of optimal stability is crucial for viral replication. *J. Virol.* 2002;76(11):5667-5677.

[34] von Schwedler UK, Stray KM, Garrus JE, Sundquist WI. Functional surfaces of the human immunodeficiency virus type 1 capsid protein. *J. Virol.* 2003;77(9):5439-5450.

[35] Carteau S, Gorelick RJ, Bushman FD. Coupled integration of human immunodeficiency virus type 1 cDNA ends by purified integrase In vitro: stimulation by the viral nucleocapsid protein. *J. Virol.* 1999;73(8):6670-6679.

[36] Gorelick RJ, Gagliardi TD, Bosche WJ, Wiltrout TA, Coren LV, Chabot DJ, Lifson JD, Henderson LE, Arthur LO. Strict conservation of the retroviral nucleocapsid protein zinc finger is strongly influenced by its role in viral infection processes: characterization of HIV-1 particles containing mutant nucleocapsid zinc-coordinating sequences. *Virology* 1999;256(1):92-104.

[37] Guo J, Wu T, Anderson J, Kane BF, Johnson DG, Gorelick RJ, Henderson LE, Levin JG. Zinc finger structures in the human immunodeficiency virus type 1 nucleocapsid protein facilitate efficient minus- and plus-strand transfer. *J. Virol.* 2000;74(19):8980-8988.

[38] Buckman JS, Bosche WJ, Gorelick RJ. Human immunodeficiency virus type 1 nucleocapsid zn(2+) fingers are required for efficient reverse transcription, initial

integration processes, and protection of newly synthesized viral DNA. *J. Virol.* 2003;77(2):1469-1480.

[39] Schwartz MD, Fiore D, Panganiban AT. Distinct functions and requirements for the Cys-His boxes of the human immunodeficiency virus type 1 nucleocapsid protein during RNA encapsidation and replication. *J. Virol.* 1997;71(12):9295-9305.

[40] Pierson TC, Doms RW. HIV-1 entry and its inhibition. Curr. Top. Microbiol. *Immunol.* 2003;281:1-27.

[41] Moulard M, Decroly E. Maturation of HIV envelope glycoprotein precursors by cellular endoproteases. Biochim. Biophys. *Acta* 2000;1469(3):121-132.

[42] Orenstein JM, Meltzer MS, Phipps T, Gendelman HE. Cytoplasmic assembly and accumulation of human immunodeficiency virus types 1 and 2 in recombinant human colony-stimulating factor-1-treated human monocytes: an ultrastructural study. *J. Virol.* 1988;62(8):2578-2586.

[43] Gelderblom HR. Assembly and morphology of HIV: potential effect of structure on viral function. *AIDS* 1991;5:617-638.

[44] Gelderblom H. Fine Structure of HIV and SIV. Korber B, Foley B, Leitner T, McCutchan F, Hahn B, Mellors J, Myers G, Kuiken C, editors. Los Alamos: Los Alamos National Laboratory; 1997.

[45] Ott DE. Potential roles of cellular proteins in HIV-1. *Rev. Med. Virol.* 2002;12(6):359-374.

[46] Berkowitz R, Fisher J, Goff SP. RNA packaging. *Curr. Top. Microbiol. Immunol.* 1996;214:177-218.

[47] Swanstrom R, Wills JW. Synthesis, assembly, and processing of viral proteins. In: Coffin J, Hughes S, Varmus H, editors. Retroviruses. Plainview, N. Y.: *Cold Spring Harbor Laboratory Press;* 1997. p 263-334.

[48] Rein A. Retroviral RNA packaging: A review. *Arch. Virol. Suppl.* 1994;9:513-522.

[49] Amarasinghe GK, De Guzman RN, Turner RB, Chancellor KJ, Wu ZR, Summers MF. NMR structure of the HIV-1 nucleocapsid protein bound to stem-loop SL2 of the psi-RNA packaging signal. Implications for genome recognition. *J. Mol. Biol.* 2000;301(2):491-511.

[50] Berkowitz RD, Goff SP. Analysis of binding elements in the human immunodeficiency virus type 1 genomic RNA and nucleocapsid protein. *Virology* 1994;202(1):233-246.

[51] Clever J, Sassetti C, Parslow TG. RNA secondary structure and binding sites for gag gene products in the 5' packaging signal of human immunodeficiency virus type 1. *J. Virol.* 1995;69(4):2101-2109.

[52] Clavel F, Orenstein JM. A mutant of human immunodeficiency virus with reduced RNA packaging and abnormal particle morphology. *J. Virol.* 1990;64(10):5230-5234.

[53] Luban J, Goff SP. Mutational analysis of cis-acting packaging signals in human immunodeficiency virus type 1 RNA. *J. Virol.* 1994;68(6):3784-3793.

[54] Dorman N, Lever A. Comparison of viral genomic RNA sorting mechanisms in human immunodeficiency virus type 1 (HIV-1), HIV-2, and Moloney murine leukemia virus. *J. Virol.* 2000;74(23):11413-11417.

[55] Butsch M, Boris-Lawrie K. Destiny of unspliced retroviral RNA: ribosome and/or virion? *J. Virol.* 2002;76(7):3089-3094.

[56] Butsch M, Boris-Lawrie K. Translation is not required to generate virion precursor RNA in human immunodeficiency virus type 1-infected T cells. *J. Virol.* 2000;74(24):11531-11537.

[57] Levin JG, Rosenak MJ. Synthesis of murine leukemia virus proteins associated with virions assembled in actinomycin D-treated cells: Evidence for persistence of viral messenger RNA. *Proc. Natl. Acad. Sci.* U S A 1976;73(4):1154-1158.

[58] Messer LI, Levin JG, Chattopadhyay SK. Metabolism of viral RNA in murine leukemia virus-infected cells; evidence for differential stability of viral message and virion precursor RNA. *J. Virol.* 1981;40(3):683-690.

[59] Ott DE, Coren LV, Johnson DG, Kane BP, Sowder RC, II,, Kim YD, Fisher RJ, Zhou XZ, Lu KP, Henderson LE. Actin-binding cellular proteins inside human immunodeficiency virus type 1. *Virology* 2000;266(1):42-51.

[60] Poon DT, Chertova EN, Ott DE. Human immunodeficiency virus type 1 preferentially encapsidates genomic RNAs that encode pr55(Gag): functional linkage between translation and RNA packaging. *Virology* 2002;293(2):368-378.

[61] Lever AM, Kaye JF, McCann E, Chadwick D, Dorman N, Thomas J, Zhao J. Lentivirus vectors for gene therapy. *Biochem. Soc. Trans.* 1999;27(6):841-847.

[62] Kaye JF, Lever AM. Human immunodeficiency virus types 1 and 2 differ in the predominant mechanism used for selection of genomic RNA for encapsidation. *J. Virol.* 1999;73(4):3023-31.

[63] Hirsch RC, Lavine JE, Chang LJ, Varmus HE, Ganem D. Polymerase gene products of hepatitis B viruses are required for genomic RNA packaging as well as for reverse transcription. *Nature* 1990;344(6266):552-555.

[64] Bartenschlager R, Junker-Niepmann M, Schaller H. The P gene product of hepatitis B virus is required as a structural component for genomic RNA encapsidation. *J. Virol.* 1990;64(11):5324-5332.

[65] Tritel M, Resh MD. Kinetic analysis of human immunodeficiency virus type 1 assembly reveals the presence of sequential intermediates. *J. Virol.* 2000;74(13):5845-5855.

[66] Ono A, Huang M, Freed EO. Characterization of human immunodeficiency virus type 1 matrix revertants: effects on virus assembly, Gag processing, and Env incorporation into virions. *J. Virol.* 1997;71(6):4409-4418.

[67] Hermida-Matsumoto L, Resh MD. Localization of human immunodeficiency virus type 1 Gag and Env at the plasma membrane by confocal imaging. *J. Virol.* 2000;74(18):8670-8679.

[68] Sandefur S, Smith RM, Varthakavi V, Spearman P. Mapping and characterization of the N-terminal I domain of human immunodeficiency virus type 1 Pr55(Gag). *J. Virol.* 2000;74(16):7238-7249.

[69] Mouland AJ, Mercier J, Luo M, Bernier L, DesGroseillers L, Cohen EA. The double-stranded RNA-binding protein Staufen is incorporated in human immunodeficiency virus type 1: evidence for a role in genomic RNA encapsidation. *J. Virol.* 2000;74(12):5441-5451.

[70] Bassell G, Singer RH. mRNA and cytoskeletal filaments. *Curr. Opin. Cell Biol.* 1997;9(1):109-115.

[71] Mouland AJ, Xu H, Cui H, Krueger W, Munro TP, Prasol M, Mercier J, Rekosh D, Smith R, Barbarese E and others. RNA trafficking signals in human immunodeficiency virus type 1. *Mol. Cell. Biol.* 2001;21(6):2133-2143.
[72] Palacios IM, St Johnston D. Getting the message across: the intracellular localization of mRNAs in higher eukaryotes. *Annu. Rev. Cell Dev. Biol.* 2001;17:569-614.
[73] Kislauskis EH, Singer RH. Determinants of mRNA localization. *Curr. Opin. Cell Biol.* 1992;4(6):975-978.
[74] Canto-Nogues C, Hockley D, Grief C, Ranjbar S, Bootman J, Almond N, Herrera I. Ultrastructural localization of the RNA of immunodeficiency viruses using electron microscopy in situ hybridization and in vitroinfected lymphocytes. *Micron* 2001;32(6):579-589.
[75] Basyuk E, Galli T, Mougel M, Blanchard JM, Sitbon M, Bertrand E. Retroviral genomic RNAs are transported to the plasma membrane by endosomal vesicles. *Dev. Cell* 2003;5(1):161-174.
[76] Tang Y, Winkler U, Freed EO, Torrey TA, Kim W, Li H, Goff SP, Morse HC, 3rd. Cellular motor protein KIF-4 associates with retroviral Gag. *J. Virol.* 1999;73(12):10508-10513.
[77] Rey O, Canon J, Krogstad P. HIV-1 Gag protein associates with F-actin present in microfilaments. *Virology* 1996;220:530-534.
[78] Sasaki H, Nakamura M, Ohno T, Matsuda Y, Yuda Y, Nonomura Y. Myosin-actin interaction plays an important role in human immunodeficiency virus type 1 release from host cells. *Proc. Natl. Acad. Sci.* U S A 1995;92:2026-2030.
[79] Campbell SM, Crowe SM, Mak J. Lipid rafts and HIV-1: from viral entry to assembly of progeny virions. *J. Clin. Virol.* 2001;22(3):217-227.
[80] Simons K, Toomre D. Lipid rafts and signal transduction. *Nat. Rev. Mol. Cell Biol.* 2000;1(1):31-39.
[81] Lai EC. Lipid rafts make for slippery platforms. *J. Cell Biol.* 2003;162(3):365-370.
[82] Munro S. Lipid rafts: elusive or illusive? *Cell* 2003;115(4):377-388.
[83] Nguyen DH, Hildreth JE. Evidence for budding of human immunodeficiency virus type 1 selectively from glycolipid-enriched membrane lipid rafts. *J. Virol.* 2000;74(7):3264-3272.
[84] Holm K, Weclewicz K, Hewson R, Suomalainen M. Human immunodeficiency virus type 1 assembly and lipid rafts: Pr55(gag) associates with membrane domains that are largely resistant to Brij98 but sensitive to Triton X-100. *J. Virol.* 2003;77(8):4805-4817.
[85] Ono A, Freed EO. Plasma membrane rafts play a critical role in HIV-1 assembly and release. *Proc. Natl. Acad. Sci.* U S A 2001;98(24):13925-13930.
[86] Lindwasser OW, Resh MD. Multimerization of human immunodeficiency virus type 1 Gag promotes its localization to barges, raft-like membrane microdomains. *J. Virol.* 2001;75(17):7913-7924.
[87] Campbell S, Rein A. In vitro assembly properties of human immunodeficiency virus type 1 Gag protein lacking the p6 domain. *J. Virol.* 1999;73(3):2270-2279.
[88] Campbell S, Fisher RJ, Towler EM, Fox S, Issaq HJ, Wolfe T, Phillips LR, Rein A. Modulation of HIV-like particle assembly in vitro by inositol phosphates. *Proc. Natl. Acad. Sci.* U S A 2001;98(19):10875-10879.

[89] Ding L, Derdowski A, Wang JJ, Spearman P. Independent segregation of human immunodeficiency virus type 1 Gag protein complexes and lipid rafts. *J. Virol.* 2003;77(3):1916-1926.

[90] Liao Z, Graham DR, Hildreth JE. Lipid rafts and HIV pathogenesis: virion-associated cholesterol is required for fusion and infection of susceptible cells. *AIDS Res. Hum. Retroviruses* 2003;19(8):675-687.

[91] Graham DR, Chertova E, Hilburn JM, Arthur LO, Hildreth JE. Cholesterol depletion of human immunodeficiency virus type 1 and simian immunodeficiency virus with beta-cyclodextrin inactivates and permeabilizes the virions: evidence for virion-associated lipid rafts. *J. Virol.* 2003;77(15):8237-8248.

[92] Campbell SM, Crowe SM, Mak J. Virion-associated cholesterol is critical for the maintenance of HIV-1 structure and infectivity. *AIDS* 2002;16(17):2253-2261.

[93] Campbell S, Vogt VM. Self-assembly in vitro of purified CA-NC proteins from Rous sarcoma virus and human immunodeficiency virus type 1. *J. Virol.* 1995;69(10):6487-6497.

[94] Muriaux D, Mirro J, Harvin D, Rein A. RNA is a structural element in retrovirus particles. *Proc. Natl. Acad. Sci.* USA 2001;98(9):5246-5251.

[95] Cimarelli A, Sandin S, Hoglund S, Luban J. Basic residues in human immunodeficiency virus type 1 nucleocapsid promote virion assembly via interaction with RNA. *J. Virol.* 2000;74(7):3046-3057.

[96] Wang SW, Noonan K, Aldovini A. Nucleocapsid-RNA Interactions Are Essential to Structural Stability but Not to Assembly of Retroviruses. *J. Virol.* 2004;78(2):716-723.

[97] Wang SW, Aldovini A. RNA incorporation is critical for retroviral particle integrity after cell membrane assembly of Gag complexes. *J. Virol.* 2002;76(23):11853-11865.

[98] Morikawa Y, Hinata S, Tomoda H, Goto T, Nakai M, Aizawa C, Tanaka H, Omura S. Complete inhibition of human immunodeficiency virus Gag myristoylation is necessary for inhibition of particle budding. *J. Biol. Chem.* 1996;271(5):2868-2873.

[99] Gottlinger HG, Sodroski JG, Haseltine WA. Role of capsid precursor processing and myristoylation in morphogenesis and infectivity of human immunodeficiency virus type 1. *Proc. Natl. Acad. Sci.* USA 1989;86(15):5781-5785.

[100] Paillart J-C, Gottlinger HG. Opposing effects of human immunodeficiency virus type 1 matrix mutations support a myristyl switch model of Gag membrane targeting. *J. Virol.* 1999;73(4):2604-2612.

[101] Lindwasser OW, Resh MD. Myristoylation as a target for inhibiting HIV assembly: unsaturated fatty acids block viral budding. *Proc. Natl. Acad. Sci.* U S A 2002;99(20):13037-13042.

[102] Spearman P, Wang JJ, Vander Heyden N, Ratner L. Identification of human immunodeficiency virus type 1 Gag protein domains essential to membrane binding and particle assembly. *J. Virol.* 1994;68(5):3232-3242.

[103] Yuan X, Yu X, Lee TH, Essex M. Mutations in the N-terminal region of human immunodeficiency virus type 1 matrix protein block intracellular transport of the Gag precursor. *J. Virol.* 1993;67(11):6387-6394.

[104] Zhou W, Parent LJ, Wills JW, Resh MD. Identification of a membrane-binding domain within the amino-terminal region of human immunodeficiency virus type 1 Gag protein which interacts with acidic phospholipids. *J. Virol.* 1994;68(4):2556-2569.

[105] Hill CP, Worthylake D, Bancroft DP, Christensen AM, Sundquist WI. Crystal structures of the trimeric human immunodeficiency virus type 1 matrix protein: implications for membrane association and assembly. *Proc. Natl. Acad. Sci.* USA 1996;93(7):3099-3104.

[106] Ono A, Orenstein JM, Freed EO. Role of the Gag matrix domain in targeting human immunodeficiency virus type 1 assembly. *J. Virol.* 2000;74(6):2855-2866.

[107] Liang C, Hu J, Whitney JB, Kleiman L, Wainberg MA. A structurally disordered region at the C terminus of capsid plays essential roles in multimerization and membrane binding of the gag protein of human immunodeficiency virus type 1. *J. Virol.* 2003;77(3):1772-1783.

[108] Sandefur S, Varthakavi V, Spearman P. The I domain is required for efficient plasma membrane binding of human immunodeficiency virus type 1 Pr55Gag. *J. Virol.* 1998;72(4):2723-32.

[109] Pearce-Pratt R, Malamud D, Phillips DM. Role of the cytoskeleton in cell-to-cell transmission of human immunodeficiency virus. *J. Virol.* 1994;68:2898-2905.

[110] Pelchen-Matthews A, Kramer B, Marsh M. Infectious HIV-1 assembles in late endosomes in primary macrophages. *J. Cell Biol.* 2003;162(3):443-455.

[111] Nguyen DG, Booth A, Gould SJ, Hildreth JE. Evidence that HIV budding in primary macrophages occurs through the exosome release pathway. *J. Biol. Chem.* 2003.

[112] Roche PA. Intracellular protein traffic in lymphocytes: "how do I get THERE from HERE"? *Immunity* 1999;11(4):391-398.

[113] Katzmann DJ, Odorizzi G, Emr SD. Receptor downregulation and multivesicular-body sorting. *Nat. Rev. Mol. Cell Biol.* 2002;3(12):893-905.

[114] Stoorvogel W, Kleijmeer MJ, Geuze HJ, Raposo G. The biogenesis and functions of exosomes. *Traffic* 2002;3(5):321-330.

[115] Kleijmeer M, Ramm G, Schuurhuis D, Griffith J, Rescigno M, Ricciardi-Castagnoli P, Rudensky AY, Ossendorp F, Melief CJ, Stoorvogel W and others. Reorganization of multivesicular bodies regulates MHC class II antigen presentation by dendritic cells. *J. Cell Biol.* 2001;155(1):53-63.

[116] Marsh M, Thali M. HIV's great escape. *Nat. Med.* 2003;9(10):1262-1263.

[117] Nydegger S, Foti M, Derdowski A, Spearman P, Thali M. HIV-1 egress is gated through late endosomal membranes. *Traffic* 2003;4(12):902-910.

[118] Goff A, Ehrlich LS, Cohen SN, Carter CA. Tsg101 control of human immunodeficiency virus type 1 Gag trafficking and release. *J. Virol.* 2003;77(17):9173-9182.

[119] Sherer NM, Lehmann MJ, Jimenez-Soto LF, Ingmundson A, Horner SM, Cicchetti G, Allen PG, Pypaert M, Cunningham JM, Mothes W. Visualization of retroviral replication in living cells reveals budding into multivesicular bodies. *Traffic* 2003;4(11):785-801.

[120] Nguyen DG, Booth A, Gould SJ, Hildreth JE. Evidence That HIV Budding in Primary Macrophages Occurs through the Exosome Release Pathway. *J. Biol. Chem.* 2003;278(52):52347-52354.

[121] Gould SJ, Booth AM, Hildreth JE. The Trojan exosome hypothesis. *Proc. Natl. Acad. Sci.* U S A 2003;100(19):10592-10597.

[122] von Schwedler UK, Stuchell M, Muller B, Ward DM, Chung HY, Morita E, Wang HE, Davis T, He GP, Cimbora DM and others. The protein network of HIV budding. *Cell* 2003;114(6):701-713.

[123] Thery C, Zitvogel L, Amigorena S. Exosomes: composition, biogenesis and function. *Nat. Rev. Immunol.* 2002;2(8):569-579.

[124] Gluschankof P, Mondor I, Gelderblom HR, Sattentau QJ. Cell membrane vesicles are a major contaminant of gradient-enriched human immunodeficiency virus type-1 preparations. *Virology* 1997;230:125-133.

[125] Bess JW, Jr., Gorelick RJ, Bosche WJ, Henderson LE, Arthur LO. Microvesicles are a source of contaminating cellular proteins found in purified HIV-1 preparations. *Virology* 1997;230:134-144.

[126] Freed EO. Viral late domains. *J. Virol.* 2002;76(10):4679-4687.

[127] Freed EO. The HIV-TSG101 interface: recent advances in a budding field. *Trends Microbiol.* 2003;11(2):56-59.

[128] Demirov DG, Orenstein JM, Freed EO. The late domain of human immunodeficiency virus type 1 p6 promotes virus release in a cell type-dependent manner. *J. Virol.* 2002;76(1):105-117.

[129] Puffer BA, Parent LJ, Wills JW, Montelaro RC. Equine infectious anemia virus utilizes a YXXL motif within the late assembly domain of the Gag p9 protein. *J. Virol.* 1997;71:6541-6546.

[130] Parent LJ, Bennett RP, Craven RC, Nelle TD, Krishna NK, Bowzard JB, Wilson CB, Puffer BA, Montelaro RC, Wills JW. Positionally independent and exchangeable late budding functions of the Rous sarcoma virus and human immunodeficiency virus Gag proteins. *J. Virol.* 1995;69(9):5455-5460.

[131] VerPlank L, Bouamr F, LaGrassa TJ, Agresta B, Kikonyogo A, Leis J, Carter CA. Tsg101, a homologue of ubiquitin-conjugating (E2) enzymes, binds the L domain in HIV type 1 Pr55Gag. *Proc. Natl. Acad. Sci.* U S A 2001;98(14):7724-7729.

[132] Carter CA. Tsg101: HIV-1's ticket to ride. *Trends Microbiol.* 2002;10(5):203-205.

[133] Li L, Cohen SN. Tsg101: a novel tumor susceptibility gene isolated by controlled homozygous functional knockout of allelic loci in mammalian cells. *Cell* 1996;85(3):319-329.

[134] Babst M, Odorizzi G, Estepa EJ, Emr SD. Mammalian tumor susceptibility gene 101 (TSG101) and the yeast homologue, Vps23p, both function in late endosomal trafficking. *Traffic* 2000;1(3):248-258.

[135] Hewitt EW, Duncan L, Mufti D, Baker J, Stevenson PG, Lehner PJ. Ubiquitylation of MHC class I by the K3 viral protein signals internalization and TSG101-dependent degradation. *Embo J.* 2002;21(10):2418-2429.

[136] Wagner KU, Krempler A, Qi Y, Park K, Henry MD, Triplett AA, Riedlinger G, Rucker IE, Hennighausen L. Tsg101 is essential for cell growth, proliferation, and cell survival of embryonic and adult tissues. *Mol. Cell. Biol.* 2003;23(1):150-162.

[137] Pornillos O, Alam SL, Davis DR, Sundquist WI. Structure of the Tsg101 UEV domain in complex with the PTAP motif of the HIV-1 p6 protein. *Nat. Struct. Biol.* 2002;9(11):812-817.

[138] Pornillos O, Alam SL, Rich RL, Myszka DG, Davis DR, Sundquist WI. Structure and functional interactions of the Tsg101 UEV domain. *Embo J.* 2002;21(10):2397-2406.

[139] Vogt VM. Ubiquitin in retrovirus assembly: actor or bystander? *Proc Natl Acad Sci* U S A 2000;97(24):12945-12947.

[140] Dupre S, Volland C, Haguenauer-Tsapis R. Membrane transport: ubiquitylation in endosomal sorting. *Curr. Biol.* 2001;11(22):R932-R934.

[141] Garrus JE, von Schwedler UK, Pornillos OW, Morham SG, Zavitz KH, Wang HE, Wettstein DA, Stray KM, Cote M, Rich RL and others. Tsg101 and the vacuolar protein sorting pathway are essential for HIV-1 budding. *Cell* 2001;107(1):55-65.

[142] Demirov DG, Ono A, Orenstein JM, Freed EO. Overexpression of the N-terminal domain of TSG101 inhibits HIV-1 budding by blocking late domain function. *Proc. Natl. Acad. Sci.* U S A 2002;99(2):955-960.

[143] Martin-Serrano J, Zang T, Bieniasz PD. HIV-1 and Ebola virus encode small peptide motifs that recruit Tsg101 to sites of particle assembly to facilitate egress. *Nat. Med.* 2001;7(12):1313-1319.

[144] Pornillos O, Higginson DS, Stray KM, Fisher RD, Garrus JE, Payne M, He GP, Wang HE, Morham SG, Sundquist WI. HIV Gag mimics the Tsg101-recruiting activity of the human Hrs protein. *J. Cell. Biol.* 2003;162(3):425-434.

[145] Babst M, Katzmann DJ, Snyder WB, Wendland B, Emr SD. Endosome-associated complex, ESCRT-II, recruits transport machinery for protein sorting at the multivesicular body. *Dev. Cell* 2002;3(2):283-289.

[146] Babst M, Katzmann DJ, Estepa-Sabal EJ, Meerloo T, Emr SD. Escrt-III: an endosome-associated heterooligomeric protein complex required for mvb sorting. *Dev. Cell* 2002;3(2):271-282.

[147] Katzmann DJ, Babst M, Emr SD. Ubiquitin-dependent sorting into the multivesicular body pathway requires the function of a conserved endosomal protein sorting complex, ESCRT-I. *Cell* 2001;106(2):145-155.

[148] Martin-Serrano J, Zang T, Bieniasz PD. Role of ESCRT-I in retroviral budding. *J. Virol.* 2003;77(8):4794-4804.

[149] Strack B, Calistri A, Craig S, Popova E, Gottlinger HG. AIP1/ALIX is a binding partner for HIV-1 p6 and EIAV p9 functioning in virus budding. *Cell* 2003;114(6):689-699.

[150] Martin-Serrano J, Yaravoy A, Perez-Caballero D, Bieniasz PD. Divergent retroviral late-budding domains recruit vacuolar protein sorting factors by using alternative adaptor proteins. *Proc. Natl. Acad. Sci.* U S A 2003;100(21):12414-12419.

[151] Martin-Serrano J, Bieniasz PD. A bipartite late-budding domain in human immunodeficiency virus type 1. *J. Virol.* 2003;77(22):12373-12377.

[152] Goila-Gaur R, Demirov DG, Orenstein JM, Ono A, Freed EO. Defects in human immunodeficiency virus budding and endosomal sorting induced by TSG101 overexpression. *J. Virol.* 2003;77(11):6507-6519.

[153] Ott DE, Coren LV, Copeland TD, Kane BP, Johnson DG, Sowder I, R. C., Yoshinaka Y, Oroszlan S, Arthur LO, Henderson LE. Ubiquitin is covalently attached to the p6Gag proteins of human immunodeficiency virus type-1 and simian immunodeficiency virus and the p12Gag protein of Moloney murine leukemia virus. *J. Virol.* 1998;72:2962-2968.

[154] Ciechanover A. The ubiquitin-proteasome pathway: on protein death and cell life. *EMBO J.* 1998;17(24):7151-7160.

[155] Wilkinson KD. Ubiquitination and deubiquitination: targeting of proteins for degradation by the proteasome. Semin. *Cell Dev. Biol.* 2000;11(3):141-148.

[156] Adams J. The proteasome: structure, function, and role in the cell. *Cancer Treat. Rev.* 2003;29 Suppl 1:3-9.

[157] Goldberg AL. Functions of the proteasome: the lysis at the end of the tunnel. *Science* 1995;268:522-523.

[158] Thrower JS, Hoffman L, Rechsteiner M, Pickart CM. Recognition of the polyubiquitin proteolytic signal. *EMBO J.* 2000;19(1):94-102.

[159] Aguilar RC, Wendland B. Ubiquitin: not just for proteasomes anymore. *Curr. Opin. Cell Biol.* 2003;15(2):184-190.

[160] Ott DE, Coren LV, Chertova EN, Gagliardi TD, Schubert U. Ubiquitination of HIV-1 and MuLV Gag. *Virology* 2000;278(1):111-121.

[161] Strack B, Calistri A, Accola MA, Palu G, Gottlinger HG. A role for ubiquitin ligase recruitment in retrovirus release. *Proc. Natl. Acad. Sci.* U S A 2000;97(24):13063-13068.

[162] Ott DE, Coren LV, Sowder RC, II, Adams J, Nagashima K, Schubert U. Equine infectious anemia virus and the ubiquitin-proteasome system. *J. Virol.* 2002;76(6):3038-3044.

[163] Ott DE, Coren LV, Sowder RC, II, Adams J, Schubert U. Retroviruses have differing requirements for proteasome function in the budding process. *J. Virol.* 2003;77(6):3384-3393.

[164] Gonen H, Dickman D, Schwartz AL, Ciechanover A. Protein synthesis elongation factor EF-1 alpha is an isopeptidase essential for ubiquitin-dependent degradation of certain proteolytic substrates. *Adv. Exp. Med. Biol.* 1996;389:209-219.

[165] Gonen H, Smith CE, Siegel NR, Kahana C, Merrick WC, Chakraburtty K, Schwartz AL, Ciechanover A. Protein synthesis elongation factor EF-1 alpha is essential for ubiquitin-dependent degradation of certain N alpha-acetylated proteins and may be substituted for by the bacterial elongation factor EF-Tu. *Proc. Natl. Acad. Sci.* U S A 1994;91(16):7648-7652.

[166] Schubert U, Ott DE, Chertova EN, Welker R, Tessmer U, Princiotta MF, Bennink JR, Krausslich HG, Yewdell JW. Proteasome inhibition interferes with gag polyprotein processing, release, and maturation of HIV-1 and HIV-2. *Proc. Natl. Acad. Sci.* U S A 2000;97(24):13057-13062.

[167] Patnaik A, Chau V, Wills JW. Ubiquitin is part of the retrovirus budding machinery. Proc. *Natl. Acad. Sci.* U S A 2000;97(24):13069-13074.

[168] Bouamr F, Melillo JA, Wang MQ, Nagashima K, de Los Santos M, Rein A, Goff SP. PPPYEPTAP motif Is the late domain of human T-cell leukemia virus type 1 Gag and mediates its functional interaction with cellular proteins Nedd4 and Tsg101. *J. Virol.* 2003;77(22):11882-11895.

[169] Harty RN, Brown ME, McGettigan JP, Wang G, Jayakar HR, Huibregtse JM, Whitt MA, Schnell MJ. Rhabdoviruses and the cellular ubiquitin-proteasome system: a budding interaction. *J. Virol.* 2001;75(22):10623-10629.

[170] Patnaik A, Chau V, Li F, Montelaro RC, Wills JW. Budding of equine infectious anemia virus is insensitive to proteasome inhibitors. J. Virol. 2002;76(6):2641-2647.

[171] Khor R, McElroy LJ, Whittaker GR. The Ubiquitin-Vacuolar Protein Sorting System is Selectively Required During Entry of Influenza Virus into Host Cells. *Traffic* 2003;4(12):857-868.

[172] Hicke L. Protein regulation by monoubiquitin. *Nat. Rev. Mol. Cell Biol.* 2001;2(3):195-201.

[173] Kikonyogo A, Bouamr F, Vana ML, Xiang Y, Aiyar A, Carter C, Leis J. Proteins related to the Nedd4 family of ubiquitin protein ligases interact with the L domain of Rous sarcoma virus and are required for gag budding from cells. *Proc. Natl. Acad. Sci.* U S A 2001;98(20):11199-11204.

[174] Putterman D, Pepinsky RB, Vogt VM. Ubiquitin in avian leukosis virus particles. *Virology* 1990;176:633-637.

[175] Schubert U, Anton LC, Gibbs A, Norbury CC, Yewdell JW, Bennink JR. Rapid degradation of a large fraction of newly synthesized proteins by proteasomes. *Nature* 2000;404:770-774.

[176] Princiotta MF, Finzi D, Qian SB, Gibbs J, Schuchmann S, Buttgereit F, Bennink JR, Yewdell JW. Quantitating protein synthesis, degradation, and endogenous antigen processing. *Immunity* 2003;18(3):343-354.

[177] Yewdell JW. Not such a dismal science: the economics of protein synthesis, folding, degradation and antigen processing. *Trends Cell Biol.* 2001;11(7):294-297.

[178] Reits EA, Vos JC, Gromme M, Neefjes J. The major substrates for TAP in vivo are derived from newly synthesized proteins. *Nature* 2000;404(6779):774-778.

[179] Gurer C, Cimarelli A, Luban J. Specific incorporation of heat shock protein 70 family members into primate lentiviral virions. *J. Virol.* 2002;76(9):4666-4670.

[180] Brenner BG, Wainberg MA. Heat shock protein-based therapeutic strategies against human immunodeficiency virus type 1 infection. *Infect. Dis. Obstet. Gynecol.* 1999;7(1-2):80-90.

[181] Mitsui K, Nakayama H, Akagi T, Nekooki M, Ohtawa K, Takio K, Hashikawa T, Nukina N. Purification of polyglutamine aggregates and identification of elongation factor-1alpha and heat shock protein 84 as aggregate-interacting proteins. *J. Neurosci.* 2002;22(21):9267-9277.

[182] Hotokezaka Y, Tobben U, Hotokezaka H, Van Leyen K, Beatrix B, Smith DH, Nakamura T, Wiedmann M. Interaction of the eukaryotic elongation factor 1A with newly synthesized polypeptides. *J. Biol. Chem.* 2002;277(21):18545-18551.

[183] Finley D, Bartel B, Varshavsky A. The tails of ubiquitin precursors are ribosomal proteins whose fusion to ubiquitin facilitates ribosome biogenesis. *Nature* 1989;338(6214):394-401.

[184] Zimmerman C, Klein KC, Kiser PK, Singh AR, Firestein BL, Riba SC, Lingappa JR. Identification of a host protein essential for assembly of immature HIV-1 capsids. *Nature* 2002;415(6867):88-92.

[185] Ott DE. Cellular proteins in HIV. *Rev. Med. Virol.* 1997;7:167-180.

[186] Tremblay MJ, Fortin JF, Cantin R. The acquisition of host-encoded proteins by nascent HIV-1. *Immunol Today* 1998;19(8):346-351.

[187] Franke EK, Yuan HEH, Luban J. Specific incorporation of cyclophilin A into HIV-1 virions. *Nature* (London) 1994;372:359-362.

[188] Gamble TR, Vajdos FF, Yoo S, Worthylake DK, Houseweart M, Sundquist WI, Hill CP. Crystal structure of human cyclophilin A bound to the amino-terminal domain of HIV-1 capsid. *Cell* 1996;87:1285-1294.
[189] Thali M, Bukovsky A, Kondo E, Rosenwirth B, Walsh CT, Sodroski J, Gottlinger HG. Functional association of cyclophilin A with HIV-1 virions. *Nature* 1994;372(6504):363-365.
[190] Braaten D, Franke EK, Luban J. Cyclophilin A is required for an early step in the life cycle of human immunodeficiency virus type 1 before the initiation of reverse transcription. *J. Virol.* 1996;70:3551-3560.
[191] Luban J. Absconding with the chaperone: essential cyclophilin-Gag interaction in HIV-1 virions. *Cell* 1996;87:1157-1159.
[192] Wiegers K, Rutter G, Schubert U, Grattinger M, Krausslich HG. Cyclophilin A incorporation is not required for human immunodeficiency virus type 1 particle maturation and does not destabilize the mature capsid. *Virology* 1999;257(1):261-274.
[193] Trubey CM, Chertova E, Coren LV, Hilburn JM, Hixson CV, Nagashima K, Lifson JD, Ott DE. Quantitation of HLA class II protein incorporated into human immunodeficiency type 1 virions purified by anti-CD45 immunoaffinity depletion of microvesicles. *J. Virol.* 2003;77(23):12699-12709.
[194] Poon DT, Coren LV, Ott DE. Efficient incorporation of HLA class II onto human immunodeficiency virus type 1 requires envelope glycoprotein packaging. *J. Virol.* 2000;74(8):3918-3923.
[195] Cantin R, Fortin J-F, Lamontagne G, Tremblay M. The presence of host-derived HLA-DR1 on human immunodeficiency virus type 1 increases viral infectivity. *J. Virol.* 1997;71:1922-1930.
[196] Arthur LO, Bess JW, Jr., Sowder RC, II, Benveniste RE, Mann LD, Chermann J-C, Henderson LE. Cellular proteins bound to immunodeficiency viruses: implications for pathogenesis and vaccines. *Science* 1992;258:1935-1938.
[197] Greenfield EA, Nguyen KA, Kuchroo VK. CD28/B7 costimulation: a review. *Crit. Rev. Immunol.* 1998;18(5):389-418.
[198] Finkel TH, Tudor-Williams G, Banda NK, Cotton MF, Curiel T, Monks C, Baba TW, Ruprecht RM, Kupfer A. Apoptosis occurs predominantly in bystander cells and not in productively infected cells of HIV- and SIV-infected lymph nodes. *Nat. Med.* 1995;1(2):129-134.
[199] Rossio JL, Bess JW, Jr., Henderson LE, Cresswell P, Arthur LO. HLA class II on HIV-1 particles is functional in superantigen presentation to human T cells: implications for HIV pathogenesis. *AIDS Res. Hum. Retroviruses* 1995;11:1433-1439.
[200] Esser MT, Bess JW, Jr., Suryanarayana K, Chertova E, Marti D, Carrington M, Arthur LO, Lifson JD. Partial activation and induction of apoptosis in CD4(+) and CD8(+) T lymphocytes by conformationally authentic noninfectious human immunodeficiency virus type 1. *J. Virol.* 2001;75(3):1152-1164.
[201] Wedekind JE, Dance GS, Sowden MP, Smith HC. Messenger RNA editing in mammals: new members of the APOBEC family seeking roles in the family business. *Trends Genet.* 2003;19(4):207-216.
[202] Goff SP. Death by deamination: a novel host restriction system for HIV-1. *Cell* 2003;114(3):281-283.

[203] Lecossier D, Bouchonnet F, Clavel F, Hance AJ. Hypermutation of HIV-1 DNA in the absence of the Vif protein. *Science* 2003;300(5622):1112.

[204] Harris RS, Bishop KN, Sheehy AM, Craig HM, Petersen-Mahrt SK, Watt IN, Neuberger MS, Malim MH. DNA deamination mediates innate immunity to retroviral infection. *Cell* 2003;113(6):803-809.

[205] Mangeat B, Turelli P, Caron G, Friedli M, Perrin L, Trono D. Broad antiretroviral defence by human APOBEC3G through lethal editing of nascent reverse transcripts. *Nature* 2003;424(6944):99-103.

[206] Shindo K, Takaori-Kondo A, Kobayashi M, Abudu A, Fukunaga K, Uchiyama T. The enzymatic activity of CEM15/Apobec-3G is essential for the regulation of the infectivity of HIV-1 virion but not a sole determinant of its antiviral activity. *J. Biol. Chem.* 2003;278(45):44412-6.

[207] Kao S, Khan MA, Miyagi E, Plishka R, Buckler-White A, Strebel K. The human immunodeficiency virus type 1 Vif protein reduces intracellular expression and inhibits packaging of APOBEC3G (CEM15), a cellular inhibitor of virus infectivity. *J. Virol.* 2003;77(21):11398-11407.

[208] Marin M, Rose KM, Kozak SL, Kabat D. HIV-1 Vif protein binds the editing enzyme APOBEC3G and induces its degradation. *Nat. Med.* 2003;9(11):1398-1403.

[209] Stopak K, de Noronha C, Yonemoto W, Greene WC. HIV-1 Vif blocks the antiviral activity of APOBEC3G by impairing both its translation and intracellular stability. *Mol. Cell* 2003;12(3):591-601.

[210] Conticello SG, Harris RS, Neuberger MS. The Vif protein of HIV triggers degradation of the human antiretroviral DNA deaminase APOBEC3G. *Curr. Biol.* 2003;13(22):2009-2013.

[211] Yung E, Sorin M, Pal A, Craig E, Morozov A, Delattre O, Kappes J, Ott D, Kalpana GV. Inhibition of HIV-1 virion production by a transdominant mutant of integrase interactor 1. *Nat. Med.* 2001;7(8):920-926.

[212] Clague MJ, Urbe S. Hrs function: viruses provide the clue. *Trends Cell Biol.* 2003;13(12):603-606.

[213] Bestman-Smith J, Desormeaux A, Tremblay MJ, Bergeron MG. Targeting cell-free HIV and virally-infected cells with anti-HLA-DR immunoliposomes containing amphotericin B. *AIDS* 2000;14(16):2457-2465.

In: Progress in Virus Research
Editor: Sabine Thebault, pp. 101-127
ISBN 1-59454-438-7

Chapter IV

The Interactions of Immunodeficiency Viruses with Antigen Presenting Cells

Stuart G. Turville*
Center For Virus Research, Westmead Millennium Institute,
Sydney, NSW 2145, Australia

Abstract

Recently several reports have focussed on the molecular aspects of immunodeficiency viral pathogenesis within myeloid antigen presenting cells such as the dendritic cells and macrophage. From the receptors that are hypothesised to attach immunodeficiency viruses in the mucosae during HIV transmission, mechanisms of viral harbouring and transfer and the use of the exosome synthesis machinery during APC infection. Herein we will address the concepts of viral attachment, infection and viral transfer in light of current observations of APC subsets. In relation to viral transfer, we will more specifically address the uptake compartments in DC, the internalised half-life of virus and finally the exit pathways for either "old" regurgitated virus versus "new," *de novo* produced virus within the DC.

Introduction

HIV's reliance on cells from the myeloid antigen-presenting cell (APC) lineage is no exception to the lentiviral family. Slow, tissue associated viral replication is a clinical outcome in many lentiviruses including, HIV, SIV, maedi-visna virus, feline immunodeficiency virus, equine infectious anaemia virus and ovine lentiviruses (Bingen et al., 2002, Brodie et al., 1995, Gendelman et al., 1986, Gendelman et al., 1989, Harrold et al.,

* Corresponding author: Dr Stuart G Turville, Westmead Millennium Institute, PO Box 412 Darcy Road, Westmead, NSW 2145, AUSTRALIA.,Phone#: +61-2-9845-9113Fax#: +61-2-9845-9103, E-mail: stuart_turville@wmi.usyd.edu.au

2000, Hirsch et al., 1991, Narayan et al., 1982, Oaks et al., 1998, Singh et al., 1999). The common theme in all of these lentiviruses is a need for infection of the APC for persistence *in vivo* and the need for this growth and spread may enable survival in the hostile environment of the innate and developing immune system (or after the introduction of antiretroviral therapy). With specific respect to HIV, the viral genome is a good indicator into the importance of APC in its life cycle. Certain viral like VPR are essential to infect APC and their retention is a mark of respect for this lineage. Other viral genes appear equally as important in both lymphoid and APC lineage, although their function in each cellular environment maybe different. For instance Nef can induce responses in macrophages that attract lymphocytes for viral transfer (Swingler et al., 2003). HIV gag also appears to function very differently in macrophages allowing viral budding to occur into vacuolar compartments in contrast to the plasma membrane on lymphocytes. Such vesicular budding may impart the virus a phenotype which may temporarily evade an immune response (Gendelman et al., 1988, Pelchen-Matthews et al., 2003, Raposo et al., 2002). Thus it appears that viral proteins have evolved multifaceted nature in both APC and lymphoid derived lineages.

What stage HIV needs or benefits from APC can vary. Early on, they are linked to, and recent investigations have supported their role in HIV transmission. The fact that their home is tissue and one of their jobs is surveillance logically puts them in the first contact points when virus meets the vaginal mucosae or penile epithelium. Although their role in immunodeficiency viral pathogenesis is far more than being at the right place at the right time. The defining feature of a major subset of myeloid APC, the dendritic cell (DC) and HIV, is the ability not only to catalyse transfer to CD4 lymphocytes, but also drive further viral replication within them (Cameron et al., 1992b, McDonald et al., 2003, Pope et al., 1994, Turville et al., 2004). DCs also have inherent migratory capacity and do so to increase their probability of interacting with an appropriate lymphocyte at lymphoid organs or within inflamed peripheral tissue. Therefore it is not therefore surprising they are now seen as one of main cellular disseminators of immunodeficiency virus during anogenital transmission. In later stages of disease, virus levels also rely heavily upon myeloid APC. They are seen as active viral reservoirs, secreting and maintaining viral loads when CD4 lymphocytes have been heavily depleted or when antiretroviral therapy commences. In the later case the inaccessibility of compounds to tissue sites which bear these infected cells types provides a difficult situation in terms of effectively eliminating all active viral replication with current therapeutic approaches. Thus the study of the fundamental interactions between immunodeficiency viruses and myeloid APC, like DC and macrophages, is key to understanding and intervening on the life cycle of immunodeficiency viruses.

Splitting Hairs

To address the early interaction with HIV and APC subsets, it is essential to define and outline the terminology within the APC divisions. From the broad division of macrophages and DC through to the multiple antigenic and functional sub-divisions of the DC, it is essential to emphasise their diversity rather than their homogeneity. As it is the appreciation of diversity which clarifies the contradicting reports involving the interactions between APC and immunodeficiency viruses.

Myeloid APC can be simply split into either professional APC or DC and amateur APC or macrophages. The functional delineation between the two major subsets as "amateur," and "professional," is that DC can stimulate naïve T cells. ie. DC can carry out a primary immune response, where as macrophages are more a tune to stimulating secondary immune responses. Some argue that DCs are simply another macrophage subset as several cellular markers are shared by both. Others rightly point out that DC represent a unique leukocyte population that deserves its own lineage placement (Hart, 1997, Steinman, 1991). Even if DC were a macrophage subset, their highly effective antigen uptake capacity, selective means of phagocytosis and potent stimulatory capacity would indeed make them a very unique macrophage subset. Although with current markers it is easy to fall into the trap of mistaking a macrophage for a DC and *vice versa*. One current and important point to note is that functional assays are far more informative than so call "DC," or "macrophage" specific markers. For instance CD14 is often associated with macrophages, but is expressed on immature DC (Engering et al., 2002, Turville et al., 2002). DC-SIGN/CD209 is often associated with "DC" restriction (it's name Dendritic cell specific ICAM-3 grabbing non-integrin has a lot to do with this) although is also expressed on macrophage subsets (Soilleux et al., 2002). Therefore it is easy to be misled into what is and isn't a DC, based on so called DC or macrophage restricted markers. Functional assays do exist for DCs and macrophage division and the most commonly used method is the use of the mix leukocyte reaction (MLR). Briefly the MLR is an indicator of naïve lymphocyte proliferation that is efficiently driven by low numbers of professional APC present in the culture like DCs. Other methods such as latex particle uptake are also used but used with less frequency compared with that of the MLR.

Dendritic Cell Family

Immature Versus Mature DC Functional Division

Although this functional division has been covered in previous other DC related reviews, it warrants revision on the basis that immature and mature DCs interact with HIV and/or SIV (will be collectively referred to as immunodeficiency viruses) in a very different manner. More specifically with respect to the DC family, the broad division between immature versus mature is related to the paradox of the immune response being initiated within the secondary lymphoid organs, but exposure and recognition of pathogens occurring in peripheral tissues. In peripheral tissue, DCs are predominantly seen as immature phenotypes. ie. They have a remarkable ability to uptake antigen, yet their ability to stimulate responding T lymphocytes is somewhat limited. On the contrary, the mature DC is believed to primarily reside in lymphoid tissue or in areas of tissue where DCs are promoting secondary immune responses, ie. Their ability to uptake antigen is somewhat impaired, but their ability to stimulate lymphocytes is remarkably efficient. In terms of the differences between the "immature," and "mature," phenotypes, there are both functional and antigenic means whereby to distinguish the two. The process of differentiation from an immature to a mature DC serves to distinctly highlight their differences on the basis of their contrasting function. For instance, the immature DC needs to acquire antigen at very low concentrations, therefore expresses high levels of receptors that give them the unique ability ,to not only recognise pathogens, but also distinguish between self and non-self. C-Type Lectins Receptors (CLRs), eg. CD209, CD207, CD206 and Toll-like Receptors (TLRs) eg TLR2, TLR4 and TLR9, are just a few examples of cell surface bearing receptors that serve to do such recognition in peripheral tissue (Mitsui et al., 2004, Turville et al., 2002). In terms of functional assays, a ligand for CD206, FITC-Dextran, is commonly used to demonstrate the ability of the DC in question to uptake this molecule and can be easily accessed by flow cytometry. In terms of maturation, mature DCs don't need the equivalent uptake capacity of immature DCs and down-regulate many of their uptake receptors. Specifically in terms of CD206, you often find large quantities of this receptor in LAMP-2 positive compartments, indicating it's imminent removal from the cell through the endo-lysosomal degradation pathway. In contrast, other membrane bound receptors, which help a DC play with a lymphocyte, become readily available on the cell surface. For instance large quantities of peptide bearing MHC II is redistributed and stabilised on the cell surface, and the lymphocyte adhesion molecules ,CD83 and CD86, also appears on the surface as well (Turley et al., 2000) (although not to the amazing degree of MHC II). In addition the cell undertakes a dramatic and somewhat beautiful transformation from a pseudopod bearing immature DC to a spike ridden mature DC. Functionally the mature DCs display their T lymphocyte "playing," power through the induction of proliferation of lymphocytes and can be measured, as before mentioned, in stimulating allogeneic CD8 and CD4 peripheral lymphocyte in the MLR.

In Vivo and in Vitro Subset Divisions

The specific consideration of DC and HIV with respect to DC subsets has received sporadic attention. Although confusing in nature, it is best to outline the main subsets that will be addressed with respect to HIV behaviour in later sections. The main underlying reason for confusion regarding DC subsets, is firstly they are very hard to obtain, secondly new subsets divisions are being resolved over time and thirdly new isolation methods can now obtain fresh *ex vivo* DCs with true precursor or immature functional status. In broad terms, all DC subsets have the capacity to mature and stimulate naïve lymphocytes. Although they share this common feature, they do not exist as a homogeneous population. Starting at peripheral tissue it is possible to split immature DCs into three subsets on the basis of CLR expression such as CD209 (DC-SIGN), CD206 (MR) and CD207 (Langerin) and other pathogen recognition molecules like CD1a and CD14 (McLellan et al., 1998, Nestle et al., 1993, Turville et al., 2002). Langerhans cells are the easiest to pinpoint given their exclusive habitation above the basement membrane within the epithelium and high levels of CD207 and CD1a expression (Valladeau et al., 2000). The dermis under the epithelium is somewhat more complicated. On the basis of CD14, CD1a and CD209 expression, it is possible to find two subsets: CD1a moderate expression, or CD14 and CD209 expressing DC (McLellan et al., 1998, Turville et al., 2002). The mature versions of these two are a little difficult to pinpoint as CD14 and CD209 are rapidly down regulated in skin explant cultures (McLellan et al., 1998, Nestle et al., 1993, Turville et al., 2002), although cervical explants have shown persistence of CD209 expression on a subset of DCs (Hu et al., 2004). To note, explant cultures have been a good model for obtaining mature DCs, as it is convenient to obtain them from their migration out of tissue sections into culture supernatant. In addition, their maturation does not require addition of pathogenic products like LPS or pro-inflammatory cytokines, but rather due to the result of tissue trauma (Lukas et al., 1996).

Moving away from peripheral tissue are the DC precursors in blood (Cells that are believed to seed the peripheral tissue with immature DC). For three main examples of blood DC precursors, there are i) lineage negative (negative for CD19, CD3, CD14, CD56) CD1b/c, CD11c and HLA-DR positive ii) lineage negative, CD123, BDCA-2 and HLA-DR positive and the lineage low (more specifically CD14 low), CD11c, CD209 and HLA-DR positive. For a more complete and thorough look all blood DC precursor and characterisation please refer to the work by Mac Donald and colleagues (MacDonald et al., 2002a, MacDonald et al., 2002b). The principle feature about the precursor DCs is that they do not resemble the immature pseudopod bearing DCs and are very similar in morphology to a peripheral blood monocyte for the CD11c restricted, to the more plasmacytoid appearance of a lymphocyte in the CD123 BDC2A restricted subset. Nevertheless brief culture periods have profound effects on these cells types, with extensive dendrites appearing overnight when they obtain the mature DC phenotype (they skip the immature stage without appropriate stimulation).

In vitro DC subsets have also been found by the pioneering work by Romani and colleagues (Romani et al., 1994). The key was to use peripheral blood monocytes and simply incubate GM-CSF and IL-4 for several days. Initially incubations involved as long as 6 to 7 days resulting in an immature DC with high levels of CD1a, CD209 and CD206. Others have shown that shorter incubations (around 24 to 48 hours) can generate immature DCs in antigenic profile to the CD14 positive dermal/Lamina propria DC (Dauer et al., 2003) and unpublished observations). Given the large numbers and ease which the *in vitro* DCs can be generated, the 6 day MDDC have largely served as the model for DC-HIV interactions. Convenient yes maybe, but their major failing is that their equivalent *in vivo* DC has yet to be identified and are therefore seen as a unique DC subset in their own right.

Molecular Considerations with Respect to APC-Immunodeficiency Virus Attachment, Infection and Transfer

Although evidence *in vivo* is in support of myeloid APC in transmission, the molecular mechanisms whereby it does so have been rather cloudy. Which and how each DC does its part in *in vivo* viral dissemination is not well known. Several investigators have looked at not only different subsets, but also with respect to maturational status. It is these two features that we will concentrate on. As in general, differences in immunodeficiency virus attachment are at the level DC subsets and the capacity of virus to infect DCs is at the level of maturation status (Figure 1. and Table I). Viral transfer from DCs to CD4 lymphocytes is a little more complicated as it can occur either via a process of viral regurgitation or through active viral budding of newly synthesised virus. Which transfer predominates is more reflective on the maturation status, but there are a few exceptions to this rule. Herein we will address sequentially evidence for viral attachment, infection and transfer to CD4 lymphocytes in the main DC subsets to date.

Figure 1. Handling of immunodeficiency viruses in immature and mature DC subsets over time. Cartoon highlighting the early time points in (1) where immature DCs attach virus for digestion (hostile situation) or the virus has escaped digestion by fusion through CD4 and appropriate chemokine receptors (virus entering on the left of the DC face). For mature DCs, the predominate handling is via uptake into vacuolar compartments believed to be multivesicular bodies. This pathway appears less hostile over time (friendly situation).In immature DCs in (2) virus is degraded with an approximate half-life of about 4 hours (eating DC). For mature DCs, virus is also degraded but more intact virions are maintained in vesicular compartments (eating DC with a full mouth).Finally in (3) immature DCs after 24 hours (48 to 72 hours), virus reappears from an infected immature DC (multiple virions erupting virions from the DC face to the contacting T cell below). In mature DCs the virus is regurgitated from vesicular stores with little or no virus produced *de novo* (vomiting DC onto T cell below).

Table I. Summary of immunodeficiency viral attachment, infection, regurgitation *and de novo* viral transfer in macrophages and dendritic cell subsets. Main subsets are highlighted, although other subsets maybe absent due to lack of observations. Scoring for receptors is either present (+), absent (-) or intermediate (-/+). Scoring for processes is high/predominate (+++ to ++), medium to low (+), (-/+) intermediate to undetectable and undetectable (-). Cases where observations are unknown are marked by (?).

Attachment: C-Type Lectins: CD206	CD207	CD209	CD4		Infection	Regurgitation	*de novo*
+	-	+/-		Macrophage	+++	+	+++
				Dendritic Cells:			
				LC Family			
-	+	-	+	ImLC	+	?	+
-	+	-	+	MatLC	?	?	?
+	+	+	+	(MDLC)*	+	?	+
				Dermal Family			
+	-	+	+	ImCD14+ve	+	?	+
+	-	-	+	ImCD1a+ve	+	?	+
-	-	-	+	Mat -CD1a+ve	+	?	+
-	-	-	+	Mat -CD14+ve/-ve	+	?	+
				MDDC Family			
+	-	+	+	ImMDDC	+	+	+
-/+	-	+	+	MatMDDC	-/+	++	-
				Blood DC Family			
-	-	-	+	Lin-ve/CD11c/ CD1b/c	+	?	+
-	-	-	+	Lin-ve/CD123/ BDCA-2	++	?	+
?	-	+	+	Lindim/CD14dim/ CD11c	+	?	?
-	-	-	+	Mat Blood DC	-	++	-

*MDLC represents an *in vivo* LC equivalent that is differentiated from monocytes in the presence of IL-4, GM-CSF and TGF-β.

The Macrophage

The syndecans, a member of the heparan sulfate proteoglycan (HSPGs) family, have been shown to serve as the major class of attachment receptors for HIV-1 on macrophages (Saphire et al., 2001). Although others have implicated the CLR CD206 in attachment (Nguyen and Hildreth, 2003). CD4 is present at far lower levels on macrophages compared to lymphocytes, yet still provides a restricting factor for infectious fusion of HIV.

Infection of macrophages with HIV has been known for sometime both *in vivo* and *in vitro* (Gartner et al., 1986a, Gartner et al., 1986b, Koenig et al., 1986, Salahuddin et al., 1986, Wiley et al., 1986). Overall R5 isolates can infect macrophages, although there are incidences where primary X4 isolates or dual X4R5 isolates have also been observed to infect (Gorry et al., 2002, Naif et al., 2002, Yi et al., 2003, Yi et al., 1998). One feature on infection strikingly different from CD4 lymphocytes is the ability to bud virus into intracellular (Gendelman et al., 1988, Pelchen-Matthews et al., 2003, Raposo et al., 2002) prior to release.

Transfer of virus from macrophages can occur independent of macrophage infection, at least in the short term. Transfer via this type of regurgitation is thought to be largely carbohydrate dependent, through the CLR CD206 and only present within the first 24 hours after the macrophage viral pulse. Although the likely long term and more efficient means of transfer is still reliant on *de novo* viral production in macrophages. In addition vesicular budding in macrophages is thought be associated with restricted and directed release of contacting CD4 lymphocytes (Carr et al., 1999, Pelchen-Matthews et al., 2003, Raposo et al., 2002).

The Immature MDDC

Viral attachment on immature DCs is predominantly mediated by C-Type lectin receptors (CLR), although CD4 binding still contributes (Turville et al., 2001) (Table I). Apparent controversy still exists as to which CLR mediates this binding. Geijtenbeek *et al*, initially supported CD209 as the sole receptor on immature DCs (Geijtenbeek et al., 2000). Yet it was clear from other receptors like CD4 and CD206 being present that the situation was a little more complex (Turville et al., 2001). The relative role of each receptor is somewhat confusing and is largely dependent on the given the assay conditions and the heterogeneity in MDDC produced, not only from each donor, but also in each laboratory. In the former point, low viral titres are routinely used to see a CD209 mediated effect, given it has the highest affinity for HIV gp120. Although to what extent these models are reflective on *in vivo* transmission is not known. Lack of effective anti CD206 blocking mAbs also compounds the issue of to what extent this CLR plays in attachment.

The immature MDDC subsets have been covered extensively with respect to HIV infection. The general consensus now is that they can be infected by R5 HIV (CCR5 using HIV isolates) on the basis that they express CD4 in conjunction with CCR5 rather than CXCR-4 (Blauvelt et al., 1997). Infection however is low and maybe related to the fact that new synthesised virus leaving the cell is consumed and degraded by neighbouring DCs (Turville et al., 2003, Turville et al., 2004)

Transfer of virus from immature MDDCs to CD4 lymphocytes was initially proposed to not involve the HIV infection pathway through the DC. Therefore investigators have largely assumed the initial virus pulsed on the DCs is the virus given to the CD4 lymphocyte. This is of course the case in the short term (ie. less than 6 hours) and when looking at the majority of papers investigating transfer, you see that the DCs are pulsed for 2 hours and then mixed with CD4 lymphocytes. Therefore the majority of virus that the DCs give to the CD4 lymphocytes is in a process of viral regurgitation and therefore the virus is from the initial inoculum given to the DC. Mac Donald and colleagues have quite effectively captured and defined the mechanisms of this regurgitation in real time (McDonald et al., 2003). Of note is that the virus is concentrated to the CD4 lymphocyte contact point and the CD4 lymphocyte aggregates its CD4 and CCR5 to the same point. Thus this process significantly increases the level of probability of a CD4/CCR5 based fusion event on the CD4 lymphocytes. Some argue that it is this catalytic mechanism which enables CD4 lymphocytes to be infected *in vivo* (Engering et al., 2002). Indeed this maybe the case when using low viral inoculums, but it is one of many cases that may occur *in vivo*. The only problem with the viral regurgitation phenomena is that it is transient, as its efficiency drops significantly over 4 to 6 hours after the immature DC has met the CD4 lymphocyte (and is totally gone after 24 hours) (Turville et al., 2004). The physiological significance of this, is that DCs take their time to get to CD4 lymphocytes. In inflamed tissue DC it maybe easy for a DC to meet a lymphocyte and transfer virus in the short term (thus supporting the correlation of viral transmission and genital inflammation or co-infections with pathogens like Herpes). Although in conditions with no inflammation, a DC may need to migrate out of tissue and towards the draining lymph node to increase its probability of interacting with an appropriate CD4 lymphocyte. Thus the problem with this is that it takes time and in viral transfer via regurgitation is unlikely if this is the means whereby a DC meets a T lymphocyte on the majority of occasions (Turville et al., 2004). The other means of viral transfer from a DC to a lymphocyte is via an infected DCs. Studies observing this are somewhat limited as it has been difficult to eliminate the contribution of this phase and the regurgitation phase of transfer (McDonald et al., 2003, Turville et al., 2004). Although infection of a DC provides a very stable and long-term means of viral transfer in the immature MDDC model (Turville et al., 2004). It also is consistent with observations of tissue explant models, which will be highlighted later in Langerhans cells (LC) and dermal DC subscts. Yet this means of viral transfer is often criticised on the basis it needs a high viral inoculum for the DC to get infected. Although whether the inoculum is high or low, the regurgitation model does not occur in the long term in an immature DC setting, whereas the DC infection model does (Turville et al., 2004).

The Mature in Vitro DC

Maturation does lead to an alteration of known HIV receptors. Of significant note is that CD206 is significantly down regulated (sometimes as much as 10 fold), yet in contrast the drop in CD209 is minimal (around 2 fold) (Turville et al., 2001) (Table I). CD4 is also upregulated during maturation, so with respect to HIV gp120 binding you see more of a CD4 dependent binding event (approximately 40% of all gp120 binding) (Turville et al., 2001). Yet the majority of HIV gp120 binding is still CLR dependent and as it appears that CD209 is the only known CLR to bind HIV that is still expressed at significant levels. Therefore one would predict that it would be the predominate receptor on this DC cell type.

Infection of mature DC is where things get a little confusing. In contrast to immature, mature DCs can not be effectively infected by R5 or X4 HIV isolates. Granelli-Piperno and colleagues initially argued that high levels of virus did enter the DC by fusion and it was halted at several steps, which were overcome by CD4 lymphocyte contact (Granelli-Piperno et al., 1998, Granelli-Piperno et al., 1999). While others have further shown blocks exist at both pre and post transcription levels in mature DCs (Bakri et al., 2001, Canque et al., 1999, Granelli-Piperno et al., 1997).

With respect to viral transfer from DCs to CD4 lymphocytes, it appears now, that in mature MDDC, the regurgitation pathway predominates and lasts for a longer time period (Turville et al., 2004). In fact, intact virus although reduced in number significantly, can be transferred efficiently after 24 to 48 hours (Turville et al., 2004). This second pathway is somewhat limited by the fact that the viability of mature DCs decreases rapidly over time (Hope personal and unpublished observations) and that there is no new virus being produced within the DC (ie. *de novo* viral production). Thus to conclude, it is the recently observed regurgitation phenomena (McDonald et al., 2003, Turville et al., 2004) which is predominantly observed in mature MDDC and that is the source for viral transfer to responding CD4 lymphocytes.

Blood DCs

Lineage negative blood DCs whether mature or freshly isolated without culture, do not express CLRs that are capable of binding HIV in a carbohydrate dependent manner. Rather HIV gp120 attachment on blood DC is CD4 dependent (Turville et al., 2001, Turville et al., 2002) (Table I).

With respect to infect, early studies by Patterson and Knight (1987) identified by electron microscopy that blood DC could be infected, as indicative by the visualisation of active budding at the cell surface of DCs (Patterson and Knight, 1987). Later studies by several groups produced contrasting results and the likely reason for this was associated with the different protocols used by each investigator for DC isolation (see (Cameron et al., 1996) for a comprehensive review of this area). Although what is now becoming more apparent is the DCs isolated in certain early studies were mature blood DCs. For example precursor blood DC are now known to be permissive to infection and it is related to the level of CD4 expression. eg. Infection of CD123 positive blood DCs is far greater than CD11c positive due to the higher levels of CD4 on the former (Turville et al., 2001, Turville et al., 2002). However when using blood DC, which are prepared with protocols involving culture periods, you are observing not what we now term precursor blood DCs, but mature blood DCs. For instance the initial work on DC viral transfer to CD4 lymphocytes by Cameron *et al* (Cameron et al., 1992b) produced mature blood DCs. Unknown at the time, their isolation procedure induced maturation of the blood DCs. Thus the initial observations by Cameron and colleagues (Cameron et al., 1992b) fits quite well with what is seen in *in vitro* mature MDDC. ie. Mature blood DC are not permissive to infection, but can transfer virus (Figure 2 and 3).

Transfer of virus to CD4 lymphocytes as is with DC infection, largely determined by the maturation state of the blood DC. Precursor blood DC can transfer virus by becoming infected and passing on *de Novo* produced virions to CD4 lymphocytes (Unpublished observations). Although care must be taken when observing viral transfer from CD123 blood DCs, as exposure to this cell type leads to high levels of interferon alpha production (Ferbas et al., 1994) and can significantly dampen HIV replication in longer term co-cultures (Poli et al., 1989). In terms of viral regurgitation like immature MDDC, little data is available to date regarding the freshly isolated blood precursor DC. Like in mature MDDC, mature blood DCs infection is limited and transfer consists largely as a regurgitation mechanism of the initial viral inoculum (Cameron et al., 1992a). To clearly show this last point Cameron *et al*, used murine blood DC and showed that when using permissive human CD4 T lymphocytes as the recipient in the viral transfer assay, the ability of murine blood DC (all murine cells are non-permissive for HIV) was equal to that of the human blood DCs (Cameron et al., 1992b). The actual mechanism of viral regurgitation has not been shown to date like the MDDC model, but is eluded to by Reece *et al* using latex beads similar in size to that of HIV (Reece et al., 2001) being regurgitated in mature blood DC subsets. Of interest, given we have more effective sorting strategies and markers to cleanly and quickly obtain them, is reinvestigating the blood DC subsets in their precursor state with respect to this mechanism.

Langerhans Cells

HIV attachment as shown by gp120 binding is largely CD4 independent, but carbohydrate dependent as shown by mannan inhibition. The receptors involved in mannan inhibitable binding are Langerin and potentially other yet identified CLRs expressed on LCs (Turville et al., 2002).

In skin and mucosa of seropositive patients, LCs were observed to harbour HIV or show viral replication (Bhoopat et al., 2001, Cimarelli et al., 1994, Dusserre et al., 1992, Giannetti et al., 1993, Rappersberger et al., 1988, Tschachler et al., 1987, Zambruno et al., 1991) (Cimarelli *et al*, (Cimarelli et al., 1994) estimates at least 3% of LC harbour HIV proviral DNA). Work by Reece *et al* (Reece et al., 1998) and Kawamura *et al*, (Kawamura et al., 2000) has highlighted *in situ* infection of LCs in skin explant cultures occurs predominantly through CD4/CCR5. Work on freshly isolated LCs has yet to be done as enzymes which enable epidermal sheet dissociation, like trypsin, cleave the first domain of CD4 used for HIV envelope attachment. Therefore as with immature MDDCs, LCs can be infected, although at a very low level.

Most of the transfer studies to date involve the use of *in situ* infection of the LC in the epidermal skin sheet and then explant skin cultures to bring the LCs out by migration. The normal time for maximal emigrant LC harvests is over 72 hours (Reece et al., 1998). Naturally as what is seen with MDDC, the regurgitation phase of viral transfer only occurs efficiently if the DCs are co-cultured with CD4 lymphocytes less than 24 hours after HIV exposure. Therefore LCs within these explant models only appear to transfer virus to CD4 lymphocytes via the *de novo* synthesis of virions. Recent studies by Kawaruma and colleagues have used attachment and fusion inhibitors to try and address the regurgitation phase. When using the CLR inhibitor mannan there is no reduction of viral transfer, whereas when using CCR5 inhibitors there is total block of not only LC infection, but also viral transfer to CD4 lymphocytes (Kawamura et al., 2003). Thus in terms of the skin explant model, it is apparent that LCs require infection for transfer to CD4 lymphocytes. Whether this is the same in freshly isolated LCs has yet to be determined.

Dermal DCs

Attachment of HIV gp120 to dermal DCs is mannan inhibitable and therefore carbohydrate dependent when dermal DCs are in their immature state (Turville et al., 2002). Mature dermal DCs can only mediate HIV gp120 binding via CD4, as CLRs are down regulated, whereas in contrast CD4 is upregulated after maturation (Turville et al., 2002). Although recent studies on cervical explants, have show that a significant portion CD209 can persist on a subset of lamina propria DC, that are also bear the maturation marker CD83 (Hu et al., 2004). The receptors responsible for binding are CD209 and CD206 on the CD14 positive dermal DC subset and CD206 only on the CD1a dermal DC subset (Turville et al., 2002).

From *in vivo* observations both in humans and macaques it is apparent that dermal/lamina propria DCs are infected (Bhoopat et al., 2001, Hu et al., 1998). The resolution as to whether the CD1a or the CD14 subset is infected has yet to be determined. Explant cultures of dermal skin sections have been infected *in situ*, but the level of infection was not determined (rather the capacity to transfer to CD4 lymphocytes). Mature dermal DCs isolated from skin explant cultures have been shown to be infected by the fact *de novo* synthesis is required for the virus to be transfer to CD4 lymphocytes (Pope et al., 1995).

To date no study has freshly isolated dermal DCs and used them in transfer assays. In parallel studies epidermal skin sections have been infected *in situ* and the LCs in conjunction with dermal DCs have been investigated in terms of viral transfer to CD4 lymphocytes. Dermal DCs as LCs could effectively transfer R5 virus to CD4 lymphocytes, although dermal DCs could also transfer X4 viruses albeit at lower efficiency (Reece et al., 1998). Therefore infection prior to transfer appears to be the likely model again in the skin explant cultures, give DCs are isolated 72 hours after viral exposure. Dermal DCs were treated in this study as one population given the potential for resolution in the DC subsets after emigration is relatively difficult given the down regulation of CD14 and CLRs which split the two subsets distinctly apart (Reece et al., 1998). Although recent studies have shown mature CD83 DC emigrants from cervical explants also have a regurgitation phase, largely dependent on CD209 attachment, in addition to a potential CD4 dependent phase *de novo* phase (Hu et al., 2004).

Incoming and Outgoing Virus in APCs: Intracellular Reservoirs for Both De Novo and Regurgitated Virus

The normal cellular vesicular pathway has been recently shown to be intimately involved with incoming and outgoing immunodeficiency virions in APC. For immunodeficiency viruses, entry into DC and macrophages and DCs after attachment has been observed to occur through either viral host cell membrane fusion through CD4/CCR5 and cytoplasmic entry, endocytosis into coated pits and/or entry into vacuolar compartments (Frank et al., 2002, Kwon et al., 2002, Turville et al., 2004). In terms of exiting the DC, this can occur via vesicular regurgitation to the cell surface of the original viral inoculum the APC has seen or via *de novo* viral production and viral exit via budding. Although with macrophages HIV viral budding does not occur at the plasma membrane surface like it does in CD4 lymphocytes (see early works by (Gendelman et al., 1988)). Rather virions are accumulated into vacuolar compartments where they then exit the cell to infect neighbouring cell types (Gendelman et al., 1988, Pelchen-Matthews et al., 2003, Raposo et al., 2002). Such observations, although only recently observed in macrophages likely represent what may be happening in DCs as well.

Although what vesicular compartments are the virions going to in each DC subset? Like scenario with DC infection, subsets and maturation status play an influential role with incoming virions. Although to date there have been only a few studies which have specifically observed this process in only MDDC and LCs. In the later case of LCs, several studies using gp120 have looked at the early internalisation events by electron microscopy. In LCs internalisation of HIV gp120 is linked to the formation of birbeck granules at the cells surface (Dezutter-Dambuyant, 1995). Birbeck granules are structurally pockets of two limiting membranes, separated by leaflets with periodic striations and are induced to form by the presence of the CLR langerin/CD207 (Valladeau et al., 2003, Valladeau et al., 2000). Recent studies by Mc Dermott and colleagues (Mc Dermott et al., 2002) suggest that birbeck granules are subdomains of the endosomal recycling compartment where Langerin has accumulated. Therefore is likely that HIV initially is loaded into an early recycling endosome in LCs via Langerin. In immature and mature MDDCs several investigations have used chemically inactivated HIV/SIV or VPR-GFP fluorescent virion constructs (Frank et al., 2002, Kwon et al., 2002, McDonald et al., 2003, Turville et al., 2004). In immature MDDCs it is clear that early viral internalisation (less than 15 minutes) is through the transferin positive early endosomal pathway (Kwon et al., 2002). Later trafficking (greater than 45 minutes) results in significant loss of virus, but intact virions persist in a non-lysosomal, small peripheral vesicular compartments just under the membrane (Frank et al., 2002, Kwon et al., 2002, Turville et al., 2004). Mature MDDC appear to be using quite a different mechanism, as they appear to actively load incoming virions into deeper and quite large vesicular compartments. These vesicles lack early endosomal and lysosomal markers, but contain late endosomal/ MVB markers like CD81 and CD63 ((Frank et al., 2002, Turville et al., 2004) and Hope T personal communication).

In terms of outgoing virus, CD4 lymphocyte (or other cells for that matter; see (McDonald et al., 2003)) contact directed viral regurgitation of the initial inoculum is seen in both immature and mature MDDC (McDonald et al., 2003, Turville et al., 2004). Outgoing *de novo* produced virus in macrophages (and likely DCs) is also contact driven release of vesicular budding virus (Carr et al., 1999). Although contact zones between DC and lymphocytes are also seen as a very active site of polarised viral budding at the cell surface (Hladik et al., 1999). Recently several investigators have attempted to identify not only the vesicular compartment which contains the actively budding virus but also, at protein level, of how this process occurs (Pelchen-Matthews et al., 2003, Raposo et al., 2002, von Schwedler et al., 2003). What is becoming apparent from recent studies, it that the virus appears to be using the exosomal synthesis and release pathway within the MVB. Most of the work has been currently focussed on macrophages *in vitro*, but with the efficient ability that DCs generate exosomes, it would be no surprise as to whether the virus uses the same pathway in DCs as recently observed in macrophages. Thus outgoing virus appears to be regulated by the MVB and the key to understanding what is potentially going on lies with the basic biology of this cellular compartment.

Biological Mechanisms Responsible for Outgoing Viruses

The Multivesicular Body

One feature that is both common to the above mentioned mature DC subsets, and to somewhat of a degree macrophages, is the intracellular reservoirs of virus produced *de novo* or sequestered from the initial viral inoculum, use the MVB compartment for short term lodging prior re-release to extracellular environment. Although what defines the MVB? Within the early vesicular transport system from the plasma membrane are the vesicles now known as the early endosome and the late endosome. The early endosomes are derived from the plasma membrane and typically contain proteins like EEA-1 and receptors, which are cycling between vesicles and the plasma membrane (recycling receptors). Late endosomes appear to be derived from early endosomes (Stoorvogel et al., 1991) in a process referred to as vesicular maturation. Structurally late, compared to early endosomes, increase in size and have vesicles inwardly budding into them from their limiting membrane. It is this later feature which has coined the other name for the late endosome as the MVB. Thus MVB are simply large vesicles matured from early endosomes and contain several inwardly budding internal vesicles. So what is their function? MVB can undertake a maintenance role by removing targeted receptors by fusing themselves with lysosomes (Futter et al., 1996, Mullock et al., 1998). The second major role is seen as a storage target. For instance in immature DCs, there is a need to have a high capacity to store MHC-II, which they do within the MVB (Kleijmeer et al., 2001). Then when immature DCs meet pathogens, or other stimuli that induces maturation, the MHC-II vesicles fuse with the MVB limiting membrane resulting in a beautiful tubular like structure. This has been recently demonstrated in EGFP-MHC-II transgenic mice with the tubular structure stretching to the plasma membrane and thus permitting contact with neighbouring lymphocytes (Boes et al., 2002). The third and final role of MVB is seen when the limiting membrane fuses with the plasma membrane. When this happens, the internal vesicles can escape into the extracellular environment. These later, now extracellular vesicles are now known as exosomes and warrant further discussion in detail as they structurally and biologically share many features with immunodeficiency virions.

Exosomes

An interesting and common feature with exosomes and immunodeficiency virions is not only their mechanism of synthesis, but also their lipid and protein composition. Recent observations have highlighted this with respect to virus budding into the MVB and will be discussed here in detail. Although what are exosomes? As highlighted in the definition of the MVB, this compartment is responsible for the production of exosomes and the release of these vesicles into the extracellular environment. Structurally exosomes range from 40 to 100nm is size and are enriched for specific proteins like Tsg101, tetraspanins (CD9, CD63, CD81 and CD82), MHC-II and MHC-I (see (Stoorvogel et al., 2002) for review). They are also enriched in lipid composition for cholesterol and sphingomyelin. Their function depends on the cell type secreting them. Exosomes from cytotoxic CD8 lymphocytes containing perforin and granzymes (Peters et al., 1991)) and thus help mediate the cytolytic immune responses, whereas maturing reticulocytes shed their obsolete receptors like CD71 through this pathway (Johnstone et al., 1991). In DCs, exosomes are enriched for MHC-II and can mediate MHC-II transfer between DCs and/or provide additional MHC-II/peptide complexes and augment CD4 lymphocyte stimulation when other DCs are present (Thery et al, 2002). Exosomes can also stimulate CD4 or CD8 lymphocytes in a MHC-II or MHC-I manner, but at levels at least 10 to 20 fold lower than compared to when whole cells are present (Raposo et al., 1996). Follicular DCs do not secrete exosomes or synthesise MHC-II, but are known to acquire peptide loaded MHC-II from donour cells (Gray et al., 1991). Given follicular DCs also bear numerous exosomes on their surface with MHC-II it is likely that exosomal capture is the means of their MHC-II acquisition. Transfer of surface receptors is not limited to MHC-II, as CCR5 has been observed by Mack and colleagues (Mack et al., 2000) to be transferred from transfected cell lines through microparticles/exosomes to PBMCs which lack CCR5 and this permitting R5 HIV isolate entry into these once non-viral permissive cell types. Thus with respect to APCs, exosomes provide directed fluid receptor transfer and augmentation in a MHC-II and MHC-I mediated immune response (see (Denzer et al., 2000) for further review).

Exosomes Formation and Viral Budding

What is amazing and has received little attention until late is the similarities exosomes have with immunodeficiency virions (see (Gould et al., 2003) for a recent review). Firstly look at sites and means of exosomal and viral production. It is now established that HIV in macrophages buds into the MVB where exosome precursors also form (Denzer et al., 2000, Pelchen-Matthews et al., 2003, Raposo et al., 2002). Virions in T lymphocytes bud predominantly at the cell surface and maybe akin to exosomal formation at the plasma membrane at endosomal regions at the cell surface (called intermediate exosome biogenesis) (Gould et al., 2003). Exosomal production is also shut down in mature DCs (Thery et al., 1999) and so to is viral production seen in mature DCs (although there are pre and post transriptional blocks as mentioned in the preceding sections, the later block relevant when compared to shutting down exosomal synthesis). The synthesis machinery for exosomes and immunodeficiency virions provide further similarities. As intraluminal vesicles in the MVB serve as precursors for exosomes, the process of MVB formation is key to understanding their production as well as immunodeficiency virions. The majority of the work on MVB formation has been carried out in yeast and has led to the discovery of a network of Vacuolar Protein Sorting (vps) proteins. The class E vps mutants belong to a class of 17 protein members, which form class E compartments in yeast (Katzmann et al., 2002, Raiborg et al., 2003). Class E compartments in yeast are structurally characterised as multilameller compartments, abnormally enlarged as they cannot form intraluminal vesicles (Katzmann et al., 2001). The role of vps proteins is to be recruited from the cytoplasm and sequentially form complexes known as Endosomal Sorting Complexes Required for Transport (ESCRTs), for formation of vesicular bodies at the site of the MVB (Babst et al., 2002a, Babst et al., 2002b, Babst et al., 2000, Bache et al., 2003, Katzmann et al., 2003, Odorizzi et al., 2003).

Studies largely confined to the gag region of HIV (as gag alone is sufficient to form virus-like particles (Accola et al., 2000) have also shown how retroviruses can also use the same vesicular apparatus that is used to form exosomes. Indeed the mutation of several different human class E vps homologs, including TSG101, vps28, CHMP2A, CHMP3, CHMP4B, CHMP4C, vps4A, vps4B and AIP1 (Garrus et al., 2001, Martin-Serrano et al., 2003a, Martin-Serrano et al., 2001, Martin-Serrano et al., 2003b, Strack et al., 2003, von Schwedler et al., 2003) all result in arresting of HIV budding at the late stage. That is, virions appear to undergo normal budding, but fail to complete the process of "pinching" off from the limiting membrane and thus resemble round stalk like structures similar to the shape of tennis rackets (see (Strack et al., 2003, von Schwedler et al., 2003) for examples of electron micrographs). Therefore the HIV gag protein uses the exosomal/vesicle machinery in the MVB and results in HIV budding akin to that of a normal exosome.

Structural and Biological Similarities with Exosomes and Immunodeficiency Virions

Given similar synthesis machinery and location off virions and exosomes in APC, it is no surprise that they share common biological behaviour and composition. For instance the lipid composition of exosomes and HIV are enriched for cholesterol and sphingomyelin (Ono and Freed, 2001, Stoorvogel et al., 2002). Host derived cell proteins sorted into this lipid "patches," or "rafts," are also acquired by budding virions and exosomes. Of note the receptors CD55 and CD59 are membrane regulators of complement, are selectively acquired by exosomes and HIV (Clayton et al., 2003, Saifuddin et al., 1997) and interestingly acquisition which imparts the biological phenotype of complement resistance. Follicular DCs (FDC) also share a large capacity to bind and harbour exosomes derived from other cell types and bind and harbour (Denzer et al., 2000, Schmitz et al., 1994). The mechanism whereby exosomes are attached to FDC is largely unknown, although HIV is known to attach via an immune complex pathway and a CD54/ICAM-1 and CD11a/LFA1 interaction (Fujiwara et al., 1999). Therefore given acquisition of CD54 by exosomes, it is likely that their binding mechanism is similar to that seen on FDC with HIV.

Exosomal and Viral Transfer by "Backfusion": Potential Role for Tetraspanins

Transfer of significant amounts of transmembrane proteins, first seen by the action of exosomes, is also a biological phenotype seen in immunodeficiency virions. When tracking transfer of AT2 chemically inactivated virus from DC to CD4 lymphocytes, large segments of SIV envelope positive membrane could be seen transferred readily between DCs and CD4 lymphocytes and later between the lymphocytes by themselves ((Turville et al., 2004) and unpublished observations). Observations like these are similar to those seen when significant amounts of MHC-II is transferred between MHC-II positive and negative DCs (Thery et al., 2002), between exosomes and FDC and in studies observing CCR5 transfer from CCR5 transfectants to wild type CCR5 negative lymphocytes (Mack et al., 2000). How exosomes achieve this is only recently been address in a mechanism known as "back-fusion". For instance the majority of MHC-II resides in immature MDDC as internal vesicles (Murk et al., 2003). Therefore to come to the cell surface via the MHC-II tubules seen in EGFP transgenic mice DC, the MHC-II must switch from the internal vesicles to the outer MVB membrane. Therefore the only mechanism whereby MHC II can do this is back fusing from the internal free vesicles to the external MVB membrane. Selective accumulation of a family of tetraspanins may provide a mechanistic clue into how, not only exosomes may "back-fuse," but also how virions may do so independent of viral envelope. The recent review by Gould and colleagues highlights many instances whereby retroviruses may infected cells, albeit inefficiently, independently of their envelope glycoprotein. To further this discussion, it appears the tetraspanin family may not only be partially responsible for this mechanism in exosomes, but also in several cases where viral infection is occurring. Indeed FIV spread (budding and cell to cell transfer) is inhibited by antibodies to the tetraspanin CD9 (CD9 is also implicated in sperm-egg fusion (Miyado et al., 2000)). Hepatitis C is also inhibited by antibodies and specific depletion of the tetraspanin CD81 by siRNAs (Zhang et al., 2004). Recently the tetraspanin CD63 has been demonstrated to block HIV infection specifically in macrophages (von Lindern et al., 2003). Although it is unclear as to the mechanism by which CD63 antibodies do this. The cell to cell spread of virus maybe a mechanism which may need the use of tetraspanins like CD63. Indeed studies on both FIV and HTLV have observed the need for CD9 and CD82 respectively, for viral cell to cell transfer (de Parseval et al., 1997, Pique et al., 2000). Thus in the specific case of blocking the tetraspanin CD63 in macrophages, this may interfere with release of virus at several levels. Firstly it may inhibit the release of virus from the MVB as seen with FIV. This may occur if the predominate means of viral transfer from an infected cell to another contacting cell type involves prior back fusion in the MVB and transfer to the plasma membrane akin to MHC-II tubule formation (Boes et al., 2002, Chow et al., 2002). Studies using the monocytoid cell line U937, may have provided the first observed precedent for this, as U937s contacting epithelial cells not only project microvilli in the cell to cell contact zone but also have HIV budding from such microvillic structures (Bourinbaiar and Phillips, 1991). Secondly the inhibition maybe at the level of infected macrophages fusing to form multinucleated giant cells and CD63 maybe key in this cell to cell fusion process, as other tetraspanins are in syncytia formation (Pique et al., 2000). Although it is likely CD63 promotes both postulated viral transfer pathways: ie. may be essential for giant cell syncytia formation and cell to cell transfer.

Conclusions

The continued investigations with respect to immunodeficiency viruses and the myeloid APC lineages indeed warrants future studies. Although the role CD4 lymphocytes is important, the impact of APC must not be underestimated. As before mentioned the virus does not forget the role of the APC and its importance in its pathogenesis. Indeed if it had, it would have lost its long term memory of the APC by purging those genes now seen important for its successful interaction with this cell type.

A wealth of knowledge has been forged of late, with particular emphasis on DC and immunodeficiency viruses. The confusing nature and the apparent controversies in such a field are slowly becoming clarified and the extent and complexity with which the virus interacts with this cell type being often astounding. Although care must be taken when trying to throw blanket theories over the interactions of DCs and immunodeficiency viruses. In doing so you may appear to simplify an area for easy digestion, although in reality subtleties may be overlooked and actual be incredibly important for the design of a therapeutic drugs or microbicides. The controversy of whether DCs are infected or not infected is one example where people may or may not take into account de novo viral production as a source of virus that DCs can transfer to lymphoid tissue. In addition the recent focus on CD209 may indeed be warranted, but also must be taken into a broader perspective. For instance studies purely focussed on this molecule in microbicide development maybe missing the crucial role of CD4 (and other lectins and/or attachment molecules for that matter) (see (Davis and Doms, 2004, Hu et al., 2004) for a recent example). With respect to the other C-Type Lectins, CD206, CD207 and CD209L, we know that they have a remarkable capacity to bind viral envelope and virions. They may or may not fit in the paradigm of the lectin CD209, where they bind and allow viral transfer to CD4 lymphocytes. If they do not handle virions like CD209, research on why two receptors handle ligands differently may yield some fascinating observations. Even if they do bind large quantities of virus and then degrade it, it would be a promising avenue of investigation to look out how to enhance this process and thus provide a more natural mechanism to reduce viral loads in vivo. In terms of a microbicide, fierce competition by these lectins and subsequent trafficking for degradation may be enough to further reduce the probability of those few viral particles successfully disseminating in peripheral tissue.

References

Accola, M. A., Strack, B. and Gottlinger, H. G. (2000). Efficient particle production by minimal Gag constructs which retain the carboxy-terminal domain of human immunodeficiency virus type 1 capsid-p2 and a late assembly domain. *J Virol* 74, 5395-402.

Babst, M., Katzmann, D. J., Estepa-Sabal, E. J., Meerloo, T. and Emr, S. D. (2002a). Escrt-III: an endosome-associated heterooligomeric protein complex required for mvb sorting. *Dev Cell* 3, 271-82.

Babst, M., Katzmann, D. J., Snyder, W. B., Wendland, B. and Emr, S. D. (2002b). Endosome-associated complex, ESCRT-II, recruits transport machinery for protein sorting at the multivesicular body. *Dev Cell* 3, 283-9.

Babst, M., Odorizzi, G., Estepa, E. J. and Emr, S. D. (2000). Mammalian tumor susceptibility gene 101 (TSG101) and the yeast homologue, Vps23p, both function in late endosomal trafficking. *Traffic* 1, 248-58.

Bache, K. G., Brech, A., Mehlum, A. and Stenmark, H. (2003). Hrs regulates multivesicular body formation via ESCRT recruitment to endosomes. *J Cell Biol* 162, 435-42.

Bakri, Y., Schiffer, C., Zennou, V., Charneau, P., Kahn, E., Benjouad, A., Gluckman, J. C. and Canque, B. (2001). The maturation of dendritic cells results in postintegration inhibition of HIV-1 replication. *J Immunol* 166, 3780-8.

Bhoopat, L., Eiangleng, L., Rugpao, S., Frankel, S. S., Weissman, D., Lekawanvijit, S., Petchjom, S., Thorner, P. and Bhoopat, T. (2001). In vivo identification of Langerhans and related dendritic cells infected with HIV-1 subtype E in vaginal mucosa of asymptomatic patients. *Mod Pathol* 14, 1263-9.

Bingen, A., Nonnenmacher, H., Bastien-Valle, M. and Martin, J. P. (2002). Tissues rich in macrophagic cells are the major sites of feline immunodeficiency virus uptake after intravenous inoculation into cats. *Microbes Infect* 4, 795-803.

Blauvelt, A., Asada, H., Saville, M. W., Klaus-Kovtun, V., Altman, D. J., Yarchoan, R. and Katz, S. I. (1997). Productive infection of dendritic cells by HIV-1 and their ability to capture virus are mediated through separate pathways. *J Clin Invest* 100, 2043-53.

Boes, M., Cerny, J., Massol, R., Op den Brouw, M., Kirchhausen, T., Chen, J. and Ploegh, H. L. (2002). T-cell engagement of dendritic cells rapidly rearranges MHC class II transport. *Nature* 418, 983-8.

Bourinbaiar, A. S. and Phillips, D. M. (1991). Transmission of human immunodeficiency virus from monocytes to epithelia. *J Acquir Immune Defic Syndr* 4, 56-63.

Brodie, S. J., Pearson, L. D., Zink, M. C., Bickle, H. M., Anderson, B. C., Marcom, K. A. and DeMartini, J. C. (1995). Ovine lentivirus expression and disease. Virus replication, but not entry, is restricted to macrophages of specific tissues. *Am J Pathol* 146, 250-63.

Cameron, P., Pope, M., Granelli-Piperno, A. and Steinman, R. M. (1996). Dendritic cells and the replication of HIV-1. *J Leukoc Biol* 59, 158-71.

Cameron, P. U., Forsum, U., Teppler, H., Granelli-Piperno, A. and Steinman, R. M. (1992a). During HIV-1 infection most blood dendritic cells are not productively infected and can induce allogeneic CD4+ T cells clonal expansion. *Clin Exp Immunol* 88, 226-36.

Cameron, P. U., Freudenthal, P. S., Barker, J. M., Gezelter, S., Inaba, K. and Steinman, R. M. (1992b). Dendritic cells exposed to human immunodeficiency virus type-1 transmit a vigorous cytopathic infection to CD4+ T cells. *Science* 257, 383-7.

Canque, B., Bakri, Y., Camus, S., Yagello, M., Benjouad, A. and Gluckman, J. C. (1999). The susceptibility to X4 and R5 human immunodeficiency virus-1 strains of dendritic cells derived in vitro from CD34(+) hematopoietic progenitor cells is primarily determined by their maturation stage. *Blood* 93, 3866-75.

Carr, J. M., Hocking, H., Li, P. and Burrell, C. J. (1999). Rapid and efficient cell-to-cell transmission of human immunodeficiency virus infection from monocyte-derived macrophages to peripheral blood lymphocytes. *Virology* 265, 319-29.

Chow, A., Toomre, D., Garrett, W. and Mellman, I. (2002). Dendritic cell maturation triggers retrograde MHC class II transport from lysosomes to the plasma membrane. *Nature* 418, 988-94.

Cimarelli, A., Zambruno, G., Marconi, A., Girolomoni, G., Bertazzoni, U. and Giannetti, A. (1994). Quantitation by competitive PCR of HIV-1 proviral DNA in epidermal Langerhans cells of HIV-infected patients. *J Acquir Immune Defic Syndr* 7, 230-5.

Clayton, A., Harris, C. L., Court, J., Mason, M. D. and Morgan, B. P. (2003). Antigen-presenting cell exosomes are protected from complement-mediated lysis by expression of CD55 and CD59. *Eur J Immunol* 33, 522-31.

Dauer, M., Obermaier, B., Herten, J., Haerle, C., Pohl, K., Rothenfusser, S., Schnurr, M., Endres, S. and Eigler, A. (2003). Mature dendritic cells derived from human monocytes within 48 hours: a novel strategy for dendritic cell differentiation from blood precursors. *J Immunol* 170, 4069-76.

Davis, C. W. and Doms, R. W. (2004). HIV Transmission: Closing All the Doors. *J Exp Med* 199, 1037-40. Epub 2004 Apr 12.

de Parseval, A., Lerner, D. L., Borrow, P., Willett, B. J. and Elder, J. H. (1997). Blocking of feline immunodeficiency virus infection by a monoclonal antibody to CD9 is via inhibition of virus release rather than interference with receptor binding. *J Virol* 71, 5742-9.

Denzer, K., Kleijmeer, M. J., Heijnen, H. F., Stoorvogel, W. and Geuze, H. J. (2000). Exosome: from internal vesicle of the multivesicular body to intercellular signaling device. *J Cell Sci* 113, 3365-74.

Dezutter-Dambuyant, C. (1995). In vivo and in vitro infection of human Langerhans cells by HIV-1. *Adv Exp Med Biol* 378, 447-51.

Dusserre, N., Dezutter-Dambuyant, C., Mallet, F., Delorme, P., Philit, F., Ebersold, A., Desgranges, C., Thivolet, J. and Schmitt, D. (1992). In vitro HIV-1 entry and replication in Langerhans cells may clarify the HIV-1 genome detection by PCR in epidermis of seropositive patients. *J Invest Dermatol* 99, 99S-102S.

Engering, A., Van Vliet, S. J., Geijtenbeek, T. B. and Van Kooyk, Y. (2002). Subset of DC-SIGN(+) dendritic cells in human blood transmits HIV-1 to T lymphocytes. *Blood* 100, 1780-6.

Ferbas, J. J., Toso, J. F., Logar, A. J., Navratil, J. S. and Rinaldo, C. R., Jr. (1994). CD4+ blood dendritic cells are potent producers of IFN-alpha in response to in vitro HIV-1 infection. *J Immunol* 152, 4649-62.

Frank, I., Piatak, M., Jr., Stoessel, H., Romani, N., Bonnyay, D., Lifson, J. D. and Pope, M. (2002). Infectious and whole inactivated simian immunodeficiency viruses interact similarly with primate dendritic cells (DCs): differential intracellular fate of virions in mature and immature DCs. *J Virol* 76, 2936-51.

Fujiwara, M., Tsunoda, R., Shigeta, S., Yokota, T. and Baba, M. (1999). Human follicular dendritic cells remain uninfected and capture human immunodeficiency virus type 1 through CD54-CD11a interaction. *J Virol* 73, 3603-7.

Futter, C. E., Pearse, A., Hewlett, L. J. and Hopkins, C. R. (1996). Multivesicular endosomes containing internalized EGF-EGF receptor complexes mature and then fuse directly with lysosomes. *J Cell Biol* 132, 1011-23.

Garrus, J. E., von Schwedler, U. K., Pornillos, O. W., Morham, S. G., Zavitz, K. H., Wang, H. E., Wettstein, D. A., Stray, K. M., Cote, M., Rich, R. L., Myszka, D. G. and Sundquist, W. I. (2001). Tsg101 and the vacuolar protein sorting pathway are essential for HIV-1 budding. *Cell* 107, 55-65.

Gartner, S., Markovits, P., Markovitz, D. M., Betts, R. F. and Popovic, M. (1986a). Virus isolation from and identification of HTLV-III/LAV-producing cells in brain tissue from a patient with AIDS. *Jama* 256, 2365-71.

Gartner, S., Markovits, P., Markovitz, D. M., Kaplan, M. H., Gallo, R. C. and Popovic, M. (1986b). The role of mononuclear phagocytes in HTLV-III/LAV infection. *Science* 233, 215-9.

Geijtenbeek, T. B., Kwon, D. S., Torensma, R., van Vliet, S. J., van Duijnhoven, G. C., Middel, J., Cornelissen, I. L., Nottet, H. S., KewalRamani, V. N., Littman, D. R., Figdor, C. G. and van Kooyk, Y. (2000). DC-SIGN, a dendritic cell-specific HIV-1-binding protein that enhances trans-infection of T cells. *Cell* 100, 587-97.

Gendelman, H. E., Narayan, O., Kennedy-Stoskopf, S., Kennedy, P. G., Ghotbi, Z., Clements, J. E., Stanley, J. and Pezeshkpour, G. (1986). Tropism of sheep lentiviruses for monocytes: susceptibility to infection and virus gene expression increase during maturation of monocytes to macrophages. *J Virol* 58, 67-74.

Gendelman, H. E., Orenstein, J. M., Baca, L. M., Weiser, B., Burger, H., Kalter, D. C. and Meltzer, M. S. (1989). The macrophage in the persistence and pathogenesis of HIV infection. *Aids* 3, 475-95.

Gendelman, H. E., Orenstein, J. M., Martin, M. A., Ferrua, C., Mitra, R., Phipps, T., Wahl, L. A., Lane, H. C., Fauci, A. S., Burke, D. S. and et al. (1988). Efficient isolation and propagation of human immunodeficiency virus on recombinant colony-stimulating factor 1-treated monocytes. *J Exp Med* 167, 1428-41.

Giannetti, A., Zambruno, G., Cimarelli, A., Marconi, A., Negroni, M., Girolomoni, G. and Bertazzoni, U. (1993). Direct detection of HIV-1 RNA in epidermal Langerhans cells of HIV-infected patients. *J Acquir Immune Defic Syndr* 6, 329-33.

Gorry, P. R., Zhang, C., Wu, S., Kunstman, K., Trachtenberg, E., Phair, J., Wolinsky, S. and Gabuzda, D. (2002). Persistence of dual-tropic HIV-1 in an individual homozygous for the CCR5 Delta 32 allele. *Lancet* 359, 1832-4.

Gould, S. J., Booth, A. M. and Hildreth, J. E. (2003). The Trojan exosome hypothesis. *Proc Natl Acad Sci U S A* 100, 10592-7. Epub 2003 Aug 28.

Granelli-Piperno, A., Chen, D., Moser, B. and Steinman, R. M. (1997). The HIV-1 life cycle is blocked at two different points in mature dendritic cells. *Adv Exp Med Biol* 417, 415-9.

Granelli-Piperno, A., Delgado, E., Finkel, V., Paxton, W. and Steinman, R. M. (1998). Immature dendritic cells selectively replicate macrophagetropic (M-tropic) human immunodeficiency virus type 1, while mature cells efficiently transmit both M- and T-tropic virus to T cells. *J Virol* 72, 2733-7.

Granelli-Piperno, A., Finkel, V., Delgado, E. and Steinman, R. M. (1999). Virus replication begins in dendritic cells during the transmission of HIV-1 from mature dendritic cells to T cells. *Curr Biol* 9, 21-9.

Gray, D., Kosco, M. and Stockinger, B. (1991). Novel pathways of antigen presentation for the maintenance of memory. *Int Immunol* 3, 141-8.

Harrold, S. M., Cook, S. J., Cook, R. F., Rushlow, K. E., Issel, C. J. and Montelaro, R. C. (2000). Tissue sites of persistent infection and active replication of equine infectious anemia virus during acute disease and asymptomatic infection in experimentally infected equids. *J Virol* 74, 3112-21.

Hart, D. N. (1997). Dendritic cells: unique leukocyte populations which control the primary immune response. *Blood* 90, 3245-87.

Hirsch, V. M., Zack, P. M., Vogel, A. P. and Johnson, P. R. (1991). Simian immunodeficiency virus infection of macaques: end-stage disease is characterized by widespread distribution of proviral DNA in tissues. *J Infect Dis* 163, 976-88.

Hladik, F., Lentz, G., Akridge, R. E., Peterson, G., Kelley, H., McElroy, A. and McElrath, M. J. (1999). Dendritic cell-T-cell interactions support coreceptor-independent human immunodeficiency virus type 1 transmission in the human genital tract. *J Virol* 73, 5833-42.

Hu, J., Pope, M., Brown, C., O'Doherty, U. and Miller, C. J. (1998). Immunophenotypic characterization of simian immunodeficiency virus-infected dendritic cells in cervix, vagina, and draining lymph nodes of rhesus monkeys. *Lab Invest* 78, 435-51.

Hu, Q., Frank, I., Williams, V., Santos, J. J., Watts, P., Griffin, G. E., Moore, J. P., Pope, M. and Shattock, R. J. (2004). Blockade of Attachment and Fusion Receptors Inhibits HIV-1 Infection of Human Cervical Tissue. *J Exp Med* 199, 1065-75. Epub 2004 Apr 12.

Johnstone, R. M., Mathew, A., Mason, A. B. and Teng, K. (1991). Exosome formation during maturation of mammalian and avian reticulocytes: evidence that exosome release is a major route for externalization of obsolete membrane proteins. *J Cell Physiol* 147, 27-36.

Katzmann, D. J., Babst, M. and Emr, S. D. (2001). Ubiquitin-dependent sorting into the multivesicular body pathway requires the function of a conserved endosomal protein sorting complex, ESCRT-I. *Cell* 106, 145-55.

Katzmann, D. J., Odorizzi, G. and Emr, S. D. (2002). Receptor downregulation and multivesicular-body sorting. *Nat Rev Mol Cell Biol* 3, 893-905.

Katzmann, D. J., Stefan, C. J., Babst, M. and Emr, S. D. (2003). Vps27 recruits ESCRT machinery to endosomes during MVB sorting. *J Cell Biol* 162, 413-23.

Kawamura, T., Cohen, S. S., Borris, D. L., Aquilino, E. A., Glushakova, S., Margolis, L. B., Orenstein, J. M., Offord, R. E., Neurath, A. R. and Blauvelt, A. (2000). Candidate microbicides block HIV-1 infection of human immature langerhans cells within epithelial tissue explants. *J Exp Med* 192, 1491-500.

Kawamura, T., Gulden, F. O., Sugaya, M., McNamara, D. T., Borris, D. L., Lederman, M. M., Orenstein, J. M., Zimmerman, P. A. and Blauvelt, A. (2003). R5 HIV productively infects Langerhans cells, and infection levels are regulated by compound CCR5 polymorphisms. *Proc Natl Acad Sci U S A* 100, 8401-6.

Kleijmeer, M., Ramm, G., Schuurhuis, D., Griffith, J., Rescigno, M., Ricciardi-Castagnoli, P., Rudensky, A. Y., Ossendorp, F., Melief, C. J., Stoorvogel, W. and Geuze, H. J. (2001). Reorganization of multivesicular bodies regulates MHC class II antigen presentation by dendritic cells. *J Cell Biol* 155, 53-63.

Koenig, S., Gendelman, H. E., Orenstein, J. M., Dal Canto, M. C., Pezeshkpour, G. H., Yungbluth, M., Janotta, F., Aksamit, A., Martin, M. A. and Fauci, A. S. (1986). Detection of AIDS virus in macrophages in brain tissue from AIDS patients with encephalopathy. *Science* 233, 1089-93.

Kwon, D. S., Gregorio, G., Bitton, N., Hendrickson, W. A. and Littman, D. R. (2002). DC-SIGN-mediated internalization of HIV is required for trans-enhancement of T cell infection. *Immunity* 16, 135-44.

Lukas, M., Stossel, H., Hefel, L., Imamura, S., Fritsch, P., Sepp, N. T., Schuler, G. and Romani, N. (1996). Human cutaneous dendritic cells migrate through dermal lymphatic vessels in a skin organ culture model. *J Invest Dermatol* 106, 1293-9.

MacDonald, K. P. A., Munster, D. J., Clark, G. J., Dzionek, A., Schmitz, J. and Hart, D. N. J. (2002a). Peripheral blood dendritic cell heterogeneity. *Blood* In press.

MacDonald, K. P. A., Munster, D. J., Clark, G. J., Vuckovic, S. and Hart, D. N. J. (2002b). Peripheral blood dendritic cell subset analysis. In *Leucocyte Typing VII*. Edited by D. Mason. Oxford: Oxford University Press.

Mack, M., Kleinschmidt, A., Bruhl, H., Klier, C., Nelson, P. J., Cihak, J., Plachy, J., Stangassinger, M., Erfle, V. and Schlondorff, D. (2000). Transfer of the chemokine receptor CCR5 between cells by membrane-derived microparticles: a mechanism for cellular human immunodeficiency virus 1 infection. *Nat Med* 6, 769-75.

Martin-Serrano, J., Yarovoy, A., Perez-Caballero, D. and Bieniasz, P. D. (2003a). Divergent retroviral late-budding domains recruit vacuolar protein sorting factors by using alternative adaptor proteins. *Proc Natl Acad Sci U S A* 100, 12414-9. Epub 2003 Sep 30.

Martin-Serrano, J., Zang, T. and Bieniasz, P. D. (2001). HIV-1 and Ebola virus encode small peptide motifs that recruit Tsg101 to sites of particle assembly to facilitate egress. *Nat Med* 7, 1313-9.

Martin-Serrano, J., Zang, T. and Bieniasz, P. D. (2003b). Role of ESCRT-I in retroviral budding. *J Virol* 77, 4794-804.

Mc Dermott, R., Ziylan, U., Spehner, D., Bausinger, H., Lipsker, D., Mommaas, M., Cazenave, J. P., Raposo, G., Goud, B., de la Salle, H., Salamero, J. and Hanau, D. (2002). Birbeck granules are subdomains of endosomal recycling compartment in human epidermal Langerhans cells, which form where Langerin accumulates. *Mol Biol Cell* 13, 317-35.

McDonald, D., Wu, L., Bohks, S. M., KewalRamani, V. N., Unutmaz, D. and Hope, T. J. (2003). Recruitment of HIV and its receptors to dendritic cell-T cell junctions. *Science* 300, 1295-7.

McLellan, A. D., Heiser, A., Sorg, R. V., Fearnley, D. B. and Hart, D. N. (1998). Dermal dendritic cells associated with T lymphocytes in normal human skin display an activated phenotype. *J Invest Dermatol* 111, 841-9.

Mitsui, H., Watanabe, T., Saeki, H., Mori, K., Fujita, H., Tada, Y., Asahina, A., Nakamura, K. and Tamaki, K. (2004). Differential expression and function of Toll-like receptors in Langerhans cells: comparison with splenic dendritic cells. *J Invest Dermatol* 122, 95-102.

Miyado, K., Yamada, G., Yamada, S., Hasuwa, H., Nakamura, Y., Ryu, F., Suzuki, K., Kosai, K., Inoue, K., Ogura, A., Okabe, M. and Mekada, E. (2000). Requirement of CD9 on the egg plasma membrane for fertilization. *Science* 287, 321-4.

Mullock, B. M., Bright, N. A., Fearon, C. W., Gray, S. R. and Luzio, J. P. (1998). Fusion of lysosomes with late endosomes produces a hybrid organelle of intermediate density and is NSF dependent. *J Cell Biol* 140, 591-601.

Murk, J. L., Humbel, B. M., Ziese, U., Griffith, J. M., Posthuma, G., Slot, J. W., Koster, A. J., Verkleij, A. J., Geuze, H. J. and Kleijmeer, M. J. (2003). Endosomal compartmentalization in three dimensions: implications for membrane fusion. *Proc Natl Acad Sci U S A* 100, 13332-7. Epub 2003 Nov 3.

Naif, H. M., Cunningham, A. L., Alali, M., Li, S., Nasr, N., Buhler, M. M., Schols, D., de Clercq, E. and Stewart, G. (2002). A human immunodeficiency virus type 1 isolate from an infected person homozygous for CCR5Delta32 exhibits dual tropism by infecting macrophages and MT2 cells via CXCR4. *J Virol* 76, 3114-24.

Narayan, O., Wolinsky, J. S., Clements, J. E., Strandberg, J. D., Griffin, D. E. and Cork, L. C. (1982). Slow virus replication: the role of macrophages in the persistence and expression of visna viruses of sheep and goats. *J Gen Virol* 59, 345-56.

Nestle, F. O., Zheng, X. G., Thompson, C. B., Turka, L. A. and Nickoloff, B. J. (1993). Characterization of dermal dendritic cells obtained from normal human skin reveals phenotypic and functionally distinctive subsets. *J Immunol* 151, 6535-45.

Nguyen, D. G. and Hildreth, J. E. (2003). Involvement of macrophage mannose receptor in the binding and transmission of HIV by macrophages. *Eur J Immunol* 33, 483-93.

Oaks, J. L., McGuire, T. C., Ulibarri, C. and Crawford, T. B. (1998). Equine infectious anemia virus is found in tissue macrophages during subclinical infection. *J Virol* 72, 7263-9.

Odorizzi, G., Katzmann, D. J., Babst, M., Audhya, A. and Emr, S. D. (2003). Bro1 is an endosome-associated protein that functions in the MVB pathway in Saccharomyces cerevisiae. *J Cell Sci* 116, 1893-903. Epub 2003 Mar 18.

Ono, A. and Freed, E. O. (2001). Plasma membrane rafts play a critical role in HIV-1 assembly and release. *Proc Natl Acad Sci U S A* 98, 13925-30.

Patterson, S. and Knight, S. C. (1987). Susceptibility of human peripheral blood dendritic cells to infection by human immunodeficiency virus. *J Gen Virol* 68, 1177-81.

Pelchen-Matthews, A., Kramer, B. and Marsh, M. (2003). Infectious HIV-1 assembles in late endosomes in primary macrophages. *J Cell Biol* 162, 443-55. Epub 2003 Jul 28.

Peters, P. J., Borst, J., Oorschot, V., Fukuda, M., Krahenbuhl, O., Tschopp, J., Slot, J. W. and Geuze, H. J. (1991). Cytotoxic T lymphocyte granules are secretory lysosomes, containing both perforin and granzymes. *J Exp Med* 173, 1099-109.

Pique, C., Lagaudriere-Gesbert, C., Delamarre, L., Rosenberg, A. R., Conjeaud, H. and Dokhelar, M. C. (2000). Interaction of CD82 tetraspanin proteins with HTLV-1 envelope glycoproteins inhibits cell-to-cell fusion and virus transmission. *Virology* 276, 455-65.

Poli, G., Orenstein, J. M., Kinter, A., Folks, T. M. and Fauci, A. S. (1989). Interferon-alpha but not AZT suppresses HIV expression in chronically infected cell lines. *Science* 244, 575-7.

Pope, M., Betjes, M. G., Romani, N., Hirmand, H., Cameron, P. U., Hoffman, L., Gezelter, S., Schuler, G. and Steinman, R. M. (1994). Conjugates of dendritic cells and memory T lymphocytes from skin facilitate productive infection with HIV-1. *Cell* 78, 389-98.

Pope, M., Gezelter, S., Gallo, N., Hoffman, L. and Steinman, R. M. (1995). Low levels of HIV-1 infection in cutaneous dendritic cells promote extensive viral replication upon binding to memory CD4+ T cells. *J Exp Med* 182, 2045-56.

Raiborg, C., Rusten, T. E. and Stenmark, H. (2003). Protein sorting into multivesicular endosomes. *Curr Opin Cell Biol* 15, 446-55.

Raposo, G., Moore, M., Innes, D., Leijendekker, R., Leigh-Brown, A., Benaroch, P. and Geuze, H. (2002). Human macrophages accumulate HIV-1 particles in MHC II compartments. *Traffic* 3, 718-29.

Raposo, G., Nijman, H. W., Stoorvogel, W., Liejendekker, R., Harding, C. V., Melief, C. J. and Geuze, H. J. (1996). B lymphocytes secrete antigen-presenting vesicles. *J Exp Med* 183, 1161-72.

Rappersberger, K., Gartner, S., Schenk, P., Stingl, G., Groh, V., Tschachler, E., Mann, D. L., Wolff, K., Konrad, K. and Popovic, M. (1988). Langerhans' cells are an actual site of HIV-1 replication. *Intervirology* 29, 185-94.

Reece, J. C., Handley, A. J., Anstee, E. J., Morrison, W. A., Crowe, S. M. and Cameron, P. U. (1998). HIV-1 selection by epidermal dendritic cells during transmission across human skin. *J Exp Med* 187, 1623-31.

Reece, J. C., Vardaxis, N. J., Marshall, J. A., Crowe, S. M. and Cameron, P. U. (2001). Uptake of HIV and latex particles by fresh and cultured dendritic cells and monocytes. *Immunol Cell Biol* 79, 255-63.

Romani, N., Gruner, S., Brang, D., Kampgen, E., Lenz, A., Trockenbacher, B., Konwalinka, G., Fritsch, P. O., Steinman, R. M. and Schuler, G. (1994). Proliferating dendritic cell progenitors in human blood. *J Exp Med* 180, 83-93.

Saifuddin, M., Hedayati, T., Atkinson, J. P., Holguin, M. H., Parker, C. J. and Spear, G. T. (1997). Human immunodeficiency virus type 1 incorporates both glycosyl phosphatidylinositol-anchored CD55 and CD59 and integral membrane CD46 at levels that protect from complement-mediated destruction. *J Gen Virol* 78, 1907-11.

Salahuddin, S. Z., Rose, R. M., Groopman, J. E., Markham, P. D. and Gallo, R. C. (1986). Human T lymphotropic virus type III infection of human alveolar macrophages. *Blood* 68, 281-4.

Saphire, A. C., Bobardt, M. D., Zhang, Z., David, G. and Gallay, P. A. (2001). Syndecans serve as attachment receptors for human immunodeficiency virus type 1 on macrophages. *J Virol* 75, 9187-200.

Schmitz, J., van Lunzen, J., Tenner-Racz, K., Grossschupff, G., Racz, P., Schmitz, H., Dietrich, M. and Hufert, F. T. (1994). Follicular dendritic cells retain HIV-1 particles on their plasma membrane, but are not productively infected in asymptomatic patients with follicular hyperplasia. *J Immunol* 153, 1352 9.

Singh, D. K., Chebloune, Y., Mselli-Lakhal, L., Karr, B. M. and Narayan, O. (1999). Ovine lentivirus-infected macrophages mediate productive infection in cell types that are not susceptible to infection with cell-free virus. *J Gen Virol* 80, 1437-44.

Soilleux, E. J., Morris, L. S., Leslie, G., Chehimi, J., Luo, Q., Levroney, E., Trowsdale, J., Montaner, L. J., Doms, R. W., Weissman, D., Coleman, N. and Lee, B. (2002). Constitutive and induced expression of DC-SIGN on dendritic cell and macrophage subpopulations in situ and in vitro. *J Leukoc Biol* 71, 445-457.

Steinman, R. M. (1991). The dendritic cell system and its role in immunogenicity. *Annu Rev Immunol* 9, 271-96.

Stoorvogel, W., Kleijmeer, M. J., Geuze, H. J. and Raposo, G. (2002). The biogenesis and functions of exosomes. *Traffic* 3, 321-30.

Stoorvogel, W., Strous, G. J., Geuze, H. J., Oorschot, V. and Schwartz, A. L. (1991). Late endosomes derive from early endosomes by maturation. *Cell* 65, 417-27.

Strack, B., Calistri, A., Craig, S., Popova, E. and Gottlinger, H. G. (2003). AIP1/ALIX is a binding partner for HIV-1 p6 and EIAV p9 functioning in virus budding. *Cell* 114, 689-99.

Swingler, S., Brichacek, B., Jacque, J. M., Ulich, C., Zhou, J. and Stevenson, M. (2003). HIV-1 Nef intersects the macrophage CD40L signalling pathway to promote resting-cell infection. *Nature* 424, 213-9.

Thery, C., Duban, L., Segura, E., Veron, P., Lantz, O. and Amigorena, S. (2002). Indirect activation of naive CD4+ T cells by dendritic cell-derived exosomes. *Nat Immunol* 3, 1156-62.

Thery, C., Regnault, A., Garin, J., Wolfers, J., Zitvogel, L., Ricciardi-Castagnoli, P., Raposo, G. and Amigorena, S. (1999). Molecular characterization of dendritic cell-derived exosomes. Selective accumulation of the heat shock protein hsc73. *J Cell Biol* 147, 599-610.

Tschachler, E., Groh, V., Popovic, M., Mann, D. L., Konrad, K., Safai, B., Eron, L., diMarzo Veronese, F., Wolff, K. and Stingl, G. (1987). Epidermal Langerhans cells--a target for HTLV-III/LAV infection. *J Invest Dermatol* 88, 233-7.

Turley, S. J., Inaba, K., Garrett, W. S., Ebersold, M., Unternaehrer, J., Steinman, R. M. and Mellman, I. (2000). Transport of peptide-MHC class II complexes in developing dendritic cells. *Science* 288, 522-7.

Turville, S., Wilkinson, J., Cameron, P., Dable, J. and Cunningham, A. L. (2003). The role of dendritic cell C-type lectin receptors in HIV pathogenesis. *J Leukoc Biol* 74, 710-8. Epub 2003 Sep 2.

Turville, S. G., Arthos, J., Mac Donald, K., Lynch, G., Naif, H., Clark, G., Hart, D. and Cunningham, A. L. (2001). HIV gp120 receptors on human dendritic cells. *Blood* 98, 2482-8.

Turville, S. G., Cameron, P. U., Handley, A., Lin, G., Pohlmann, S., Doms, R. W. and Cunningham, A. L. (2002). Diversity of receptors binding HIV on dendritic cell subsets. *Nat Immunol* 3, 975-83.

Turville, S. G., Santos, J. J., Frank, I., Cameron, P. U., Wilkinson, J., Miranda-Saksena, M., Dable, J., Stossel, H., Romani, N., Piatak, M., Jr., Lifson, J. D., Pope, M. and Cunningham, A. L. (2004). Immunodeficiency virus uptake, turnover, and 2-phase transfer in human dendritic cells. *Blood* 103, 2170-9. Epub 2003 Nov 20.

Valladeau, J., Dezutter-Dambuyant, C. and Saeland, S. (2003). Langerin/CD207 sheds light on formation of birbeck granules and their possible function in Langerhans cells. *Immunol Res* 28, 93-107.

Valladeau, J., Ravel, O., Dezutter-Dambuyant, C., Moore, K., Kleijmeer, M., Liu, Y., Duvert-Frances, V., Vincent, C., Schmitt, D., Davoust, J., Caux, C., Lebecque, S. and Saeland, S. (2000). Langerin, a novel C-type lectin specific to Langerhans cells, is an endocytic receptor that induces the formation of Birbeck granules. *Immunity* 12, 71-81.

von Lindern, J. J., Rojo, D., Grovit-Ferbas, K., Yeramian, C., Deng, C., Herbein, G., Ferguson, M. R., Pappas, T. C., Decker, J. M., Singh, A., Collman, R. G. and O'Brien, W. A. (2003). Potential role for CD63 in CCR5-mediated human immunodeficiency virus type 1 infection of macrophages. *J Virol* 77, 3624-33.

von Schwedler, U. K., Stuchell, M., Muller, B., Ward, D. M., Chung, H. Y., Morita, E., Wang, H. E., Davis, T., He, G. P., Cimbora, D. M., Scott, A., Krausslich, H. G., Kaplan, J., Morham, S. G. and Sundquist, W. I. (2003). The protein network of HIV budding. *Cell* 114, 701-13.

Wiley, C. A., Schrier, R. D., Nelson, J. A., Lampert, P. W. and Oldstone, M. B. (1986). Cellular localization of human immunodeficiency virus infection within the brains of acquired immune deficiency syndrome patients. *Proc Natl Acad Sci U S A* 83, 7089-93.

Yi, Y., Chen, W., Frank, I., Cutilli, J., Singh, A., Starr-Spires, L., Sulcove, J., Kolson, D. L. and Collman, R. G. (2003). An unusual syncytia-inducing human immunodeficiency virus type 1 primary isolate from the central nervous system that is restricted to CXCR4, replicates efficiently in macrophages, and induces neuronal apoptosis. *J Neurovirol* 9, 432-41.

Yi, Y., Rana, S., Turner, J. D., Gaddis, N. and Collman, R. G. (1998). CXCR-4 is expressed by primary macrophages and supports CCR5-independent infection by dual-tropic but not T-tropic isolates of human immunodeficiency virus type 1. *J Virol* 72, 772-7.

Zambruno, G., Mori, L., Marconi, A., Mongiardo, N., De Rienzo, B., Bertazzoni, U. and Giannetti, A. (1991). Detection of HIV-1 in epidermal Langerhans cells of HIV-infected patients using the polymerase chain reaction. *J Invest Dermatol* 96, 979-82.

Zhang, J., Randall, G., Higginbottom, A., Monk, P., Rice, C. M. and McKeating, J. A. (2004). CD81 is required for hepatitis C virus glycoprotein-mediated viral infection. *J Virol* 78, 1448-55.

In: Progress in Virus Research
Editor: Sabine Thebault, pp. 129-167

ISBN 1-59454-438-7

Chapter V

Mechanisms of Retrovirus-Host Interaction Revealed Through Studies of Mouse Genetic Models

Laure K. Case and Tatyana V. Golovkina
The Jackson Laboratory, 600 Main Street, Bar Harbor, ME, 04609

Abstract

An efficient host defense against viral pathogens operates at two levels: innate and adaptive immunity resulting in long-lasting protective immunity. Viruses, however, have developed mechanisms to evade both innate and adaptive immunity, and in some cases certain aspects of normal protective host responses have been redirected or modified to the advantage of these viruses. Susceptibility to retroviral pathogenesis is controlled by the genetic make-up of the host and can be studied in mice from different inbred backgrounds, because allelic differences enable cloning of the genes responsible for the phenotype and subsequent understanding of the mechanism of their action. Using mice from distinct genetic backgrounds we learned that mouse mammary tumor virus (MMTV) has developed strategies to both exploit and subvert the immune system. For example, MMTV activates profound T-cell proliferation necessary for virus propagation via virally-encoded superantigen (SAg). At the same time, MMTV can stimulate production of the immunosuppressive cytokine IL-10 by subverting innate immune signaling. This allows the virus to avoid the ongoing anti-virus immune response and to replicate without restraint. In addition to their unique ability to evade immune responses, retroviruses are also very efficient in causing tumors. It is well accepted that most retroviruses cause tumors by acting as insertional mutagens activating expression of cellular protoocogenes. We found, however, that MMTV activation of oncogenes is a necessary but insufficient step in mammary tumor induction. By using different viral variants and mice from distinct genetic backgrounds, we discovered that virally-encoded genes must co-operate with oncogenes for efficient tumor induction in the MMTV model. Identification of novel pathways utilized by retroviruses to induce tumors will increase our understanding of the signal transduction pathways involved in tumorigenesis in general.

This chapter will focus on the usefulness of mouse strains with distinct virus-susceptible or virus-resistant phenotypes to gain understanding of the fundamental mechanisms of virus-host interactions.

MMTV Pathogenesis

MMTV was the first transmissible agent shown to cause cancer in mammals and, for almost 75 years, the MMTV model has been widely used to investigate different aspects of virus-host interaction.

MMTV is transmitted as an exogenous virus [1] or as a stably integrated provirus (*Mtv*). Exogenous MMTV is spread via the milk of infected females and is acquired by suckling pups [2]. The primary targets for exogenous MMTV are T and B cells located in the Peyer's patches of the gastrointestinal tract of neonatally infected pups (Fig. 1) [3, 4]. MMTV gains access to these cells by passing from the apical to the basal surface of M cells located in the follicular associated epithelium of Peyer's patches [5]. After amplification within cells of the immune system, MMTV travels to mammary gland cells, which secrete virus into milk during lactation. The majority of endogenous *Mtv*s do not produce infectious viral particles because of mutations in either their transcriptional regulatory or coding regions [6]. Endogenous viruses however, can recombine with exogenous MMTV to generate new exogenous virus variants with altered biological activities [7-9].

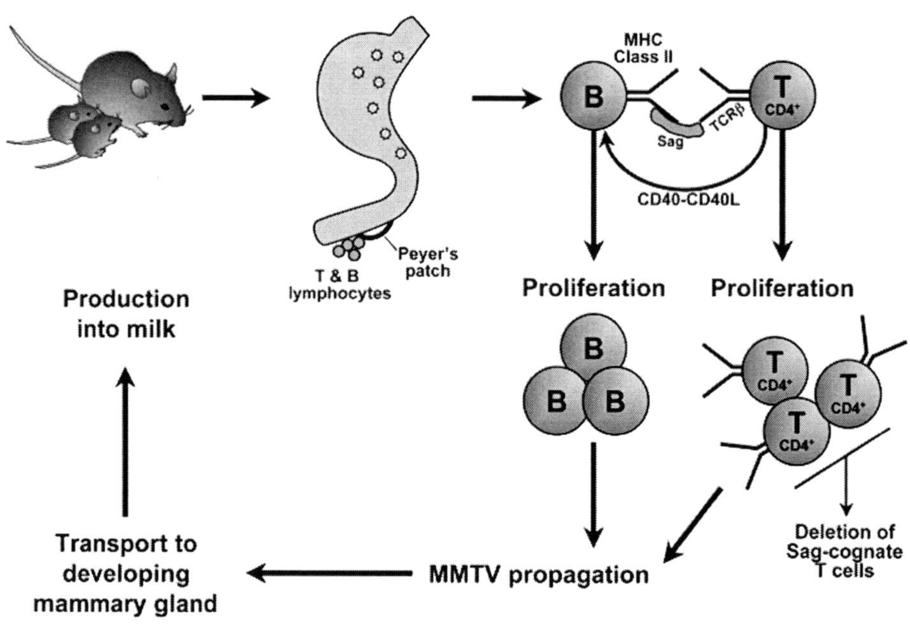

Figure 1. Life cycle of exogenous MMTV. See explanations in the text.

MMTV represents the prototype of the B-type retroviruses. Like all replication-competent retroviruses, MMTV encodes two structural genes, *gag* (the non-glycosylated proteins of the capsid), and *env* (the glycosylated proteins of the viral envelope), as well as two enzymes, *pol* (the reverse transcriptase), and *pro* (viral protease) [10].

The Gag protein is the precursor to the internal structural proteins of all retroviruses. Gag is organized into regions which are proteolitically liberated within assembled virions during viral maturation. The *pro* gene product cleaves the Gag polyprotein into the following mature proteins: matrix (p21MA), which lines the virion; capsid (p27CA), which forms the virion core; and nuclear capsid (p14NC) zinc-finger protein, which coats the genomic RNA.

MMTV contains two major Env glycoproteins, gp52 surface (SU) and gp36 transmembrane (TM) [11]. Like other retroviruses, MMTV matures by budding from the surface membrane of its host cells. The area of cell surface at which budding occurs is specifically enriched in viral glycoproteins. During infection, gp52SU binds to the cellular receptor expressed on susceptible cells, whereas gp36TM facilitates viral-cell membrane fusion.

The long terminal repeat (LTR) of exogenous and endogenous MMTVs encodes a superantigen (SAg) [12], which plays an important role in MMTV transmission. Conventional antigens are presented by major histocompatibility (MHC) class II molecules expressed on MMTV-infected antigen-presenting cells and are recognized by both the Vβ and the Vα chains of the T cell receptor (TCR), resulting in stimulation of a relatively small percentage of T cells. SAgs are also presented by MHC class II molecules. However, in contrast to conventional antigens, SAg recognition relies predominately on the Vβ domain of the TCR, resulting in stimulation of approximately 10 to 30 percent of all T cells [13, 14]. Overall, the LTR sequences of different MMTVs are highly conserved [15]. However, the region encoding the C-terminal segment of SAg is more diverse and is known as the hypervariable region [15]. The amino acid sequence of this region contacts the Vß-chain of the TCR and determines which T cells are affected [16]. These specific Vβ-expressing T cells proliferate and stimulate proliferation of B cells via interaction between the CD40 receptor expressed on B cells and the CD40 ligand expressed on T cells [17]. SAg presentation by infected cells results in a pool of proliferating cells which are essential for efficient MMTV infection [18]. This occurs because, as a retrovirus, MMTV requires dividing cells to deliver the pre-integrated complex into nuclei of infected cells [19]. Ultimately the virus will be transported to the mammary glands and deposited into the milk during lactation, therefore completing the replication cycle. Thus, unlike many microorganisms that must evade the immune system in order to persist, MMTV relies upon SAg to stimulate the immune system for its survival. SAg-activated T cells undergo subsequent deletion, and the deletion of these SAg-cognate T cells is used as a readout to determine whether mice are MMTV-infected [20].

A requirement for SAg function during MMTV infection was demonstrated in studies in which we generated and analyzed transgenic mice carrying a new MMTV provirus that had a frameshift mutation in the *sag* region of an exogenous MMTV virus. This resulted in premature termination of the SAg [7]. Even though transgenic females produced non-infectious virus, all mice fostered on their milk became infected. Sequence analysis demonstrated that the SAg coding regions of all newly acquired viruses were restored because of recombination between the mutant virus and an endogenous *Mtv1* [7]. In the absence of retroviral recombination, viruses without functional SAgs were not pathogenic [21].

Once integrated into a chromosome, expression of proviral DNA is regulated by specific sequences within the LTR that cause increased viral transcription in response to glucocorticoid receptor/steroid hormone complexes [22]. The large group of retroviruses that induce malignancies do not carry oncogenes in their genome, and proviral integration near a proto-oncogene – a phenomenon termed proviral insertional mutagenesis – plays a central role in tumor induction. Proviral integration occurs either at the 3' or the 5' end of the protooncogene in contrast to transcription orientation, suggesting the usage of viral enhancer rather than promoter [23]). Like all retroviral integrations, insertions near oncogenes occur by chance and require multiple rounds of re-infection/re-integration. This is why multiparous females develop mammary tumors with higher incidence and shorter latency compared to virgin females.

MMTV-induced mammary adenocarcinomas develop from hyperplastic alveolar nodules (HANs) [2]. HANs are focal proliferations of lobuloalveolar epithelium that contain immortal cells. They arise from a single cell which contains newly integrated provirus next to the protoncogene. Mammary carcinomas are postulated to arise from HANs by progressive clonal selection of variants, called hyperplastic outgrowth (HOG), with increased growth autonomy.

Based on the clonal nature of MMTV-induced tumors, viral genome was used as a tag for identifying genes activated by the MMTV insertion. *Int1* was the first protooncogene to be cloned following activation by viral insertion in mammary tumors of C3H/He mice [24]. Its initial name, *int1,* was subsequently changed to *Wnt1* because of its homology to the *Drosophila* Wingless gene [25]. The gene encodes for a secreted protein which determines the fate of surrounding cells. In mice, its expression is restricted to the developing nervous system, where it is expressed at day 8-14 [26]. Insertional activation leading to up-regulation of *Wnt1* expression occurs in ~70 percent of mammary tumors of C3H/He mice [25].

The receptors for *Wnt* have been identified as a class of seven transmembrane proteins known as Frizzled [27]. In the absence of *Wnt1,* the constitutively active glycogen synthase kinase 3 (GSK3) phosphorylates ß-catenin and targets it for degradation. After binding the *Wnt* ligand, Frizzle transmits a signal to cytoplasmic phosphoproteins in the disheveled family (DV1). DV1 inhibits activity of GSK3 and allows ß-catenin's translocation into the nucleus. ß-catenin associates with several DNA-binding proteins, such as members of the TCF/LEF family. Heterodimers of ß-catenin and transcription factors transactivate a number of cellular protocogenes, including c-myc and cyclin D1 [27, 28].

Using a similar approach, Peters and colleagues have cloned another gene affected by MMTV insertion, initially called *int2* [29]. *Int2/Fgf3* is a member of the fibroblast growth factor (FGF) family and like *Wnt1*, *Int2/Fgf3* is not normally expressed in the cells of mammary glands but is insertionally upregulated in 50 percent of mammary tumors of BR6 mice [29]. In 1987 another common MMTV integration site was cloned from MMTV-induced mammary tumors in CzechII mice [30] and called *int3.* Later identified as a member of the Notch family, *int3/Notch4* [30] encodes transmembrane receptor proteins that determine cell fate during development. In contrast to *Wnt1* and *int2/Fgf3*, *int3/Notch4* was found to be rearranged by MMTV insertion [31]. This integration caused constitutive overexpression of the portion of the gene encoding the intracellular domain of the protein (gain of function mutation) [31]. Oncogenic properties of the *Wnt1*, *int2/Fgf3* and *int3/Notch4* genes were proven in transgenic mice. *Wnt1*, *int2/Fgf3 and int3/Notch4* transgenic mice developed mammary tumors when transgenic expression was directed by MMTV LTR [32-35]

To identify other genes affected by MMTV integration, *Wnt1* transgenic mice were infected with MMTV to accelerate tumorigenesis. Tumors that developed in these transgenic mice were used to clone other MMTV integrations. Other members of the Wnt and Fgf families were found, as well as novel unrelated genes (translation initiation factor 3) among genes affected by MMTV insertions [33, 36, 37].

Thus, many genes and pathways were discovered using MMTV-induced tumors, but none of them could fully explain a high rate of mammary tumors in the MMTV system. Indeed, even though all cells in the mammary gland express the *Wnt1* gene, only a few tumors develop per mouse [32]. In addition, these mice have a relatively delayed onset of tumor development. Furthermore, *Wnt1/int2* bitrasgenic mice demonstrated only slight acceleration in tumor development [36], suggesting that co-operation between oncogenes also does not provide a convincing explanation for the high rate of tumors in MMTV-infected mice. We found that in addition to the insertional mutagenesis, MMTV-encoded genes also contribute directly to the transformation of mammary epithelial cells.

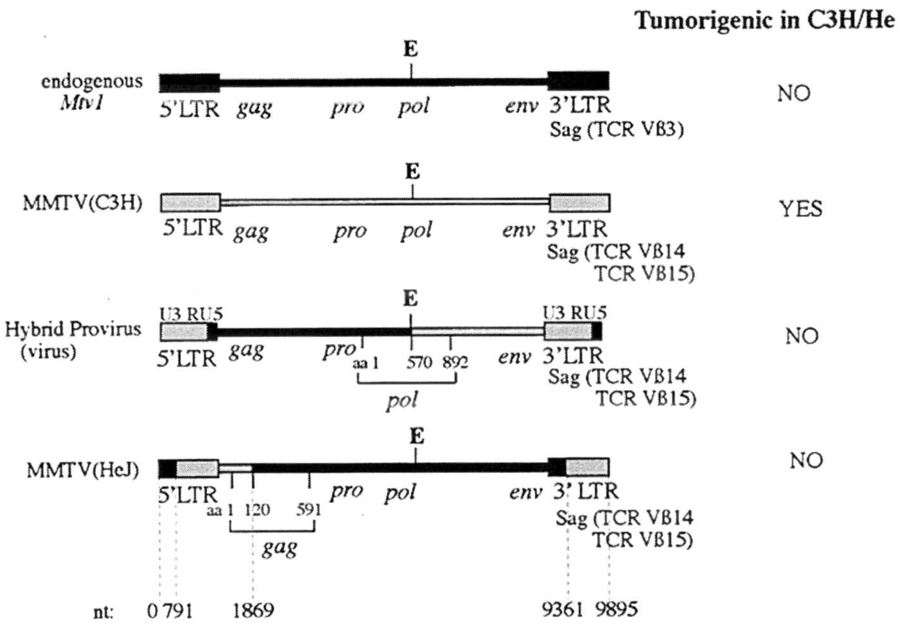

Figure. 2. Structure and tumorigenic features of different MMTVs. Exogenous MMTV(C3H), MMTV(HeJ), HP and endogenous *Mtv1* proviruses were cloned and sequenced. The HP *pol* gene is chimeric, that has the first 570 a. a. derived from *Mtv1* and the rest from exogenous MMTV(C3H). The recombination break point in MMTV(HeJ) occurred in the *gag* gene between nt position 1845-1869 [nt 1845 is specific for MMTV(C3H), whereas nt 1869 is specific for *Mtv1*] and in LTR between nt position 9333-9361 [nt 9333 is specific for *Mtv1*, whereas nt 9361 is specific for MMTV(C3H)]. The hypervariable region of *sag*, nt 9430-9530. As a result, the first 120 a.a of the *gag* gene and the hypervariable region of the *sag* gene in MMTV(HeJ) are derived from MMTV(C3H). Although the original DNA construct used to make HP transgenic mice has a 5'LTR derived from *Mtv1* and a 3'LTR derived from MMTV(C3H), upon infection of cells with the resulting virus, MMTV(C3H)-specific information in the U3 region of the 3'LTR is present in the newly synthesized 5'LTR. (from Hook et al., *J. Virol.* 74, 8876, 2000, with permission).

A number of years ago we produced C3H/HeN transgenic mice that had the genetically engineered hybrid MMTV provirus (HP) in which the 5' half – including the 5' LTR, the *gag* gene, and part of the *pol* gene – were derived from the endogenous *Mtv1* locus, and the 3' half – including the rest of the *pol* gene, the *env* gene, and the 3' LTR – were derived from the exogenous MMTV(C3H) virus (Fig. 2) [38]. Transgenic females shed the HP virus into their milk, and non-transgenic mice foster-nursed by them became infected with the virus, showing that it was infectious [38]. We also showed that HP amplification within the mammary gland was similar to that of the exogenous wild-type MMTV(C3H) by analyzing viral RNA production in the mammary glands after each subsequent pregnancy [39]. However, when we monitored mice infected with HP for mammary gland tumors, we discovered that this virus was not tumorigenic in C3H/HeN mice. While 100 percent of C3H/HeN females infected with MMTV(C3H) developed tumors after approximately 300 days, none of the C3H/HeN mice infected with HP developed tumors over a period of 350 days [39].

Another infectious MMTV variant with attenuated tumorigenicity has been identified as naturally occurring virus in C3H/HeJ mice, called MMTV(HeJ) [39]. While C3H/HeJ mice are highly susceptible to MMTV(C3H)-induced mammary tumors with 100 percent females developing tumors after approximately 250 days, only 10 percent of the MMTV(HeJ)-infected C3H/HeJ females developed tumors after 350 days [39].

In more than 80 percent of C3H/He MMTV-induced mammary tumors, at least one of the additional MMTV genomes is integrated near the cellular *Wnt1* gene resulting in transcription activation of the gene [24]. In approximately 10 percent of the tumors, proviral insertion up-regulates expression of the *Int2/Fgf3* gene [40, 41]. It is possible that the enhancer elements in MMTV(HeJ) and HP were not capable of upregulating transcription of cellular proto-oncogenes, resulting in the failure of these viruses to cause tumors. To test this possibility, we took advantage of the fact that BALB/cJ mice appear to be sensitive to tumors caused by all viruses [39]. BALB/cJ mice were foster-nursed on milk containing either HP or MMTV(HeJ). HP-infected BALB/cJ mice developed mammary tumors at approximately the same rate as MMTV(C3H)-infected BALB/cJ mice [39]. RNA isolated from these BALB/cJ mammary tumors and from rare MMTV(HeJ)-induced mammary tumors in C3H/HeJ mice was analyzed for expression of *Wnt1* and *Int2/Fgf3* transcripts. Expression of *Wnt1* was detected in 71 percent of MMTV(HeJ)-induced C3H/HeJ mammary tumors and in 58 percent of HP-induced BALB/cJ mammary tumors [39]. Expression of the *Int2/Fgf3* gene was detected in 25 percent of the rare MMTV(HeJ)-induced C3H/HeJ tumors and in 16 percent of BALB/cJ mammary tumors induced by HP [39]. Thus, both HP and MMTV(HeJ) were capable of up-regulating cellular proto-oncogenes in tumors they induced.

Both C3H/HeN and C3H/HeJ originated from MMTV(C3H)-infected, mammary tumor prone progenitor strain, called C3H/St. Since recombination is a driving force leading to generation of new viral variants, we hypothesized that MMTV(HeJ) is a recombinant virus generated from wild-type MMTV(C3H) and one of the endogenous MMTVs. Of five different endogenous MMTVs present in the C3H/He genome, *Mtv1* is the only one that can be co-packaged with exogenous MMTV in the mammary gland to give rise to a new recombinant [8]. We cloned and sequenced *Mtv1*, MMTV (C3H), MMTV(HeJ) and HP using a panel of overlapping plasmids and found that the latter is a recombinant virus (Fig. 2) [39]. Almost the entire genome of MMTV(HeJ) was derived from *Mtv1* except for the SAg hypervariable region and the first 360 bp of the *gag* region, which were from MMTV(C3H). Knowing the sequences of different viruses, it was possible to compare the phenotypes they produced and to deduce which gene might be responsible for loss of tumorigenicity (Fig. 2). Two genes (*gag* and part of *pol*) in non-tumorigenic HP were derived from endogenous *Mtv1*. Since the *pol* gene encodes a highly conserved enzyme whose function is required for retroviral replication, and HP was replicating normally, we hypothesized that the *Mtv1 gag* gene must be responsible for the failure of HP and MMTV(HeJ) to cause tumors (Fig. 2) [39]. We found five non-conservative amino acid changes between the Gag of non-tumorigenic *Mtv1* and the Gag of tumorigenic MMTV(C3H); four are located within two putative bipartite nuclear localization domains of the MA protein (I 182 N, G 185 H, V 235 D and Q 240 L) and one is located in the NC protein (S 551 P) [39].

Gag-encoded proteins orchestrate assembly and uncoating of the retroviral virion [42]. Gag also recruits other viral components required for infectivity, including genomic RNA and Env glycoprotein, by direct interaction during assembly [42]. After binding with its receptor, the virus fuses with the plasma membrane and delivers the ribonucleoprotein core into the cytoplasm, where Pol initiates reverse transcription of viral genomic RNA. Gag-encoded proteins then dissociate from the complex, allowing reverse transcription to begin. Gag protein also participates in the transport of the DNA copy of the viral genome to the nucleus. Finally, MA protein demonstrates a nuclear export activity important for transporting unspliced viral RNA to the plasma membrane [43]. Our data suggest that in addition to known functions of Gag, it also has a novel function in mammary tumorigenesis. Experiments are ongoing which will help us to identify the mechanism by which Gag contributes to mammary tumors.

Mechanisms of Resistance to MMTV Pathogenesis

Like other retroviruses, MMTV is not efficiently eliminated by the immune system of susceptible mice. MMTV can efficiently infect both newborn and adult mice [21], suggesting that the virus has evolved strategies to suppress or subvert immune system attack and does not rely on tolerance in newborn mice alone. Although $CD4^+$-mediated T-cell help to B cells, and $CD8^+$-T-cell responses against the virus are detectable, they do not clear infection and do not prevent the virus from infecting its final targets, the mammary gland cells [44, 45]. The above findings suggesting that MMTV can avoid immune system attack prompted us to further investigate the underlying mechanisms. A large number of genetically well-characterized inbred mouse strains, which differ in their susceptibility to MMTV pathogenesis, serves as a great tool enabling researchers to dissect mechanisms of viral evasion.

There are two well-documented mechanisms of resistance to MMTV infection and subsequent tumorigenesis. One type of resistance is due to the virus' inability to achieve SAg-dependent amplification in infected hosts, whereas another type of resistance is independent of the virus' capacity to stimulate SAg-dependent amplification.

MMTV Resistant Mechanisms Associated with SAg Function

MHC genes. One type of resistance to MMTV relies on MHC class II proteins, which are essential for presenting foreign antigenic peptides to $CD4^+$ T cells. In mice, there are two types of class II isotypic heterodimeric MHC proteins, termed $A\alpha A\beta$ (I-A) and $E\alpha E\beta$ (I-E).

Inbred strains of either the b, f, q, or s MHC haplotype do not express I-E molecules because of mutations in either the Eα or Eβ gene. It appears that most but not all MMTV SAgs can be presented much more efficiently by the I-E molecules than by I-A molecules [46)] As a result, mice lacking the I-E molecule are relatively resistant to MMTV infection and MMTV-induced mammary tumors [47, 48]. Some resistant mouse strains without the I-E molecule include C57BL/6J, which has the $H-2^b$ haplotype, SJL/J with the $H-2^s$ haplotype, and FVB/NJ with the $H-2^q$ haplotype.

Endogenous Mtvs. Another type of resistance to MMTV infection and MMTV-induced mammary tumors occurs when endogenous Mtvs with SAg of the same T cell specificity are present in the genome of germ line cells. These endogenous Mtvs cause negative selection of SAg cognate T cells possessing specific Vβ elements [49-51]. There are more than 40 different endogenous Mtvs encoding different Vβ specific SAgs, and the presence of a particular endogenous Mtv in the genome of a mouse can be determined by the absence of specific T cells bearing the Vβ element that interacts with the SAg of that particular Mtv. Mice that lack SAg-cognate T cells due to the expression of SAg-transgenes or endogenous proviruses were shown to be resistant to exogenous viruses bearing SAgs of the same Vß specificity [18, 52]. Because almost all of the endogenous Mtv loci have maintained functional sag genes, in contrast to the genes encoding for structural proteins, it has been proposed that endogenous viruses are retained in the mouse genome as an antiviral defense mechanism.

Is SAg function indispensable from the virus life cycle? Even though mice lacking SAg-cognate T cells were found to be resistant to exogenous viruses bearing SAgs of the same Vß specificity [18, 52], they still developed MMTV-induced mammary tumors, albeit with much longer latency compared to mice having SAg-cognate T cells [53]. In addition, viruses without functional SAgs were only able to establish initial infection in vivo which did not spread beyond cells of the immune system [21]. Together, these data suggested that there may be two stages in MMTV replication: SAg-independent and SAg-dependent.

To elucidate the role of SAg at the different steps of virus infection, we took advantage of exogenous MMTV(LA) virus produced by BALB/cLA mice. BALB/cT mice contained two novel exogenous MMTVs: BALB2, with Vβ2 specific SAg, and BALB14, with Vβ14-specific SAg. After crossing BALB/cT mice with AKR/J mice, a new exogenous virus with Vβ6- and Vβ8.1-specific SAg was detected in the milk of these F_1 mice [54]. This new virus, termed BALBLA, was a result of recombination between exogenous BALB14 in BALB/cT mice and endogenous *Mtv7* in AKR/J mice [9]. Therefore, MMTV(LA) consisted of three exogenous MMTVs, including BALB2, BALB14, and BALBLA.

We reasoned that, unless infection in neonates is SAg-independent, *Mtv7*-positive mice should be resistant to BALBLA, since SAg-cognate T cells used by this virus are deleted in these mice at around birth [55]. To test this hypothesis, *Mtv7*-positive CBA/J and DBA/2J mice – and *Mtv7*-negative BALB/cJ, CBA/CaJ, and C3H/HeN mice – were infected with MMTV(LA) by three different routes. Young adult mice aged 3 to 4 weeks were injected with syngeneic MMTV(LA)-infected splenocytes, while newborn mice were either foster-nursed by viremic BALB/cLA females or injected with syngeneic MMTV(LA)-infected splenocytes. All mice were bred, and RNA isolated from their milk was subjected to an RNase T_1 protection assay with probes specific for BALB2, BALB14, and BALBLA. As anticipated, all adult *Mtv7*-negative mice became infected with all three viruses, while *Mtv7*-positive mice were resistant to BALBLA and transmitted only BALB2 and BALB14 viruses into their milk (Table 1) (Pobezinskya et al., J. Immunol. 2004). Surprisingly, newborn mice of both *Mtv7*-positive and negative strains were equally permissive for all three viruses regardless of the route of infection. Thus, the lack of SAg-cognate T cells and SAg-mediated T cell activation confers resistance to MMTV infection in adult mice, but not in neonatal mice.

Table 1. Age-dependent susceptibility to MMTV infection in different strains of mice

Strain	MHC	I-E	*Mtv7*	Infected as neonates			Infected as adults		
	haplotype			BALB2	BALB14	BALBLA	BALB2	BALB14	BALBLA
BALB/cJ	H-2^d	Yes	N	+	+	+	+	+	+
CBA/CaJ	H-2^k	Yes	No	+	+	+	+	+	+
C3H/HeN	H-2^k	Yes	No	+	+	+	+	+	+
DBA/2J	H-2^d	Yes	Yes	+	+	+	+	+	-
CBA/J	H-2^k	Yes	Yes	+	+	+	+	+	-
LP/J	H-2^b	No	No	+	+	+	-	-	+
D1.LP/J	H-2^b	No	Yes	+	+	+		-	-

RNA isolated from milk or lactating mammary glands of different mice either foster-nursed by BALB/cLA females or injected with syngeneic MMTV-infected splenocytes was subjected to RNAse T_1 protection analysis with probes specific for BALB2, BALB14 and BALBLA viruses.

Expression of MHC class II I-E molecules and endogenous *Mtv7* by infected mice is indicated by "Yes" or "No".

+ or -; presence or absence of viral RNA.

Some MMTV variants encode I-E-independent SAgs, including endogenous *Mtv7* [56] and exogenous BALBLA [57]. It was shown that newborn mice from the I-E-negative genetic background are susceptible to I-E-dependent MMTV variants if infected concomitantly with an I-E-independent MMTVs. Therefore, it was possible that BALBLA was incapable of infecting Vβ6$^+$ T cell-deficient mice on its own but acquired infectivity upon mixing with BALB2 and BALB14 viruses in I-E-positive mice. To investigate this possibility, we used two strains of mice with I-E-negative H-2^b haplotypes, specifically LP/J and D1.LP/J. They are identical to one another with the exception of the *Mtv7* locus, which is present in D1.LP/J but not in LP/J mice. As a result, T cells cognate for BALBLA SAg are absent in D1.LP/J mice due to negative selection, but are present in LP/J mice. Importantly, due to I-E deficiency in D1.LP/J and LP/J mice, T cells cognate for I-E-dependent SAgs encoded by BALB2 and BALB14 cannot be activated. If concomitant infection underlies virus susceptibility in neonatal mice, newborn D1.LP/J mice should be resistant to all three viruses, as BALB2 and BALB14 SAgs are unable to activate cognate T cells, and BALBLA SAg lacks its target T cells. Newborn and adult D1.LP/J and LP/J mice were infected with MMTV(LA) via foster-nursing or by injection with infected splenocytes, respectively. The newborn mice from both strains became infected with all three viruses, indicating that infection was SAg-independent (Table 1). At the same time, adult D1.LP/J mice were completely resistant to all these viruses, while adult LP/J mice were susceptible to BALBLA, because its SAg was presented to T cells by the I-A^b molecule. Therefore, the susceptibility of neonatal mice to MMTV infection cannot be explained by concomitant infection with the three different viruses and does not require SAg presentation. We concluded that this susceptibility is instead a property of a developmental stage in mice.

To determine the time point at which mice lacking SAg-cognate Vß6$^+$/Vß8.1$^+$ T cells become resistant to BALBLA, we foster-nursed several groups of DBA/2J mice on

BALB/cLA mothers for different three-day periods between 15 and 20 days postpartum and then returned them to their uninfected mothers. The foster-nursed mice were subsequently bred and their milk was tested for the presence of MMTV. It became clear that the DBA/2J mice were permissible for all three viruses up until day 17 postpartum but became resistant to BALBLA after day 17 (Table 2).

Table 2. Neonatal mice are susceptible to all MMTVs until day 17 *postpartum*

			Virus in the milk			
Mouse strain	Fostered between days	Mice (n)	BALB2	BALB14	BALBLA	cell proliferation *in vitro*
DBA/2J	15-18	8	+	+	+	+
	16-19	5	+	+	+	+
	17-20	10	+	+	-	-
BALB/cJ	1-3...17-19	8	+	+	+	

DBA/2J mice and control BALB/cJ mice were foster-nursed by BALB/cLA females for different three-day periods and then returned to uninfected mothers. The foster-nursed mice were bred, and RNA isolated from their milk was subjected to RNase T_1 protection analysis with probes specific for BALB2, BALBLA, and BALB14 viruses.
+, presence of the virus or cell proliferation in culture
-, no detectable virus or absence of cell proliferation in cell culture

These studies suggest that developmental changes determine whether the host is susceptible or resistant to retroviruses. As previously mentioned, most retroviruses require the target cells to actively cycle in order to allow integration of the reverse-transcribed retroviral genome into the host chromosome. We hypothesized that MMTV might use the neonatal period as an opportunity to invade the host even without functional SAg but relies on the SAg function after this proliferative stage has been completed.

It has been recognized that the thymocytes and splenocytes of neonatal mice proliferate in vitro [58-60]. Thus, we sought to determine whether the ability of these cells from newborn mice to proliferate spontaneously is developmentally regulated. Splenocytes from DBA/2J mice of different ages were labeled with CFSE, cultured in vitro for 4 days, and analyzed by FACS (fluorescence-activated cell sorter). In contrast to cultured adult splenocytes, cells from newborn mice divided vigorously as indicated by the dilution of CFSE label and the appearance of additional peaks. Notably, the number of proliferating cells decreased significantly after day 17 postpartum, which correlated well with the observed resistance to MMTV infection (Table 2) (Pobezinskaya et al., manuscript submitted).

These results indicate that initial infection does not require viral SAg, because MMTV is infecting cells that are proliferating naturally as a result of their developmental program. In contrast, SAg-dependent activation of T cells with consequent activation of B-cell proliferation via the CD40 receptor is necessary for MMTV amplification during the later stages of infection [17]. Therefore, the survival strategy of MMTV is much more complex than has been indicated by previous research, since the discovery of viral SAgs more than a decade ago.

MMTV Resistant Mechanisms Independent of SAg Function

Innate and adaptive immunity evolved as host's defense mechanisms against different pathogens. The major targets of innate immune recognition are pathogen-associated molecular patterns (PAMPs) [61]. PAMPs represent conserved molecular structures that are produced only by pathogens, but not by the host organisms, and are essential for the survival of the pathogen. For example, peptidoglycan (PGN) and polysaccharide (LPS) are produced by Gram-positive and Gram-negative bacteria, respectively, and not by eukaryotic hosts, and are absolutely required for survival of the pathogen. PAMPs are recognized by the germ line-encoded Toll-like innate receptors (TLR) [62]. The mammalian TLRs represent a family of homologous proteins characterized by an extracellular leucine-rich domain and a cytoplasmic domain that is responsible for signal transduction. TLR signaling induces NF-κB dependent gene expression and results in production of proinflammatory cytokines, such as TNF-α and IL-12, and co-stimulatory molecules, such as CD80 (B7-1) and CD86 (B7-2) [62, 63]. Co-stimulatory molecules are necessary for antigen presentation via MHC class I and II molecules to naive T cells and their subsequent differentiation to T_H1 and T_H2 subsets. Therefore, innate immunity ultimately contributes to induction of the adaptive immune response.

All successful virus pathogens have evolved to specifically evade both innate and adaptive immune responses. Genetic variations as the means to escape antibody and cytotoxic lymphocyte responses were among the first of the strategies to be described [64]. In addition, viruses downregulate MHC class I and MHC class II expression of antigen-presenting cells, block natural killer (NK) activation, and interfere with apoptosis induced in the infected host's cell, with cytokine effector function, and with functions of complement [64], and with TLR4 signaling [65].

Despite viral evasion of the immune response, most viral infections result in virus clearance and induce long lasting protective immunity, or are adequately controlled by an adaptive immune response. However, infection with retroviruses rarely, if ever, leads to an immune response that can prevent ongoing virus replication. This is primarily the result of rapid retroviral replication and high mutation rates [66]. In addition to this rapid accumulation of minor genetic changes, retroviruses can also recombine at a high rate, generating large genetic alterations [67]. Continuous production of viral mutants results in adaptation to most environmental changes (anti-retrovirus immune response, anti-retroviral therapy) and subsequent selection of the most fit variants [68].

The best examples of immune system subversion by retroviruses have come from studies of the human and simian immunodeficiency lentiviruses (HIV and SIV, respectively). HIV and SIV undermine immune system control by infecting key $CD4^+$ T immune cells, thereby impairing the response of both the infected $CD4^+$ T cells and uninfected $CD8^+$ T cells [69]. Furthermore, both viruses have evolved other mechanisms to evade immune system attack, including downregulation of MHC class I expression [64].

Studies have also shown that evasion of the immune response by retroviruses occurs in the murine leukemia virus (MuLV) system. The host immune response to infection is dependent on the infection protocol and the genetic background of the mouse. Mice infected at birth by MuLV generally show little evidence of antiviral humoral and cellular immunity

[70]. However, infection of adult mice with most MuLV isolates produces a stronger antiviral response that can influence the efficiency of infection and pathogenesis, and it was concluded that the lack of cellular and humoral immunity in neonatally infected mice should be the result of tolerization. Thus, by infecting newborn mice, the virus has evolved a mechanism to avoid immune system recognition.

In this part of our chapter, we will illustrate how MMTV evolved to evade adaptive immune response (I/LnJ mice) and subvert innate immune response (C3H/HeJ mice) for its survival. In the final part, we will describe a novel dominant mechanism of resistance to MMTV infection inherited by YBR/Ei mice.

I/LnJ mice. It has been documented that mice from the inbred mouse strain I/LnJ exhibit resistance to MMTV-induced mammary tumors. Andervont was among the first scientists to extensively study I/LnJ's resistance to MMTV-induced tumors. His data clearly indicated that MMTV was not capable of passing from the infected mother to the nursing pups and that the virus was lost within a few generations [71]. Andervont suggested that I/LnJ mice are capable of either destroying or inactivating MMTV. In the same series of experiments, it was established that the tumor resistance mechanism inherited by I/LnJ mice is recessive, because hybrid females obtained from crosses between susceptible C3H/HeN and resistant I/LnJ mice were as susceptible to MMTV-induced mammary tumors as C3H/HeN mice. Experiments performed in our laboratory have confirmed these results (Fig. 3).

Early work with transplanted mammary glands indicated that resistance to MMTV-induced tumor growth in I/LnJ mice was controlled at the level of the host. Like I/LnJ mice, C57BL/6 mice are resistant to MMTV-induced mammary tumors. Because a different mechanism underlies resistance in each strain, hybrid progeny of crosses between resistant C57BL/6 and I/LnJ mice are susceptible to MMTV-induced mammary tumors [72, 73]. Mammary tissue fragments from mice of the two parental strains transplanted into mammary parenchyma-free fat pads of MMTV-infected F_1 mice were susceptible to HAN and tumor development (Fig. 4). Nodules arising in C57BL/6 mammary transplants gave rise to HOGs and mammary tumors following secondary transplantation into mammary parenchyma-free fat pads of uninfected C57BL/6 or F_1 mice. As with C57BL/6 transplants, I/LnJ transplants produced tumors on secondary transplantation into mammary parenchyma-free fat pads of F_1 hosts. However, I/LnJ transplants never produced tumors after secondary transplantation into syngeneic uninfected I/LnJ hosts. Also, tumors of I/LnJ origin (propagated in an F_1 host) never grew upon transplantation into I/LnJ mice (Fig. 4) [72, 73]. Thus, it appears that I/LnJ mice fail to support tumor growth. These data suggest that enhanced immune surveillance of tumor cells might underlie the mechanism of resistance to MMTV-induced mammary tumors.

Studies from our laboratory have shown that I/LnJ mice have the H-2^J haplotype and express the I-E molecule necessary for SAg presentation [74]. Uninfected I/LnJ mice also have a normal percentage of SAg-cognate T cells that are Vβ14$^+$. Therefore, I/LnJ mice possess T cells necessary for MMTV(C3H) infection [74]. From these data, we concluded that I/LnJ mice have a novel mechanism of resistance to MMTV infection and MMTV-induced mammary tumors unrelated to MHC genes or endogenous *Mtv*s.

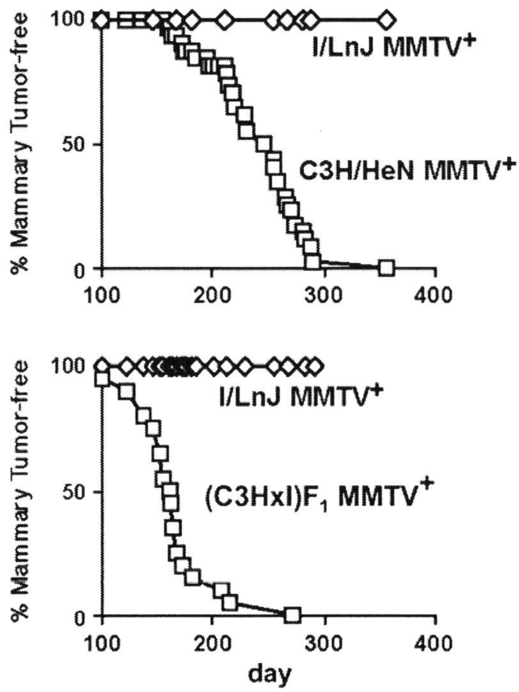

Figure 3. Mammary gland tumorigenesis in MMTV(C3H)-infected I/LnJ, C3H/HeN and F_1 mice. Mice were infected by foster nursing on viremic C3H/HeN females, bred and monitored for mammary tumors.

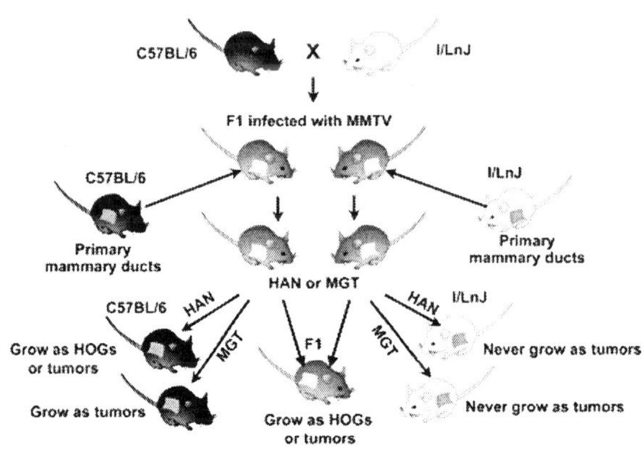

Figure 4. Resistance of I/LnJ mice to MMTV-induced mammary tumors exists at the organismal level. See explanations in the text. MGT, mammary gland tumors; HOGs, hyperplastic outgrowths; HAN, hyperplastic alveolar nodules.

First, it needed to be determined whether I/LnJ mice are capable of becoming MMTV-infected. To address this question, newborn I/LnJ mice as well as susceptible BALB/cJ mice were foster-nursed on MMTV(C3H)-infected BALB/cJ females, bred and their pups or pups from other susceptible strains were allowed to suckle on their milk. To determine whether mice from both generations became MMTV-infected, we analyzed percentages of their SAg-cognate T cells in the periphery and looked at integrated proviruses in lymphocytes and at virus secretion into the milk. These experiments showed that infection with MMTV(C3H) exogenous virus did not progress to the mammary glands of I/LnJ mice, even though mice responded to viral SAg and their lymphocytes were productively infected [74]. Furthermore, MMTV-infected I/LnJ lymphocytes failed to deliver virus to the mammary gland of other susceptible syngeneic mice [74].

To test whether resistance to mammary gland infection and subsequent tumorigenesis applied to other MMTV viruses in addition to MMTV(C3H), we infected I/LnJ mice with MMTV(LA) [9]. MMTV(LA) viruses are produced at much higher titers compared to MMTV(C3H) [9] and Pobezinskya et al., manuscript submitted). I/LnJ females foster-nursed by BALB/cLA mice were bred and tested for the presence of virus in their milk (Fig. 5). All animals became mammary gland-infected, since they produced all three MMTVs in their milk (Fig. 5) [45]. Thus, in contrast to MMTV(C3H), infection with MMTV(LA) progressed to the mammary glands of I/LnJ mice.

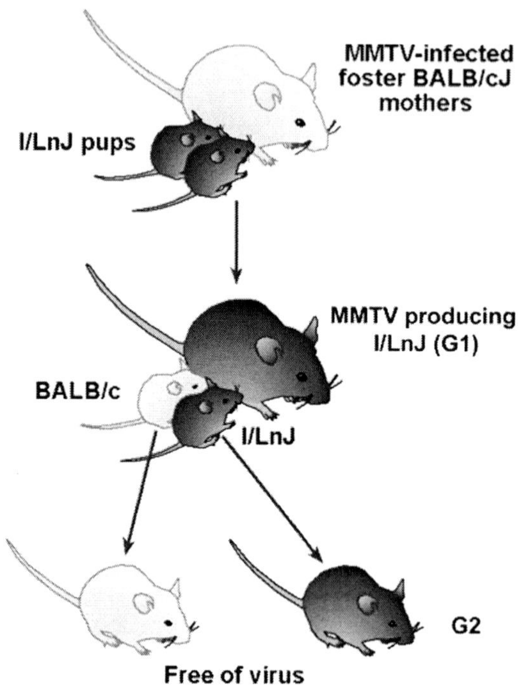

Figure 5. MMTV(LA) infection progresses to the mammary gland of I/LnJ mice, but infectious virus is eliminated in the G_2 generation. I/LnJ females fostered by BALB/cLA females produced virus into the milk as determined by RNase T_1 protection analysis with MMTV-specific probes. However, female offspring from infected I/LnJ mice (G_2 generation) as well as BALB/cJ mice fostered by the same mothers were found to be virus free.

To determine whether viruses secreted by infected I/LnJ mice were passed to the offspring, we produced G_2 generation animals (offspring of fostered I/LnJ females) and tested their milk for the virus (Fig. 5). Almost all animals either eliminated viruses or produced them at much-reduced titers compared to their mothers (Fig. 5) [45]. Like the I/LnJ offspring, BALB/cJ mice that were nursed on the same mothers did not produce virus into the milk [45]. Thus, although I/LnJ mice became infected with MMTV(LA) when foster-nursed by

BALB/cLA females, viruses they produced were uninfectious even in susceptible BALB/cJ mice.

To explain the loss of virus in subsequent generations of MMTV-infected I/LnJ mice, we hypothesized that MMTV virions produced by infected I/LnJ mice are coated with neutralizing antibodies (Fig. 6). These antibodies bind to surface glycoproteins on the virus, which blocks their binding to the plasma membrane receptors expressed on the host cell and ultimately prevents viral entry. To test this hypothesis, milk was collected from MMTV(LA)-infected I/LnJ or BALB/cJ females and virus particles were pelleted and viral proteins bound to plastic. Secondary antibodies against different immunoglobulin isotypes conjugated with alkaline phosphatase (AP) were used to develop an ELISA (enzyme-linked immunoabsorbent assay) (Fig. 7). These experiments showed that in contrast to viral particles isolated from MMTV-infected BALB/cJ milk, viral particles isolated from MMTV-infected I/LnJ milk were found to be associated with IgG2a immunonoglubulins (Fig. 7) [45]. Another ELISA was performed to determine whether anti-MMTV antibodies of the IgG2a isotype are present in the sera of MMTV-infected I/LnJ mice. In this ELISA, viral proteins were bound to plastic, incubated with sera from infected and uninfected I/LnJ mice and control BALB/cJ mice, and developed with either polyvalent antibodies or anti-IgG2a antibodies conjugated with AP. From this experiment, we found that only MMTV-infected I/LnJ mice have circulating antibodies against MMTV in their sera [45].

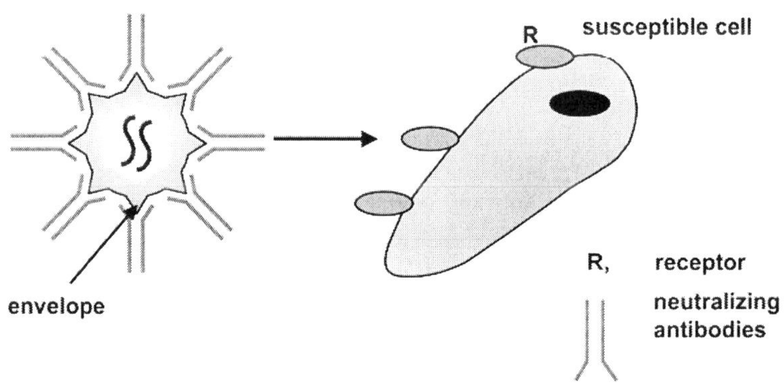

Figure 6. Hypothetical model. I/LnJ mice produce virus-neutralizing antibodies directed against the SU *env* gene product which coat cell-free viral particles and thus, affect receptor recognition.

In order to confirm our hypothesis, we needed to determine whether these anti-MMTV antibodies produced by infected I/LnJ mice were capable of neutralizing MMTV. An *in vitro* neutralizing assay was performed, where virus isolated from the milk of MMTV-infected BALB/cJ mice was incubated with serum samples from either uninfected or infected I/LnJ and BALB/cJ mice and then injected into the mammary glands of susceptible, uninfected BALB/cJ mice. To test whether injected mice became MMTV-infected, the percent of SAg cognate T cells was measured five weeks after injection. We found that the BALB/cJ recipients did not become MMTV-infected when the virus was incubated with sera obtained from infected I/LnJ mice (Table 3) [45]. Therefore, MMTV-infected I/LnJ mice produce antivirus neutralizing antibodies of the IgG2a isotype.

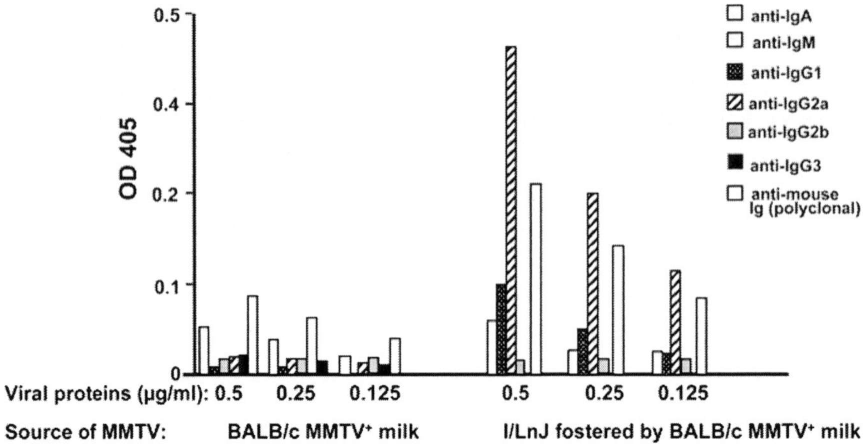

Figure 7. Viral particles isolated from the milk of I/LnJ mice fostered by viremic BALB/cLA mothers are coated with antibodies of the IgG2a isotype. MMTV(LA) was pelleted from the skim milk of I/LnJ or BALB/cJ mice fostered by BALB/cLA females and bound to the plate. Secondary antibodies against different mouse immunoglobulins conjugated with alkaline phosphatase (AP) were used at the second step. Background obtained with the secondary antibody alone was subtracted (from Purdy et al., *J. Exp. Med.* 197, 233, 2003, with permission).

Table 3. MMTV-infected I/LnJ mice produce anti-virus neutralizing antibodies

Sera used	$CD4^+/V\beta6^+$ (%)
I/lnJ	5.5±2.3
I/LnJ MMTV$^+$	11.5±1.1
BALB/cJ	3.4±1.7
BALB/cJ	5.7±1.3

MMTV was isolated from viremic BALB/cJ females and incubated with different sera followed by injection into the mammary gland of 3 to 4-week-old BALB/cJ females. Mice were bled 6 to 9 weeks later and percentages of SAg-cognate T cells were determine in the peripheral blood. Uninfected BALB/cJ mice have 11.0±1.1 $CD4^+/V\beta6^+$ T cells. 6 – 10 mice were used per group.

Similarly, an *in vivo* neutralization assay was performed where sera from either infected I/LnJ mice or uninfected I/LnJ control mice were injected into MMTV-infected adult or newborn BALB/cJ mice nursing on viremic mothers. Both groups of mice were bred and RNA was isolated from their lactating mammary glands to be analyzed for the presence of MMTV-specific transcripts. The data showed that both MMTV-infected adult and newborn BALB/cJ mice were protected from MMTV infection when injected with sera from MMTV-infected I/LnJ mice [45]. Therefore, anti-MMTV antibodies produced by infected I/LnJ mice are capable of neutralizing MMTV infection *in vitro* and *in vivo*.

With this conclusion firmly established, the next step was to identify the viral proteins recognized by the anti-MMTV antibodies. This was analyzed by western blot. Viral proteins were separated on a gel, blotted to nitrocellulose, incubated with sera from either MMTV-infected or uninfected I/LnJ mice followed by anti-mouse antibodies against IgG2a or IgG1 immunoglobulins.

There are three major proteins within the MMTV virion: gp52SU and gp36TM, both products of the *env* gene, and p27CA, a product of the *gag* gene [11]. A pool of sera from three uninfected I/LnJ mice showed no reactivity against any of these MMTV proteins (Fig. 8). In contrast, all six MMTV-infected I/LnJ mice produced anti-gp52SU and anti-gp36TM antibodies as well as antibodies against p27CA (Fig. 8). Furthermore, all the reactive antibodies were of the IgG2a and not of the IgG1 isotype, except for mouse #3 (Fig. 8) [45]. The appearance of anti-p27CA- and anti-gp36 TM-specific antibodies is surprising, since these proteins are localized within infected cells. It is possible that infected cells are coated by anti-gp52SU antibodies and thus facilitate virus uptake into phagocytic cells through interaction with the Fc receptor or through the complement pathway. Phagocytic antigen-presenting cells then might present digested proteins to T cells and stimulate the T-cell dependent B-cell response against internal viral proteins.

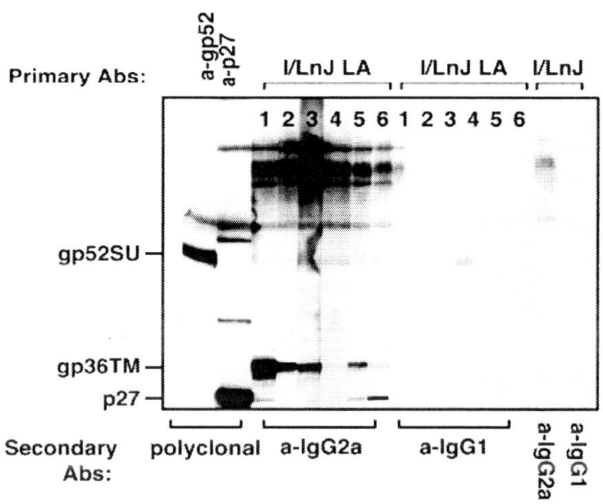

Figure 8. Antibodies produced by MMTV-infected I/LnJ mice recognize major virion proteins. Western blot analysis of MMTV virion proteins with sera of MMTV(LA)-infected I/LnJ mice. MMTV(LA) purified from the milk of BALB/cLA females was run on a 10% denaturing acrylamide gel, blotted to nitrocellulose, and incubated with mouse monoclonal antibodies against gp52 (a-gp52), p27 (a-p27), gp36 (not shown), or sera from infected I/LnJ mice (I/LnJ LA) or uninfected I/LnJ mice (I/LnJ). Anti-mouse polyclonal, or anti-mouse IgG2a-specific or IgG1-specific antibodies coupled to horseradish peroxidase were used at the second step. A 1:100 dilution of each serum was used. 1 - 6, individual mice (from Purdy et al., *J. Exp. Med.* 197, 233, 2003, with permission).

The fact that the antibodies of the IgG2a isotype produced by MMTV-infected I/LnJ mice were capable of neutralizing MMTV suggested that these antibodies might also prevent MMTV-induced mammary tumors by blocking the virus re-infection/re-integration action necessary for tumor induction. To confirm this hypothesis, we monitored MMTV(LA)-infected I/LnJ and control BALB/cJ mice for mammary tumors. None of the infected I/LnJ mice developed mammary tumors, while 80 percent of the infected BALB/cJ mice developed mammary tumors within 1 year.

In order to understand when the immune response is initiated after infection with MMTV, sera from both MMTV-infected susceptible BALB/cJ and resistant I/LnJ mice were tested for reactivity against MMTV proteins at different time points. These experiments showed that both BALB/cJ and I/LnJ mice began to produce polyvalent immunoglobulins against MMTV at three weeks of age (Fig. 9). However, only infected I/LnJ mice undergo a class switch to immunoglobulins of the IgG2a isotype and demonstrate a steady and increasing production of these antibodies beginning at about six weeks of age (Fig. 9). Susceptible BALB/cJ mice were incapable of mounting a long-lasting antibody response against MMTV and did not class switch to the IgG2a isotype (Fig. 9).

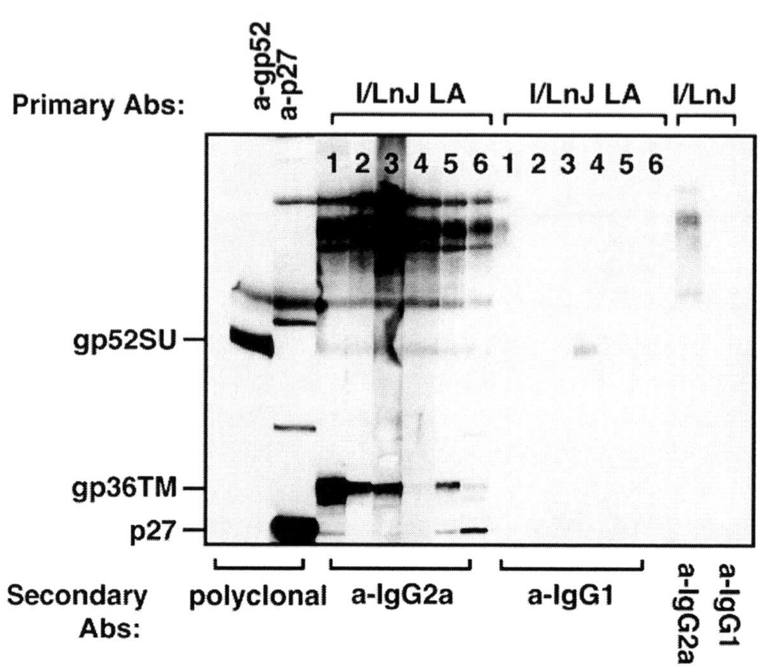

Figure 9. Kinetics of anti-MMTV antibody appearance in neonatally infected I/LnJ and BALB/cJ mice. Three mice of each strain were bled weekly from 3 to 8 weeks of age, and the presence of antibodies against MMTV virion proteins in their sera was detected by an ELISA. A 2.5x10-2 dilution of sera was used in this experiment. Readings obtained with sera from age-matched uninfected mice were subtracted (from Purdy et al., J. Exp. Med. 197, 233, 2003, with permission).

In general, the immune response in I/LnJ mice was not skewed towards the IgG2a isotype, because, as with BALB/cJ mice, I/LnJ mice immunized with MMTV virion proteins in complete Freund adjuvant (CFA) produced mostly IgG1 specific antibodies against MMTV antigens [45]. These data clearly established that the production of anti-MMTV antibodies of the IgG2a isotype by infected I/LnJ mice is a specialized response to retroviral infection.

To determine the cellular basis of I/LnJ resistance, we took advantage of the fact that F_1 progeny of a cross between susceptible MHC-congenic C3.JK and resistant I/LnJ mice were susceptible to MMTV infection [74] and performed bone marrow transfer experiments. Adult MMTV-infected susceptible (C3.JK x I/LnJ)F_1 mice were lethally irradiated and injected with bone marrow isolated from I/LnJ mice. The chimeric mice were tested for their ability to produce anti-MMTV antibodies 70 days after transfer. All I/LnJ to F_1 chimeras showed production of antibodies against MMTV in their sera (Table 4) (Case et al., manuscript in preparation). These results show that cells of I/LnJ bone marrow origin are sufficient to convey the ability to produce neutralizing antibodies to susceptible mice. Therefore, the stromal cells of the secondary lymphoid organs and follicular dendritic cells can be eliminated as possible cell types that contribute to the sustained production of MMTV-neutralizing antibodies in I/LnJ mice.

Table 4. Antibody production in bone marrow chimeras

Donor to Recipient			MMTV Infected[1]	Percent Chimerism[2]	Antibody production[3]
I	to	F_1	+	48%	+
I	to	F_1	+	39%	+
I	to	F_1	+	44%	+
I	to	F_1	+	56%	+
I	to	F_1	+	62%	+
I	to	F_1	+	50%	+
F_1	to	F_1 (5)	+	-	-

[1]Determined by $CD4^+/V\beta14^+$ T cell deletion.
[2]Percentage of CD5-1 $^+$/CD5-1.1$^-$ cells over all CD5 $^+$ cells. I/LnJ mice have CD5-1 allele, whereas C3.JK mice have CD5-1.1 allele.
[3]ELISA assay 13 weeks after bone marrow transfer.

Next we sought to determine which subset of bone marrow-derived cells is responsible for production of neutralizing antibodies by I/LnJ mice. To address this question, we performed chimera experiments in which some cell populations were negatively selected from I/LnJ and residual cells were injected into adult MMTV-infected IFN-γ deficient I/LnJ mice, which were sub-lethally irradiated. The recipient mice were tested for antibody production two months after injection. All recipient mice receiving either unmanipulated I/LnJ splenocytes or I/LnJ splenocytes lacking T, B, MHC class positive cells, CD11C or CD11B cells began to produce anti-MMTV antibodies. In contrast, recipient mice receiving I/LnJ splenocytes lacking CD4 positive cells were unable to produce virus-neutralizing antibodies. Therefore, preliminary data suggests that IFN-γ secreting CD4 positive T cells are responsible for the production of neutralizing antibodies in I/LnJ mice (Case et al., manuscript submitted).

The mechanism of resistance to MMTV infection in I/LnJ mice is recessive, since infected F_1 mice obtained from crosses between resistant I/LnJ mice and susceptible C3H/HeN mice were susceptible to MMTV infection and MMTV-induced mammary tumors [45, 74]. When these infected F_1 females were backcrossed to I/LnJ males producing an N_2 generation, 75 percent of these mice were found to be resistant because they produced antibodies against MMTV, whereas 25 percent of the mice were found to be susceptible, because they did not produce antibodies against MMTV and succumbed to mammary tumors (Fig. 10). Subsequent analysis of mice from generations N_3 and N_4 demonstrated that two phenocopying genes control the mechanism of resistance inherited by I/LnJ mice. We refer to these gene loci as *vic1* and *vic2*, which stands for ***v**irus **i**nfectivity **c**ontroller* 1 and 2.

To map the location of the *vic* genes, a genome-wide screen was performed with 50 susceptible and 50 resistant N_2 mice with markers at every 10 to 15 cM. *Vic*1 and *vic*2 have been mapped to chromosome 17 and chromosome 15, respectively (Case, manuscript in preparation).

MMTV-infected I/LnJ mice demonstrate a steady and increasing production of anti-MMTV antibodies of the IgG2a isotype after an initial poly-isotypic response (Fig. 9). It has been well documented that class switching to the IgG2a isotype is induced primarily by interferon γ (IFN-γ) [75]. We hypothesized that IFN-γ is required for anti-MMTV antibody production in resistant I/LnJ mice. To investigate this hypothesis, I/LnJ mice with a targeted mutation of IFN-γ were produced by 10 repetitive backcrosses of C57BL6/J IFN-γ knockout (KO) mice to I/LnJ mice and tested for susceptibility to MMTV infection. All IFN-γ KO I/LnJ mice became MMTV-infected; they experienced deletion of SAg-cognate T cells and contained newly integrated MMTV proviruses, establishing that IFN-γ is not required for MMTV infection [45]. To test whether IFN-γ is necessary for the production of virus-neutralizing antibodies by I/LnJ mice, sera from MMTV-infected IFN-γ-sufficient and IFN-γ-deficient I/LnJ mice were tested for reactivity against MMTV by ELISA.

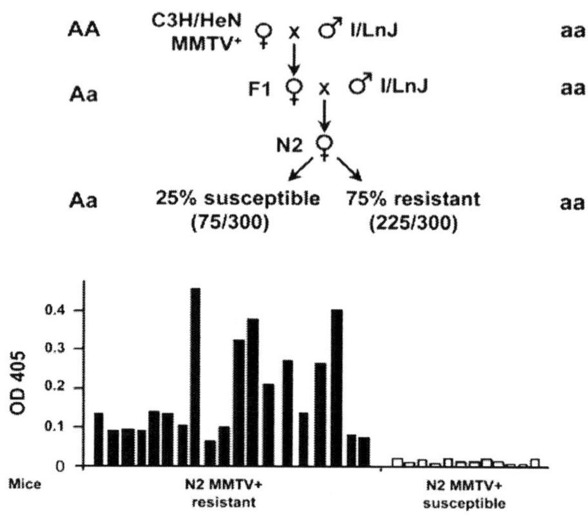

Figure 10. Mechanism of resistance to MMTV infection in I/LnJ mice is recessive and controlled by two redundant genes. Serum samples from MMTV(C3H)-infected resistant and susceptible N_2 females, I/LnJ, C3H/HeN and F_1 hybrid mice were tested for reactivity against MMTV virion proteins by ELISA. Anti-mouse IgG2a antibodies coupled to alkaline phosphatase were used at the second step. Sera were diluted from 1:20 to 1:3000, dilution 1:400 is shown. Antibody-negative mice did not show any reactivity at all dilutions tested. Either MMTV(C3H) virus isolated from the milk of C3H/HeN MMTV$^+$ mice, or MMTV(LA) virus isolated from the milk of BALB/cLA mice were attached to the plastic [(MMTV(C3H) is shown]. Background obtained with the secondary antibody alone was subtracted.

We found that, unlike virus-infected I/LnJ mice, I/LnJ mice with a targeted mutation of INF-γ were incapable of producing virus-neutralizing antibodies and secreted infectious virus into their milk. As a result, mice fostered on their milk all became MMTV-infected [45]. These data show that IFN-γ KO I/LnJ mice no longer have the ability to block virus transmission.

IFN-γ production is tightly regulated by a number of molecules that modulate signaling responses. These negative regulators of IFN-γ include the SH2-containing tyrosine phosphatases (SHP-1 and SHP-2) and inositol phosphatase (SHIP1 and SHIP2), protein inhibitors of activated STATs (PIAS), as well as suppressors of cytokine signaling (SOCS-1-

7) [76-78]. Since the resistant mechanism exhibited by I/LnJ mice is recessive and dependent on IFN-γ, we have hypothesized that the *vic* genes signal within the negative loop controlling IFN-γ production (Fig. 11). In this scenario, both MMTV-infected susceptible and resistant mice initiate the production of a poly-isotypic humoral response against the virus while secreting IFN-γ. However, in susceptible mice, IFN-γ production is quickly suppressed due to a negative regulator, which favors virus replication and results in long-life virus persistence (Fig. 11). In contrast, resistant I/LnJ mice, which fail to properly control IFN-γ due to recessive alleles of the *vic* genes, continue to secrete the cytokine, and thus class switch to the IgG2a isotype response. IgG2a immunoglobulins are most potent in neutralizing virus infections [79], and this results in complete clearance of MMTV in I/LnJ mice (Fig. 11). Therefore, retroviruses might use a timely negative regulation of IFN-γ production as a way to evade antivirus immune response.

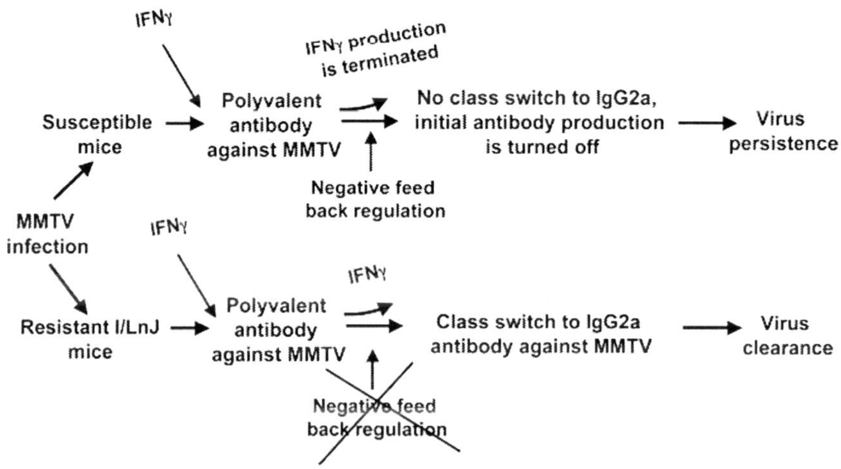

Figure 11. Hypothetical model of the *vic* genes functioning. See the text for explanations.

Humoral immune response, which includes the production of antiviral antibodies, is a mainline defense against viral infections. However, retroviruses like HIV are generally able to escape humoral immune responses because of their rapid mutation rate, leading to the selection of immune escape variants [80]. These immune escape variants begin to flourish and will ultimately defeat the immune responses that were mounted against them. Most of the mutations found in the immune escape variants were mapped within the SU domain of the *env* gene, which encodes for the less conserved extracellular domain of Env. As a result, the importance of neutralizing antibodies in controlling HIV infection still remains a subject of controversy.

Intriguingly, not all individuals infected with HIV had developed acquired immunodeficiency syndrome (AIDS), even though they demonstrated low virus titer in the blood [81]. These data suggested that a humoral immune response mounted against HIV infection might protect these individuals from the devastating disease. Indeed, virus-neutralizing antibodies have been detected in the sera of these non-progressors, and the majority of these antibodies were of the IgG fraction recognizing different domains of Env [82]. We believe that the humoral immune response fails to control HIV infection in humans because it is not robust and sustained. Infected I/LnJ mice, on the other hand, are capable of rapid production of retrovirus-neutralizing antibodies and are able to sustain this response throughout their lifetime. Furthermore, these neutralizing antibodies are directed against multiple viral proteins that cross-react with different MMTV variants, preventing the selection of escape variants [45]. Elucidation of the mechanism of resistance to retroviral infection is of fundamental importance, as it will ultimately lead to increased knowledge about the antivirus immune response in general, and about variations in susceptibility to viral infections in humans. In addition, we hope that it would have major implications in the development of effective vaccines against HIV and other viruses.

In addition to their resistance to virally-induced tumors, mice of the I/LnJ strain are also highly resistant to propagable tumors, DMBA-induced mammary tumors [83-85], and MCA-induced fibrosarcomas [86]. No spontaneous tumors have been documented in I/LnJ mice. The ability of I/LnJ mice to block the growth of syngeneic tumors upon transplantation [72, 73] indicates that this strain can recognize tumor-specific antigens. Resistance of I/LnJ mice to MMTV-induced mammary tumors results from an IFN-γ-dependent, robust, and sustained humoral antivirus immune response. Therefore, I/LnJ mice continue to recognize virus-specific antigens as foreign and maintain the immune response throughout their life. It appears that I/LnJ mice are always aware of "non-self" antigens. Since tumor cells express tumor-specific antigens that distinguish them from normal cells, we speculate that tumor suppression in I/LnJ mice might also be associated with the unique ability of their immune system to discriminate between normal and tumor cells. Identification of the *vic* genes and mechanisms of their action will help to address this question.

C3H/HeJ mice. More than 50 years ago, the high-tumor-incidence, MMTV(C3H)-infected C3H mouse strain was divided between the National Institutes of Health (C3H/HeN) and The Jackson Laboratory (C3H/HeJ) (Fig. 12). In 1973, it was reported that C3H/HeJ mice demonstrated a drastic decrease in mammary tumor incidence compared to MMTV-infected C3H/HeN mice [87]. Since we learned that C3H/HeJ mice did not lose exogenous virus [39], we reasoned that such attenuation of mammary tumorigenesis in C3H/HeJ mice could be due to either differences in the genetic background between C3H/HeN and C3H/HeJ, or occurrence of a novel MMTV variant.

To discriminate between these two possibilities, we fostered pups from the same C3H/HeN or C3H/HeJ MMTV$^-$ litters either on C3H/HeN MMTV$^+$ milk or on C3H/HeJ MMTV$^+$ milk (Fig. 13). After mice reached puberty, they were bred and monitored for mammary gland tumors. Fifty percent of the C3H/HeN mice fostered on C3H/HeN MMTV$^+$ milk developed tumors after approximately 250 days, whereas mice from the same substrain fostered on C3H/HeJ MMTV$^+$ milk developed no tumors within the same period of time, even though all mice became MMTV infected [39]. In contrast, C3H/HeJ mice fostered by C3H/HeN MMTV$^+$ females developed tumors at the same rate as C3H/HeN MMTV$^+$ mice, even though C3H/HeJ mice infected with their own virus were tumor-resistant [39]. Therefore, exogenous MMTV naturally occurring in C3H/HeJ mice was non-tumorigenic even in tumor-susceptible C3H/HeN mice, and C3H/HeJ mice were susceptible to MMTV(C3H)-induced mammary tumors. We concluded that attenuation of mammary tumors in C3H/HeJ mice must be due to a novel MMTV variant, which we called MMTV(HeJ).

To determine the structure of MMTV(HeJ), this virus was cloned and sequenced together with MMTV(C3H). As mentioned above, MMTV(HeJ) was found to be a recombinant between endogenous *Mtv1* present in all C3H/He strains and exogenous wild-type

MMTV(C3H) with almost the entire genome of the virus derived from *Mtv1* except for the hypervariable region of SAg (Fig.14) [39]. Since both C3H/HeJ and C3H/HeN mice were derived from a mouse strain infected with highly tumorigenic virus we hypothesized that selective pressure against the wild-type MMTV(C3H) in C3H/HeJ mice eliminated this variant and promoted survival of the recombinant MMTV(HeJ). Coincidently, *Mtv1* sequences made the virus incapable of causing mammary tumors.

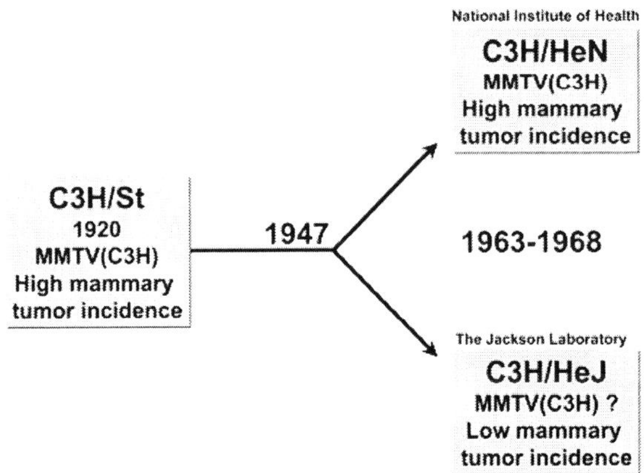

Figure 12. Origin of C3H/HeN and C3H/HeJ strains. See the text for explanations.

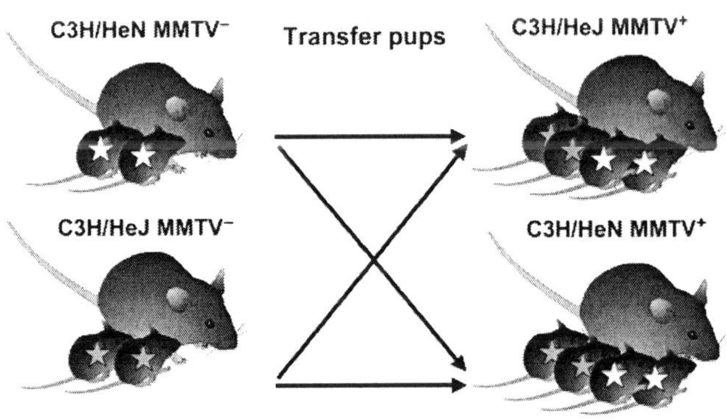

Figure 13. Foster-nursing experiments. See the text for explanations.

To follow the generation of a new recombinant virus, progeny of the MMTV-negative C3H/HeJ mice were infected with exogenous MMTV(C3H) by foster-nursing on C3H/HeN $MMTV^+$ milk. These MMTV(C3H)-infected C3H/HeJ females (generation 1 or G_1) then were mated to C3H/HeJ males to produce infected generation 2 (G_2) offspring (Fig. 15). The infected females from each generation were mated to C3H/HeJ males to produce subsequent generations. Mice from each generation were sacrificed at 8 to 10 weeks of age and their splenic DNA was tested for newly integrated exogenous proviruses using semi-quantitative PCR that specifically distinguished MMTV(C3H) from recombinant MMTV(HeJ) [88]. We found that MMTV(C3H) virus was the only virus detectable in the foster-nursed G_1 and G_2 C3H/HeJ animals. However, the recombinant MMTV(HeJ) became evident starting with G_3 animals, dominated by G_6 and completely replaced the wild-type MMTV(C3H) by G_9 (Fig.

15). At the same time, the recombinant virus was never detected in more than 25 tested MMTV(C3H)-infected C3H/HeN mice [88]. This is a clear example of retroviral evolution and suggests a selective pressure applied by the C3H/HeJ genome against MMTV(C3H).

To confirm that the newly selected virus has a structure similar to MMTV(HeJ), we cloned and sequenced virus present in G_{10} of the MMTV(C3H)-infected C3H/HeJ mouse pedigree. Like MMTV(HeJ), newly selected recombinant MMTV is recombinant between the wild-type MMTV(C3H) and endogenous *Mtv1*. In contrast to MMTV(HeJ), however, only the *gag* gene of the newly selected recombinant virus was derived from *Mtv1* (Fig. 14). These data suggest that selection of recombinant MMTV(HeJ)-like viruses from wild-type MMTV(C3H) is reproducible and that the selective pressure against wild-type MMTV(C3H) in C3H/HeJ mice is directed against the *gag* gene product.

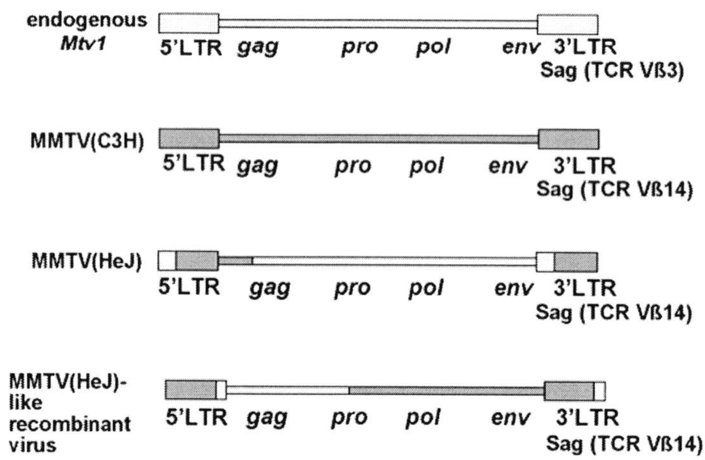

Figure 14. Structure of *Mtv1*, MMTV(C3H), MMTV(HeJ), and a newly selected recombinant MMTV. The recombination break point in the newly selected recombinant MMTV occurred in the *pro* gene, whereas MMTV(HeJ) consisted primarily of endogenous sequences derived from endogenous *Mtv1*. *Mtv1 gag* gene sequences were conserved in all recombinant viruses.

Why is the recombinant MMTV selected in C3H/HeJ mice but not in C3H/HeN mice? The best known genetic difference between C3H/HeJ and C3H/HeN mice is within the lypopolysaccharide sensitivity (*lps*) locus, which determines cellular response to LPS [89, 90]. C3H/HeN mice are LPS-sensitive, whereas C3H/HeJ mice are LPS-resistant. Both C3H/HeJ and C3H/HeN strains were derived from the LPS-sensitive C3H/St mouse strain between 1945-1947 (Fig. 16). Later, three strains were derived from the C3H/HeJ strain: C3Heb/FeJ in 1949, C3H/DiSn in 1953, and C3H/A^{vy} in 1960. All three of these strains are LPS-responsive, indicating that the original C3H/HeJ strain must have been LPS-responsive as late as 1960.

In 1968 the C3H/Bts strain was derived from the C3H/HeJ strain. C3H/Bts mice are as unresponsive as C3H/HeJ mice to LPS, demonstrating that a gene mutation causing complete unresponsiveness to LPS occurred after 1960 and by 1968 was homozygous in the C3H/HeJ colony. Like C3H/HeN mice, the parental LPS-sensitive C3H/St mice were characterized by a high incidence of mammary tumors, whereas C3H/HeJ mice demonstrated a dramatic decrease in mammary tumor incidence. Thus, the loss of LPS sensitivity in C3H/HeJ mice coincided with the attenuation of MMTV-induced tumorigenicity in this strain in the late 1960s, suggesting a link between *Lps* mutation and selection of the recombinant MTV(HeJ) [39] (Fig. 16).

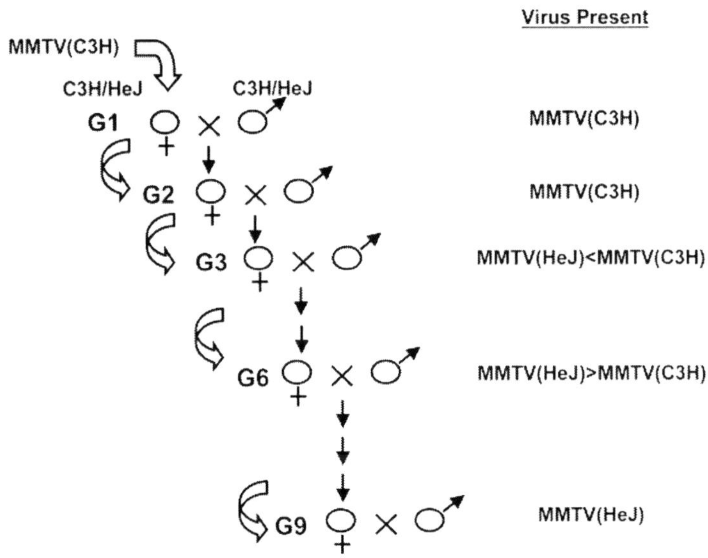

Figure. 15. Recombinant MMTVs are selected in C3H/HeJ mice infected with MMTV(C3H). C3H/HeJ mice free of MMTV were fostered by MMTV(C3H)-infected C3H/HeN females (G1), and crossed to C3H/HeJ males to generate G2 mice. Females from each generation were crossed to C3H/HeJ males to generate subsequent generations (G2 - G9). Mice from each generation were sacrificed and their splenic DNA was subjected to the semi-quantitative PCR analysis specific for either the MMTV(C3H) or MMTV(HeJ) virus. PCR products were separated on agarose gel, transferred to a nylon membrane, hybridized with the LTR probe, and intensity of resulting bands was compared using a PhosphoImager densitometric analysis. Recombinant MMTV becomes detectable at G3 in the MMTV(C3H)-infected C3H/HeJ mouse pedigree, dominates by G7, and completely replaces MMTV(C3H) in G9.

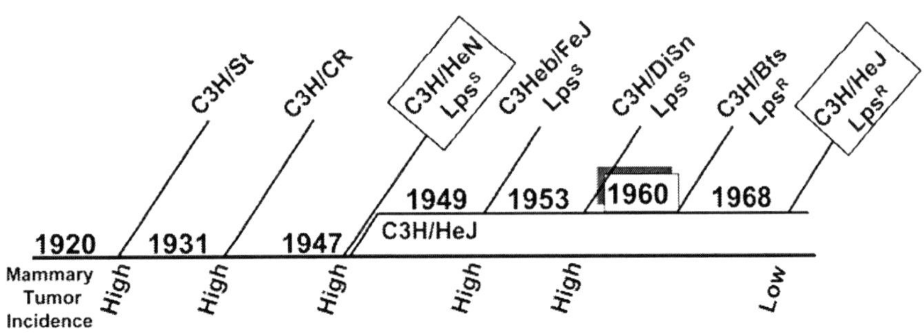

Figure 16. Tumor attenuation in C3H/HeJ mice coincided with a loss of responsiveness to lypopolysaccharide (Lps). C3H/HeJ and C3H/HeN strains were derived from the LPS-sensitive C3H/St mouse strain between 1945-1947. Later, three strains were derived from the C3H/HeJ strain: C3Heb/FeJ in 1949, C3H/DiSn in 1953, and C3H/A^{vy} in 1960. All three of these strains are LPS-responsive, indicating that the original C3H/HeJ strain must have been LPS-responsive as late as 1960. In 1968 the C3H/Bts strain was derived from the C3H/HeJ strain. C3H/Bts are as unresponsive as the C3H/HeJ mice to LPS demonstrating that a gene mutation causing complete unresponsiveness to LPS occurred after 1960 and by 1968 was homozygous in the C3H/HeJ colony. Like C3H/HeN mice, the parental LPS-sensitive C3H/St mice were characterized by a high incidence of mammary tumors, whereas C3H/HeJ mice demonstrated a dramatic decrease in mammary tumor incidence.

Recently it was shown that the innate immune Toll-like receptor (*Tlr*) 4 is the product of the *Lps* locus [62]. TLR4 is a member of the Toll family proteins, which recognize specific patterns associated with microbial pathogens. TLR4 recognizes LPS produced by Gram-negative bacteria. C3H/HeJ mice contain a dominant negative missense mutation of the *Tlr4* gene, whereas C57BL10/ScNcR mice are homozygous for a null mutation at the gene [91].

Our data suggested that the loss of normal TLR4 functioning may be a factor determining selection of recombinant MMTV(HeJ). To investigate whether the dominant negative allele of the *Tlr4* gene is such a factor, we designed the following breeding scheme. C57B10/ScNCr ($Tlr4^{0/0}$) mice were crossed either to C3H/HeJ ($Tlr4^{dn/dn}$) mice or C3H/HeN ($Tlr4^{+/+}$) mice. F_1 mice were intercrossed and the resultant F_2 mice were fostered by MMTV(C3H)-infected C3H/HeN females (Fig. 17). These F_2 mice were also intercrossed to produce generations of mice with 4 different genotypes (Fig. 17). Semi-quantitative PCR specific for MMTV(C3H) and recombinant MMTV(HeJ) was performed with the splenic DNA isolated from mice of 4 genotypes after three virus passages. This experiment showed that 29 out of 30 mice deficient in TLR4 signaling ($Tlr4^{0/0}$, 20/20 $Tlr4^{dn/dn}$ and 33/35 $Tlr4^{dn/0}$) produced recombinant virus, whereas all mice with functional TLR4 signaling did not show any evidence for recombinant virus (Table 5).

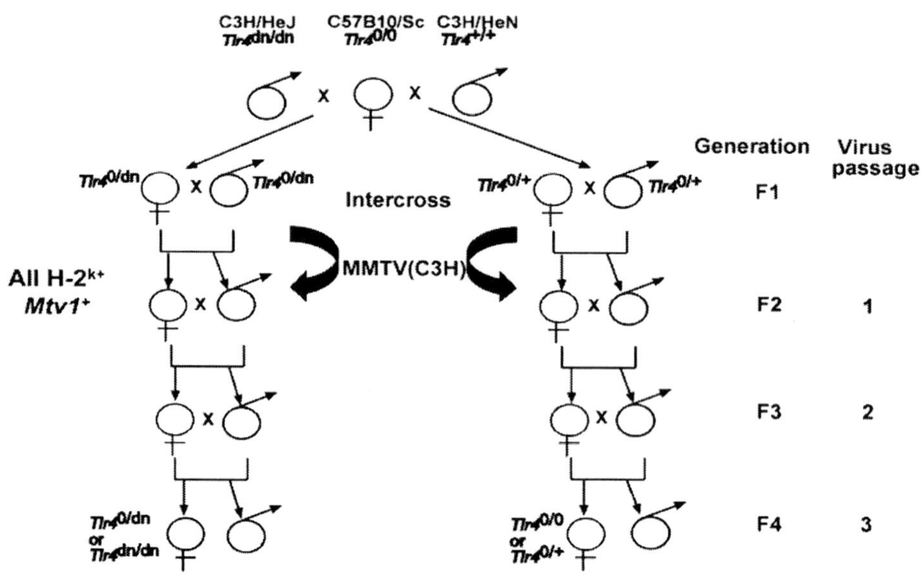

Figure 17. Breeding scheme for testing the effect of dominant-negative and null mutations of TLR4 on the selection of recombinant MMTV. C57B10/ScNCr mice were crossed either to C3H/HeJ or C3H/HeN mice. F_1 mice were intercrossed and the resultant F_2 mice with permissive for MMTV infection H-2k$^{+/+}$ MHC haplotype were fostered by MMTV(C3H)-infected C3H/HeN females. MMTV-infected F_2 mice were intercrossed again to produce generation F_3 animals. Females in each generation were screened for the *Tlr4* gene and *Mtv1*. Only *Mtv1* positive mice were used in the following crosses. Semi-quantitative PCR specific for MMTV(C3H) and recombinant MMTV(HeJ) was performed with the splenic DNA isolated from the F_4 mice. +, the wild-type allele of *Tlr4*; dn, the dominant negative allele of *Tlr4*; 0, the null allele of *Tlr4*.

Table 5. Recombinant MMTV is selected only in TLR4 deficient mice

Genotype	Total mice tested	Mice with recombinant MMTV
$Tlr4^{0/+}$	25	0
$Tlr4^{+/+}$	15	0
$Tlr4^{0/0}$	30	29
$Tlr4^{0/dn}$	35	33
$Tlr4^{dn/dn}$	20	20

Summary of PCR data from mice differing in TLR4 expression infected with wild-type MMTV (C3H). 0, null allele of *Tlr4*; dn, dominant negative allele of TLR4.

Therefore, we established a definitive genetic link between the mutation in the *Tlr4* gene and selection of the recombinant virus. Our data also suggested that MMTV could activate the TLR4 signaling system. To determine whether MMTV is capable of transducing TLR4 signaling, we co-transfected 293 cells with the plasmid containing MMTV provirus and the plasmid containing the mouse TLR4, and the NF-κB luciferase reporter gene. If binding of MMTV proteins results in NF-kB signaling, we should detect luciferase activity in transfected cells. Cells were harvested 24 hours post-transfection, and luciferase activity was measured. MMTV provirus greatly increased activation of NF-κB in cells expressing mouse TLR4, but not in TLR-4-negative cells (Fig. 18).

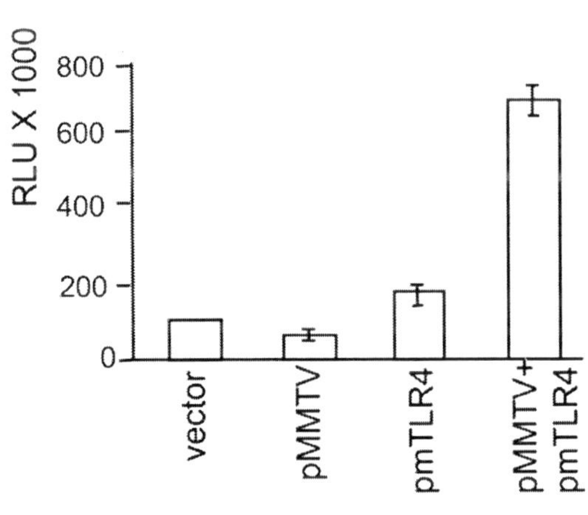

Figure 18. MMTV proteins interact with the TLR 4 signaling system. Human embryonic kidney 293 cells were transfected with a plasmid containing hybrid MMTV (pHP, 1.8 μg) and/or a plasmid containing the mouse *Tlr4* gene (mTLR4, 0.2μg), filler DNA to 2 μg , and NF-κB luciferase reporter gene (0.1μg). For the luciferase reporter assay, cells were harvested 24h post-transfection and lysed. RLU, relative luminescence units. pcDNA3, control vector alone.

Thus MMTV interacts with the TLR4 signaling system. Why does MMTV need TLR4? We hypothesize that virus/TLR4 interaction is needed to escape an antivirus immune response and that the selected recombinant virus is an immune escape variant. This hypothesis predicts that immunodeficient C3H/HeJ mice should not restrict replication of the wild-type MMTV(C3H) and should not select the recombinant virus. To determine whether immune response controls viral replication in C3H/HeJ mice, we used C3H/HeJ mice which lack ß2 macroglobulin (ß2M). These animals do not express MHC class I and thus do not generate cytotoxic CD8 T cells [92]. C3H/HeJ ß2m-deficient and C3H/HeJ ß2m-sufficient mice were fostered by C3H/HeN MMTV$^+$ females and bred. Virus fate was followed in C3H/HeJ ß2m-deficient and C3H/HeJ control mice after three virus passages. While all G_3 control C3H/HeJ mice produced recombinant MMTV, none of the C3H/HeJ ß2m-deficient mice produced the recombinant virus (Fig.19). Notably, the titers of wild-type MMTV(C3H) were higher in G_3 MMTV(C3H)-infected ß2m-deficient C3H/HeJ mice than in G_3 ß2m-sufficient C3H/HeJ mice [88]. Thus, it is adaptive immune response that controls MMTV replication in TLR4-deficient C3H/HeJ mice.

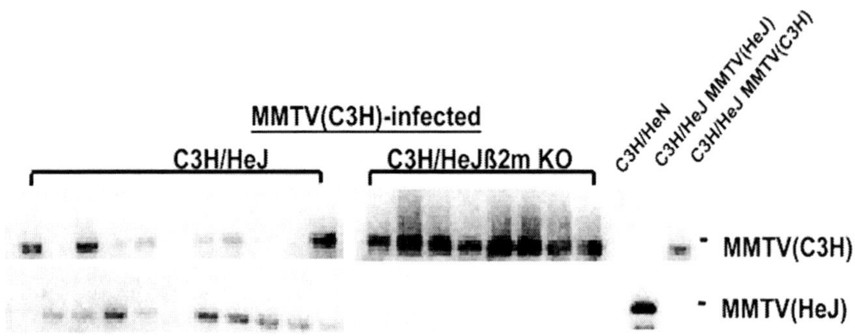

Figure 19. Recombinant MMTV is not selected in C3H/HeJ mice with a targeted mutation of $ß_2M$. $ß_2M$-sufficient and $ß_2M$-deficient C3H/HeJ mice were infected with MMTV(C3H) and bred to produce subsequent generations. At G_3, the presence of wild-type and recombinant viruses was tested in 8-week-old mice. Controls: C3H/HeJ mice infected with MMTV(C3H) (G_1); C3H/HeJ mice infected with MMTV(HeJ); uninfected C3H/HeN mice. (from Jude et al., *Nature Immunology.* 4, 573, 2003, with permission)

How does the virus evade the immune response in TLR4-sufficient C3H/HeN mice? One possibility is that MMTV stimulates proliferation of its primary target cells (B cells) through TLR4 signaling [93], leading to increased virus titers sufficient to withstand an ongoing immune response. However, replication of recombinant viruses in C3H/HeJ mice was not affected by TLR4 deficiency [88]. Another possibility is that MMTV/TLR4 engagement induces immunosuppressive signals. IL-10 is a primary candidate for such a signal, as it inhibits a broad spectrum of activated macrophage and monocyte functions, including IL-12 synthesis, production of nitric oxide, and expression of MHC class II and co-stimulatory molecules [94] and downregulates immunity to several pathogens [94-96], including retroviruses [97]. It is not surprising that some viruses, like poxviruses and cytomegaloviruses encode an IL-10 homolog that by blocking IL-10 signaling contributes to immune evasion during these viruses infection [98, 99]. In addition, it was recently shown that another pathogen, *Yersinia enterocolitica*, exploits TLR2 to stimulate IL-10-mediated immunosuppression [100]. We hypothesize that MMTV/TLR4 interaction leads to suppression of anti-virus immune response via stimulation of IL-10 production.

To examine whether MMTV/TLR4 interaction results in IL-10 production, splenocytes from C3H/HeN and C3H/HeJ mice as well as C57B10/J and C57BL10/ScNCr mice were incubated with intact MMTV virions *in vitro,* and culture supernatants were assayed for IL-10 production after 16 hours. IL-10 production was induced in TLR4-sufficient mice, but not in TLR4-deficient mice in response to MMTV (Table 6). IL-10 production required intact virions, because virions heated for 10 min at 85^{0}C failed to stimulate IL-10 production [88].

To determine the cellular basis of MMTV/TLR4-induced up-regulation of IL10-production, we performed a similar experiment with mice that lack either B cells or T cells. Our results showed that B cells are absolutely required for MMTV/TLR4-induced up-regulation of IL-10 production, because splenocytes from B cell-less mice were not capable of inducing IL-10 production (Table 6). In contrast, splenocytes from T cell-less mice stimulated IL-10 production in response to MMTV (Table 6) [88].

Table 6. Live MMTV stimulates TLR4-dependent IL-10 production by B cells

Mice	IL-10 production
C3H/HeN	+
$TLR4^{+/+}$	
C3H/HeJ	-
$TLR4^{dn/dn}$	
B10/ScNCr	-
$TLR4^{0/0}$	
B10/J	+
B6TCRβδKO	+
$TLR4^{+/+}$	
Igh-6	-
$TLR4^{+/+}$	
B10/ScNCr + Igh-6	+
B10/ScNCr + M∅	+
B10/ScNCr + DC	+
M∅	-
DC	-

Splenocytes were isolated from indicated mice and incubated *in vitro* with live MMTV virions. Sixteen hours later culture supernatans were assayed for IL-10 production by ELISA.

B6TCRβδ - splenocytes from TCRβδKO mice

Igh-6, splenocytes from B6 B-less mice

M, macrophages; DC, dendritic cells from RAG1 KO mice

B10/ScNCr + Igh-6, splenocytes from TLR4-deficient B10ScNCr mice were mixed with splenocytes from TLR4-sufficient B-cell deficient mice.

B10/ScNCr + M∅ splenocytes from TLR4-deficient B10ScNCr mice were mixed with M∅ from TLR4-sufficient T-cell and B-cell deficient RAG1 KO mice.

B10/ScNCr + DC, splenocytes from TLR4-deficient B10ScNcr mice were mixed with DC from TLR4-sufficient T-cell and B-cell deficient RAG1 KO mice.

To identify the cells that were activated through TLR4 and provided signaling to trigger IL-10 production by B cells, TLR4-deficient splenocytes were mixed with either TLR4-sufficient peritoneal macrophages or dendritic cells isolated from recombinase activating gene 1 (RAG1)-deficient mice (lacking T and B cells) and were exposed to MMTV. Both macrophages and dendritic cells were able to stimulate production of IL-10 by TLR4-deficient splenocytes (Table 6). Thus, MMTV stimulates IL-10 production by B cells through dendritic cell and macrophage activation mediated by TLR4 signaling [88]. The nature of the signaling to B cells by macrophages and dendritic cells remains to be elucidated.

Based on our data, we propose the following model. In C3H/HeN mice, TLR4/MMTV interaction leads to immunomodulation of antivirus immune response via production of IL-10 and to successful viral replication (Fig 20A). Absence of functional TLR4 in C3H/HeJ mice renders the virus unable to stimulate production of IL-10 and suppress antivirus immune responses and results in selection of an immune escape variant (Fig 20B). Thus, MMTV is the first retrovirus shown to subvert signaling through pattern recognition innate receptors for its own benefit.

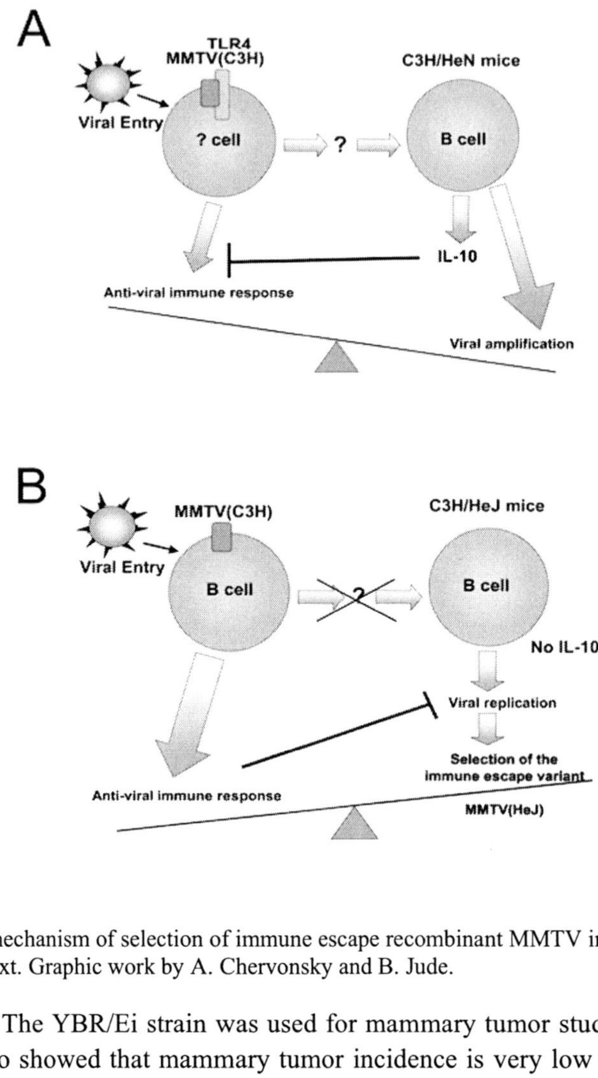

Figure 20. Possible mechanism of selection of immune escape recombinant MMTV in C3H/HeJ mice. See explanations in the text. Graphic work by A. Chervonsky and B. Jude.

YBR/Ei mice. The YBR/Ei strain was used for mammary tumor studies in the laboratory of Andervont, who showed that mammary tumor incidence is very low in breeding females, with or without MMTV infection [71]. However, the nature of MMTV infection in YBR/Ei mice or whether these mice are even susceptible to MMTV infection remained undetermined, because only mammary tumors were monitored. To identify the mechanism of resistance to MMTV-induced mammary tumors, we first needed to determine whether YBR/Ei mice are infectable with MMTV.

YBR/Ei mice have the H-2^d haplotype and express I-E molecules necessary to SAg presentation. Also, YBR/Ei mice have 8.8±0.5 percent (n=10) of SAg-cognate $CD4^+$/Vß14^+ T cells, suggesting that they do not inherit endogenous *Mtv* capable of deleting these T cells. To test whether MMTV can establish primary infection in YBR/Ei mice, we fostered newborn pups from this strain together with control susceptible C3H/HeN mice by viremic C3H/HeN females and analyzed the percentage of $CD4^+/Vß14^+$ SAg-cognate T cells in the periphery of the infected mice at different ages. A slow, progressive loss of $CD4^+/Vß14^+$ T cells was seen in fostered YBR/Ei mice as they aged, similar to that observed in C3H/HeN mice, albeit with slow kinetics. Approximately thirty percent of $CD4^+/Vß14^+$ T cell were deleted in both MMTV-infected YBR/Ei mice at 10 weeks after birth (Case et al., unpublished data).

To determine whether the mammary glands of YBR/Ei mice became MMTV-infected and the virus is present in the milk, fostered mice were bred, and RNA isolated from their lactating mammary glands after the first and second pregnancies was subjected to MMTV(C3H)-specific RNase T_1 protection analysis. Virus-specific RNA was detected in lactating mammary glands of infected YBR/Ei mice (Fig. 21), and the virus was present in the milk (not shown). We have also investigated whether virus amplification within the mammary glands of YBR/Ei mice was similar to that in C3H/HeN mice by analyzing viral RNA production in the mammary glands after each subsequent pregnancy. MMTV(C3H) did not amplify in the mammary gland of YBR/Ei mice compared to the mammary glands of C3H//HeN, suggesting that virus production was affected in YBR/Ei cells (Fig. 21). It is unlikely that provirus expression was affected in infected YBR/Ei mice, because expression of endogenous *Mtvs* did not appear to be decreased in these mice (Fig. 21).

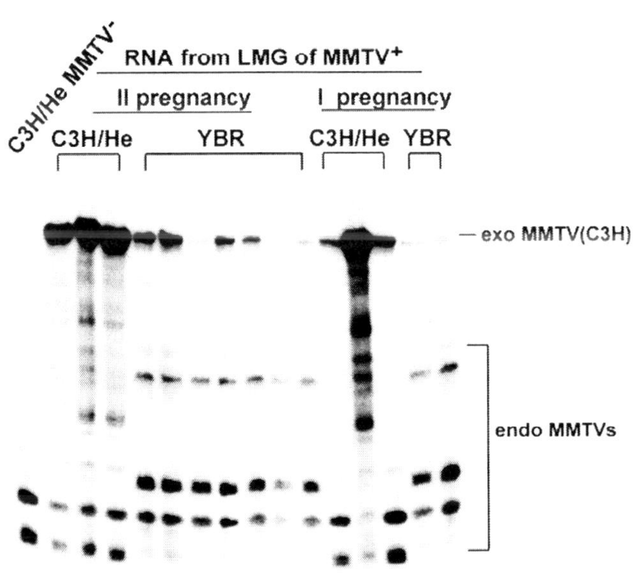

Figure 21. Virus-resistant mechanism inherited by YBR/Ei mice is due to attenuated virus production. YBR/Ei females were fostered by MMTV(C3H)-infected C3H/HeN females, bred and RNA isolated from their lactating mammary gland subsequent pregnancies was subjected to RNase T_1 protection analysis with MMTV(C3H)-specific probe. exo MMTV(C3H), full length protection fo exogenous MMTV(C3H); endo MMTVs, partial protection of endogenous *Mtv* transcripts.

In agreement with originally published data [71], MMTV-infected YBR/Ei mice were highly resistant to MMTV-induced tumors when compared to C3H/HeN mice (Fig. 22). Whereas 100 percent of C3H/HeN females infected with MMTV(C3H) developed tumors after approximately 300 days, none of the YBR/Ei mice infected with the virus developed tumors over a period of 350 days (Fig. 22).

Next we sought to determine whether the resistant mechanism inherited by YBR/Ei is recessive or dominant. Accordingly, we produced hybrid mice between susceptible C3H/HeN mice and resistant YBR/Ei and infected them with MMTV via foster-nursing. Mice were bred and RNA isolated from their lactating mammary glands was tested for virus-specific transcripts after the first pregnancy. Like MMTV-infected YBR/Ei mice, MMTV-infected F_1 mice produced severely reduced amounts of virus-specific transcripts in the mammary glands (not shown). When F_1 mice were monitored for mammary tumors, they demonstrated the same resistant phenotype as YBR/Ei mice, and none of the mice succumbed to tumors even at 400 day (Fig. 22). Therefore, in contrast to I/LnJ mice, the resistance mechanism inherited by YBR/Ei mice is dominant. We are currently investigating the mechanism of resistance to MMTV infection in YBR/Ei mice.

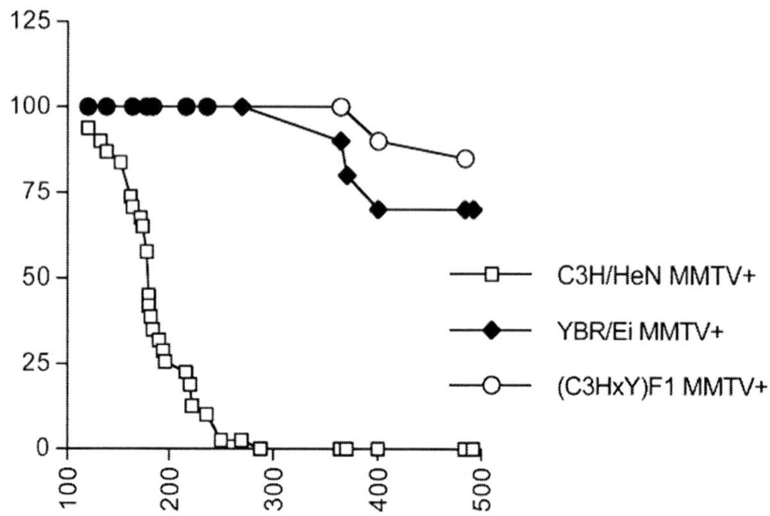

Figure 22. Mammary tumor incidence in MMTV(C3H)-infected YBR/Ei, C3H/HeN and F_1 females. Mice were infected via fostering, bred and monitored for mammary tumors.

Conclusions

The life cycle of MMTV is tightly linked to the immune system. A strong immune response induced by viral SAg is required for productive infection and virus transmission to the next generation. Despite strong stimulation of the immune system, the virus remains unrecognized and persists. Thus, MMTV has developed a delicate balance between its need to stimulate the immune system (subversion), and its needs to escape elimination (evasion). Our new data suggest that MMTV has evolved to manipulate both the innate and adaptive immune responses.

MMTV is also known for its ability to induce a high frequency of mammary tumors in susceptible mice by what was thought to be activation of cellular protooncogenes via insertional mutagenesis. However, our data indicate that MMTV activation of oncogenes is a necessary but insufficient step in mammary tumorigenesis, and that the virally-encoded *gag* gene is required for transformation of mammary gland cells. Identification of the mechanism by which Gag contributes to virally-induced mammary tumors will subsequently delineate the signal transduction pathway involved in tumorigenesis in general.

In summary, these studies demonstrated the usefulness of inbred mouse strains to gain understanding of the fundamental mechanisms of virus-host interactions.

Acknowledgments

We are thankful to Drs. A. Chervonsky, D. Roopenian, S. Ross and all members of the laboratory for helpful discussion, and to S. Williamson for the graphic work.

References

[1] Coffin, J. M. 1996. Retroviridae: The viruses and their replication. In *Fundamental Virology*. B. N. Fields, P. M. Howley, and D. M. Knipe, eds. Raven Press, New York, p. 763.

[2] Nandi, S., and C. M. McGrath. 1973. Mammary neoplasia in mice. *Adv. Cancer Res. 17:353.*

[3] Beutner, U., E. Kraus, D. Kitamura, K. Rajewsky, and B. T. Huber. 1994. B cells are essential for murine mammary tumor virus transmission, but not for presentation of endogenous superantigens. *J. Exp. Med. 179:1457.*

[4] Held, H., A. N. Shakhov, S. Izui, G. A. Waanders, L. Scarpellino, H. R. MacDonald, and H. Acha-Orbea. 1993. Superantigen-reactive CD4+ T cells are required to stimulate B cells after infection with mouse mammary tumor virus. *J. Exp. Med. 177:359.*

[5] Golovkina, T. V., M. Shlomchik, L. Hannum, and A. Chervonsky. 1999. Organogenic role of B lymphocytes in mucosal immunity. *Science 286:1965.*

[6] Kozak, C., G. Peters, R. Pauley, V. Morris, R. Michalides, J. Dudley, M. Green, M. Davisson, O. Prakash, A. Vaidya, J. Hilgers, A. Verstraeten, N. Hynes, H. Diggelmann, D. Peterson, J. C. Cohen, C. Dickson, N. Sarkar, R. Nusse, H. Varmus, and R. Callahan. 1987. A standardized nomenclature for endogenous mouse mammary tumor viruses. *J. Virol. 61:1651.*

[7] Golovkina, T. V., J. P. Dudley, A. B. Jaffe, and S. R. Ross. 1995. Mouse mammary tumor viruses with functional superantigen genes are selected during *in vivo* infection. *Proc. Natl. Acad. Sci. USA 92:4828.*

[8] Golovkina, T. V., A. Jaffe, and S. R. Ross. 1994. Coexpression of exogenous and endogenous mouse mammary tumor virus RNA *in vivo* results in viral recombination and broadens the virus host range. *J. Virol. 68:5019.*

[9] Golovkina, T. V., I. Piazzon, I. Nepomnaschy, V. Buggiano, M. de Olano Vela, and S. R. Ross. 1997. Generation of a tumorigenic milk-borne mouse mammary tumor virus by recombination between endogenous and exogenous viruses. *J. Virol. 71:3895.*

[10] Coffin, J. M. 1984. Structure of retroviral genome. In *RNA Tumor Viruses: Molecular Biology of Tumor Viruses*. R. Weiss, N. Teich, H. Varmus, and J. Coffin, eds. Cold Spring Harbor Laboratory, Cold Spring Harbor, New York, p. 261.

[11] Dickson, C., R. Eisenman, H. Fan, E. Hunter, and N. Teich. 1984. Protein biosynthesis and assembly. In *RNA tumor viruses*. R. Weiss, N. Teich, H. E. Varmus, and J. M. Coffin, eds. Cold Spring Harbor Laboratory, Cold Spring Harbor, New York, p. 513.

[12] Choi, Y., J. W. Kappler, and P. Marrack. 1991. A superantigen encoded in the open reading frame of the 3' long terminal repeat of the mouse mammary tumor virus. *Nature 350:203.*

[13] MacDonald, H. R., R. Schneider, R. K. Lees, R. C. Howe, H. Acha Orbea, H. Festenstein, R. M. Zinkernagel, and H. Hengartner. 1988. T-cell receptor V beta use predicts reactivity and tolerance to Mlsa-encoded antigens. *Nature 332:40.*

[14] MacDonald, H. R., A. L. Glasebrook, R. Schneider, R. K. Lees, H. Pircher, T. Pedrazzini, O. Kanagawa, J. F. Nicolas, R. C. Howe, R. M. Zinkernagel, and et al. 1989. T-cell reactivity and tolerance to Mlsa-encoded antigens. *Immunol. Rev. 107:89.*

[15] Brandt-Carlson, C., J. S. Butel, and D. Wheeler. 1993. Phylogenetic and structural analysis of MMTV LTR ORF sequences of exogenous and endogenous origins. *Virology 185:171.*

[16] Yazdanbakhsh, K., C. G. Park, G. M. Winslow, and Y. Choi. 1993. Direct evidence for the role of COOH terminus of mouse mammary tumor virus superantigen in determining T cell receptor Vb specificity. *J. Exp. Med. 178:737.*

[17] Chervonsky, A. V., J. Xu, A. K. Barlow, M. Khery, R. A. Flavell, and C. Janeway, Jr. 1995. Direct physical interaction involving CD40 ligand on T cells and CD40 on B cells is required to propagate MMTV. *Immunity 3:139.*

[18] Held, W., G. Waanders, A. N. Shakhov, L. Scarpellino, H. Acha-Orbea, and H. R. MacDonald. 1993. Superantigen-induced immune stimulation amplifies mouse mammary tumor virus infection and allows virus transmission. *Cell 74:529.*

[19] Varmus, H. E., T. Padgett, S. Heasley, G. Simon, and J. M. Bishop. 1977. Cellular functions are required for the synthesis and integration of avian sarcoma virus-specific DNA. *Cell 11:307.*

[20] Marrack, P., E. Kushnir, and J. Kappler. 1991. A maternally inherited superantigen encoded by mammary tumor virus. *Nature 349:524.*

[21] Golovkina, T. V., J. P. Dudley, and S. R. Ross. 1998. B and T cells are required for mouse mammary tumor virus spread within the mammary gland. *J. Immunol. 161:2375.*

[22] Yamamoto, K. 1985. Steroid receptor regulated transcription of specific genes and gene networks. *Annu. Rev. Genet. 19:209.*

[23] Peters, G., and C. Dickson. 1987. On the mechanism of carcinogenesis by mouse mammary tumor virus. In *Cellular and molecular biology of breast cancer*. D. Medina, W. Kidwell, G. Heppner, and E. Anderson, eds. Plenum Publishing Corp., New York, p. 307.

[24] Nusse, R., and H. Varmus. 1982. Mammary tumors induced by the mouse mammary tumor virus contain a provirus integrated in the same region of the host genome. *Cell 31:99.*

[25] Nusse, R. 1988. The *int* genes in mammary tumorigenesis and in normal development. *Trends Genet. 4:291.*

[26] Katoh, M. 2002. WNT and FGF gene clusters (Review). *Int J Oncol 21:1269.*

[27] Miller, J. R., and R. T. Moon. 1996. Signal transduction through β-catenin and specification of cell fate during embryogenesis. *Genes Dev. 10:2527.*

[28] Peters, G., A. E. Lee, and C. Dickson. 1984. Activation of cellular gene by mouse mammary tumour virus may occur early in mammary tumour development. *Nature 309:273.*

[29] Gallahan, D., C. Kozak, and R. Callahan. 1987. A new common integration region (int-3) for mouse mammary tumor virus on mouse chromosome 17. *J Virol 61:218.*

[30] Callahan, R., and G. H. Smith. 2000. MMTV-induced mammary tumorigenesis: gene discovery, progression to malignancy and cellular pathways. *Oncogene 19:992.*

[31] Tsukamoto, A. S., R. Grosschedl, R. C. Guzman, T. Parslow, and H. E. Varmus. 1988. Expression of the int-1 gene in transgenic mice is associated with mammary gland hyperplasia and adenocarcinomas in male and female mice. *Cell 55:619.*

[32] Lee, F. S., T. F. Lane, A. Kuo, G. M. Shackleford, and P. Leder. 1995. Insertional mutagenesis identifies a member of the Wnt gene family as a candidate oncogene in the mammary epithelium of int-2/Fgf-3 transgenic mice. *Proc. Natl. Acad. Sci. USA 92:2268.*

[33] Smith, G. H., D. Gallahan, F. Diella, C. Jhappan, G. Merlino, and R. Callahan. 1995. Constitutive expression of a truncated INT3 gene in mouse mammary epithelium impairs differentiation and functional development. *Cell Growth Differ 6:563.*

[34] Jhappan, C., D. Gallahan, C. Stahle, E. Chu, G. H. Smith, G. Merlino, and R. Callahan. 1992. Expression of an activated Notch-related int-3 transgene interferes with cell differentiation and induces neoplastic transformation in mammary and salivary glands. *Genes Dev 6:345.*

[35] Shackleford, G. M., C. A. MacArthur, H. C. Kwan, and H. E. Varmus. 1993. Mouse mammary tumor virus infection accelerates mammary carcinogenesis in Wnt-1 transgenic mice by insertional activation of int-2/Fgf-3 and hst/Fgf-4. *Proc. Natl. Acad. Sci. USA 90:740.*

[36] MacArthur, C. A., D. B. Shankar, and G. M. Shackleford. 1995. Fgf-8, activated by proviral insertion, cooperates with the Wnt-1 transgene in murine mammary tumorigenesis. *J. Virol. 69:2501.*

[37] Golovkina, T. V., A. Chervonsky, J. C. Prescott, C. A. Janeway, and S. R. Ross. 1994. The mouse mammary tumor virus envelope gene product is required for superantigen presentation to T cells. *J. Exp. Med. 179:439.*

[38] Hook, L. M., Y. Agafonova, S. R. Ross, S. J. Turner, and T. V. Golovkina. 2000. Genetics of mouse mammary tumor virus-induced mammary tumors: linkage of tumor induction to the gag gene. *J Virol 74:8876.*

[39] Gallahan, D., and R. Callahan. 1987. Mammary tumorigenesis in feral mice: identification of a new int locus in mouse mammary tumor virus (Czech II)-induced mammary tumors. *J Virol 61:66.*

[40] Peters, G., A. E. Lee, and C. Dickson. 1986. Concerted activation of two potential proto-oncogenes in carcinomas induced by mouse mammary tumour virus. *Nature 320:628.*

[41] Swanstrom, R., and J. W. Wills. 1997. Synthesis, assembly, and processing of viral proteins. In *Retroviruses*. J. M. Coffin, S. H. Hughes, and H. E. Varmus, eds. Cold Spring Harbor Press, New York, p. 263.

[42] Dupont, S., N. Sharova, C. DeHoratius, C. M. Virbasius, X. Zhu, A. G. Bukrinskaya, M. Stevenson, and M. R. Green. 1999. A novel nuclear export activity in HIV-1 matrix protein required for viral replication [In Process Citation]. *Nature 402:681.*

[43] Acha-Orbea, H., D. Finke, A. Attinger, S. Schmid, N. Wehrli, S. Vacheron, I. Xenarios, L. Scarpellino, K. M. Toellner, I. C. MacLennan, and S. A. Luther. 1999. Interplays between mouse mammary tumor virus and the cellular and humoral immune response. *Immunol Rev 168:287.*

[44] Purdy, A., L. Case, M. Duvall, M. Overstrom-Coleman, N. Monnier, A. Chervonsky, and T. Golovkina. 2003. Unique resistance of I/LnJ mice to a retrovirus is due to sustained interferon gamma-dependent production of virus-neutralizing antibodies. *J Exp Med 197:233.*

[45] Held, W., G. A. Waanders, H. R. MacDonald, and H. Acha Orbea. 1994. MHC class II hierarchy of superantigen presentation predicts efficiency of infection with mouse mammary tumor virus. *Int. Immunol. 6:1403.*

[46] Pucillo, C., R. Cepeda, and R. J. Hodes. 1993. Expression of a MHC class II transgene determines both superantigenicity and susceptibility to mammary tumor virus infection. *J. Exp. Med. 178:1441.*

[47] Wrona, T., and J. P. Dudley. 1996. Major histocompatibility complex class II I-E-independent transmission of C3H mouse mammary tumor virus. *J. Virol. 70:1246.*

[48] Frankel, W. N., C. Rudy, J. M. Coffin, and B. T. Huber. 1991. Linkage of Mls genes to endogenous mammary tumour viruses of inbred mice. *Nature 349:526.*

[49] Woodland, D. L., M. P. Happ, K. J. Gollob, and E. Palmer. 1991. An endogenous retrovirus mediating deletion of alpha beta T cells? *Nature 349:529.*

[50] Dyson, P. J., A. M. Knight, S. Fairchild, E. Simpson, and K. Tomonari. 1991. Genes encoding ligands for deletion of V beta 11 T cells cosegregate with mammary tumour virus genomes. *Nature 349:531.*

[51] Golovkina, T. V., A. V. Chervonsky, J. P. Dudley, and S. R. Ross. 1992. Transgenic mouse mammary tumor virus superantigen expression prevents viral infection. *Cell 69:637.*

[52] Golovkina, T., J. Prescott, and S. Ross. 1993. Mouse mammary tumor virus-induced tumorigenesis in sag transgenic mice: a laboratory model of natural selection. *J. Virol. 67:7690.*

[53] Piazzon, I., A. Goldman, S. Torello, I. Nepomnaschy, A. Deroche, and G. Dran. 1994. Transmission of an Mls-1a-like superantigen to BALB/c mice by foster-nursing on F1 Mls-1bxa mothers. Sex-influenced onset of clonal deletion. *J. Immunol. 153:1553.*

[54] Webb, S. R., and J. Sprent. 1993. Factors controlling the reactivity of immature and mature T cells to Mls antigens in vivo. *Immunol Rev 131:169.*

[55] Golovkina, T., Y. Agafonova, D. Kazansky, and A. Chervonsky. 2001. Diverse repertoire of the MHC class II-peptide complexes is required for presentation of viral superantigens. *J Immunol 166:2244.*

[56] Buggiano, V., A. Goldman, I. Nepomnaschy, P. Bekinschtein, P. Berguer, G. Lombardi, A. Deroche, M. V. Francisco, and I. Piazzon. 1999. Characterization of two infectious mouse mammary tumour viruses: superantigenicity and tumorigenicity. *Scand. J. Immunol. 49:269.*

[57] Howe, M. L., A. L. Goldstein, and J. R. Battisto. 1970. Isogeneic lymphocyte interaction: recognition of self antigens by cells of the neonatal thymus. *Proc. Natl. Acad. Sci. U.S.A. 67:613.*

[58] Von Boehmer, H., K. Shortman, and P. Adams. 1972. Nature of the stimulating cell in the syngeneic and the allogeneic mixed lymphocyte reaction in mice. *J Exp Med 136:1648.*

[59] Riggs, J. E., G. R. Sirken, L. G. Prior, and M. V. Hobbs. 1995. The murine autologous mixed lymphocyte response: distribution of stimulator cells. *J Autoimmun 8:21.*

[60] Janeway, C. A. 1989. Approaching the asymptote? Evolution and revolution in immunology. *Cold Spring Harb Symp Quant Biol 54:1.*

[61] Medzhitov, R., P. Preston-Hurlburt, and C. A. Janeway. 1997. A human homologue of the Drosophila Toll protein signals activation of adaptive immunity. *Nature 388:394.*

[62] Medzhitov, R. 2001. Toll-like receptors and innate immunity. *Nature Rev Immunol 1:135.*

[63] Tortorella, D., B. E. Gewurz, M. H. Furman, D. J. Schust, and H. L. Ploegh. 2000. Viral subversion of the immune system. *Annu Rev Immunol 18:861.*

[64] Bowie, A., E. Kiss-Toth, J. A. Symons, G. L. Smith, S. K. Dower, and L. A. O'Neill. 2000. A46R and A52R from vaccinia virus are antagonists of host IL-1 and toll-like receptor signaling. *Proc Natl Acad Sci U S A 97:10162.*

[65] Preston, B. D., and J. P. Dougherty. 1996. Mechanisms of retroviral mutation. *Trends Microbiol 4:16.*

[66] Hu, W. S., and H. M. Temin. 1990. Genetic consequences of packaging two RNA genomes in one retroviral particle: pseudodiploidy and high rate of genetic recombination. *Proc. Natl. Acad. Sci. USA 87:1556.*

[67] Domingo, E., and J. J. Holland. 1997. RNA virus mutations and fitness for survival. *Annu Rev Microbiol 51:151.*

[68] Fauci, A. S., and R. C. Desrosiers. 1997. Pathogenesis of HIV and SIV. J. M. Coffin, S. H. Highes, and H. E. Varmus, eds. Cold Spring Harbor Laboratory Press, Retroviruses, p. 587.

[69] Chieco-Bianchi, L., D. Collavo, and G. Biasi. 1988. Immunologic unresponsiveness to murine leukemia virus antigens: mechanisms and role in tumor development. *Adv Cancer Res 51:277.*

[70] Andervont, H. B. 1940. The influence of foster nursing upon the incidence of spontaneous mammary cancer in resistant and susceptible mice. *J. Nat. Cancer Inst. 1:147.*

[71] Nandi, S., M. Handin, A. Robinson, D. R. Pitelka, and L. E. Webber. 1966. Susceptibility of mammary tissues of "genetically resistant" strains of mice to mammary tumor virus. *J. Natl. Cancer Inst. 36:783.*

[72] Nandi, S., S. Haslam, and C. Helmich. 1972. Mechanisms of resistance to mammary tumor development in C57BL and I strains of mice. I. Noduligenesis, tumorigenesis, and characteristics of nodules and tumors. *J. Natl. Cancer Inst. 48:1005.*

[73] Golovkina, T. V. 2000. A novel mechanism of resistance to mouse mammary tumor virus infection. *J Virol 74:2752.*

[74] Snapper, C. M., and W. E. Paul. 1987. Interferon-gamma and B cell stimulatory factor-1 reciprocally regulate Ig isotype production. *Science 236:944.*

[75] Yasukawa, H., A. Sasaki, and A. Yoshimura. 2000. Negative regulation of cytokine signaling pathways. *Annu Rev Immunol 18:143.*

[76] Krystal, G. 2000. Lipid phosphatases in the immune system [In Process Citation]. *Semin Immunol 12:397.*

[77] Zhang, J., A. K. Somani, and K. A. Siminovitch. 2000. Roles of the SHP-1 tyrosine phosphatase in the negative regulation of cell signalling [In Process Citation]. *Semin Immunol 12:361.*

[78] Coutelier, J. P., J. T. van der Logt, F. W. Heessen, G. Warnier, and J. Van Snick. 1987. IgG2a restriction of murine antibodies elicited by viral infections. *J Exp Med 165:64.*

[79] Wei, X., J. M. Decker, S. Wang, H. Hui, J. C. Kappes, X. Wu, J. F. Salazar-Gonzalez, M. G. Salazar, J. M. Kilby, M. S. Saag, N. L. Komarova, M. A. Nowak, B. H. Hahn, P. D. Kwong, and G. M. Shaw. 2003. Antibody neutralization and escape by HIV-1. *Nature 422:307.*

[80] Sattentau, Q. J. 1996. Neutralization of HIV-1 by antibody. *Curr Opin Immunol 8:540.*

[81] Burrer, R., D. Salmon-Ceron, S. Richert, G. Pancino, G. Spiridon, S. Haessig, V. Roques, F. Barre-Sinoussi, A. M. Aubertin, and C. Moog. 2001. Immunoglobulin G (IgG) and IgA, but also nonantibody factors, account for in vitro neutralization of human immunodeficiency virus (HIV) type 1 primary isolates by serum and plasma of HIV-infected patients. *J Virol 75:5421.*

[82] Andervont, H. B. 1937. The use of pure strain animals in studies on natural resistance to transplantable tumors. *Public Health Rep. 52:1885.*

[83] Andervont, H., and H. Stewart. 1937. Adenomatous lesion in the stomach of strain I mice. *Science 86:566.*

[84] Thomas, P. E., J. J. Hutton, and B. A. Taylor. 1973. Genetic relationship between aryl hydrocarbon hydroxylase inducibility and chemical carcinogen induced skin ulceration in mice. *Genetics 74:655.*

[85] Strong, L. 1952. Differences in response among mice of fifteen inbred strains to the subcutaneous injection of methylcholanthrene. *Yale J. bIol. Med. 25:34.*

[86] Outzen, H. C., D. Corrow, and L. D. Shultz. 1985. Attenuation of exogenous murine mammary tumor virus virulence in the C3H/HeJ mouse substrain bearing the *Lps* mutation. *J. Natl. Cancer Inst. 75:917.*

[87] Jude, B. A., Y. Pobezinskaya, J. Bishop, S. Parke, R. M. Medzhitov, A. V. Chervonsky, and T. V. Golovkina. 2003. Subversion of the innate immune system by a retrovirus. *Nat Immunol 4:573.*

[88] Glode, L. M., and D. L. Rosenstreich. 1976. Genetic control of B cell activation by bacterial lipopolysaccharide is mediated by multiple distinct genes or alleles. *J Immunol 117:2061.*

[89] Vogel, S. N., D. Johnson, P. Y. Perera, A. Medvedev, L. Lariviere, S. T. Qureshi, and D. Malo. 1999. Cutting edge: functional characterization of the effect of the C3H/HeJ defect in mice that lack an Lpsn gene: in vivo evidence for a dominant negative mutation. *J Immunol 162:5666.*

[90] Poltorak, A., X. He, I. Smirnova, M. Y. Liu, C. V. Huffel, X. Du, D. Birdwell, E. Alejos, M. Silva, C. Galanos, M. Freudenberg, P. Ricciardi-Castagnoli, B. Layton, and B. Beutler. 1998. Defective LPS signaling in C3H/HeJ and C57BL/10ScCr mice: mutations in Tlr4 gene. *Science 282:2085.*

[91] Koller, B. H., P. Marrack, J. W. Kappler, and O. Smithies. 1990. Normal development of mice deficient in beta 2M, MHC class I proteins, and CD8+ T cells. *Science 248:1227.*

[92] Rassa, J. C., J. L. Meyers, Y. Zhang, R. Kudaravalli, and S. R. Ross. 2002. Murine retroviruses activate B cells via interaction with toll-like receptor 4. *Proc Natl Acad Sci U S A 99:2281.*

[93] Moore, K. W., R. de Waal Malefyt, R. L. Coffman, and A. O'Garra. 2001. Interleukin-10 and the interleukin-10 receptor. *Annu Rev Immunol 19:683.*

[94] Sing, A., A. Roggenkamp, A. M. Geiger, and J. Heesemann. 2002. Yersinia enterocolitica evasion of the host innate immune response by V antigen-induced IL-10 production of macrophages is abrogated in IL-10-deficient mice. *J Immunol 168:1315.*

[95] Eaton, K. A., M. Mefford, and T. Thevenot. 2001. The role of T cell subsets and cytokines in the pathogenesis of Helicobacter pylori gastritis in mice. *J Immunol 166:7456.*

[96] Akridge, R. E., L. K. Oyafuso, and S. G. Reed. 1994. IL-10 is induced during HIV-1 infection and is capable of decreasing viral replication in human macrophages. *J Immunol 153:5782.*

[97] Fleming, S. B., C. A. McCaughan, A. E. Andrews, A. D. Nash, and A. A. Mercer. 1997. A homolog of interleukin-10 is encoded by the poxvirus orf virus. *J Virol 71:4857.*

[98] Spencer, J. V., K. M. Lockridge, P. A. Barry, G. Lin, M. Tsang, M. E. Penfold, and T. J. Schall. 2002. Potent immunosuppressive activities of cytomegalovirus-encoded interleukin-10. *J Virol 76:1285.*

[99] Sing, A., D. Rost, N. Tvardovskaia, A. Roggenkamp, A. Wiedemann, C. J. Kirschning, M. Aepfelbacher, and J. Heesemann. 2002. Yersinia V-antigen exploits toll-like receptor 2 and CD14 for interleukin 10-mediated immunosuppression. *J Exp Med 196:1017.*

In: Progress in Virus Research
Editor: Sabine Thebault, pp. 169-189

ISBN 1-59454-438-7

Chapter VI

Coxsackievirus B4 and Type 1 Diabetes

Fabienne Brilot[1], Vincent Geenen[1] and Didier Hober[2,*]

[1] University of Liege, Center of Immunology, Institute of Pathology CHU-B23, B-4000 Liege-Sart-Tilman, Belgium.

[2]UPRES EA3610 "Pathogenèse virale du diabète de type 1" Université de Lille 2 , Faculté de Médecine et Service de Virologie, Bât. Paul Boulanger, CHRU, 59037 Lille, France

Abstract

Enteroviruses are small, naked viruses whose genome is a positive-strand RNA. Type 1 diabetes is an autoimmune pathology characterized by a loss of insulin due to destruction of insulin-secreting β-cells in pancreas Langerhans islets. Epidemiological studies on identical twin pairs pointed out that even though genetic factors are important factors in disease susceptibility, environmental factors, like virus infection, also play a role. Among viruses, coxsackieviruses B4 (CVB4) are suggested to be involved in disease emergence. Although the mechanism responsible for virus-induced diabetes is unknown, two mechanisms are thought to act at different levels of the immune system. Molecular mimicry is based on sequence homology between host glutamic acid decarboxylase 65 (GAD 65) and viral protein 2C (P2C). Virus can also act at the pancreas level with an islet direct infection followed by bystander activation. Recent research has demonstrated that CVB4 is able to infect thymic epithelial and lymphoid compartments of human thymus and to induce severe thymic dysfunction.

Keywords: Coxsackievirus B4, Diabetes

* Corresponding author: Didier Hober, MD, PhD, § UPRES EA3610 "Pathogenèse virale du diabète de type 1" Université de Lille 2 , Faculté de Médecine et Service de Virologie, Bât. Paul Boulanger, CHRU, 59037 Lille, France . Tel: 33 3 20 44 66 88, Fax: 33 3 20 44 52 81, E-mail: dhober@chru-lille.fr

Coxsackievirus B4 Classification and Structure

Late in the 1940s, a virus was isolated from a child with a flaccid paralysis who lived in Coxsackie a small village located on the Hudson River south of Albany in New York state. This isolate and others "parapoliomyelitis viruses" constitued a new group of enteroviruses named coxsackie viruses, originally two words but now combined into one.

Enteroviruses are small icosahedric virus (30 nm) (Figure 1), belonging to *picornaviridae* family. Traditionally, picornavirus have been defined in terms of serotypes grouped in genera. Recently, a change has been introduced with the concept of a picornavirus species consisting of several serotypes. According to the current taxonomic situation (see Table 1), the six coxsackievirus B serotypes in addition to echovirus serotypes, EV-69 and SVDV (swine vesicular disease virus) belong to the Human enterovirus B (HEV-B) species of the Enterovirus genus within the Picornaviridae family [102].

Table 1 Classification of enterovirus[a]

Family	Genus[b]	Species (abbreviation)	Serotype[c]
Picornaviridae	Enterovirus	Poliovirus	PV serotype 1,2,3
		Human Enterovirus A (HEV-A)	CVA serotypes 2-8, 10, 12, 14 EV-71
		Human Enterovirus B (HEV-B)	CVB serotypes 1-6; CVA9; EV-69 Echovirus 1-7; 8; 9; 11-21; 24-27; 29-33; SVDV
		Human Enterovirus C (HEV-C)	CVA serotypes 1; 11; 13; 15; 17-22,24
		Human Enterovirus D (HEV-D)	EV-68; EV-70

[a] In addition to these enterovirus involved in human pathology, there are animal enteroviruses: Bovine enterovirus (BEV) (BEV-1; BEV-2); Porcine enterovirus A (PEV-A) (PEV-8); Porcine enterovirus B (PEV-B) (PEV-9; PEV-10).

[b] In addition to enterovirus, the Picornaviridae family includes the following genus (species): Rhinovirus (Human rhinovirus A; Human rhinovirus B), Cardiovirus (Encephalomyocarditis virus; Theilovirus), Aphtovirus (Foot-and-mouth disease virus; Equine rhinitis A virus), Hepatovirus (Hepatitis A virus; Avian encephalomyelitis-like virus), Parechovirus (Human parechovirus), Erbovirus (Equine rhinitis B virus), Kobuvirus (Aichi virus), Teschovirus (Procine teschovirus).

[c] PV poliovirus; CVA coxsackievirus A; CVB coxsackievirus B; EV enterovirus; SVDV swine vesicular disease virus (variant of CVB5).

(Figure 3). Crystallography studies showed that protomers display 60 structures constituted in surface-exposed VP1, VP2, VP3 and RNA-associated intern VP4 [63, 112, 113].

The three-dimensional structure of coxsackievirus B4 has not been determined, it has been inferred from X-ray cristallography studies of coxsackievirus B3 and other enteroviruses that resemble coxsackievirus B4 [39, 80].

Coxsackievirus B4 Viral Cycle

CVB4 cycle occurs in host cell cytoplasm (Figure 4). The first event is virus attachment to virus receptor on cell surface. A receptor called CAR (Coxsackievirus and Adenovirus receptor) has been isolated and has been thought to be the common receptor for coxsackieviruses B and adenoviruses 2 and 5. All 6 CVB serotypes bind CAR extracellular domain, whose function in regular cell cycle is not known [14]. CAR contains an extracellular domain with two disulphuro bridges, a hydrophobic transmembrane domain and 107-aa intracytoplasm domain. Murine CAR homolog is also used as a CVB receptor [15, 106]. Both receptors have 91 % homology in the extracellular domain that allows attachment and virus entry into host cells [111]. Human CAR is expressed on β-cells [24, 117]. The second kind of identified receptor for CVB is DAF (Decay Accelerator Factor, also called CD55). CVB1, 3 and 5 bind to human DAF and to CAR either or both as receptors. In pig cells, Spiller and coll. reported that CVB did bind cells expressing pig or human CAR or bind human DAF but not pig DAF [100]. Although DAF role in infection is still uncertain, interaction with DAF on the apical surface of polarized epithelial cells facilitates infection by CVB3 and 5 [98]. CVB4 does not bind DAF [54, 75]. Other unknown surface molecules are thought to be receptors. For example, several CVB4 strains like V89-4557 and T318 replicate in CHO cell lines that do not express CAR at their surface [38]. Mice Langerhans islet cells do not express CAR but are easily infected by CVB4, meaning that CVB4 may bind another receptor to entry in those cells. Thus, Verdugo and coll. have suggested the existence of a third receptor. They isolated a 100 KDa protein that all CVB serotypes bind to [31]. Receptor function is double; it helps to set the virus at a good distance from cellular membrane and it induces capsid conformation change (eclipse phenomenom) that will have capsid uncoating in host cell cytoplasm as a final consequence [93].

The interaction between virus and receptor is essential for infection. Microinjection of virus in susceptible cells is not followed by an infection [78]. The eclipse phenomenom is a conformational change characterized by a loss of VP4, release of uninfectious “A” particles and empty capsid in infected cells supernatants [70]. Uncoating occurs either from cytoplasm membrane, or endosomes. For example, the use of endocytosis inhibitors like chloroquine prevents CVB4 infection in monocytes [49]. The uncoating final step is infection, that is viral RNA injection in the host cell cytoplasm where translation and replication take place. A hypothetical mechanism for translocation of enteroviral RNA across the cell membrane inferred from poliovirus studies suggest that the capsid undergoes conformational changes including the externalisation of VP4 and the N terminus of VP1. Myristoyl VP4 might facilitate the formation of a channel by five copies of the N terminus of VP1 inserted into the cell membrane. The viral RNA may pass through this channel [52].

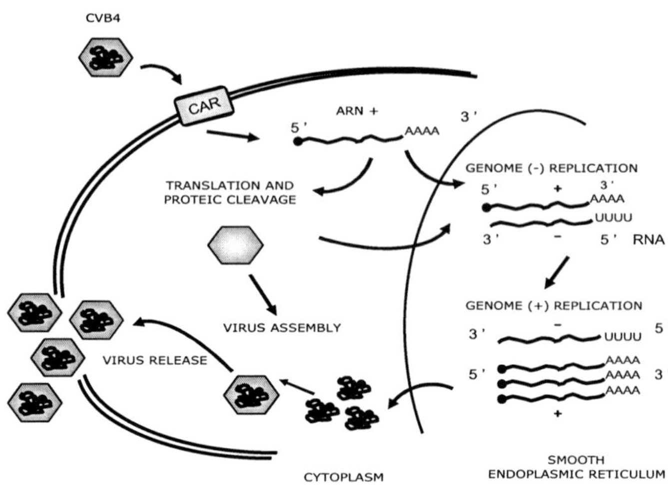

Figure 4. CVB4 viral cycle. *Adapted from Melnik et al., 1996.*

Viral proteins are translated from positive viral RNA and translation products are cleaved in P1, P2 and P3 by viral 2A and $3C^{pro}$ proteases. Viral proteins are then synthetized [93]. The first step of viral replication happens quickly. Viral positive RNA is transcripted by a template- and primer-dependent RNA polymerase $3D^{pol}$ into a negative replication RNA that will be used as matrix to synthetize numerous positive genomic RNA [93]. The viral proteins most directly involved in RNA synthesis are $3D^{pol}$ [34], the terminal protein VPg [3A] [87], and $3CD^{pro}$ [90], a proteinase and an important RNA-binding protein [3, 65]. Other viral proteins containing 2B, 2C, 2BC, 3A sequences bind to and induce alterations in membrane morphology [2, 25, 32, 104, 107, 109]. All nonstructural proteins and viral RNA become associated with the membranes to form an active replication complex. It is unclear how RNA or proteins associate with and are maintained within these areas but it has been recently shown that intact membranes are required for viral RNA synthesis [35]. Negative-strand RNA synthesis is then initiated by the formation of an uridylylated VPg [87]. To recognize its own template during replication, virus uses *cis*-acting signals ensuring efficient specific initiation of the complementary RNA strands [1]. The cis-replicating elements include three main structures. A cloverleaf structure of 88 nucleotides at the 5‘ end of the genome interacts with the virus 3CproDpol and either the VPg-containing precursor 3AB or the cellular poly-C binding type 2 (PCBP2) [3, 4, 48, 85]. This 5’UTR ribonucleoprotein complex is required for replication. RNA sequences and structures within the 3’ UTR presumed to be involved in replication are much less well defined. Signals might be located within the 3’UTR heteropolymeric region and include elements with conserved secondary and tertiary “kissing” structure [88]. Although there is a degree of flexibility in the 3’UTR sequences and structures, this region of the genome plays a role in virus replication by influencing the efficiency of the process rather than being an obligatory component [87]. In addition, an important 61-nucleotide RNA hairpin (*cre*(2C)) is located in the poliovirus-coding sequence of protein $2C^{ATPase}$ [45]. This RNA element is similar in size and shape to rhinoviral cre elements such as HRV2 *cre*(2A) (44) and HRV14 *cre*(VP1) [76] and covariance analysis suggested the sequence is well conserved in the enterovirus genus. Because both plus- and minus-strand RNA are VPg-linked *in vivo*, it is generally accepted that the uridylylated VPg is the primer used by $3D^{pol}$ for the synthesis of both RNA [82]. The model would be that negative-strand synthesis initiates with VPg uridylylated in the 3' poly(A) tail in virion RNA and positive-strand synthesis initiates with VPgpUpU synthetized on the *cre*(2C) hairpin. The pool of excess VPgpUpU synthesized on the *cre*(2C) hairpin shoud support high levels of positive-strand synthesis and thereby provided promote the asymmetric replicartion of enterovirus RNA [79].

The viral genome is protected by an icosahedrally symmetric capsid that is made of 60 copies of capsid proteins VP1 and VP3, and an average of 58 to 59 copies of capsid proteins VP2 and VP4, plus 1 to 2 copies of their capsid precursor, VP0. Morphogenesis of the poliovirus virion is initiated by proteolytic processing of the myristoylayed capsid precursor protein, to form the cleaved protomer. This cleaved protomer is a heterotrimer composed of VP0-VP3-VP1. The self-assembly of 5 protomers leads to the formation of 14S pentamers, that are an obligate intermediate in the virion assembly pathway [51]. Encapsidation of the genomic RNA by 12 pentamers forms the provirion with the final cleavage of VP0, to yield VP2 and VP4, occurring during the formation and maturation of the provirion. VP0 cleavage, also called maturation cleavage, is hypothesized to be an autocatalytic cleavage that is dependent on viral encapsidation [8, 51, 92] and is crucial for virion stability and infectivity [6, 12, 33]. Recent study has shown that poliovirus mutants at histidine 195 of VP2 do not cleave VP0 into VP2 and VP4 [6, 12, 33]. Immature pentamers package positive RNA to form uninfectious provirions [33, 71]. Infectious particles are then released in supernatants after lysis of host cell [59, 60, 93]. Data argue that enterovirus can be released from infected cells without overt cellular lysis however the mechanism of this nonlytic release is not understood [83, 94].

Epidemiology and Pathogenesis of Coxsackievirus B4

Human CVB4 have their reservoir only in human, their natural host. Virus can be spread by feco-oral and respiratory routes, contacts between humans and CVB4-contamined food. Enterovirus sexual transmissions have not been reported. Incubation periods last from 7 to 14 days, sometimes 2 to 35-day long. Initial multiplication occurs in satellite lymphoid tissues of intestine. Then, infection leads to viremia and subsequent transports to target organs like spinal cord, brain, meninges, skin, liver, pancreas and myocardium [77]. Accurate infections are the most common and are mostly silent, mild, or subclinical. Until now, three kinds of chronical infections have been associated to CVB4 infections; chronic fatigue syndrome (also known as postviral fatigue syndrome) [73], amyotrophic lateral sclerosis [11], and principally type 1 diabetes.

Type 1 Diabetes

Type 1 diabetes is the common diabetes in Caucasian children and young adults. The clinical manifestations of diabetes become apparent when there is a deficiency in insulin secretion resulting from an autoimmune-mediated destruction of insulin secreting ß-cells in pancreas. Type 1 diabetes is a pathology in which genetic, immunologic and environmental factors are implied. Genetic factors determine susceptibility or resistance to the disease. They control the disease inheritance and risk but are not sufficient for diabetes development in individuals. Immunologic factors are considered to provide the autoimmune character of the disease. Diabetes is induced by a transfer of T lymphocytes from diabetic animal models like NOD (Non Obese Diabetic) mice to syngenic nondiabetic mice [13] and self-directed autoantibodies are detected long before the disease observation.

Environmental Factors in the Pathogenesis of Type 1 Diabetes

Only 10% of the individuals bearing a genetic predisposition develop type1 diabetes and the contribution of an environmental role was suspected for a long time. Indeed, more than 50% of identical twins are discordants for the disease [10, 64, 72]. Assertion that environmental factors contribute to the development of diabetes came first from epidemiological studies. They have documented seasonal differences in the incidence of the disease, with most cases occurring in the fall and the winter [64, 95]. Moreover, geographic localization also seems to determine important variations in pathology incidence. A North-South gradient appears when type 1 diabetes incidences in high-rated Northern European countries such as Finland (36 to 40 new cases/100.000 inhabitants per year) and in low-rated Southern countries such as Venezuela (0.1 new case/100,000 inhabitants per year) are compared [68]. Even though epidemiological studies among migrants have highlighted the important role of genetic background, they also have documented an environmental crucial role in geographic zones. Thus, migrant populations moving from low-incidence countries tend to acquire the same risk than inhabitants of the welcoming countries [86, 96]. For example, diabetes annual incidence among Pakistani children living in United Kingdom is identical to the one which is observed among English children, e.g. 11.7 new cases/100,000 inhabitants per year against 1 new case/100,000 inhabitants per year in Pakistan [16, 101]. Type 1 diabetes appears in 12 to 40 % of patients with Congenial Rubella Syndrome after a 5- to 20-year-long incubation period [36]. Congenial Rubella Syndrome is responsible for only a small proportion of type 1 diabetes but these observations show that viruses can play a role in the disease. Viral infection was one of the first suggested hypothesis in order to explain environmental influence on Type 1 diabetes development [66]

Coxsackievirus and Type 1 Diabetes

Early epidemiologic studies have suggested that CVB might be involved in the aetology of human type 1 diabetes. Using neutralizing antibody, several teams showed that newly type 1 diabetes patients were more positive for CVB4 than were the control subjects [40, 41].

Later, other studies using RT-PCR detection of the virus genome reinforced the association between CVB4 and type 1 diabetes. Clements and coll. studied a group of very young children at the onset of their diabetes and showed that 64% were positive for enteroviruses as opposed to 4% of the control group [26]. In a second study, they gathered a larger group and confirmed their results [81]. In one of our study, EV genome was detected in five of 12 (42%) newly diagnosed IDDM patients and in one of 12 (8%) patients in the course of IDDM. By contrast, none of the 12 non-IDDM patients and none of the 15 healthy adults had EV sequences in whole blood [5]. Additionnal studies in Sweden and in Australia confirmed those results [27, 115]. CVB genome, and alpha-interferon have been detected by RT-PCR in peripheral blood of type 1 diabetes patients [24]. It has been reported that CVB4 infection of PBMC and CVB4-induced synthesis of IFN-α by these cells in vitro can be enhanced through interactions between CVB4 and specific antibodies isolated from plasma of healthy subjects, segregated from neutralizing antibodies and a receptor for CVB called CAR (coxsackievirus and adenovirus receptor) [22]. In addition, it has been demonstrated that CVB4, through interactions with circulating and/or cell-bound IgGs, can strongly induce the production of IFN-α by PBMC from patients with type 1 diabetes [50]. The difference of CVB4 response in these studies between patients and controls can be explained by a history of CVB4 infections responsible for the presence of higher levels of IFN-α-enhancing anti-CVB4 antibodies in patients than in healthy subjects. A high proportion of T cell response to CVB4 antigen, as evidenced by an enhanced production of interferon-γ, was observed in patients with type 1 diabetes [110].

Overall, the average proportion of enteroviral RNA-positive patients in various studies was 33% compared to 4% of control subjects (56). An additional advantage of these studies was that RT-PCR viral detection could be extended to delineate virus nucleotide sequences in order to identify diabetes-inducer virus. Thus, a wide range of enteroviruses of the coxsackie type are clearly associated with type 1 diabetes, with CVB4 especially common.

Additional evidences for a role of CVB in the etiopathogenesis of type1 diabetes are provided by the finding of insulitis and ß-cell destruction on post-mortem examination of the pancreases from a few children who have succumbed to CVB infection [62, 118]. CVB4 E2 was isolated from the pancreas of a child with ketoacidosis [118]. CVB4 E2 sequencing showed that the virus shared 94% homology with 111 amino acid substitutions, whose 82 were located in the non structural genome (8 in P2C area) and 29 in capsid proteins, mostly VP1 and VP2 [67]. CVB5 also isolated from a patients with rapid onset diabetes [20]. Those two isolates are particularly important because both strains were able to induce insulitis, ß-cell destruction and overt diabetes when injected into mice [20, 119, 120]. CVB4 E2 was also shown to persistently infect α and β cells from islets in human (23). Furthermore, a new isolated CVB4 variant, VD2921, causes an islet cells persistent infection with consequences on insulin secretion and proinsulin synthesis [116]. Recently, Yin and coll. detected CVB4 E2 and VD2921 genomes in diabetic patients by RT-PCR. Interestingly, six out of 7 control individuals were positive for CVB4 but had been infected by a different phyllogenetic branch of CVB4 than the one detected in diabetic patients [115].

Recent prospective studies have provided evidence for a role of CVB in diabetes development, whereas infections occur before the onset of the disease several years before the diabetes diagnostic [57]. Moreover, enterovirus infection during pregnancy may be a cause of diabetes among children. CVB-directed IgM and IgG antibodies are found more frequently among mothers of diabetic children than among others [28, 30]. Clinical studies among Swedish mothers of diabetic children reported an increased risk during pregnancy three first months. In this cohort, six mothers among 85 were positive for CVB infection against one among 172 in control mothers [28-30]. Altogether these observation and others suggest that enterovirus infections, especially CVB infections during pregnancy could cause type 1 diabetes in children [58].

Potential Mechanisms of Virus-Induced Type 1 Diabetes

Among enteroviruses, CVB4 has received the most attention with respect to a viral etiology for type 1 diabetes. This is in part because of the molecular homology between the 28-50 amino acids sequence of P2C, a structural viral protein of the replicative complex, and amino acids 250-273 of the ß-cell antigen glutamic acid decarboxylase (GAD) (Figure 5) [9, 89, 105]. The homologous sequence carries an epitope recognized by GAD 65-directed T lymphocytes in diabetic patients [9]. This suggests that the mechanism of molecular mimicry induction of autoimmune diabetes by CVB could apply. Molecular mimicry hypothesis lays on epitopes shared between virus proteins and host self-antigen. Viral antigens have to be different from host protein to generate an immune response but similar enough to induce a cross-reaction against the host [9, 84, 114]. Studies with NOD mice have confirmed that the immunization with GAD 65 and P2C-derivated peptides is able to induce cross-reactive immune responses [105].

```
GAD 65, aa 250-273   A M M I   A R F K M F P E V K E K G M A A L P R L
                     . .     . . . . | . . | | | | | |     .   |     |
CVB4-2C, aa 28-50    F I   E W L K V K I L P E V K E K H E F - L S R L
```

Figure 5. Sequence homology between human GAD65 and CVB4 protein 2C. Lines (in red) show identical amino acids (aa). Dots (in blue) show aa whose charge, polarity or hydrophobicity are similar

Other molecular homologies have been found between viral components and self-antigens. Tyrosine phosphatases (IA-2, IAR) represent major target autoantigens in type 1 diabetes. Both humoral and cellular immune responses are directed to the carboxy-terminal (C-terminal) part of the protein. This region has a 5-amino acid sequence identity, followed by 5 amino acid similarity with the conservative motif in enterovirus VP1 (PALTAVETGA/HT), which is a highly immunogenic B- and T-cell epitope in enterovirus infection-induced immune responses. The results obtained with human paired sera, collected during enterovirus infection, indicated that enterovirus infection in humans may also occasionally induce a humoral response that cross-reacts with IA-2/IAR [46]. In NOD mice, an autoimmune model of type 1 diabetes, one of the reactive β-cell autoantigens is the heat shock protein 60 (HSP60). These studies were prompted by sequence homology discovered between the immunogenic region in HSP60 and two regions in enterovirus capsid proteins, one in VP1 and the other in VP0, the precursor of VP2 and VP4 proteins. The HSP60/65 family of proteins is highly conserved and there is a striking degree of homology between bacterial and human HSP. Rabbit antibodies to HSP65 of Mycobacterium bovis that reacted with human HSP60 were also found to recognise capsid protein VP1 of coxsackievirus A9, VP1, and/or VP2 of CVB4. Both viruses were also recognised by antisera raised against HSP60 of Chlamydia pneumoniae. In addition to the capsid proteins derived from native virions, antisera to both bacterial HSP proteins recognised expression protein VP1 of coxsackievirus A9. The cross-reactivity was also demonstrated the other way around; antisera to purified virus particles reacted with the HSP60/65 proteins to some extent [47]. These two last studies suggest that, apart from the well-documented sequence homology between P2C and the β-cell autoantigen GAD, there are other motifs in picornavirus proteins homologous to islet cell autoantigens, which might induce cross-reacting immune responses during picornavirus infections.

CVB direct tropism for pancreas has been observed during post-mortem examination of pancreases from children who died from CVB infection [62, 108]. Moreover, CVB4 E2 variant was isolated from a pancreas of a child who succumbed to acute diabetes [118]. Since then, fundamental studies provided evidence of murine and human ß-cell *in vitro* infection. Insulin secretion perturbations in CVB4-infected human ß-cell have been demonstrated accompanying by cellular death nuclear pycnosis [91]. Chehadeh and coll. showed that CVB4 persistent infection in human islets cells leads to α- and β-cell infection and to IFN-α secretion by β-cells [23]. Thus CVB4 was able to induce the synthesis of IFN-α in human β-cells, which mimicked the expression of IFN-α in β-cells but not other islet cells in diabetic patients at autopsy [37].

CVB4-induced IFN-α secretion played a role in the initiation and/or maintenance of chronic CVB4 infection in human islets in our *in vitro* system [22]. The production of IFN-α by β-cells can play a role in the pathogenesis of type 1 diabetes since transgenic mice with IFN-α–expressing β-cells develop the disease with activation of autoreactive T-cells [19, 103]. CVB4 infection of islet ß-cells can be followed by a release of specific ß-cell autoantigens and a bystander activation of autoreactive T-cells. This is one explanation of the relationship between type 1 diabetes incidence and CVB4 infection in addition to the former hypothesis of molecular mimicry between CVB4 and GAD65 sequences. Recent data show an islet cell tropism of enterovirus in the human pancreas and detection by in situ hybridization of enterovirus-positive islet cells in autopsy specimen of type 1 diabetic patients but not in control pancreas [117].

Bystander activation is another mechanism suggested to explain viral-induced diabetes. The concept of bystander activation requires destruction of specific tissue, release of sequestered antigen and increased local immune inflammation. Lymphocytes would be recruited to the tissue and those reactive to the released, sequestered self-antigen would in turn be restimulated in the inflammatory response. Thus, autoreactive lymphocytes would gain access to the target tissue without being involved in the initial viral insult or reactive to the viral antigens [55]. CVB4 ability to directly infect pancreas and to induce a strong local inflammatory response transforms the analysis of CVB4-induced mechanisms in diabetes pathogenesis into a dual non-exclusive phenomenon: molecular mimicry or/and bystander activation. Recently, studies have attempted to discriminate both mechanisms. Three different strains of mice were infected with CVB4 [53]. NOD and B10.H2g7 mice expressed I-A^{g7}, diabetes major susceptibility haplotype potentially able to cross-react with GAD 65. Diabetes develops spontaneously in NOD mice but not faster neither in greater percentages in CVB4-infected animals compared to negative controls. No cross-reaction was observed. B10.H2g7 mice do not develop diabetes spontaneously neither after CVB4 infection. Those results suggest that the predominant response against the virus contains other viral epitopes than the GAD 65 homolog epitopes. The third used strain is the transgenic BDC2.5 mouse that does not develop spontaneous diabetes but over-expressed an islet-directed diabetogenic TCR. The antigen is different from GAD 65 and does not react against CVB4. BDC2.5 mice develop diabetes 2 to 4 weeks after CVB4 injection. In this model, CVB4 infection and following pancreas inflammation induced an auto-reactive T cells activation by a bystander effect. Other authors showed that CVB4 can speed up diabetes development in NOD mice through a bystander effect but only when a certain number of auto-reactive T cells were present [97]. The importance of pre-existing auto-reactive cells and the time of infection was showed to be important.

CVB4 and Thymus

The pathogenic mechanism by which CVB4 can contribute to the development of an autoimmune disease like type 1 diabetes has not been elucidated. As for other organ- or cell-specific autoimmune diseases, the development of type 1 diabetes corresponds to the absence or to the breakdown of immune self-tolerance at central and/or peripheral level [99]. The induction of immune self-tolerance is an active multi-step process initiated within the thymus network during T-cell ontogeny [42, 43, 74]. Thymus fullfills its function by deleting self-directed T cells and by generating regulatory T cells. In the thymic stoma, thymic epithelial cells (TEC) play a critical role in the differentiation of T-cell precursors, providing a microenvironment with a unique capacity to generate functional and self-tolerant T cells. The role of a virus in the induction and/or triggering of type 1 diabetes could result from interference with the central tolerogenic role of the thymus through an interaction between the virus and thymic cells. Interestingly, we recently demonstrated that TEC can be persistently infected with CVB4 *in vitro*, which results in an increased production of IL-6 (interleukine-6), LIF (leukaemia inhibitory factor) and GM-CSF (granulocyte macrophage-colony stimulating factor) by these cells. We hypothesized that this persistence could interfere with thymus function and contribute to the development of autoimmunity [17]. Using human fetal thymus organ culture, we have recently shown that human thymus in culture is infected by CVB4 and that cortical $CD4^+$ $CD8^+$ thymocytes are the principal target of infection. CVB4 infection of these cells lead to a marked and progressive depletion and upregulation of type I major histocompatibily complex (MHC I) molecules, that seems to be the direct consequence of CVB4 infection of $CD4^+$ $CD8^+$ thymocytes rather than an cytokine-mediated indirect effect [18]. Diabetogenic CVB4 can reach the thymus in the course of systemic infection of mice, which results in a significant increase in $CD4^-$ $CD8^-$ thymic T lymphocytes together with insulitis and hyperglycemia [21]. The nature of infected cells within the thymus was not elucidated in the previous study, but these data suggest that CVB4 can induce abnormal T lymphocyte maturation, which may contribute to development of diabetes. Studies are in progress in our laboratory to investigate whether, *in vivo*, CVB4 can infect the epithelial network of the thymus, as suggested by our *in vitro* studies. It will be of interest to investigate whether CVB4-infected TEC and the resulting increase in cytokine levels in the microenvironment of TEC play a role in the impairment of central T-cell self-tolerance.

Conclusions

Direct epidemiological associations between viral infections and autoimmune pathologies are difficult to establish for a few reasons. First, each individual already has undergone multiple virus infections at the time when diabetes is observed, meaning that most retrospective epidemiological studies, as far as causality is concerned, are very difficult to understand. Second, immune reactions leading to the disease development may occur when all diabetes inducing virus have been cleared from the organism, whereas virus detection at the time of diabetes diagnostic may represent a concomitant infection without any meanings in regard to the pathology etiology. Nevertheless the role of CVB4 in the etiology of type 1 diabetes is strongly suspected. Indeed, recent data have highlighted that enterovirus infections are a major trigger for type 1 diabetes. Evidences were detection of antisera, viral RNA in patients, as well as isolation of CVB4 strains from pancreas of diabetic patients. Studies using mice confirmed later that CVB4 was able to induce diabetes, bringing further evidences of CVB4 involvement. Although implicated mechanisms are still unknown, new studies have suggested that direct infection of pancreas with IFN-α secretion and CVB4 direct infection of thymus might interfere with function of the immune system and promotes autoimmunity.

Acknowledgments

F. Brilot is scientific collaborator at the "Fond National de la recherche scientifique" (FNRS, Belgium) and was supported by FNRS grants n°3.4508.04 and 9.4566.04. V. Geenen is research director at FNRS. This work was supported by the Fondation Leon Fredericq (Liege Medical School, Belgium), the Fondation Vaugrenier for Tolerance Research (Switzerland), the European Association for the Study of Diabetes (EASD, Germany). D. Hober was supported by Centre Hospitalier Régional et Universitaire de Lille and by Ministère de l'Education Nationale de la Recherche et de la Technologie ,UPRES EA3610 "Pathogenèse virale du diabète de type 1" Université Lille 2 (France), ALFEDIAM. Both laboratories belong to Euro-Thymaide FP6 Integrated Project LSHB-CT-2003-503410.

References

[1] Agol, V. I., A. V. Paul, and E. Wimmer. 1999. Paradoxes of the replication of picornaviral genomes. *Virus Res* 62:129-47.

[2] Aldabe, R., and L. Carrasco. 1995. Induction of membrane proliferation by poliovirus proteins 2C and 2BC. *Biochem Biophys Res Commun* 206:64-76.

[3] Andino, R., G. E. Rieckhof, P. L. Achacoso, and D. Baltimore. 1993. Poliovirus RNA synthesis utilizes an RNP complex formed around the 5'-end of viral RNA. *EMBO J* 12:3587-98.

[4] Andino, R., G. E. Rieckhof, and D. Baltimore. 1990. A functional ribonucleoprotein complex forms around the 5' end of poliovirus RNA. *Cell* 63:369-80.

[5] Andreoletti, L., D. Hober, C. Hober-Vandenberghe, S. Belaich, M. C. Vantyghem, J. Lefebvre, and P. Wattre. 1997. Detection of coxsackie B virus RNA sequences in whole blood samples from adult patients at the onset of type I diabetes mellitus. *J Med Virol* 52:121-7.

[6] Ansardi, D. C., and C. D. Morrow. 1995. Amino acid substitutions in the poliovirus maturation cleavage site affect assembly and result in accumulation of provirions. *J Virol* 69:1540-7.

[7] Argos, P., G. Kamer, M. J. Nicklin, and E. Wimmer. 1984. Similarity in gene organization and homology between proteins of animal picornaviruses and a plant comovirus suggest common ancestry of these virus families. *Nucleic Acids Res* 12:7251-67.

[8] Arnold, E., M. Luo, G. Vriend, M. G. Rossmann, A. C. Palmenberg, G. D. Parks, M. J. Nicklin, and E. Wimmer. 1987. Implications of the picornavirus capsid structure for polyprotein processing. *Proc Natl Acad Sci* U S A 84:21-5.

[9] Atkinson, M. A., M. A. Bowman, L. Campbell, B. L. Darrow, D. L. Kaufman, and N. K. Maclaren. 1994. Cellular immunity to a determinant common to glutamate decarboxylase and coxsackie virus in insulin-dependent diabetes. *J Clin Invest* 94:2125-9.

[10] Barnett, A. H., C. Eff, R. D. Leslie, and D. A. Pyke. 1981. Diabetes in identical twins. A study of 200 pairs. *Diabetologia* 20:87-93.

[11] Bartfeld, H., C. Dham, H. Donnenfeld, R. A. Ollar, M. T. deMasi, and R. Kascsak. 1989. Enteroviral-related antigen in circulating immune complexes of amyotrophic lateral sclerosis patients. *Intervirology* 30:202-12.

[12] Basavappa, R., R. Syed, O. Flore, J. P. Icenogle, D. J. Filman, and J. M. Hogle. 1994. Role and mechanism of the maturation cleavage of VP0 in poliovirus assembly: structure of the empty capsid assembly intermediate at 2.9 A resolution. *Protein Sci* 3:1651-69.

[13] Bendelac, A., C. Carnaud, C. Boitard, and J. F. Bach. 1987. Syngeneic transfer of autoimmune diabetes from diabetic NOD mice to healthy neonates. Requirement for both L3T4+ and Lyt-2+ T cells. *J Exp Med* 166:823-32.

[14] Bergelson, J. M., J. A. Cunningham, G. Droguett, E. A. Kurt-Jones, A. Krithivas, J. S. Hong, M. S. Horwitz, R. L. Crowell, and R. W. Finberg. 1997. Isolation of a common receptor for Coxsackie B viruses and adenoviruses 2 and 5. *Science* 275:1320-3.

[15] Bergelson, J. M., A. Krithivas, L. Celi, G. Droguett, M. S. Horwitz, T. Wickham, R. L. Crowell, and R. W. Finberg. 1998. The murine CAR homolog is a receptor for coxsackie B viruses and adenoviruses. *J Virol* 72:415-9.

[16] Bodansky, H. J., A. Staines, C. Stephenson, D. Haigh, and R. Cartwright. 1992. Evidence for an environmental effect in the aetiology of insulin dependent diabetes in a transmigratory population. *BMJ* 304:1020-2.

[17] Brilot, F., W. Chehadeh, C. Charlet-Renard, H. Martens, V. Geenen, and D. Hober. 2002. Persistent infection of human thymic epithelial cells by coxsackievirus B4. *J. Virol.* 76:5260-5.

[18] Brilot, F., V. Geenen, D. Hober, and C. A. Stoddart. 2004. Coxsackievirus B4 infection of human fetal thymus. *J. Virol. in press.*

[19] Chakrabarti, D., B. Hultgren, and T. A. Stewart. 1996. IFN-alpha induces autoimmune T cells through the induction of intracellular adhesion molecule-1 and B7.2. *J Immunol* 157:522-8.

[20] Champsaur, H., E. Dussaix, D. Samolyk, M. Fabre, C. Bach, and R. Assan. 1980. Diabetes and Coxsackie virus B5 infection. *Lancet* 1:251.

[21] Chatterjee, N. K., J. Hou, P. Dockstader, and T. Charbonneau. 1992. Coxsackievirus B4 infection alters thymic, splenic, and peripheral lymphocyte repertoire preceding onset of hyperglycemia in mice. *J. Med. Virol.* 38:124-31.

[22] Chehadeh, W., A. Bouzidi, G. Alm, P. Wattre, and D. Hober. 2001. Human antibodies isolated from plasma by affinity chromatography increase the coxsackievirus B4-induced synthesis of interferon-alpha by human peripheral blood mononuclear cells in vitro. *J. Gen. Virol.* 82:1899-907.

[23] Chehadeh, W., J. Kerr-Conte, F. Pattou, G. Alm, J. Lefebvre, P. Wattre, and D. Hober. 2000. Persistent infection of human pancreatic islets by coxsackievirus B is associated with alpha interferon synthesis in beta cells. *J. Virol.* 74:10153-64.

[24] Chehadeh, W., J. Weill, M. C. Vantyghem, G. Alm, J. Lefebvre, P. Wattre, and D. Hober. 2000. Increased level of interferon-alpha in blood of patients with insulin-dependent diabetes mellitus: relationship with coxsackievirus B infection. *J. Infect. Dis.* 181:1929-39.

[25] Cho, M. W., N. Teterina, D. Egger, K. Bienz, and E. Ehrenfeld. 1994. Membrane rearrangement and vesicle induction by recombinant poliovirus 2C and 2BC in human cells. 202:129-45.

[26] Clements, G. B., D. N. Galbraith, and K. W. Taylor. 1995. Coxsackie B virus infection and onset of childhood diabetes. *Lancet* 346:221-3.

[27] Craig, M. E., N. J. Howard, M. Silink, and A. Chan. 2000. The rising incidence of childhood type 1 diabetes in New South Wales, Australia. *J Pediatr Endocrinol Metab* 13:363-72.

[28] Dahlquist, G., G. Frisk, S. A. Ivarsson, L. Svanberg, M. Forsgren, and H. Diderholm. 1995. Indications that maternal coxsackie B virus infection during pregnancy is a risk factor for childhood-onset IDDM. *Diabetologia* 38:1371-3.

[29] Dahlquist, G. G., J. E. Boman, and P. Juto. 1999. Enteroviral RNA and IgM antibodies in early pregnancy and risk for childhood-onset IDDM in offspring. *Diabetes Care* 22:364-5.

[30] Dahlquist, G. G., S. Ivarsson, B. Lindberg, and M. Forsgren. 1995. Maternal enteroviral infection during pregnancy as a risk factor for childhood IDDM. A population-based case-control study. *Diabetes* 44:408-13.

[31] de Verdugo, U. R., H. C. Selinka, M. Huber, B. Kramer, J. Kellermann, P. H. Hofschneider, and R. Kandolf. 1995. Characterization of a 100-kilodalton binding protein for the six serotypes of coxsackie B viruses. *J Virol* 69:6751-7.

[32] Echeverri, A. C., and A. Dasgupta. 1995. Amino terminal regions of poliovirus 2C protein mediate membrane binding. *Virology* 208:540-53.

[33] Fernandez-Tomas, C. B., and D. Baltimore. 1973. Morphogenesis of poliovirus. II. Demonstration of a new intermediate, the proviron. *J Virol* 12:1122-30.

[34] 34. Flanegan, J. B., and D. Baltimore. 1977. Poliovirus-specific primer-dependent RNA polymerase able to copy poly(A). *Proc Natl Acad Sci* U S A 74:3677-80.

[35] Fogg, M. H., N. L. Teterina, and E. Ehrenfeld. 2003. Membrane requirements for uridylylation of the poliovirus VPg protein and viral RNA synthesis in vitro. *J. Virol.* 77:11408-16.

[36] Forrest, J. M., M. A. Menser, and J. A. Burgess. 1971. High frequency of diabetes mellitus in young adults with congenital rubella. *Lancet* 2:332-4.

[37] Foulis, A. K., M. A. Farquharson, and A. Meager. 1987. Immunoreactive alpha-interferon in insulin-secreting beta cells in type 1 diabetes mellitus. *Lancet* 2:1423-7.

[38] Frisk, G., T. Elfstrom, and H. Diderholm. 2001. The replication of certain Coxsackie B virus strains in CHO cells. *J Virol Methods* 98:161-5.

[39] Fry, E. E., N. J. Knowles, J. W. Newman, G. Wilsden, Z. Rao, A. M. King, and D. I. Stuart. 2003. Crystal structure of Swine vesicular disease virus and implications for host adaptation. *J Virol* 77:5475-86.

[40] Gamble, D. R., M. L. Kinsley, M. G. FitzGerald, R. Bolton, and K. W. Taylor. 1969. Viral antibodies in diabetes mellitus. *Br Med J* 3:627-30.

[41] Gamble, D. R., and K. W. Taylor. 1973. Coxsackie B virus and diabetes. *Br Med J* 1:289-90.

[42] Geenen, V., O. Kecha, F. Brilot, I. Hansenne, C. Renard, and H. Martens. 2001. Thymic T-cell tolerance of neuroendocrine functions: physiology and pathophysiology. *Cell Mol Biol* (Noisy-le-grand) 47:179-88.

[43] Geenen, V., and G. Kroemer. 1993. Multiple ways to cellular immune tolerance. Immunol. *Today.* 14:573-5.

[44] Gerber, K., E. Wimmer, and A. V. Paul. 2001. Biochemical and genetic studies of the initiation of human rhinovirus 2 RNA replication: identification of a cis-replicating element in the coding sequence of 2A(pro). *J Virol* 75:10979-90.

[45] Goodfellow, I., Y. Chaudhry, A. Richardson, J. Meredith, J. W. Almond, W. Barclay, and D. J. Evans. 2000. Identification of a cis-acting replication element within the poliovirus coding region. *J Virol* 74:4590-600.

[46] Harkonen, T., H. Lankinen, B. Davydova, T. Hovi, and M. Roivainen. 2002. Enterovirus infection can induce immune responses that cross-react with beta-cell autoantigen tyrosine phosphatase IA-2/IAR. *J Med Virol* 66:340-50.

[47] Harkonen, T., M. Puolakkainen, M. Sarvas, U. Airaksinen, T. Hovi, and M. Roivainen. 2000. Picornavirus proteins share antigenic determinants with heat shock proteins 60/65. *J Med Virol* 62:383-91.

[48] Harris, K. S., W. Xiang, L. Alexander, W. S. Lane, A. V. Paul, and E. Wimmer. 1994. Interaction of poliovirus polypeptide 3CDpro with the 5' and 3' termini of the poliovirus genome. Identification of viral and cellular cofactors needed for efficient binding. *J Biol Chem* 269:27004-14.

[49] Hober, D., W. Chehadeh, A. Bouzidi, and P. Wattre. 2001. Antibody-dependent enhancement of coxsackievirus B4 infectivity of human peripheral blood mononuclear cells results in increased interferon-alpha synthesis. *J. Infect. Dis.* 184:1098-108.

[50] Hober, D., W. Chehadeh, J. Weill, C. Hober, M. C. Vantyghem, P. Gronnier, and P. Wattre. 2002. Circulating and cell-bound antibodies increase coxsackievirus B4-induced production of IFN-alpha by peripheral blood mononuclear cells from patients with type 1 diabetes. *J Gen Virol* 83:2169-76.

[51] Hogle, J. M., M. Chow, and D. J. Filman. 1985. Three-dimensional structure of poliovirus at 2.9 A resolution. *Science* 229:1358-65.

[52] Hogle, J. M., and V. R. Racaniello. 2002. Poliovirus receptors and cell entry, p. 71-83. *In* B. L. Semler and E. Wimmer (ed.), Molecular Biology of Picarnovirus. *ASM Press,* Washington DC.

[53] Horwitz, M. S., L. M. Bradley, J. Harbertson, T. Krahl, J. Lee, and N. Sarvetnick. 1998. Diabetes induced by Coxsackie virus: initiation by bystander damage and not molecular mimicry. *Nat Med* 4:781-5.

[54] Horwitz, M. S., T. Krahl, C. Fine, J. Lee, and N. Sarvetnick. 1999. Protection from lethal coxsackievirus-induced pancreatitis by expression of gamma interferon. *J Virol* 73:1756-66.

[55] Horwitz, M. S., and N. Sarvetnick. 1999. Viruses, host responses, and autoimmunity. *Immunol. Rev.* 169:241-53.

[56] Hyoty, H. 2002. Enterovirus infections and type 1 diabetes. *Ann Med* 34:138-47.

[57] Hyoty, H., M. Hiltunen, M. Knip, M. Laakkonen, P. Vahasalo, J. Karjalainen, P. Koskela, M. Roivainen, P. Leinikki, and T. Hovi. 1995. A prospective study of the role of coxsackie B and other enterovirus infections in the pathogenesis of IDDM. Childhood Diabetes in Finland (DiMe) Study Group. *Diabetes*. 44:652-7.

[58] Hyoty, H., and K. W. Taylor. 2002. The role of viruses in human diabetes. *Diabetologia* 45:1353-61.

[59] Jacobson, M. F., J. Asso, and D. Baltimore. 1970. Further evidence on the formation of poliovirus proteins. *J Mol Biol* 49:657-69.

[60] Jacobson, M. F., and D. Baltimore. 1968. Morphogenesis of poliovirus. I. Association of the viral RNA with coat protein. *J Mol Biol* 33:369-78.

[61] Jenkins, O., J. D. Booth, P. D. Minor, and J. W. Almond. 1987. The complete nucleotide sequence of coxsackievirus B4 and its comparison to other members of the Picornaviridae. *J Gen Virol* 68 (Pt 7):1835-48.

[62] Jenson, A. B., H. S. Rosenberg, and A. L. Notkins. 1980. Pancreatic islet-cell damage in children with fatal viral infections. *Lancet* 2:354-8.

[63] Johnston, M. D., and S. J. Martin. 1971. Capsid and procapsid proteins of a bovine enterovirus. *J Gen Virol* 11:71-9.

[64] Joner, G., and O. Sovik. 1989. Increasing incidence of diabetes mellitus in Norwegian children 0-14 years of age 1973-1982. *Diabetologia* 32:79-83.

[65] Jore, J., B. De Geus, R. J. Jackson, P. H. Pouwels, and B. E. Enger-Valk. 1988. Poliovirus protein 3CD is the active protease for processing of the precursor protein P1 in vitro. *J Gen Virol* 69 (Pt 7):1627-36.

[66] Jun, H. S., and J. W. Yoon. 2003. A new look at viruses in type 1 diabetes. 19:8-31.

[67] Kang, Y., N. K. Chatterjee, M. J. Nodwell, and J. W. Yoon. 1994. Complete nucleotide sequence of a strain of coxsackie B4 virus of human origin that induces diabetes in mice and its comparison with nondiabetogenic coxsackie B4 JBV strain. *J Med Virol* 44:353-61.

[68] Karvonen, M., M. Viik-Kajander, E. Moltchanova, I. Libman, R. LaPorte, and J. Tuomilehto. 2000. Incidence of childhood type 1 diabetes worldwide. Diabetes Mondiale (DiaMond) Project Group. *Diabetes Care* 23:1516-26.

[69] Kitamura, N., B. L. Semler, P. G. Rothberg, G. R. Larsen, C. J. Adler, A. J. Dorner, E. A. Emini, R. Hanecak, J. J. Lee, S. van der Werf, C. W. Anderson, and E. Wimmer. 1981. Primary structure, gene organization and polypeptide expression of poliovirus RNA. *Nature* 291:547-53.

[70] Kramer, B., M. Huber, C. Kern, K. Klingel, R. Kandolf, and H. C. Selinka. 1997. Chinese hamster ovary cells are non-permissive towards infection with coxsackievirus B3 despite functional virus-receptor interactions. *Virus Res* 48:149-56.

[71] Lee, W. M., S. S. Monroe, and R. R. Rueckert. 1993. Role of maturation cleavage in infectivity of picornaviruses: activation of an infectosome. *J Virol* 67:2110-22.

[72] Lo, S. S., R. Y. Tun, M. Hawa, and R. D. Leslie. 1991. Studies of diabetic twins. *Diabetes Metab Rev* 7:223-38.

[73] Manian, F. A. 1994. Simultaneous measurement of antibodies to Epstein-Barr virus, human herpesvirus 6, herpes simplex virus types 1 and 2, and 14 enteroviruses in chronic fatigue syndrome: is there evidence of activation of a nonspecific polyclonal immune response? *Clin Infect Dis* 19:448-53.

[74] Martens, H., B. Malgrange, F. Robert, C. Charlet, D. De Groote, D. Heymann, A. Godard, J. P. Soulillou, G. Moonen, and V. Geenen. 1996. Cytokine production by human thymic epithelial cells: control by the immune recognition of the neurohypophysial self-antigen. *Regul Pept* 67:39-45.

[75] Martino, T. A., M. Petric, M. Brown, K. Aitken, C. J. Gauntt, C. D. Richardson, L. H. Chow, and P. P. Liu. 1998. Cardiovirulent coxsackieviruses and the decay-accelerating factor (CD55) receptor. *Virology* 244:302-14.

[76] McKnight, K. L., and S. M. Lemon. 1998. The rhinovirus type 14 genome contains an internally located RNA structure that is required for viral replication. 4:1569-84.

[77] Melnick, J. L. 1996. Enteroviruses: polioviruses, Coxsackieviruses, Echoviruses, and newer enteroviruses, Fields Virology, third edition ed. Lippincott-Raven publishers, Philadelphia.

[78] Modalsli, K. R., G. Bukholm, S. O. Mikalsen, and M. Degre. 1991. Microinjected Coxsackie B1 virus does not replicate in HEp-2 cells. *Virology* 185:888-90.

[79] Morasco, B. J., N. Sharma, J. Parilla, and J. B. Flanegan. 2003. Poliovirus cre(2C)-dependent synthesis of VPgpUpU is required for positive- but not negative-strand RNA synthesis. 77:5136-44.

[80] Muckelbauer, J. K., M. Kremer, I. Minor, G. Diana, F. J. Dutko, J. Groarke, D. C. Pevear, and M. G. Rossmann. 1995. The structure of coxsackievirus B3 at 3.5 A resolution. *Structure* 3:653-67.

[81] Nairn, C., D. N. Galbraith, K. W. Taylor, and G. B. Clements. 1999. Enterovirus variants in the serum of children at the onset of Type 1 diabetes mellitus. *Diabet Med* 16:509-13.

[82] Nomoto, A., B. Detjen, R. Pozzatti, and E. Wimmer. 1977. The location of the polio genome protein in viral RNAs and its implication for RNA synthesis. 268:208-13.

[83] Okada, Y., G. Toda, H. Oka, A. Nomoto, and H. Yoshikura. 1987. Poliovirus infection of established human blood cell lines: relationship between the differentiation stage and susceptibility of cell killing. *Virology* 156:238-45.

[84] Oldstone, M. B. 1998. Molecular mimicry and immune-mediated diseases. *FASEB J.* 12:1255-65.

[85] Parsley, T. B., J. S. Towner, L. B. Blyn, E. Ehrenfeld, and B. L. Semler. 1997. Poly (rC) binding protein 2 forms a ternary complex with the 5'-terminal sequences of poliovirus RNA and the viral 3CD proteinase. *RNA* 3:1124-34.

[86] Patrick, S. L., C. S. Moy, and R. E. LaPorte. 1989. The world of insulin-dependent diabetes mellitus: what international epidemiologic studies reveal about the etiology and natural history of IDDM. *Diabetes Metab Rev* 5:571-8.

[87] Paul, A. V., J. H. van Boom, D. Filippov, and E. Wimmer. 1998. Protein-primed RNA synthesis by purified poliovirus RNA polymerase. *Nature* 393:280-4.

[88] Pilipenko, E. V., S. V. Maslova, A. N. Sinyakov, and V. I. Agol. 1992. Towards identification of cis-acting elements involved in the replication of enterovirus and rhinovirus RNAs: a proposal for the existence of tRNA-like terminal structures. *Nucleic Acids Res* 20:1739-45.

[89] Richter, W., T. Mertens, B. Schoel, P. Muir, A. Ritzkowsky, W. A. Scherbaum, and B. O. Boehm. 1994. Sequence homology of the diabetes-associated autoantigen glutamate decarboxylase with coxsackie B4-2C protein and heat shock protein 60 mediates no molecular mimicry of autoantibodies. *J Exp Med* 180:721-6.

[90] Rieder, E., A. V. Paul, D. W. Kim, J. H. van Boom, and E. Wimmer. 2000. Genetic and biochemical studies of poliovirus cis-acting replication element cre in relation to VPg uridylylation. *J Virol* 74:10371-80.

[91] Roivainen, M., S. Rasilainen, P. Ylipaasto, R. Nissinen, J. Ustinov, L. Bouwens, D. L. Eizirik, T. Hovi, and T. Otonkoski. 2000. Mechanisms of coxsackievirus-induced damage to human pancreatic beta-cells. *J Clin Endocrinol Metab* 85:432-40.

[92] Rossmann, M. G., E. Arnold, J. W. Erickson, E. A. Frankenberger, J. P. Griffith, H. J. Hecht, J. E. Johnson, G. Kamer, M. Luo, A. G. Mosser, and et al. 1985. Structure of a human common cold virus and functional relationship to other picornaviruses. *Nature* 317:145-53.

[93] Rueckert, R. R. 1996. Picornaviridae: the viruses and their replication, Fields Virology, third edition ed. Lippincott-Raven publishers, Philadelphia.

[94] Schlegel, A., and K. Kirkegaard. 1995. Cell biology of enterovirus infection, p. 135-154. *In* H. A. Rotbart (ed.), *Human enterovirus infections.* ASM, Washington DC.

[95] Schober, E., and H. Frisch. 1988. Incidence of childhood diabetes mellitus in Austria 1979-1984. *Acta Paediatr Scand* 77:299-302.

[96] Serrano-Rios, M., A. Goday, and T. Martinez-Larrad. 1999. Migrant populations and the incidence of type 1 diabetes mellitus: an overview of the literature with a focus on the Spanish-heritage countries in Latin America. *Diabetes Metab Res Rev* 15:113-32.

[97] Serreze, D. V., E. W. Ottendorfer, T. M. Ellis, C. J. Gauntt, and M. A. Atkinson. 2000. Acceleration of type 1 diabetes by a coxsackievirus infection requires a preexisting critical mass of autoreactive T-cells in pancreatic islets. *Diabetes* 49:708-11.

[98] Shieh, J. T., and J. M. Bergelson. 2002. Interaction with decay-accelerating factor facilitates coxsackievirus B infection of polarized epithelial cells. J Virol 76:9474-80.

[99] Sinha, A. A., M. T. Lopez, and H. O. McDevitt. 1990. Autoimmune diseases: the failure of self tolerance. *Science.* 248:1380-8.

[100] Spiller, O. B., I. G. Goodfellow, D. J. Evans, S. J. Hinchliffe, and B. P. Morgan. 2002. Coxsackie B viruses that use human DAF as a receptor infect pig cells via pig CAR and do not use pig DAF. *J Gen Virol* 83:45-52.

[101] Staines, A., S. Hanif, S. Ahmed, P. A. McKinney, S. Shera, and H. J. Bodansky. 1997. Incidence of insulin dependent diabetes mellitus in Karachi, Pakistan. *Arch Dis Child* 76:121-3.

[102] Stanway, G., T. Hovi, N. J. Knowles, and T. Hyypiä. 2002. Molecular and biological basis of picornavirus taxonomy, p. 17-21. *In* B. L. Semley and E. Wimmer (ed.), Molecular Basis of Picarnavirus. *ASM Press,* Washington DC.

[103] Stewart, T. A., B. Hultgren, X. Huang, S. Pitts-Meek, J. Hully, and N. J. MacLachlan. 1993. Induction of type I diabetes by interferon-alpha in transgenic mice. *Science* 260:1942-6.

[104] Teterina, N. L., A. E. Gorbalenya, D. Egger, K. Bienz, and E. Ehrenfeld. 1997. Poliovirus 2C protein determinants of membrane binding and rearrangements in mammalian cells. *J Virol* 71:8962-72.

[105] Tian, J., P. V. Lehmann, and D. L. Kaufman. 1994. T cell cross-reactivity between coxsackievirus and glutamate decarboxylase is associated with a murine diabetes susceptibility allele. *J Exp Med* 180:1979-84.

[106] Tomko, R. P., R. Xu, and L. Philipson. 1997. HCAR and MCAR: the human and mouse cellular receptors for subgroup C adenoviruses and group B coxsackieviruses. *Proc Natl Acad Sci* U S A 94:3352-6.

[107] Towner, J. S., T. V. Ho, and B. L. Semler. 1996. Determinants of membrane association for poliovirus protein 3AB. *J Biol Chem* 271:26810-8.

[108] Ujevich, M. M., and R. Jaffe. 1980. Pancreatic islet cell damage. Its occurrence in neonatal coxsackievirus encephalomyocarditis. *Arch Pathol Lab Med* 104:438-41.

[109] van Kuppeveld, F. J., J. G. Hoenderop, R. L. Smeets, P. H. Willems, H. B. Dijkman, J. M. Galama, and W. J. Melchers. 1997. Coxsackievirus protein 2B modifies endoplasmic reticulum membrane and plasma membrane permeability and facilitates virus release. *Embo J* 16:3519-32.

[110] Varela-Calvino, R., R. Ellis, G. Sgarbi, C. M. Dayan, and M. Peakman. 2002. Characterization of the T-cell response to coxsackievirus B4: evidence that effector memory cells predominate in patients with type 1 diabetes. *Diabetes* 51:1745-53.

[111] Wang, X., and J. M. Bergelson. 1999. Coxsackievirus and adenovirus receptor cytoplasmic and transmembrane domains are not essential for coxsackievirus and adenovirus infection. *J Virol* 73:2559-62.

[112] Wetz, K., and K. O. Habermehl. 1982. Specific cross-linking of capsid proteins to virus RNA by ultraviolet irradiation of poliovirus. *J Gen Virol* 59:397-401.

[113] Wetz, K., and K. O. Habermehl. 1979. Topographical studies on poliovirus capsid proteins by chemical modification and cross-linking with bifunctional reagents. *J Gen Virol* 44:525-34.

[114] Whitton, J. L., and R. S. Fujinami. 1999. Viruses as triggers of autoimmunity: facts and fantasies. *Curr. Opin. Microbiol.* 2:392-7.

[115] Yin, H., A. K. Berg, T. Tuvemo, and G. Frisk. 2002. Enterovirus RNA is found in peripheral blood mononuclear cells in a majority of type 1 diabetic children at onset. *Diabetes.* 51:1964-71.

[116] Yin, H., A. K. Berg, J. Westman, C. Hellerstrom, and G. Frisk. 2002. Complete nucleotide sequence of a Coxsackievirus B-4 strain capable of establishing persistent infection in human pancreatic islet cells: Effects on insulin release, proinsulin synthesis, and cell morphology. *J Med Virol* 68:544-57.

[117] Ylipaasto, P., K. Klingel, A. M. Lindberg, T. Otonkoski, R. Kandolf, T. Hovi, and M. Roivainen. 2004. Enterovirus infection in human pancreatic islet cells, islet tropism in vivo and receptor involvement in cultured islet beta cells. *Diabetologia* 47:225-39.

[118] Yoon, J. W., M. Austin, T. Onodera, and A. L. Notkins. 1979. Isolation of a virus from the pancreas of a child with diabetic ketoacidosis. N. Engl. *J. Med.* 300:1173-9.

[119] Yoon, J. W., W. T. London, B. L. Curfman, R. L. Brown, and A. L. Notkins. 1986. Coxsackie virus B4 produces transient diabetes in nonhuman primates. *Diabetes* 35:712-6.

[120] Yoon, J. W., T. Onodera, and A. L. Notkins. 1978. Virus-induced diabetes mellitus. XV. Beta cell damage and insulin-dependent hyperglycemia in mice infected with coxsackie virus B4. *J Exp Med* 148:1068-80.

In: Progress in Virus Research
Editor: Sabine Thebault, pp. 191-225

ISBN 1-59454-438-7

Chapter VII

EBV Associated Malignancies: Epidemiology, Pathogenesis and Therapeutic Perspectives

***Riccardo Dolcetti*[1], *Diego Serraino and Maria G. Masucci*[3]**
[1]Immunovirology and Biotherapy Unit, Dept. of Pre-Clinical and Epidemiological Research C.R.O. – I.R.C.C.S., National Cancer Institute, Aviano (PN), Italy
[2]Dept. of Epidemiology, INMI L. Spallanzani, IRCCS, Rome, Italy.
[3]Microbiology and Tumor Biology Center, Karolinska Institutet, S-171 77 Stockholm, Sweden.

Abstract

Epstein-Barr virus (EBV), a widespread human γ-herpes virus, is a well-established risk factor for Burkitt's lymphoma, Hodgkin's lymphoma, a subset of T cell lymphomas and various tumors of epithelial cell origin with nasopharyngeal carcinoma as the best studied example. Furthermore, EBV-associated lymphoproliferative disorders are a well-recognized major complication of immune suppression. These EBV-associated lymphoid and epithelial cell malignancies are characterized by peculiar geographic distributions and distinctive epidemiologic features. EBV establishes a life-long asymptomatic infection in immunocompetent hosts by colonizing the memory B-lymphocytes compartment and hijacking cellular signaling pathways that regulate antigen-dependent B-cell activation and differentiation. This multifaceted interaction with B cells is mediated by a set of EBV-encoded proteins that are differently expressed in normal and malignant cells. The quantification of circulating EBV DNA and EBV-epitope specific T cells are valuable tools in the diagnosis, clinical monitoring and prognosis of some EBV-associated tumors. Recent advances in elucidation of the mechanisms underlying EBV-induced growth transformation and immune evasion have been instrumental in guiding the design of novel approaches for the treatment EBV-associated malignancies.

Introduction

The year 2004 marks the 40th anniversary of the discovery of Epstein-Barr virus (EBV), which was first identified using electron microscopy in cultured Burkitt's lymphoma cells when no human lymphoid cell had ever been maintained in culture (Barr, 1964). In these four decades, a huge body of information has been accumulated on the epidemiology and biology of EBV infection as well as on the pathogenetic association of this virus with human cancers. EBV is a gamma herpesvirus of humans with two major target tissues *in vivo*: B-lymphocytes and stratified epithelium. In B-cells, the infection is predominantly non-productive or "latent" and leads to the establishment of a carrier state, which is maintained for the entire life of the infected host. Spread to new hosts is ensured by intermittent reactivation and productive replication at epithelial surfaces. Transmission is usually asymptomatic and only when primary infection is delayed until adolescence or adulthood it may cause a benign lymphoproliferative disease known as infectious mononucleosis (IM). Yet, the presence of viral DNA and the regular expression of viral gene products strongly implicate EBV in the pathogenesis of a broad spectrum of malignancies. These include lymphomas of B, T and NK cell origin such as the immunoblastic lymphoma of immunosuppressed, endemic Burkitt's lymphoma, Hodgkin's Disease, anaplastic large T cell lymphoma and midline granuloma (reviewed in Deacon *et al.*, 1993), but also lymphoepitheliomas of the nasopharynx, thymus and stomach (Imai *et al.*, 1994; Leyvraz *et al.*, 1985) and leiomyosarcomas arising in organ transplant children and HIV patients (Lee *et al.*, 1995; McClain *et al.*, 1995). The association of at least some of these tumors with severe immunosuppression stresses the importance of immune responses in keeping a non-pathogenic equilibrium between the oncogenic virus and the infected host. The opposing arms of this equilibrium are the recognition and destruction of virus infected cells by host immune defense and the capacity of the virus to evade such control in order to maintain the latent infection and to allow reactivation and transmission to new hosts.

Epidemiology of EBV Infection

Epidemiological Patterns

Infection by EBV is nearly ubiquitous, being present in metropolitan areas and in remote populations where other common viral infections are absent. It has been estimated that more than 80% of all individuals above the age of 30 are infected making EBV one of the most successful viruses infecting humans (Black *et al.,* 1970; Tischendorf *et al.*, 1970; Evans and Niederman, 1989; IARC, 1997). Once seroconversion for EBV has occurred, infection persists for the individual's lifetime and most of the persistently infected people secrete EBV in their saliva. Thus, the oral route is by far the most common mode of transmission, although rare cases of EBV transmission through blood transfusion or organ transplantation have been also documented (Alfieri *et al.,* 1996; Haque *et al.,* 1996; IARC, 1997).

Since the early '70s, epidemiological studies conducted worldwide have provided consistent evidence that different age-related patterns of EBV infection are associated with socioeconomic conditions in defined geographic areas (Henle *et al.,* 1969; Black *et al.,* 1970;

Tischendorf *et al.,* 1970; Biggar *et al.,* 1978; IARC, 1997). Generally, EBV seroprevalence in young adults is inversely correlated with family size, hygienic standard and, ultimately, with socioeconomic status. In developing countries, and in disadvantaged population groups of developed areas, up to 95% of the children are infected before the age of six through intra-familial exposures. In contrast, in developed counties, EBV infection tends to occur later in life, generally during adolescence. In this group, infection is assumed to occur mainly through kissing and IM develops in 50%-74% of these cases (Macsween and Crawford, 2003).

Recent studies have taken into account new life styles and demographic changes to better delineate the epidemiologic patterns of EBV infection (IARC, 1997; Macsween and Crawford, 2003). The presence of EBV in male and female genital secretions has been recognized (Israele, *et al.,* 1991), and one study of teenaged girls in Sweden found that EBV seropositivity was associated with sexual activity (Andersson-Ellström *et al.,* 1995). An epidemiological investigation undertaken among 1006 students under age 21 attending the University of Edinburgh, United Kingdom, showed a higher prevalence of EBV seropositivity among sexually active students (82.7%) compared to those who were not (63.7%). A greater number of sex partners was a highly significant risk factor for EBV seropositivity (Crawford *et al.,* 2002), suggesting that EBV transmission may occur during sexual intercourse or closely associated behaviors.

On the basis of epidemiological evidence obtained from human and animal studies, the International Agency for Cancer Research (IARC) has classified EBV as a group 1 carcinogen because of its strong association with Burkitt's lymphoma (BL), nasopharyngeal carcinoma (NPC), Hodgkin's lymphoma (HL) and immunosuppression related lymphoproliferative disorders (IARC, 1997; Macsween *et al.,* 2003) *(Table 1).*

Table 1. EBV-associated cancers

Tumor type	At risk populations
Burkitt's lymphoma	African children HIV-positive individuals in western countries
Immunosuppression-related non-Hodgkin's lymphoma	HIV-positive individuals Individuals immune suppressed after organ transplantation or with other types of immune deficiency
Hodgkin's lymphoma	Young adults and HIV-positive individuals in western countries Children in developing countries
Nasopharyngeal carcinoma	Southern China (Cantonese populations); Chinese population groups in Hong Kong and Singapore; Mediterranean Arabs; Malays in Singapore

EBV and Burkitt's Lymphoma

In the immunocompetent population of western countries, BL is a very rare tumor with an approximate incidence of 1/1.000.000 children of 4-14 years of age. In contrast, BL accounts for 30-70% of childhood cancers in equatorial Africa, where it occurs at high frequency in a broad belt around the equator (approximately 15° N and 15° S) (IARC, 1997).

Holoendemic malaria infection appears to play an important role as a co-factor for the disease. In Uganda, where EBV infection is acquired early in life and malaria is holoendemic, about 12 children/100.000 are affected by BL, while the incidence drops to 1-2 children/100.000 in Algeria, where EBV infection is acquired early in life but malaria is not endemic.

EBV was identified in 1964 by electron microscopy in BL biopsies in an attempt to verify the hypothesis, proposed by Dennis Burkitt, that the lymphoma that now bears his name might be caused by an oncogenic infectious agent. The association between EBV and BL shows strong geographic variations. In Africa, between 50% to 95% of BL cases carry EBV DNA whereas only 10%-30% of European and American BLs are associated with EBV (IARC, 1997). Both EBV positive and EBV negative BLs carry chromosomal translocations that cause constitutive activation of the *c-myc* oncogene by juxtaposition to an immunoglobulin enhancer. Thus, activation of the oncogene appears to be the primary cause of BL while EBV acts as a co-factor that, together with other geographically distributed co-factors, such as age of primary EBV infection, malaria, socioeconomic and lifestyle factors, influences the occurrence of the disease. In 1970, a cohort of 42.000 children was followed up in the West Nile district of Uganda for 5-8 years. During the study period, 14 cases of BL were registered, and the results obtained suggested that high VCA titers represented a strong risk factor for the disease (De Thé *et al.,* 1978). However, this finding was not consistently reproduced in non-African patients, since some studies in the United States did not find higher VCA titers in BL cases compared to controls (Hirshaut *et al.,* 1973). The occurrence of primary EBV infection at an early age increases the risk of BL of nearly 20 times, while a heavy malaria burden increases the BL risk of 100-fold, giving a 200-fold higher risk in Africa than in western countries (De Thé *et al.,* 2000).

The localization of BL in different anatomic sites varies according to age and frequency of the tumor in various geographic areas. The jaw is the predominantly involved site in African young children, while abdominal involvement is more common among adults (IARC, 1997). Outside equatorial Africa, up to 80% of BL cases show abdominal involvement, with differences between patients from endemic and sporadic (i.e., Europe, Australia and North America) countries. For instance, presentation with a large tumor mass in the right iliac fossa is common in sporadic tumors but infrequent in endemic cases (IARC, 1997). Central nervous system involvement is found in nearly one third of patients at presentation in Africa, but is much less common in regions in which the disease is sporadic (Ziegler *et al.,* 1970).

Thirtyfive-70% of BL cases in AIDS patients are EBV-associated (Gaidano *et al.,* 1995; Davi *et al.,* 1998). AIDS-associated EBV-positive BL seems to be pathogenetically similar to sporadic BL, with chronic stimulation of B-lymphocytes playing a crucial role. AIDS-associated lymphomas with c-*myc* translocation tends to occur in patients with a relatively intact immune system, while mutations in the p53 gene have been detected in approximately 60% of the cases (Gaidano *et al.,* 1995). AIDS-associated BL also show an up-regulation of CD39 and CD70 activation antigens and of the CD11a/LFA-1 adhesion molecule and a lower CD4+ cell count (Davi *et al.,* 1998).

Although the c-*myc* gene is translocated to one of the immunoglobulin loci in all BLs, the breakpoints are often different in the endemic and sporadic forms of the disease. In endemic BLs the breakpoint is usually 5' of the c-*myc* gene, which remains intact, while in sporadic and AIDS-associated BLs, the breakpoint is often in the first exon or in the first intron (Magrath, 1990; Shiramizu *et al.*, 1991). In all cases the coding region located in the second and third exons, is always fully conserved. The translocation of *c-myc* to different regions within the immunoglobulin loci suggests that B cells in different stages of activation/differentiation may be involved. The finding that in sporadic and AIDS-associated BLs the breakpoint affects the class switch region is consistent with the occurrence of the translocation in mature B lymphocyte during attempted immunoglobulin class switching. In contrast, in endemic BLs, the breakpoint usually occurs at the VDJ region suggesting that the rearrangement of c-*myc* is an early event occurring in a pre-B cell during V(D)J recombination (Magrath, 1990; Shiramizu *et al.*, 1991).

EBV and Immune-Suppression-Associated Lymphoma

The relationship between immunodeficiency and the increased frequency of non-Hodgkin lymphoma (NHL) was first noted nearly 35 years ago among transplanted patients (Penn *et al.*, 1969), and became more evident when an increase of NHL was observed in parallel with the emergence of the human immunodeficiency virus (HIV) epidemic in the early 80's (Ziegler *et al.*, 1984). HIV-infected individuals show a 10 to 100-fold increased risks of three types of NHL including: primary central nervous system lymphoma (PCNSL), immunoblastic lymphoma and BL. PCNSL shows the strongest association with EBV, since the virus is present in nearly 100% of cases (IARC, 1997) as compared to an average prevalence of 40% in systemic NHL (Ambinder and Sparano, 2001). PCNSL is associated with a severe degree of immunodeficiency (i.e., less than 50 CD4+ cell/mm) and it is thousand times more frequent in HIV-infected than in HIV-seronegative individuals (International Collaboration on HIV and Cancer, 2000). After the introduction of highly active antiretroviral therapy (HAART), a dramatic decrease in the progression of HIV infection to AIDS and to death has been reported in western countries (Palella *et al.*, 1998) and this has been paralleled by a substantial reduction in the incidence of PCNSL (International Collaboration on HIV and Cancer, 2000). Moreover, a higher likelihood of response to chemotherapy, as well as a longer survival, has been found after the use of HAART in patients with PCNSL (Hoffmann *et al.*, 2001). However, even though these observations suggested that the use of HAART might be associated with improved prognosis, the survival PCNSL of these patients still remains very poor (Antinori *et al.*, 2002).

Post-transplant lymphoproliferative disease (PTLD) is increasingly recognized as an important determinant of both graft and patient survival. The EBV genome is found in more than 90% of B-cell PTLD occurring during the first year after transplantation (i.e., in the early phase), while up to 45% of late onset PTLD may be EBV negative (Cockfield *et al.*, 2001; Preikasaitis *et al.*, 2003). Thus, while the etiologic role of EBV infection in the occurrence of early PTLD has been well recognized, its role in the occurrence of late PTLD is still uncertain. Factors associated with an increased risk of early PTLD have been identified, and high risk patients may be targeted for monitoring protocols and preemptive therapy. Among such factors, primary EBV infection, the type of transplanted organ, concomitant infection with cytomegalovirus (CMV), antigenic stimulation, severity of immunodeficiency and recipient age deserve attention. In particular, transplanted patients at risk for primary EBV infection are at 10- to 70-fold higher risk of PTLD than EBV seropositive patients, while the relative risk of seropositive individuals with EBV reactivation versus those re-infected with a second strain is unknown. In addition to EBV serological status, the type of transplanted organ also influences the risk of developing PTLD, as illustrated in Figure 1. The lowest risk of PTLD is registered in recipients of renal allograft (1%), whereas the risk is highest among recipients of heart-lung allografts (9.4%). In individuals who underwent heart transplantation, the risk of PTLD in the first year after transplantation was approximately 5-fold higher than among kidney transplant recipient (Opelz *et al.,* 1993). Although these variations may be partly attributed to differences in the type and cumulative dosage of immunosuppressive treatment, it is possible that yet undefined properties of the allografted organ may also predispose to PTLD (Preikasaitis *et al.,* 2003). Current post-transplant regimens have resulted in the earlier occurrence of PTLD (within 4 months in 32% of cases) and increased involvement of the allograft and lymph nodes, whereas the involvement of the central nervous system is decreased. There is now convincing evidence demonstrating that the use of potent anti-lymphocyte antibodies is associated with an increased incidence of early PTLD (Swinnen *et al.,* 1990), although the greatest risk for PTLD in the early post-transplantation period does probably result from the cumulative effect of immunosuppressive drugs rather than from a single agent (Preikasaitis *et al.,* 2003). CMV mismatching (seropositive donor and seronegative recipient) favors the occurrence of PTLD, since the occurrence of CMV disease in patients with a primary EBV infection was associated with a 7-fold higher risk of PTLD. Infection with human herpes virus 8 (HHV8) seems to contribute to the risk of early PTLD (Dotti *et al.,* 1999) and age is also an important determinant. Pediatric recipients are at higher risk of early PTLD, probably because they are more exposed to primary EBV (and CMV) infection, whereas older recipients are at higher risk for late PTLD, with rates being 5 times lower in patients under 20 years of age than in those aged 50 years or older (Opelz *et al.,* 1993).

EBV and Hodgkin's Lymphoma

HL is the commonest malignancy in young adults. Time trends in the incidence of HL in Nordic European countries (Denmark, Sweden, Norway, and Finland) during 1978-97 showed a significant decrease in all age groups above 40 year, whereas there was a significant increase incidence of HL, primarily for the nodular sclerosis (NS) subtype, among adolescents and young adults (Hjalgrim *et al.,* 2001). The epidemiology of HL is characterized by a bimodal age incidence curve, with a first peak around the age of 15-35 and the second peak around the age of 50 (MacMahon *et al.,* 1966). This peculiar age distribution pattern was taken as suggestive of an infectious etiology of HL, possibly as a late, abnormal host response to a widely diffuse agent in childhood. Because the presentation pattern of HL is very similar to that of IM, EBV was one of the first infectious agents considered as possible pathogenetic cofactors for the development of the disease

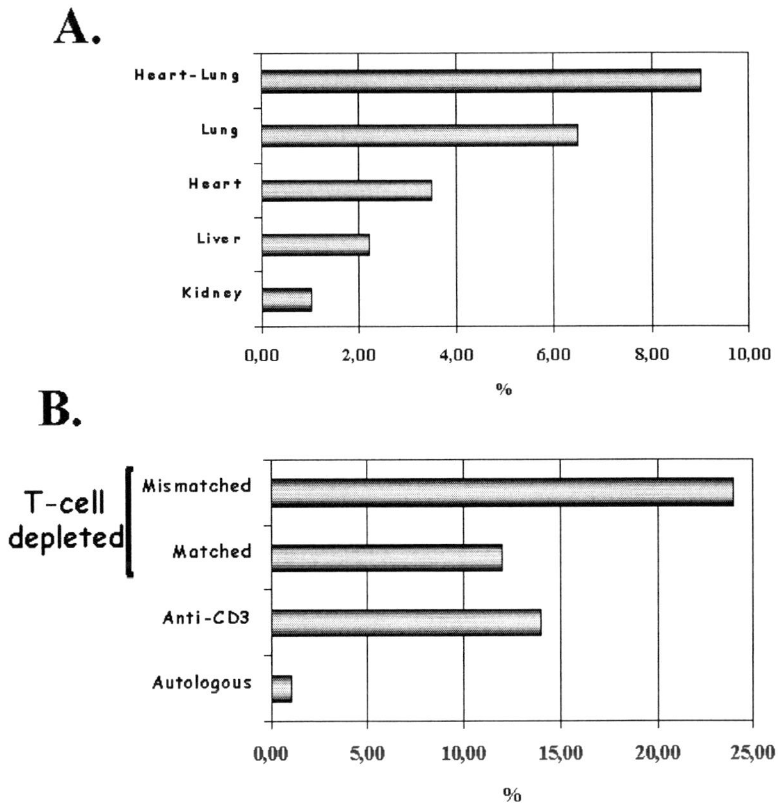

Figure 1. Incidence of PTLD in patients transplanted with different solid organs (A) and in recipients of autologous or allogeneic bone marrow transplantation (B).

The possible pathogenetic link between EBV infection and HL was subsequently strengthened by the detection of increased titers of antibodies to EBV in HL patients as well as by prospective data showing altered EBV serology before HL development (Mueller *et al.*, 1989). After the identification, in 1987, of the EBV genome in Reed-Sternberg cells, subsequent studies have consistently demonstrated the presence of EBV in approximately 50% of HL cases in western countries, varying from 80% in the mixed cellularity (MC) subtype to 20% in the nodular sclerosis (NS) subtype.

In the last three decades, consistent evidence has supported the association between EBV infection and HL. Case control studies have shown positive, although weak, increased risks for HL in individuals with a history of IM, particularly young adults and patients with the NS subtype (Serraino *et al.*, 1991). The risk of HL in patients with past IM was also evaluated by longitudinal studies, which showed a 3-fold excess risk (IARC, 1997). Serological studies on serum samples collected before the occurrence of HL also confirmed the excess risks linked to EBV infection, with a 6-fold increase of HL risk with high titers of EBNA and a decreased risk with IgM antibodies (particularly in sera collected at least three years before HL diagnosis) (IARC, 1997).

Recently, Hjalgrim and colleagues have compared the incidence rates of HL in two population-based cohorts of Danish patients tested for IM; 38,555 individuals with serologic evidence of EBV infection, and 24,614 with no such evidence. Biopsy specimens of HL occurring during follow-up were investigated for the presence of EBV. The findings strongly suggested a causal association between IM-related EBV infection and the EBV-positive subgroup of HL in young adults, since only serologically confirmed IM was associated with a persistently increased risk of HL. Sixteen of 29 tumors (55%) obtained from patients with IM had evidence of EBV infection and there was a significant 4-fold increased risk of EBV-positive HL. In contrast, there was no evidence for increased risk of EBV-negative HL after IM. The estimated median incubation time from IM to EBV-positive HL was 4 years (Hjalgrim *et al.,* 2003).

Collectively, these findings support a causal role of EBV in the development of a proportion of HL cases, while age-specific incidence curves suggest that EBV-associated and EBV-unrelated HL may be two distinct etiological entities. Alternatively, several mechanisms may lead to the development of HL. In keeping with this view, it might be possible to develop HL prevention strategies for patients affected by IM (Ambinder *et al.,* 2003).

EBV and Nasopharyngeal Carcinoma

Nasopharyngeal carcinoma (NPC) is an unusual tumor showing peculiar epidemiological characteristics. In most parts of the world, NPC is a rare malignancy, with incidence rates less than 1/100.000 in both sexes, whereas in high-risk areas the incidence rates may be up to 20-fold higher. Populations at very high risk for NPC have been identified in southern China, among Inuits and other natives in the Arctic region, while intermediate to high-risk populations have been identified in South-East Asia and in northern Africa (IARC, 1997).

The association between EBV infection and the occurrence of NPC has primarily been assessed by means of sero-epidemiological studies. NPC patients were shown to have elevated IgG and IgA antibody titers to the EBV capsid antigen (VCA) and the early antigen (EA) that are associated with EBV replication (IARC, 1997), and that these factors were predictive of tumor burden, remission and recurrence (Henle *et al.,* 1976; Traub, 2000). These findings emphasize the association between EBV reactivation/replication with the development of NPC. From an epidemiological viewpoint, EBV should be considered a major determinant of NPC, since the virus is detected in all cases of undifferentiated NPC and all tumor cells usually carry EBV infection. However, the wide geographical variations in incidence rates of NPC suggest that several other cofactors are likely to contribute to the pathogenesis of NPC, among them host, dietary, genetic, and environmental factors (IARC, 1997).

The consumption of Cantonese salted fish has been implicated in the etiology of NPC in various Chinese populations, with a 2-fold to 7-fold increased risk, particularly when such habit is acquired at an early age. A positive association with NPC was also noted for other types of salted fish, and for early exposure to other preserved foods, although the evidence is not as strong as for Cantonese salted fish (IARC, 1997). Several authors hypothesized that domestic exposure to smoke from wood fires inside chimneyless houses, or occupational exposure to smoke, dust and fumes were determinants of NPC, but these results are inconsistent (IARC, 1997), while recent epidemiological studies consistently indicate that cigarette smoking increases the risk of NPC by approximately 3-fold (IARC, 1997).

Among the genetic factors, an association between HLA haplotype and NPC risk has been determined in both Chinese and non-Chinese populations. A 2-fold increased risk of NPC was associated with the presence of both A2 and BW46 among Chinese in Singapore, Malaysia, Hong Kong and Guangzhou, while a 20-fold excess risk was noted in southern Chinese siblings. In non-Chinese populations, associations between the HLA profile and NPC risk was described for HLA A3 in Australian whites, HLA B5 in Germans and HLA A2 in the United States (for further information on cofactors for EBV-associated NPC see IARC, 1997).

EBV Pathogenesis

In spite of its association with malignancies, EBV is a largely non-pathogenic virus. B-lymphocytes are the primary targets of EBV infection *in vitro* and *in vivo* and the site of virus persistence in healthy carriers. These cells are largely non permissive for virus replication but readily express a set of viral genes that are collectively known as the "latency" genes to distinguish them from the much more numerous "lytic" genes that are expressed during productive infection. The latency genes encode for six nuclear and three membrane associated proteins known as the EBV nuclear antigens (EBNA)-1, -2, -3(or 3A), -4 (or 3B), -5 (or LP) and -6 (or 3C) and the latent membrane proteins (LMP)-1, LMP-2A and LMP-2B (reviewed in (Kieff, 1996).

Latently infected B-lymphocytes express one of four EBV latency programs that appear to reflect the adaptation of the virus to different stages of B-cell activation/differentiation *(Table 2)*. Healthy EBV carriers have between 1 to 50 EBV-infected B-lymphocytes per million cells in the peripheral blood (Babcock *et al.*, 1998 and 1999; Chen *et al.*, 1995). These EBV carrying cells are phenotypically similar to long-lived memory cells and may express a putative Latency 0, characterized by complete silencing of the viral genome, or Latency I, in which LMP-2A is expressed alone or together with EBNA-1 (Figure 2) (Miyashita *et al.*, 1997; Chen *et al.*, 1995). In the absence of immune pressure, as observed *in vitro* or *in vivo* in immunosuppressed patients, the EBV-infected B cells express a Latency III program where all nine latency proteins are regularly detected. This latency program is also known as the "growth program" (Figure 2) since it is associated with B cell proliferation *in vitro* and *in vivo*. The proliferation of infected B blasts is likely to play an important role in the economy of EBV infections since it may expand the pool of infected cells, and increase the likelihood of access to the memory B cell compartment. An intermediate form of latency characterized by the expression of EBNA-1 and the three LMPs (Latency II) is expressed in EBV infected B-lymphocytes that home to the germinal centers of lymphoid follicles (Babcock *et al.*, 2000a and b). This is known as the "rescue program" (Figure 2) since the latent membrane proteins regulate B cell activation and apoptosis and may allow the maturation of the EBV infected lymphoblasts into memory cells (reviewed in Thorley-Lawson, 2001). The relevance of these discrete latency programs is strongly supported by studies of EBV-associated malignancies (Table 2). Thus, latency III is expressed in the immunoblast-like cells of EBV carrying lymphoproliferative disorders arising in organ and bone marrow transplant recipients and HIV patients (Thomas *et al.*, 1991), while latency I is found in EBV carrying BLs that are phenotypically similar to memory B lymphocytes (Rowe DT *et al.*, 1986). In line with their germinal center origin, cells from EBV positive HL express a Latency II program (reviewed in (Niedobitek *et al.*, 2000). Latency II is also expressed in non-B-cell tumors of both hematopoietic and epithelial cell origin (reviewed in Knecht *et al.*, 2001; Dolcetti and Masucci, 2003).

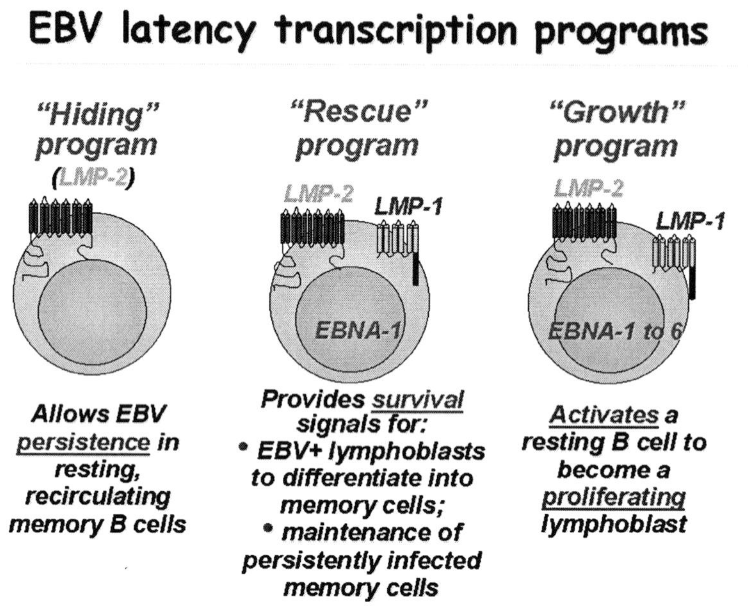

Figure 2. EBV latency transcription programs and their expression in B lymphocytes. These latency programs appear to reflect the adaptation of the virus to different stages of B-cell activation/differentiation and allow the persistence of the virus within the memory B cell compartment.

Table 2. EBV-latency programs in malignant cells

Latency program	Disease	Expressed viral genes
Latency 0	None	None
Latency I	Burkitt's Lymphoma [a]	EBNA-1
Latency II	Hodgkin's Disease, Peripheral T cell lymphoma, Nasal T/NK cell lymphoma, Nasopharyngeal carcinoma, Lymphoepithelioma (stomach, thymus)	EBNA-1, LMP-1, LMP-2A, -2B
Lantency III	AIDS-related immunoblastic B-cell lymphoma PTLD - post transplant lymphoproliferative disorder	EBNA-1, -2, -3, -4, -5, -6, LMP-1, LMP-2A, -2B

a. Some tumors were shown to express EBNA-3, -4 and –6

Regulation of Cell Growth and Apoptosis

The highly malignant B cell lymphomas arising in patients suffering from severe immune deficiencies illustrate the potential danger of the Latency III program. However, under normal circumstances, the growth potential of these cells is counteracted by the concomitant expression of many B cell activation markers and co-stimulatory molecules that enhance the activity of various components of the antigen presentation machinery, rendering the infected cells highly immunogenic and therefore easily attacked by T cell-mediated immune responses. LMP-1 plays a major role in driving this phenotype by acting as a master switch that controls the proliferation, survival and immunogenicity of EBV infected cells.

LMP-1 is the only EBV protein with recognized oncogenic activity. Transfection of LMP-1 into mouse or human fibroblasts and epithelial cells confers tumorigenicity in immunosuppressed animals (Hu *et al.*, 1993; Yang X *et al.*, 2000; Takanashi *et al.*, 1999), and LMP-1 transgenic mice develop hyperproliferations and lymphomas (Kulwichit *et al.*, 1998; Curran *et al.*, 2001). LMP-1 contains a short N-terminal cytoplasmic domain followed by six membrane-spanning domains and a large cytoplasmic C-terminal domain that is involved in signaling (reviewed in Kieff, 1996). LMP-1 acts as a constitutive CD40-like receptor and induces expression of NF-κB through activation of the tumor necrosis factor-receptor associated factor (TRAF) signaling pathway (Mosialos *et al.*, 1995; Kieser *et al.*, 1999). However, unlike CD40, LMP-1 does not target TRAF-2 and -3 for proteasomal degradation, which results in their constitutive activation and sustained activity (Brown *et al.*, 2001). Constitutive activation of NF-kB, and the consequent activation of cytokines and cytokine receptors on tumor cells and the surrounding stroma, are critical for the capacity of LMP-1 to modulate the interaction of the infected cells with the environment by regulating angiogenesis and invasiveness. Several components of the antigen presentation pathway are also up-regulated in LMP-1 expressing cells including the transporters associated with antigen processing, (TAPs) (Rowe M *et al.*, 1995; Frisan *et al.*, 1996) and the interferon-γ-regulated subunits of the proteasome core and regulatory particles, which results in enhanced enzymatic activity and changed cleavage specificity (Frisan *et al.*, 1998).

Regulation of Virus Replication

It is generally believed that production of infectious EBV particles is associated with death of the infected cells which implies that reactivation of latent infection must be tightly regulated *in vivo*. The events that lead to disruption of latency in B cells are poorly understood but studies with tumor cell lines suggest that triggering of the B-cell receptor (BCR) plays a critical role (Rowe M *et al.*, 1992). LMP-2A regulates this process through its ability to inhibit BCR signaling by interfering with the activity of BCR-associated tyrosine kinases (Miller *et al.*, 1994a and b). Recent studies have shed some light on this viral regulatory circuit. It was shown that the amino terminal domain of LMP-2A contains eight tyrosines that associate with the cellular protein tyrosine kinases *Lyn* and *Syk* via SH2-phosphotyrosine interactions and five proline-rich regions, three of which possess the PxxP core consensus sequence required for interacting with SH3 domains and two of which possess the PPxY core consensus sequence (PY motif) required for interacting with proteins containing WW modules (reviewed in Longnecker, 2000). In screening for proteins interacting with the PY domains, two laboratories have independently identified members of the Nedd4 ubiquitin ligase family including AIP4/Itchy, WWP2/AIP2 and KIAA0439 (Ikeda *et al.*, 2000; Winberg *et al.*, 2000). Binding of these ubiquitin ligases to LMP-2A correlates with ubiquitination of *Lyn* and *Syk* and accelerated degradation of *Lyn*. Ubiquitination of *Syk* does not appear to affect its turnover although accelerated degradation of a small pool of activated *Syk* cannot be excluded. LMP-2A is specifically ubiquitinated by AIP4 and WWP2 and this modification is required for modulation of BCR signaling. Although much of the involved interactions remain obscure, these data confirm that the expression of scaffold proteins that recruit components of the ubiquitin-proteasome pathway is a common viral strategy for selective activation/inactivation of cellular substrates.

Mechanism of Virus Persistence

The regular and in some cases exclusive expression of EBNA-1 in EBV associated malignancies has spurred an intensive search for specific T cell responses that could be selectively boosted in cancer patients. These efforts were initially frustrated by the failure to isolate EBV specific CTLs capable of recognizing cells infected with recombinant vaccinia or adenovirus vectors expressing EBNA-1 (reviewed in Rickinson, 1992). Later studies have demonstrated that EBNA-1 specific T cells exist but they are either of the CD4 type (reviewed in (Paludan and Munz, 2003), or recognize antigen presenting cells fed with exogenous EBNA-1 (Blake *et al.*, 1997) while the endogenous protein is not processed for MHC class I restricted presentation. More recent studies measuring T cell responses by interferon-gamma production in ELISPOT assays rather then the less sensitive target cell lysis in ^{51}Cr release assays have demonstrated that presentation of endogenous EBNA-1 does actually occur (Voo *et al.*, 2004) but appears to be less efficient than for other viral genes. The poor presentation of EBNA-1 appears to be due to inefficient processing of this viral protein, which was shown to correlate with the presence of an internal glycine-alanine repeat (GAr) domain. Removal of the GAr resulted in accelerated protein turnover and efficient presentation of peptide epitopes derived from the endogenous protein (Levitskaya *et al.*, 1995; Blake *et al.*, 1997). In line with these findings, deletion of the GAr reconstituted the capacity of EBNA-1 to trigger specific rejection responses in a mouse tumor model (Mukherjee *et al.*, 1998). The capacity of EBNA-1 to escape immune recognition may offer an important advantage in the establishment of EBV latency by permitting the continuous expression of this essential protein whereas other highly immunogenic viral products must be downregulated in order to prevent elimination of the viral reservoir.

Elucidation of the mechanism of action of the GAr is particularly interesting since this is the first example of a protein domain that affects antigen presentation and this feature could be exploited in immune- and gene-therapy settings. Using an *in vitro* processing assay it was shown that the GAr is a specific inhibitor of ubiquitin-dependent proteolysis and acts as a transferable element that, when expressed in the context of other viral or cellular proteasome substrates, abrogates or severely inhibits their degradation (Levitskaya *et al.*, 1997). Several features of this stabilization signal were revealed using a set of GAr chimeras involving IκB (Sharipo *et al.*, 1998), p53 (Heessen *et al.*, 2002), and green fluorescent protein (GFP)-based proteasome substrates in mammalian (Dantuma *et al.*, 2000a) and budding yeast (Heessen *et al.*, 2003). It was found that the activity of the GAr is independent of its location in the target protein (Levitskaya *et al.*, 1997; Sharipo, *et al.* 1998; Dantuma *et al.*, 2000a; Heessen *et al.*, 2002) and is not restricted by the type of ubiquitin ligase involved in substrate modification (Sharipo *et al.*, 1998; Heessen *et al.*, 2002; Dantuma *et al.*, 2000a). Fusions of the GAr to GFP-based reporters that are targeted for degradation with different efficiency (Dantuma *et al.*, 2000b) showed that the GAr counteracts the degradation signal in a length dependent manner (Dantuma *et al.*, 2000a). Indeed, EBNA-1 itself could be targeted for ubiquitin-dependent proteolysis using a strong degradation signal (Dantuma *et al.*, 2000a). The only restriction for the GAr activity appears to be the requirement for a sufficiently long stretch of Ala or similar small hybrophobic residues, preferably interspersed by 1, 2 or 3 Gly residues which may act by increasing solubility (Sharipo *et al.*, 2001). These findings and the demonstration that ubiquitinated GAr containing IκB cannot form stable complexes with the proteasome (Sharipo *et al.*, 1998) but still interacts with the S5a ubiquitin-binding subunit of the 19S cap (Heessen *et al.*, 2002; Leonchiks *et al.*, 2002), suggest that the highly hydrophobic GAr may compete for a putative recognition site in the proteasome. This possibility is substantiated by the demonstration that a synthetic GAr peptide inhibits the degradation of biotinylated lysozyme *in vitro* (Leonchiks *et al.*, 2002).

The importance of EBNA-1 stability in the context of viral immune escape is questioned by the demonstration that analogous repeat domains found in baboons and rhesus macaques γ-herpesviruses do not confer protection from CTL lysis in autologous systems (Blake *et al.*, 1999). Furthermore, recent evidence suggests that ubiquitin-dependent processing of EBNA-1 mRNA-derived defective ribosomal products (DRiPs) could bypass the inhibitory effect of the GAr, which points to a non-immunologic role for EBNA-1 stability. Resistance to proteolysis may be required for the capacity of EBNA-1 to act as a transcriptional regulator either alone or together with interacting cellular proteins. Indeed, using both affinity chromatography and TAP-tagging approaches Holowaty *et al.* have recently shown that EBNA-1 interacts with the isopeptidase USP7, also known as herpes virus associated ubiquitin specific protease, HAUSP (Holowaty *et al.*, 2003a). This nuclear enzyme was first identified by virtue of its interaction with the ICP0 protein of herpes simplex virus type 1, which is required for efficient initiation of the HSV-1 lytic cycle (Everett *et al.*, 1997) and was later shown to bind and deubiquitinate p53, resulting in p53 stabilization and p53-dependent growth arrest and apoptosis (Boutell *et al.*, 2003). The significance of EBNA-1:HAUSP interaction in the context of EBV infection remains unclear. USP7 appears to regulate EBNA-1 activity since disruption of USP7 binding enhanced the capacity of EBNA-1 to replicate *ori*P-containing plasmids. This may be achieved through regulation of EBNA-1 ubiquitination, which may affect its capacity for protein-protein interaction (Muller *et al.*, 2001; Spence *et al.*, 2000). Alternatively, EBNA-1 may bring USP7 to *ori*P where it could act on specific cellular substrates. It is also possible that binding to EBNA-1 may interfere with the physiologic function of USP7. By sequestering USP7, EBNA-1 may destabilize p53 with important consequences for both B cell immortalization and the development of EBV-associated tumors. This possibility was recently substantiated by the demonstration that a synthetic peptide corresponding to residues 395-450 of EBNA-1 is able to compete with a p53 peptide for binding to USP7 (Holowaty *et al.*, 2003b).

The Contribution of Cellular Oncogenes

Malignancies expressing one or more EBV antigens usually arise in relatively immunocompetent individuals suggesting that specific features of the tumor cells may promote their evasion from the host immune control. This concept is clearly illustrated by Burkitt's lymphoma. Only EBNA-1 is usually detected in EBV-positive BLs (Rowe DT *et al.*, 1986) and, although tumors expressing EBNA-1 together with the high molecular weight EBNA-3, -4 and -6 were recently described (Kelly *et al.*, 2002), EBNA-2 and LMP-1 are never expressed which explains the lack of B-cell activation markers, adhesion and co-stimulatory molecules and growth as single cell suspensions (Rowe DT *et al.*, 1986). Characteristic for the tumor cells are also the low levels of MHC class I and selective loss of certain class I alleles, HLA A11 in particular (Andersson *et al.*, 1991), which may contribute to the poor allostimulatory capacity of these cells in mixed lymphocyte cultures (Avila-Carino *et al.*, 1987). In addition, BL cells are deficient in their ability to present endogenous antigens via the MHC class I pathway (Frisan *et al.*, 1996) and express low levels of the interferon-gamma inducible subunits of the proteasome and the peptide transporters TAP1 and TAP2 (Frisan *et al.*, 1998; Rowe M *et al.*, 1995).

Studies of *in vitro* EBV-transformed cell lines that recapitulate the BL cell phenotype through regulated expression of EBNA2/LMP-1 and c-Myc have shown that overexpression of c-Myc in the absence of LMP-1 is directly responsible for the altered immunogenicity of BL cells (Staege *et al.*, 2002; Frisan *et al.*, 1996 and 1998), which also correlates with alterations of ubiquitin-dependent proteolysis (Gavioli *et al.*, 2001). BL cells were shown to be resistant to treatment with doses of proteasome inhibitors that are readily toxic for normal lymphoblasts. Furthermore, the turnover of short and long-lived proteins proceeded virtually undisturbed in BL cells in the presence of high doses of proteasome inhibitors, suggesting that other enzymatic activities participate in this critical cellular function. Indeed, deubiquitinating enzymes and the serine protease Tripeptidyl peptidase (TPP)-II, were shown to be highly expressed in tumor derived BL lines and were up-regulated upon overexpression of c-Myc in genetically normal cells.

The upregulation of TPPII and isopeptidases appears to be a direct consequence of c-Myc overexpression and does not require selection *in vivo*. The combined effect of these changes could contribute to the deregulation of a wide variety of cellular proteins involved in the control of cell cycle and apoptosis. Like many other transcription factors, the c-Myc oncoprotein is a short-lived substrate of the ubiquitin-proteasome system. Most importantly, the same protein domains regulate the turnover of c-Myc and its transcriptional activation function (Salghetti *et al.*, 1999; Bahram *et al.*, 2000). The half-life of wild type c-Myc is increased between two to six folds in Burkitt's lymphoma (Gregory *et al.*, 2000), suggesting that modulation of c-Myc ubiquitination by DUBs may play an important role in determining its turnover.

Diagnosis and Monitoring of EBV-Associated Disorders

Diagnostic Tools

Molecular tools are increasingly becoming important for the diagnostic definition of a wide spectrum of EBV-related disorders. A detailed description of the conventional tools used to diagnose EBV infection is beyond the scope of this article. It should be mentioned, however, that serology remains the gold standard to confirm acute versus past EBV infection in immunocompetent host. However, in patients with immune deficiency, humoral responses to EBV are quite variable and serology cannot provide reliable information to clinicians. Assessment of EBV seronegativity is of clinical relevance for transplanted patients since this condition is associated with a high risk to develop PTLD. Abnormal high titers against EA and IgG VCA with reduced EBNA titers are frequently observed in patients with EBV-related tumors, although this serologic pattern is not specific for malignancy. Several studies, including mass serological surveys, have clearly demonstrated that tests for serum IgA against EBV VCA and EA are clinically useful for the early diagnosis of NPC, particularly in high risk areas, as well as for monitoring the recurrence and progression of the disease (International Agency for Research on Cancer, 1997). However, accurate standardization of the sensitivity and specificity of this screening method is still lacking and a precise follow-up protocol for anti-VCA IgA positive and negative individuals has not been defined. An additional limitation to the use of anti-VCA IgA is the observation that a sizeable fraction of NPC patients, mainly the younger, remains negative for these antibodies (Sam *et al.*, 1989).

In situ hybridization of EBERs is considered the gold standard for detection and localization of latent EBV in tissue specimens (Ambinder, 1994). EBER-1 and -2 are the most abundant EBV RNAs in latently infected cells with more than 10^6 copies per cell (Howe *et al.*, 1986). Moreover, the EBERs are thought to be expressed in all forms of EBV latency whereas lesions with pure lytic EBV infection, such as oral hairy leukoplakia, are consistently EBER-negative (Gilligan *et al.*, 1990). It is generally assumed that the EBERs are uniformly expressed in all cells of EBV-related malignancies but only focal EBER expression has be detected in some instances (Pathmanathan *et al.*, 1995; Wu *et al.*, 1991). These discrepancies may be due to differences in the detection methods used or the state of EBV infection. In fact, EBV DNA may not be detected in cells carrying few EBV genomes and EBER expression may be down-regulated in cells permissive for EBV replication or within areas of cell differentiation (Pathmanathan *et al.*, 1995).

Analysis of the configuration of the EBV genome termini by Southern blotting is useful to assess the clonality of EBV-carrying cells. This method is based on the observation that, while the viral DNA is encapsidated as a linear molecule with homologous direct tandem repeats at each terminus (Raab-Traub and Flynn, 1986), the linear termini join after infection to form covalently closed circular episomes, which stably persist in the nucleus of latently infected cells. Thus, analysis of the EBV termini allows the distinction between a monoclonal infection, characterized by the presence of a single, identical EBV episome in all infected cells, and a polyclonal infection, in which multiple circular EBV molecules with fused termini of different size are detected (Raab-Traub and Flynn, 1986). This assay has revealed the presence of monoclonal EBV DNA in nearly all EBV-associated carcinomas, sarcomas,

HL and NHL (Raab-Traub and Flynn, 1986, IARC, 1997). Notably, both *in situ* and invasive NPC samples carried clonal EBV (Pathmanathan *et al.*, 1995; Raab-Traub and Flynn, 1986), indicating that a monoclonal expansion of EBV-carrying cells is present in these lesions and suggesting that clonal expansion of EBV latently infected nasopharyngeal cells is an early event in NPC development. Assessment of EBV clonality may be of particular clinical relevance in the setting of immunodeficiency. In fact, immunosuppressed patients with either oligoclonal or polyclonal proliferations of EBV-infected B cells have a better prognosis and may benefit from the sole reduction of immunosuppression (Chadburn *et al.*, 1997).

Diagnostic and Prognostic Value of EBV DNA Load

Data from several laboratories have consistently demonstrated that quantitative analysis of plasma/serum EBV DNA levels is clinically relevant in the diagnosis and monitoring of several EBV-related malignancies. Circulating cell-free EBV DNA has been detected in the plasma/serum from patients with HL (Drouet *et al.*, 1999, Gallagher *et al.*, 1999), PTLD (Limaye *et al.*, 1999), AIDS-related lymphoma (Yamamoto *et al.*, 1995), NPC (Lo *et al.*, 1999), and NK/T-cell lymphoma (Lei *et al.*, 2002). Quantitative analyses of EBV DNA in the serum/plasma of NPC patients revealed markedly higher load compared to few controls with barely detectable levels of EBV DNA (Lo *et al.*, 1999b). Moreover, advanced-stage NPC patients had higher plasma EBV DNA levels than those with early stage disease (Lo *et al.*, 1999a). Kinetics analyses carried out in NPC patients undergoing radiation therapy showed that, after an initial rise, circulating EBV DNA levels usually dropped in virtually all responsive cases (Lo *et al.*, 1999a; Lo *et al.*, 2000b). Of note, significant increases in serum EBV DNA load were observed in patients who subsequently developed tumor recurrence whereas stably low or undetectable levels of serum EBV DNA were observed in patients who remained in remission (Lo *et al.*, 1999a; Lo *et al.*, 2000b). Quantification of circulating EBV DNA may have prognostic value in this clinical setting. In fact, it has been shown that plasma/serum EBV DNA concentration is an independent prognostic factor for NPC, and is a more accurate determinant of recurrence than disease stage within the first year of follow-up (Lo *et al.*, 2000a).

More recently, significantly higher levels of cell-associated EBV DNA were detected in the blood of NPC patients compared to healthy controls (Lin *et al.*, 2001). In addition, multivariate analysis demonstrated that the EBV status in peripheral blood is an important independent prognostic factor for metastasis-free, progression-free and survival rates (Lin *et al.*, 2001). Follow-up results also indicated that the positivity for EBV DNA in PBMC was a strong predictor of subsequent distant metastasis in NPC patients (Lin *et al.*, 2001). Analysis of purified B-cell and non-B-cell fractions indicated that circulating tumor cells probably constitute the major source of EBV DNA in the blood of NPC patients (Lin *et al.*, 2001).

EBV DNA has been detected in the serum of more than 90% of patients with EBV-associated HL (Drouet *et al.*, 1999; Gallagher *et al.*, 1999), and the levels appear to correlate with serological evidence of virus reactivation (Drouet *et al.*, 1999). Moreover, analysis of a limited series of HL patients suggested that persistence of EBV DNA positivity in post-treatment serum samples could be an indicator of poor prognosis (Gallagher *et al.*, 1999). These findings await confirmation in larger series of HL patients.

The determination of EBV DNA load is of particular clinical value in solid organ and bone marrow transplant recipients since early identification and accurate monitoring of patients at high risk for PTLD is essential to reduce the morbidity and mortality associated with this disorder (Stevens *et al.*, 2002). Several studies have clearly shown that the development of PTLD is in most cases accompanied by a significant increase in the level of circulating EBV DNA (Riddler *et al.*, 1994; Savoie *et al.*, 1994; Kenagy *et al.*, 1995). Moreover, quantification of EBV DNA load was proven to be useful for monitoring the responses to treatment in patients with overt PTLD, as the amount of EBV DNA decreases to normal or undetectable levels in patients undergoing remission (Riddler *et al.*, 1994; Stevens *et al.*, 2001; Wagner *et al.*, 2001). Among the different analytical methods used, real-time PCR turned out to be the more appropriate for a precise, frequent and high-throughput monitoring of EBV load dynamics, which is crucial in high-risk patients (Rowe *et al.*, 2001; Stevens *et al.*, 2002; van Esser *et al.*, 2001). However, much remains to be done to reach standardization, particularly with regards to the DNA purification method, the EBV DNA fragment to be amplified, the clinical specimens used and the definition of cut-off values of clinical relevance. The choice of the starting biologic material (serum/plasma, PBMCs, or whole blood) is also a matter of debate, although recent findings indicate that assessment of the EBV load per ml of unfractionated whole blood may be preferable since it allows determination of EBV DNA load, irrespective of fluctuations in cell numbers (Stevens *et al.*, 2001, and 2002).. Available evidence indicates that the increased EBV DNA load observed in PTLD patients most likely represents an increase in the number of circulating EBV-carrying B cells, although it remains to be defined whether these are tumor cells or normal EBV-infected B lymphocytes. A strong correlation was found between EBV load and the numbers of circulating EBV-carrying cells, supporting the hypothesis that the number of EBV-infected cells increases as a consequence of immunosuppression rather than the number of viral genomes per cell (Yang *et al.*, 2000). Recent findings indicate that, in both healthy individuals and PTLD patients, the majority of circulating EBV-carrying B cells harbor less than 50 EBV copies per cell (Yang *et al.*, 2000). Moreover, analysis of viral gene expression in blood samples from PTLD patients at diagnosis revealed transcripts of EBERs, LMP-2, EBNA-6 and gp350 in 45% of the cases, consistently with an unrestricted pattern of latency with lytic replication (Hopwood *et al.*, 2002). Longitudinal studies carried out after transplant disclosed a progressive switch from the restricted pattern of latency seen in healthy carriers to broader patterns of EBV latency coupled with lytic replication (Hopwood *et al.*, 2002). This is probably the result of inadequate T-cell control of both latently and lytically infected cells, which may favor the proliferation of pre-existing EBV-carrying B lymphocytes as well as the infection of new B cells. The demonstration of active EBV replication in transplant patients with and without PTLD may at least partially account for the detection of elevated EBV DNA loads in serum/plasma by some studies (van Esser *et al.*, 2001a and b; Barkholt *et al.*, 1996). In these cases the viral DNA could well represent EBV virions, although possible interference due to DNA released from damaged cells cannot be excluded.

From a clinical point of view, the presence of high or increasing levels of EBV DNA has been used as a criterion to modulate the strength of immunosuppression or to start pre-emptive (CTL-based) or antitumor therapy (Baldanti *et al.*, 2000; Kogan *et al.*, 1999; Rooney *et al.*, 1995; Gustafsson *et al.*, 2000). Nevertheless, several studies have convincingly demonstrated that the detection of increased EBV load alone is not always predictive of impending PTLD (Rowe *et al.*, 1997; Savoie *et al.*, 1994; Lucas *et al.*, 1998; Hopwood *et al.*,

2002). The occurrence of high EBV load without PTLD development can be explained by the concomitant increase in EBV-specific cellular responses, as indeed demonstrated in EBV-seronegative children who underwent liver transplantation (Smets *et al.*, 2002). These findings suggest that the combined determination of EBV DNA load and EBV-specific T-cell count could allow a more precise prediction of the risk of PTLD development, at least in EBV-seronegative recipients.

Tetramer-Based Monitoring of EBV-Specific Immune Responses

Functional characterization and tracking of antigen-specific T cells is of crucial relevance for understanding and monitoring host immune responses against EBV-infected cells, particularly in patients with EBV-related malignancies. The recent development of MHC class I tetramers has allowed a direct and precise quantification of epitope-specific T lymphocyte subpopulations while avoiding extensive manipulation of cells in tissue culture. This approach is based on the use of soluble fluorescent tetramers composed of four identical MHC class I molecules folded around a known epitope peptide, which constitutes the natural target of T lymphocytes *in vivo*. Besides visualization and enumeration of epitope-specific T cells, staining with MHC-peptide tetramers also allows the *ex vivo* immunophenotyping of antigen-specific T lymphocytes (Pittet *et al.*, 2001, Appay *et al.*, 2002). Thus, antigen-specific T cells have been characterized with regards to their differentiation/activation state, TCR usage, susceptibility to apoptosis, homing properties, and expression of inhibitory receptors (Altman *et al.*, 1996; Ogg *et al.*, 1998; Tan *et al.*, 1999; Pittet *et al.*, 2001; Appay *et al.*, 2002). Moreover, following their isolation, the functional properties of epitope-specific T lymphocytes may be further characterized using techniques such as intracellular cytokine analysis, ELISPOT or quantitative RT-PCR. Results obtained using recombinant MHC-peptide tetramers have led to a re-evaluation of the nature of both primary and memory responses to several viruses, including EBV. In fact, the use of MHC-peptide tetramers has allowed the demonstration of a large expansion of activated antigen-specific CD8+ T cells during the primary response to EBV in patients with IM (Callan *et al.*, 1998). The abundance of this EBV-specific CTL response significantly exceeded previous estimates of EBV-specific CTL precursor frequencies determined using limiting-dilution cytotoxicity assays (Steven *et al.*, 1996). After clinical recovery from acute IM, the frequency of circulating antigen-specific CD8+ CTLs decreased but still remained detectable (Callan *et al.*, 1998). These studies also showed that, in peripheral blood of healthy carriers, CD8+ T cells carrying receptors specific for EBV lytic cycle proteins were usually found at higher numbers than those specific for latent EBV proteins (Callan *et al.*, 1998; Tan *et al.*, 1999), consistently with previous work with functional assays (Steven *et al.*, 1996). The use of EBV-specific tetramers in conjunction with antigen-specific activation assays allowed the identification of two main populations of CD8+ T cells specific for EBV peptides in individuals persistently infected by EBV. One subset expresses the lymph node homing receptor CD62L and do not rapidly produce interferon (IFN)-γ whereas another sub-population lacks CD62L and can readily make IFN-γ upon stimulation (Tussey *et al.*, 2000). These functional and migratory properties are similar to those of central and effector memory T cells, respectively, as recently defined by Sallusto *et al.* (1999), suggesting that this distinction can be applied also to EBV-specific T lymphocytes. More recently, it has been shown that, in both healthy donors and HIV-infected individuals, CD8+ T cells staining with EBV-specific tetramers uniformly lacked CCR7 expression and generally did not express CD62L (Chen *et al.*, 2001). These findings suggest that antigen-experienced CD8+ T cells in the setting of chronic EBV infection may be selectively excluded from the lymphoid sites, thus impairing the immune defense against a virus such as EBV that primarily targets B-lymphocytes. The combination of MHC-peptide tetramers and IFN-γ ELISPOT assay has been recently used to study EBV-specific T-cell immunity in HIV-infected patients who developed AIDS-related non-Hodgkin's lymphoma (NHL) (van Baarle *et al.*, 2001). The study showed that HIV-infected patients have normal numbers of EBV-specific CD8+ T lymphocytes although these effectors were functionally impaired when compared to those of HIV-seronegative individuals. Moreover, the development of AIDS-related NHL was not associated with a depletion of EBV-specific T lymphocytes but was related to the progressive inability of these cells to produce IFN-γ in response to EBV peptides (van Baarle *et al.*, 2001). Functional impairment of CD8+ T cells directed against EBV lytic epitopes probably accounted for an enhanced EBV reactivation, as shown by increased EBV DNA load in these patients. This may progressively enlarge the pool of EBV-infected B cells whose proliferation is poorly controlled in this setting, thus enhancing the risk of genetic alterations that may promote B cell transformation. These findings suggest that functional loss of both lytic and latent antigen-specific CD8+ T cells may contribute to the development of AIDS-related NHL. MHC-peptide tetramers in combination with IFN-γ ELISPOT assays were also used to investigate the effect of highly active antiretroviral therapy (HAART) on HIV and EBV-specific CD8+ T cells (Kostense *et al.*, 2002). Interestingly, the proportion of IFN-γ-producing T cells specific for both HIV and EBV epitopes increased in most patients, consistently with a functional restoration of virus-specific effectors in individuals successfully treated with HAART (Kostense, 2002). In this study, the improvement of antigen responsiveness of both HIV- and EBV-specific T cells was associated with an increase in the numbers of CD4+ T lymphocytes, suggesting an important role of CD4+ T cell help in the functional recovery of virus-specific immune responses. MHC-peptide tetramers have been also used to identify EBV-specific T cells in the circulation of pediatric allograft recipients (Falco *et al.*, 2002). The study demonstrated that CD8+ T cells specific for both lytic and latent EBV epitopes were detectable as early as 4 weeks after transplant in EBV-seronegative patients receiving an organ from an EBV-carrying donor (Falco *et al.*, 2002). Consistently with other reports, also in this setting the majority of tetramer reactive T cells had an activated/memory phenotype. Phenotypic and functional analysis of EBV-specific CD8+ T cells are however required to understand how iatrogenic immunesuppression affects immune responses against EBV. Moreover, combined use of MHC-peptide tetramers and functional assays may prove useful in monitoring EBV infection in the post-transplant patient and in identifying patients at risk to develop EBV-related lymphoproliferations.

New Treatment Modalities for EBV-Associated Disorders

Immunotherapy

EBV-associated malignancies constitute ideal targets for T cell-based immunotherapy approaches since immunogenic EBV antigens are expressed on tumor cells providing thus the rationale for a tumor-specific treatment. In addition, the lymphoblastoid cell lines generated by infecting normal peripheral blood B cells with EBV function as excellent antigen-presenting cells, allowing the induction/reactivation *in vitro* of EBV-specific CTLs, which may be easily cultured in large numbers. Finally, most patients are immune to EBV and EBV-specific CTL precursors persist lifelong at high frequency.

Initial treatment of patients who developed PTLD after allogeneic stem cell transplant with unmanipulated donor mononuclear cells resulted in high response rates (Papadopoulos *et al.*, 1994; O'Reilly *et al.*, 1997). However, such unselected cell populations may also contain high numbers of alloreactive T lymphocytes and induction of graft-versus-host disease (GVHD) was also observed (Papadopoulos *et al.*, 1994; O'Reilly *et al.*, 1997). Further experience in this clinical setting indicated that the administration of CTLs specific for EBV antigens may induce more potent anti-tumor effects without the risk of GVHD (Rooney *et al.*, 1995 and 1998). In particular, adoptive transfer of donor-derived EBV-specific CTLs appeared to prevent progression to PTLD in high-risk recipients of bone marrow transplantation (Rooney *et al.*, 1995 and 1998; Gustafsson *et al.*, 2000). Infused CTLs persisted *in vivo* for as long as 7 years, as shown by gene marking (Rooney *et al.*, 1995 and 1998). Notably, following adoptive transfer of EBV-specific CTLs, there is reconstitution of immunity to EBV with a rise in virus-specific cytotoxicity in peripheral blood and an increase in CTL precursor frequency (Rooney *et al.*, 1995 and 1998; Heslop *et al.*, 1996). Moreover, high EBV genome loads that existed prior to CTL administration rapidly decreased to normal levels (Heslop *et al.*, 1996; Gustafsson *et al.*, 2000). Available evidence also indicates that the efficacy of this approach in terms of PTLD prophylaxis is strictly related to the presence of a major EBV-specific component within the CTLs infused (Gustafsson *et al.*, 2000). Prophylaxis seems to have advantages over therapy of overt PTLD. In these cases, in fact, complete remissions can be also achieved, although CTL treatment of bulky disease can cause severe morbidity due to extensive inflammatory responses (Rooney *et al.*, 1998). Moreover, CTL therapy may be ineffective in some occasion, due to deletion of immunodominant epitopes in tumor cells for which the donor CTLs were specific (Gottschalk *et al.*, 2001). EBV-specific CTLs have also been administered to patients after solid organ transplant with reports of immune and clinical responses (Comoli *et al.*, 2002; Haque *et al.*, 2002).

The success of the adoptive transfer of polyclonal EBV-specific CTLs for the prophylaxis and treatment of PTLDs (Rooney *et al.*, 2001) has provided an important rationale for verify the efficacy of CTL-based therapies also in other EBV-related diseases such as HL and NPC. Although promising results were reported (Roskrow *et al.*, 1998; Comoli *et al.*, 2004), this has been a challenging task due to the low immunogenicity of LMP-1 and LMP-2 compared to the EBNAs expressed in PTLDs. The demonstration that LMP-1- and -2-specific CTL precursors can be detected in HL and NPC patients (Frisan *et al.*, 1995; Lee *et al.*, 2000; Chapman *et al.*, 2001) underscores the therapeutic potential boosting these responses. Among the new immunotherapeutic approaches is the attempt to increase the immunogenicity of LMP-1 or LMP-2 by using autologous dendritic cells for antigen presentation (Gahn *et al.*, 2001; Gottschalk *et al.*, 2003). Encouraging results have been recently reported following vaccination with LMP-2 peptide-pulsed dendritic cells in NPC patients. This vaccination strategy elicited or boosted epitope-specific CD8+ T cell responses in peripheral blood and induced partial clinical responses in a fraction of patients (Lin *et al.*, 2002). However, the effect of CTL cultures enriched in LMP-1- and LMP-2-specific effectors may be inhibited by immunosuppressive factors secreted within the microenvironment by tumor cells of both HL and NPC (Beck *et al.*, 2001). Attempts have been made to modify EBV-specific CTLs *ex vivo*, with the purpose to shield these effectors from the inhibitory effects of tumor-derived TGF-β (Bollard *et al.*, 2002).

Other Virus-Targeted Therapies

It seems reasonable to assume that persistence of the EBV genome within tumor cells may be required to maintain the transformed phenotype in several EBV associated malignancies. Thus, therapeutic approaches aimed at inducing loss of the EBV episome may be of potential clinical relevance. Chronic administration of low doses of hydroxyurea was shown to induce loss of EBV episomes in some BL and LCLs (Chodosh *et al.*, 1998). In some instances, the EBV-negative derivatives showed a decreased tumorigenic potential compared with the EBV-carrying parental cell lines (Chodosh *et al.*, 1998). Based on these findings, two patients with AIDS-related primary central nervous system lymphoma were treated with low-dose hydroxyurea, which in both cases gave a significant improvement of survival (Slobod *et al.*, 2000). However, these results need to be confirmed in larger, controlled clinical trials since integration of viral DNA into the host genome could limit the clinical efficacy of this treatment.

The almost exclusive expression of some EBV-encoded genes in tumor cells can be exploited to selectively drive the expression of cellular toxins in an EBV-dependent fashion for therapeutic purposes. Both LMP-1- and LMP-2-responsive elements were used to induce the transcription of the herpes simplex virus thymidine kinase (HSV-TK) in tumor cells (Wu *et al.*, 1998; Franken *et al.*, 1996). Expression of these constructs successfully sensitized LMP-1 or EBNA-2-expressing cells to gancyclovir-dependent killing *in vitro*. However, the clinical applicability of these strategies is limited by the restricted expression of EBNA2 to only few EBV-related malignancies and the highly heterogeneous expression of LMP-1 in others. In this respect, EBNA-1 constitutes a more attractive target, being universally expressed in all EBV-associated malignancies. Several groups have developed vectors that express toxins under the control of the oriP enhancer element, which is activated by the specific binding of EBNA-1 (Reisman *et al.*, 1985). Besides oriP-based constructs driving the expression of HSV-TK, vectors carrying the cytosine deaminase gene have been also constructed, allowing the selective killing of EBV-positive cells *in vitro* (Hirai *et al.*, 1997; Kenney *et al.*, 1998). More recently, the use of EBNA-1-driven transcriptional activation has been developed using adenoviral vectors. This resulted in the induction of oriP-driven reporters or of the p53 tumor suppressor by 1000-fold in EBV-positive cells (Li *et al.*, 2002). This adenoviral-based strategy may be also improved through the use of more cytotoxic

genes. Recent evidence indicates that recombinant adenoviral vectors mediating the expression of a mutant non-cleavable form of FasL induced significant cytotoxicity in EBV-positive NPC cell lines. The findings that the cytotoxic effect of this therapy could be enhanced by combination with ionizing radiation therapy, inducing regression of NPC xenografts in SCID mice (Li *et al.*, 2003), seems promising in view of a possible application in clinical settings. However, the clinical use of these approaches will be possible only if a marked improvement of the currently available gene delivery methods will be achieved.
A strategy that is currently being actively investigated is based on the possibility to induce the EBV lytic cycle in tumor cells. This seems however hard to achieve *in vivo* since the promoters that drive transcription of major lytic gene, BZLF1 and BRLF1, are tightly regulated. Although progress is being made, our understanding of the viral and cellular factors that regulate BZLF1 and BRLF1 gene transcription is still limited, hampering thus the clinical applicability of this strategy.

Concluding Remarks

The recent advances in our understanding of the mechanisms by which EBV induces cell transformation and escapes host immune control provide the rational background to design new strategies of intervention for EBV-related malignancies. The success of the adoptive transfer of polyclonal EBV-specific CTLs for the prophylaxis and treatment of patients with PTLDs (Rooney *et al.*, 2001) provides an important basis to verify the efficacy of CTL-based therapies also in other EBV-related diseases such as HL and NPC. The first results are promising but also indicate that boosting T-cell responses to the poorly immunogenic LMP-1 and LMP-2 is required to increase the response rates. The demonstration that CTL therapy may be useful also to prevent PTLD development is encouraging, although the indications for intervention remain to be defined. Monitoring EBV DNA load can be used for this purpose, but the predictive value of this assay is limited and additional parameters allowing the assessment of EBV-specific T cell responses are probably necessary. Considering the crucial role of EBNA-1 in the pathogenesis of all EBV-associated disorders, strategies able to enhance EBNA-1 processing and degradation *in vivo* would be highly effective since they would trigger CTL responses to all EBV-carrying cells. The identification of EBNA-1-specific CTLs and the demonstration that covalent linking of ubiquitin to EBNA-1 restores degradation of the protein suggest that this approach may be feasible. Agents able to selectively block the interaction between LMP-2A and the Lyn/Syk kinases could release the LMP-2A-mediated inhibition of BCR signaling, which may allow the induction of lytic virus replication upon physiological and/or pharmacological stimulation. Moreover, the ability of LMP-1 and LMP-2 to activate the PI3-K/Akt pathway in both lymphoid and epithelial cells suggests that this signaling cascade may constitute a potentially relevant target for innovative therapies (Vivanco and Sawyers, 2002). Finally, the demonstration that BL cells may be re-sensitized to CTL lysis by proteasome inhibitors suggests that drugs able to modulate components of the ubiquitin-proteasome system and/or associated proteolytic machineries may be of therapeutic relevance for EBV related malignancies.

References

Alfieri C, Tanner J, Carpentier L, Perpete C, Savoie A, Paradis K, Delage G, Joncas J. Epstein-Barr virus transmission from a blood donor to an organ transplant recipient with recovery of the same virus strain from the recipient's blood and oropharynx. *Blood*, 87:812-817, 1996.

Altman JD, Moss PA, Goulder PJ, Barouch DH, McHeyzer-Williams MG, Bell JI, McMichael AJ, Davis MM. Phenotypic analysis of antigen-specific T lymphocytes. *Science*, 274:94-96, 1996.

Ambinder RF, Mann RB. Epstein-Barr-encoded RNA in situ hybridization: diagnostic applications. *Hum Pathol*, 25:602-605, 1994.

Ambinder RF, Sparano JA. Primary central nervous system lymphoma. In: *HIV and HTLV-I associated malignancies. Edt.* Sparano JA, Kluwere Academic Publishers, 231-246, 2001.

Ambinder RF. Infection and Lymphoma. *N Engl J Med*, 349:1309-1311, 2003.

Andersson ML, Stam NJ, Klein G, Ploegh HL, Masucci MG. Aberrant expression of HLA class-I antigens in Burkitt lymphoma cells. *Int J Cancer*, 47:544-550, 1991.

Andersson-Ellström A, Svennerholm B, Forssman L. Prevalence of antibodies to herpes simplex virus types 1 and 2, Epstein-Barr virus and cytomegalovirus in teenage girls. *Scand J Infect Dis*, 27:315-318, 1995.

Antinori A, Cingolani A, Lorenzini P, Giancola ML, Uccella I, Bossolasco S, Grisetti S, Moretti F, Vigo B, Bongiovanni M, Del Grosso B, Arcidiacono MI, Fibbia GC, Mena M, Finazzi MG, Guaraldi G, Ammassari A, d'Arminio Monforte A, Cinque P, De Luca A; Italian Registry Investigative Neuro AIDS Study Group. Clinical epidemiology and survival of progressive multifocal leukoencephalopathy in the era of highly active antiretroviral therapy: data from the Italian Registry Investigative Neuro AIDS (IRINA). *J Neurovirol*, 9:47-53, 2003.

Appay V, Rowland-Jones SL. The assessment of antigen-specific CD8+ T cells through the combination of MHC class I tetramer and intracellular staining. *J Immunol Methods*, 268:9-19, 2002.

Avila-Carino J, Torsteinsdottir S, Ehlin-Henriksson B, Lenoir G, Klein G, Klein E, Masucci MG. Paired Epstein-Barr virus (EBV)-negative and EBV-converted Burkitt lymphoma lines: stimulatory capacity in allogeneic mixed lymphocyte cultures. *Int J Cancer*, 40:691-697, 1987.

Babcock GJ, Decker LL, Volk M, Thorley-Lawson DA. EBV persistence in memory B cells in vivo. *Immunity*, 9:395-404, 1998.

Babcock GJ, Decker LL, Freeman RB, Thorley-Lawson DA. Epstein-barr virus-infected resting memory B cells, not proliferating lymphoblasts, accumulate in the peripheral blood of immunosuppressed patients. *J Exp Med*, 190:567-576, 1999.

Babcock GJ, Hochberg D, Thorley-Lawson AD. The expression pattern of Epstein-Barr virus latent genes in vivo is dependent upon the differentiation stage of the infected B cell. *Immunity*, 13:497-506, 2000.

Babcock GJ and Thorley-Lawson DA. Tonsillar memory B cells, latently infected with Epstein-Barr virus, express the restricted pattern of latent genes previously found only in Epstein-Barr virus-associated tumors. *Proc Natl Acad Sci USA*, 97:12250-12255, 2000.

Baldanti F, Grossi P, Furione M, Simoncini L, Sarasini A, Comoli P, Maccario R, Fiocchi R, Gerna G. High levels of Epstein-Barr virus DNA in blood of solid-organ transplant recipients and their value in predicting posttransplant lymphoproliferative disorders. *J Clin Microbiol*, 38:613-619, 2000.

Bahram F, von der Lehr N, Cetinkaya C, Larsson LG. c-Myc hot spot mutations in lymphomas result in inefficient ubiquitination and decreased proteasome-mediated turnover. *Blood*, 95:2104-2110, 2000.

Barkholt LM, Dahl H, Enbom M, Linde A. Epstein-Barr virus DNA in serum after liver transplantation--surveillance of viral activity during treatment with different immunosuppressive agents. *Transpl Int*, 9:439-445, 1996.

Beck A, Pazolt D, Grabenbauer GG, Nicholls JM, Herbst H, Young LS, Niedobitek G. Expression of cytokine and chemokine genes in Epstein-Barr virus-associated nasopharyngeal carcinoma: comparison with Hodgkin's disease. *J Pathol*, 194:145-151, 2001.

Biggar RJ, Henle G, Bőcker J, Lennette ET, Fleisher G, Henle W. Primary Epstein-Barr virus infection in African Infants. II. Clinical and serological observations during seroconversion. *Int J Cancer*, 22:244-250, 1978.

Birx DL, Redfield RR, Tosato G. Defective regulation of Epstein-Barr virus infection in patients with acquired immunodeficiency syndrome (AIDS) or AISD-related disorders. *N Engl J Med,* 314:874-879, 1986.

Black FL, Woodall JP, Evans AS, Liebhaber H, Henle G. Prevalence of antibody against viruses in the Tiriyo, an isolated Amazon tribe. *Am J Epidemiol*, 91:430-438, 1970.

Blake N, Lee S, Redchenko I, Thomas W, Steven N, Leese A, Steigerwald-Mullen P, Kurilla MG, Frappier L, Rickinson A. Human CD8+ T cell responses to EBV EBNA1: HLA class I presentation of the (Gly-Ala)-containing protein requires exogenous processing. *Immunity*, 7:791-802, 1997.

Blake NW, Moghaddam A, Rao P, Kaur A, Glickman R, Cho YG, Marchini A, Haigh T, Johnson RP, Rickinson AB, Wang F. Inhibition of antigen presentation by the glycine/alanine repeat domain is not conserved in simian homologues of Epstein-Barr virus nuclear antigen 1. *J Virol*, 73:7381-7389, 1999.

Bollard CM, Rossig C, Calonge MJ, Huls MH, Wagner HJ, Massague J, Brenner MK, Heslop HE, Rooney CM. Adapting a transforming growth factor beta-related tumor protection strategy to enhance antitumor immunity. *Blood*, 99:3179-3187, 2002.

Brown KD, Hostager BS, Bishop GA. Differential signaling and tumor necrosis factor receptor-associated factor (TRAF) degradation mediated by CD40 and the Epstein-Barr virus oncoprotein latent membrane protein 1 (LMP1). *J Exp Med*, 193:943-954, 2001.

Boutell C, Everett R. The herpes simplex virus type 1 (HSV-1) regulatory protein ICP0 interacts with and Ubiquitinates p53. *J Biol Chem*, 278:36596-36602, 2003.

Callan MF, Tan L, Annels N, Ogg GS, Wilson JD, O'Callaghan CA, Steven N, McMichael AJ, Rickinson AB. Direct visualization of antigen-specific CD8+ T cells during the primary immune response to Epstein-Barr virus In vivo. *J Exp Med*, 187:1395-1402, 1998.

Chadburn A, Cesarman E, Knowles DM. Molecular pathology of posttransplantation lymphoproliferative disorders. *Semin Diagn Pathol*, 14:15-26, 1997.

Chapman AL, Rickinson AB, Thomas WA, Jarrett RF, Crocker J, Lee SP. Epstein-Barr virus-specific cytotoxic T lymphocyte responses in the blood and tumor site of Hodgkin's disease patients: implications for a T-cell-based therapy. *Cancer Res*, 61:6219-6226, 2001.

Chen F, Zou J-Z, di Renzo L, Winberg G, Hu L-F, Klein E, Klein G and Ernberg I. A subpopulation of normal B cells latently infected with Epstein-Barr virus resembles Burkitt lymphoma cells in expressing EBNA-1 but not EBNA-2 or LMP1. *J Virol*, 69:3752-3758, 1995.

Chen G, Shankar P, Lange C, Valdez H, Skolnik PR, Wu L, Manjunath N, Lieberman J. CD8 T cells specific for human immunodeficiency virus, Epstein-Barr virus, and cytomegalovirus lack molecules for homing to lymphoid sites of infection. *Blood*, 98:156-164, 2001.

Chodosh J, Holder VP, Gan YJ, Belgaumi A, Sample J, Sixbey JW. Eradication of latent Epstein-Barr virus by hydroxyurea alters the growth-transformed cell phenotype. *J Infect Dis*, 177:1194-1201, 1998.

Cockfield SM. Identifying the patient at risk for post-transplant lymphoproliferative disorder. *Transplant Infect Dis*, 3:70-78, 2001.

Comoli P, Labirio M, Basso S, Baldanti F, Grossi P, Furione M, Vigano M, Fiocchi R, Rossi G, Ginevri F, Gridelli B, Moretta A, Montagna D, Locatelli F, Gerna G, Maccario R. Infusion of autologous Epstein-Barr virus (EBV)-specific cytotoxic T cells for prevention of EBV-related lymphoproliferative disorder in solid organ transplant recipients with evidence of active virus replication. *Blood*, 99:2592-2598, 2002.

Comoli P, De Palma R, Siena S, Nocera A, Basso S, Del Galdo F, Schiavo R, Carminati O, Tagliamacco A, Abbate GF, Locatelli F, Maccario R, Pedrazzoli P. Adoptive transfer of allogeneic Epstein-Barr virus (EBV)-specific cytotoxic T cells with in vitro antitumor activity boosts LMP2-specific immune response in a patient with EBV-related nasopharyngeal carcinoma. *Ann Oncol*, 15:113-117, 2004.

Crawford DH, Swerdlow AJ, Higgins C, McAulay K, Harrison N, Williams H, Britton K, Macsween KF. Sexual History and Epstein-Barr Virus Infection. *J Infect Dis*, 186:731-736, 2002.

Curran JA, Laverty FS, Campbell D, Macdiarmid J, Wilson JB. Epstein-Barr virus encoded latent membrane protein-1 induces epithelial cell proliferation and sensitizes transgenic mice to chemical carcinogenesis. *Cancer Res*, 61:6730-6738, 2001.

Dantuma N, Heessen S, Lindsten K, Jellne M, Masucci MG. Inhibition of proteasomal degradation by the gly-Ala repeat of Epstein-Barr virus is influenced by the length of the repeat and the strength of the degradation signal. *Proc Natl Acad Sci USA*, 97:8381-8385, 2000a.

Dantuma N, Lindsten K, Glas R, Jellne M, Masucci MG. Short-lived green fluorescent proteins for quantifying ubiquitin/proteasome-dependent proteolysis in living cells. *Nat Biotechnol*, 18:538-543, 2000b.

Davi F, Delecluse HJ, Guiet P, Gabarre J, Fayon A, Gentilhomme O, Felman P, Bayle C, Berger F, Audouin J, Bryon PA, Diebold J, Raphael M. Burkitt-like lymphomas in ADS patients: characterization within a series of 103 human immunodeficiency virus-associated non-Hodgkin's lymphomas. Burkitt's Lymphoma Study Group. *J Clin Oncol*, 16:3788-3795, 1998.

de-The G, Geser A, Day NE, Tukei PM, Williams EH, Beri DP, Smith PG, Dean AG, Bronkamm GW, Feorino P, Henle W. Epidemiological evidence for a causal relationship between EBV and Burkitt's lymphoma: results of the Ugandan prospective study. *Nature,* 274:756-761, 1978.

De Thé G. Epstein-Barr virus and Burkitt's lymphoma. In: *Infectious causes of Cancer: targets for intervention.* Goedert JJ Edt., Humana Press, 77-92, 2000.

Deacon EM, Pallesen G, Niedobitek G, Crocker J, Brooks L, Rickinson AB, Young LS. Epstein-Barr virus and Hodgkin's disease: transcriptional analysis of virus latency in the malignant cells. *J Exp Med* 177:339-349, 1993.

Dolcetti R and Masucci MG. Epstein-Barr virus: induction and control of cell transformation. *J Cell Physiol*, 196:207-218, 2003.

Dotti G, Fiocchi R, Motta T. Primary effusion lymphoma after heart transplant: a new entity associated with human herpes virus 8. *Leukemia*, 13:664-670, 1999.

Drouet E, Brousset P, Fares F, Icart J, Verniol C, Meggetto F, Schlaifer D, Desmorat-Coat H, Rigal-Huguet F, Niveleau A, Delsol G. High Epstein-Barr virus serum load and elevated titers of anti-ZEBRA antibodies in patients with EBV-harboring tumor cells of Hodgkin's disease. *J Med Virol*, 57:383-389, 1999.

Epstein MA, Achong BG, Barr YM. Virus particles in cultured lymphoblasts from Burkitt's lymphoma. *Lancet* 15:702-703, 1964.

Evans AS, Niederman JC. Epstein-Barr virus. In: Evans AS, ed. *Viral Infections of Humans*, 3rd Ed. New York, Plenum, pp. 265-292, 1989.

Everett RD, Meredith M, Orr A, Cross A, Kathorai M and Parkinson J. A novel ubiquitin-specific protease is dynamically associated with the PML nuclear domain and binds to a herpesvirus regulatory protein. *EMBO J*, 16:1519-1530, 1997.

Falco DA, Nepomuceno RR, Krams SM, Lee PP, Davis MM, Salvatierra O, Alexander SR, Esquivel CO, Cox KL, Frankel LR, Martinez OM. Identification of Epstein-Barr virus-specific CD8+ T lymphocytes in the circulation of pediatric transplant recipients. *Transplantation*, 74:501-510, 2002.

Franken M, Estabrooks A, Cavacini L, Sherburne B, Wang F, Scadden DT. Epstein-Barr virus-driven gene therapy for EBV-related lymphomas. *Nat Med*, 2:1379-1382, 1996.

Frisan T, Sjoberg J, Dolcetti R, Boiocchi M, De Re V, Carbone A, Brautbar C, Battat S, Biberfeld P, Eckman M, Öst Å, Christensson B., Sundström C., Björkholm M., Pisa P., Masucci M.G. Local suppression of Epstein-Barr virus (EBV)-specific cytotoxicity in biopsies of EBV-positive Hodgkin's disease. *Blood*, 86:1493-1501, 1995.

Frisan T, Zhang QJ, Levitskaya J, Coram M, Kurilla MG, Masucci MG. Defective presentation of MHC class I-restricted cytotoxic T-cell epitopes in Burkitt's lymphoma cells. *Int J Cancer*, 68:251-258, 1996.

Frisan T, Levitsky V, Polack A, Masucci MG. Phenotype-dependent differences in proteasome subunit composition and cleavage specificity in B cell lines. *J Immunol*, 160:3281-3289, 1998.

Gahn B, Siller-Lopez F, Pirooz AD, Yvon E, Gottschalk S, Longnecker R, Brenner MK, Heslop HE, Aguilar-Cordova E, Rooney CM. Adenoviral gene transfer into dendritic cells efficiently amplifies the immune response to LMP2A antigen: a potential treatment strategy for Epstein-Barr virus--positive Hodgkin's lymphoma. *Int J Cancer*, 93:706-713, 2001.

Gaidano G, Dalla-Favera R. Molecular pathogenesis of AIDS-related lymphomas. *Adv Cancer Res*, 67:113-153, 1995.

Gallagher A, Armstrong AA, MacKenzie J, Shield L, Khan G, Lake A, Proctor S, Taylor P, Clements GB, Jarrett RF. Detection of Epstein-Barr virus (EBV) genomes in the serum of patients with EBV-associated Hodgkin's disease. *Int J Cancer*, 84:442-448, 1999.

Gavioli R, Frisan T, Vertuani S, Bornkamm GW, Masucci MG. c-myc overexpression activates alternative pathways for intracellular proteolysis in lymphoma cells. *Nat Cell Biol*, 3:283-288, 2001.

Gilligan K, Rajadurai P, Resnick L, Raab-Traub N. Epstein-Barr virus small nuclear RNAs are not expressed in permissively infected cells in AIDS-associated leukoplakia. *Proc Natl Acad Sci U S A*, 87:8790-8794, 1990.

Gottschalk S, Ng CY, Perez M, Smith CA, Sample C, Brenner MK, Heslop HE, Rooney CM. An Epstein-Barr virus deletion mutant associated with fatal lymphoproliferative disease unresponsive to therapy with virus-specific CTLs. *Blood*, 97:835-843, 2001.

Gottschalk S, Edwards OL, Sili U, Huls MH, Goltsova T, Davis AR, Heslop HE, Rooney CM. Generating CTLs against the subdominant Epstein-Barr virus LMP1 antigen for the adoptive immunotherapy of EBV-associated malignancies. *Blood*, 101:1905-1912, 2003.

Gregory MA, Hann SR. c-Myc proteolysis by the ubiquitin-proteasome pathway: stabilization of c-Myc in Burkitt's lymphoma cells. *Mol Cell Biol*, 20:2423-2435, 2000.

Gustafsson A, Levitsky V, Zou JZ, Frisan T, Dalianis T, Ljungman P, Ringden O, Winiarski J, Ernberg I, Masucci MG. Epstein-Barr virus (EBV) load in bone marrow transplant recipients at risk to develop posttransplant lymphoproliferative disease: prophylactic infusion of EBV-specific cytotoxic T cells. *Blood*, 95:807-814, 2000.

Haque T, Thomas JA, Falk KI Parratt R, Hunt BJ, Yacoub M, Crawford DH. Transmission of donor Epstein-Barr virus (EBV) in transplant organs causes lymphoproliferative disease in EBV-seronegative recipients. *J Gen Virol*, 77:1169-1172, 1996.

Haque T, Wilkie GM, Taylor C, Amlot PL, Murad P, Iley A, Dombagoda D, Britton KM, Swerdlow AJ, Crawford DH. Treatment of Epstein-Barr-virus-positive post-transplantation lymphoproliferative disease with partly HLA-matched allogeneic cytotoxic T cells. *Lancet*, 360:436-442, 2002.

Henle G, Henle W, Clifford P, Diehl V, Kafuko GW, Kirya BG, Klein G, Morrow RH, Munube GM, Pike P, Tukei PM, Ziegler JL. Antibodies to Epstein-Barr virus in Burkitt's lymphoma and control groups. *J Nat Cancer Inst*, 43:1147-1157, 1969.

Henle G, Henle W. Epstein-Barr virus-specific IgA serum antibodies as an outstanding feature of nasopharyngeal carcinoma. *Int J Cancer*, 17: 1-17, 1976.

Heslop HE, Ng CY, Li C, Smith CA, Loftin SK, Krance RA, Brenner MK, Rooney CM. Long-term restoration of immunity against Epstein-Barr virus infection by adoptive transfer of gene-modified virus-specific T lymphocytes. *Nat Med*, 2:551-555, 1996.

Hirai H, Satoh E, Osawa M, Inaba T, Shimazaki C, Kinoshita S, Nakagawa M, Mazda O, Imanishi J. Use of EBV-based Vector/HVJ-liposome complex vector for targeted gene therapy of EBV-associated neoplasms. *Biochem Biophys Res Commun*, 241:112-118, 1997.

Heessen S, Leonchiks A, Isseva N, Sharipo A, Selivanova G, Masucci MG, Dantuma N. Functional p53 chimeras containing the Epstein-Barr virus Gly-Ala repeat are protected from Mdm2- and HPV-E6-induced proteolysis. *Proc Natl Acad Sci USA*, 99:1532-1537, 2002.

Heessen S, Dantuma N, Tessarz P, Jellne M, Masucci MG. Inhibition of ubiquitin/proteasome-dependent proteolysis in Saccharomyces cerevisiae by a Gly-Ala repeat. *FEBS Lett*, 555:397-404, 2003.

Hirshaut Y, Cohen MH, Stevens DA. Epstein-Barr antibodies in American and African Burkitt's lymphoma. *Lancet*, ii:114-116, 1973.

Hjalgrim H, Askling J, Pukkala E, Hansen S, Munksgaard L, Frisch M. Incidence of Hodgkin's disease in Nordic countries. *Lancet*, 358:297-298, 2001.

Hjalgrim H, Askling J, Rostgaard K, Hamilton-Dutoit S, Frisch M, Song Zhang J, Madsen M, Rosdahl N, Bossen Konradsen H, Storm H, Melbye M. Characteristics of Hodgkin's Lymphoma after Infectious Mononucleosis. *N Engl J Med*, 349:1324-1332, 2003.

Hoffmann C, Tabrizian S, Wolf E, Eggers C, Stoehr A, Plettenberg A, Buhk T, Stellbrink HJ, Horst HA, Jager H, Rosenkranz T. Survival of AIDS patients with primary central nervous system lymphoma is dramatically improved by HAART-induced immune recovery. *AIDS*, 15:2119-2127, 2001.

Holowaty MN, Sheng Y, Nguyen T, Arrowsmith C, Frappier L. Protein interaction domains of the ubiquitin-specific protease, USP7/HAUSP. *J Biol Chem*, 278:47753-47761, 2003a.

Holowaty MN, Zeghouf M, Wu H, Tellam J, Athanasopoulos V, Greenblatt J, Frappier L. Protein profiling with Epstein-Barr nuclear antigen-1 reveals an interaction with the herpesvirus-associated ubiquitin-specific protease HAUSP/USP7. *J Biol Chem*, 278:29987-29994, 2003b.

Hopwood PA, Brooks L, Parratt R, Hunt BJ, Bokhari M, Alero Thomas J, Yacoub M, Crawford DH, Maria B, Alero TJ, Magdi Y. Persistent Epstein-Barr virus infection: unrestricted latent and lytic viral gene expression in healthy immunosuppressed transplant recipients. *Transplantation*, 74:194-202, 2002.

Howe JG, Steitz JA. Localization of Epstein-Barr virus-encoded small RNAs by in situ hybridization. *Proc Natl Acad Sci U S A*, 83:9006-9010, 1986.

Hu LF, Chen F, Zheng X, Ernberg I, Cao SL, Christensson B, Klein G, Winberg G. Clonability and tumorigenicity of human epithelial cells expressing the EBV encoded membrane protein LMP1. *Oncogene*, 8:1575-1583, 1993.

IARC Monographs on the evaluation of carcinogenic risks to humans. *Epstein-Barr virus and Kaposi's sarcoma herpesvirus/Human herpesvirus* 8. Volume 70, IARC, Lyon, 1997.

Ikeda A, Caldwell RG, Longnecker R, Ikeda M. Itchy, a Nedd4 ubiquitin ligase, downregulates latent membrane protein 2A activity in B-cell signaling. *J Virol*, 77:5529-5534, 2003.

Imai S, Koizumi S, Sugiura M, Tokonuga M, Umeura Y, Yamamoto N, Tanaka S, Sato E, Osato T. Gastric carcinoma: monoclonal epithelial malignant cells expressing Epstein-Barr virus latent infetion proteins. *Proc Nat Acad Sci USA* 91:9131-9135, 1994.

International Collaboration on HIV and Cancer. Highly active antiretroviral therapy and incidence of cancer in human immunodeficiency virus-infected adults. *J Natl Cancer Inst,* 92:1823-1830, 2000.

Israele V, Shirley P, Sixbey JW. Excretion of the Epstein-Barr virus from the genital tract of men. *J Infect Dis* 163:1341-1343, 1991.

Kelly G, Bell A, Rickinson A. Epstein-Barr virus-associated Burkitt lymphomagenesis selects for downregulation of the nuclear antigen EBNA2. *Nat Med*, 8:1098-1104, 2002.

Kenagy DN, Schlesinger Y, Weck K, Ritter JH, Gaudreault-Keener MM, Storch GA. Epstein-Barr virus DNA in peripheral blood leukocytes of patients with posttransplant lymphoproliferative disease. *Transplantation*, 60:547-554, 1995.

Kenney S, Ge JQ, Westphal EM, Olsen J. Gene therapy strategies for treating Epstein-Barr virus-associated lymphomas: comparison of two different Epstein-Barr virus-based vectors. *Hum Gene Ther*, 9:1131-1141, 1998.

Kersten MJ, Klein MR, Holwerda AM, Miedema F, van Oers MH. Epstein-Barr virus-specific cytotoxic T cell responses in HIV-1 infection: different kinetics in patients progressing to opportunistic infection or non-Hodgkin's lymphoma. *J Clin Invest*, 99:1525-1533, 1997.

Kieff E. *Fields Virology, 3rd edn, Vol. 2*, Fiels BN, Knipe DM, Howley PM *et al.* (eds). Lippincott, Raven Publishers: Philadelphia, pp. 2343-2396, 1996.

Kieser A, Kaiser C, Hammerschmidt W. LMP1 signal transduction differs substantially from TNF receptor 1 signaling in the molecular functions of TRADD and TRAF2. *EMBO J*, 18:2511-2521, 1999.

Knecht H, Berger C, Rothenberger S, Odermatt BF, Brousset P. The role of Epstein-Barr virus in neoplastic transformation. *Oncology*, 60:289-302, 2001.

Kogan DL, Burroughs M, Emre S, Fishbein T, Moscona A, Ramson C, Shneider BL. Prospective longitudinal analysis of quantitative Epstein-Barr virus polymerase chain reaction in pediatric liver transplant recipients. *Transplantation*, 67:1068-1070, 1999.

Kostense S, Otto SA, Knol GJ, Manting EH, Nanlohy NM, Jansen C, Lange JM, van Oers MH, Miedema F, van Baarle D. Functional restoration of human immunodeficiency virus and Epstein-Barr virus-specific CD8(+) T cells during highly active antiretroviral therapy is associated with an increase in CD4(+) T cells. *Eur J Immunol*, 32:1080-1089, 2002.

Kulwichit W, Edwards RH, Davenport EM, Baskar JF, Godfrey V, Raab-Traub N. Expression of the Epstein-Barr virus latent membrane protein 1 induces B cell lymphoma in transgenic mice. *Proc Natl Acad Sci USA*, 95:11963-11968, 1998.

Lee ES, Locker J, Nalesnik M, Reyes J, Jaffe R, Alashari M, Nour B, Tzakis A, Dickman PS. The association of Epstein-Barr virus with smooth-muscle tumors occurring after organ transplantation. *N Engl J Med*, 332:19-25, 1995.

Lee SP, Chan AT, Cheung ST, Thomas WA, CroomCarter D, Dawson CW, Tsai CH, Leung SF, Johnson PJ, Huang DP. CTL control of EBV in nasopharyngeal carcinoma (NPC): EBV-specific CTL responses in the blood and tumors of NPC patients and the antigen-processing function of the tumor cells. *J Immunol*, 165:573-582, 2000.

Lei KI, Chan LY, Chan WY, Johnson PJ, Lo YM. Quantitative analysis of circulating cell-free Epstein-Barr virus (EBV) DNA levels in patients with EBV-associated lymphoid malignancies. *Br J Haematol*, 111:239-246, 2000.

Lei KI, Chan LY, Chan WY, Johnson PJ, Lo YM. Diagnostic and prognostic implications of circulating cell-free Epstein-Barr virus DNA in natural killer/T-cell lymphoma. *Clin Cancer Res*, 8:29-34, 2002.

Leonchiks A, Stavropoulou V, Sharipo A, Masucci MG. Inhibition of ubiquitin-dependent proteolysis by a synthetic glycine-alanine repeat peptide that mimics an inhibitory viral sequence. *FEBS Lett*, 522:93-98, 2002.

Levitskaya J, Coram M, Levitsky V, Imreh S, Steigerwald-Mullen PM, Klein G, Kurilla MG, Masucci MG. Inhibition of antigen processing by the internal repeat region of the Epstein-Barr virus nuclear antigen-1. *Nature*, 375:685-688, 1995.

Levitskaya J, Sharipo A, Leonchiks A, Ciechanover A, Masucci MG. Inhibition of ubiquitin/proteasome-dependent protein degradation by the Gly-Ala repeat domain of the Epstein-Barr virus nuclear antigen 1. *Proc Natl Acad Sci USA*, 94:12616-12621, 1997.

Leyvraz S, Henle W, Chahinian AP, Perlmann C, Klein G, Gordon RE, Rosenblum M, Holland JF. Association of Epstein-Barr virus with thymic carcinoma. *N Engl J Med* 312:1296-1299, 1985.

Li JH, Chia M, Shi W, Ngo D, Strathdee CA, Huang D, Klamut H, Liu FF. Tumor-targeted gene therapy for nasopharyngeal carcinoma. *Cancer Res*, 62:171-178, 2002.

Li JH, Shi W, Chia M, Sanchez-Sweatman O, Siatskas C, Huang D, Busson P, Klamut H, Yeh WC, Richardson C, O'Sullivan B, Gullane P, Neligan P, Medin J, Liu FF. Efficacy of targeted FasL in nasopharyngeal carcinoma. *Mol Ther*, 8:964-973, 2003.

Limaye AP, Huang ML, Atienza EE, Ferrenberg JM, Corey L. Detection of Epstein-Barr virus DNA in sera from transplant recipients with lymphoproliferative disorders. *J Clin Microbiol*, 37:1113-1116, 1999.

Lin CL, Lo WF, Lee TH, Ren Y, Hwang SL, Cheng YF, Chen CL, Chang YS, Lee SP, Rickinson AB, Tam PK. Immunization with Epstein-Barr Virus (EBV) peptide-pulsed dendritic cells induces functional CD8+ T-cell immunity and may lead to tumor regression in patients with EBV-positive nasopharyngeal carcinoma. *Cancer Res*, 62:6952-6958, 2002.

Lo YM, Chan LY, Lo KW, Leung SF, Zhang J, Chan AT, Lee JC, Hjelm NM, Johnson PJ, Huang DP. Quantitative analysis of cell-free Epstein-Barr virus DNA in plasma of patients with nasopharyngeal carcinoma. *Cancer Res*, 59:1188-1191, 1999.

Longnecker R, Merchant M, Brown ME, Fruehling S, Bickford JO, Ikeda M, Harty RN. WW- and SH3-domain interactions with Epstein-Barr virus LMP2A. *Exp Cell Res*, 257:332-340, 2000.

Lucas KG, Filo F, Heilman DK, Lee CH, Emanuel DJ. Semiquantitative Epstein-Barr virus polymerase chain reaction analysis of peripheral blood from organ transplant patients and risk for the development of lymphoproliferative disease. *Blood*, 92:3977-3978, 1998.

Magrath I. The pathogenesis of Burkitt's lymphoma. *Adv Cancer Res*, 55:133-270, 1990.

McClain KL, Leach CT, Jenson HB, Joshi VV, Pollock BH, Parmley RT, DiCarlo FJ, Chadwick EG, Murphy SB. (1995) Association of Epstein-Barr virus with leiomyosarcomas in young people with AIDS. *N Engl J Med* 332:12-18, 1995.

MacMahon B. Epidemiology of Hodgkin's disease. *Cancer Res*, 26:1189-1200, 1966.

Macsween KF, Crawford DH. Epstein-Barr virus – recent advances. *Lancet Infect Dis*, 3:131-140, 2003.

Miller CL, Lee JH, Kieff E, Burkhardt AL, Bolen JB, Longnecker R. Epstein-Barr virus protein LMP2A regulates reactivation from latency by negatively regulating tyrosine kinases involved in sIg-mediated signal transduction. *Infect Agents Dis*, 3:128-136, 1994a.

Miller CL, Lee JH, Kieff E, Longnecker R. An integral membrane protein (LMP2) blocks reactivation of Epstein-Barr virus from latency following surface immunoglobulin crosslinking. *Proc Natl Acad Sci USA*, 91:772-776, 1994b.

Miyashita EM, Yang B, Babcock GJ and Thorley-Lawson DA. Identification of the site of Epstein-Barr virus persistence in vivo as a resting B cell. *J Virol*, 71:4882-4891, 1997.

Mosialos G, Birkenbach M, Yalamanchili R, VanArsdale T, Ware C, Kieff E. The Epstein-Barr virus transforming protein LMP1 engages signaling proteins for the tumor necrosis factor receptor family. *Cell*, 80:389-399, 1995.

Mueller N, Evans A, Harris NK, Comstock GW, Jellum E, Magnus K, Orentreich N, Polk BF, Vogelman J. Hodgkin's disease and Epstein-Barr virus. Altered antibody pattern before diagnosis. *N Engl J Med*, 320:689-695, 1989.

Muller S, Hoege C, Pyrowolakis G, Jentsch S. SUMO, ubiquitin's mysterious cousin. *Nat Rev Mol Cell Biol*, 2:202-210, 2001.

Mukherjee S, Trivedi P, Dorfman DM, Klein G, Townsend A. Murine cytotoxic T lymphocytes recognize an epitope in an EBNA-1 fragment, but fail to lyse EBNA-1-expressing mouse cells. *J Exp Med*, 187:445-450, 1998.

Niedobitek G, Baumann I, Brabletz T, Lisner R, Winkelmann C, Helm G, Kirchner T. Hodgkin's disease and peripheral T-cell lymphoma: composite lymphoma with evidence of Epstein-Barr virus infection. *J Pathol*, 191:394-399, 2000.

O'Reilly RJ, Small TN, Papadopoulos E, Lucas K, Lacerda J, Koulova L. Biology and adoptive cell therapy of Epstein-Barr virus-associated lymphoproliferative disorders in recipients of marrow allografts. *Immunol Rev*, 157:195-216, 1997.

Ogg GS, McMichael AJ. HLA-peptide tetrameric complexes. *Curr Opin Immunol*, 10:393-396, 1998.

Opclz G, Henderson R. Incidence of non-Hodgkin lymphoma in kidney and heart transplant recipients. *Lancet*, 342:1514-1516, 1993.

Palella FJ Jr, Delaney KM, Moorman AC, Loveless MO, Fuhrer J, Satten GA, Aschman DJ, Holmberg SD. Declining morbidity and mortality among patients with advanced human immunodeficiency virus infection. *N Engl J Med*, 338:853-860, 1998.

Papadopoulos EB, Ladanyi M, Emanuel D, Mackinnon S, Boulad F, Carabasi MH, Castro-Malaspina H, Childs BH, Gillio AP, Small TN, Young JW, Kernan NA, O'Reilly RJ. Infusions of donor leukocytes to treat Epstein-Barr virus-associated lymphoproliferative disorders after allogeneic bone marrow transplantation. *N Engl J Med*, 330:1185-1191, 1994.

Paludan C, Munz C. CD4+ T cell responses in the immune control against latent infection by Epstein-Barr virus. *Curr Mol Med*, 3:341-347, 2003.

Pathmanathan R, Prasad U, Sadler R, Flynn K, Raab-Traub N. Clonal proliferations of cells infected with Epstein-Barr virus in preinvasive lesions related to nasopharyngeal carcinoma. *N Engl J Med*, 333:693-698, 1995.

Penn I, Hammond W, Brettschneider L, Starzl TE. Malignant Lymphomas in transplanted patients. *Transplantation Proc*, 1:106-112, 1969.

Pittet MJ, Speiser DE, Valmori D, Rimoldi D, Lienard D, Lejeune F, Cerottini JC, Romero P. Ex vivo analysis of tumor antigen specific CD8+ T cell responses using MHC/peptide tetramers in cancer patients. *Int Immunopharmacol*, 1:1235-1247, 2001.

Preikasaitis JK, Cockfield SM. Epstein-Barr virus infection and lymphoproliferative disease after hemopoietic stem cell or solid organ transplantation. In: *Transplant Infections*, 2nd Edition, Bowden RA, Ljungman P, Paya CV, eds, Lippincott Williams and Wilkins, 326-349, 2003.

Raab-Traub N, Flynn K. The structure of the termini of the Epstein-Barr virus as a marker of clonal cellular proliferation. *Cell*, 47:883-889, 1989.

Reisman D, Yates J, Sugden B. A putative origin of replication of plasmids derived from Epstein-Barr virus is composed of two cis-acting components. *Mol Cell Biol*, 5:1822-1832, 1985.

Rickinson AB, Murray RJ, Brooks J, Griffin H, Moss DJ, Masucci MG. T cell recognition of Epstein-Barr virus associated lymphomas. *Cancer Surv*, 13:53-80, 1992.

Riddler SA, Breinig MC, McKnight JL. Increased levels of circulating Epstein-Barr virus (EBV)-infected lymphocytes and decreased EBV nuclear antigen antibody responses are associated with the development of posttransplant lymphoproliferative disease in solid-organ transplant recipients. *Blood*, 84:972-984, 1994.

Rooney CM, Smith CA, Ng CY, Loftin S, Li C, Krance RA, Brenner MK, Heslop HE. Use of gene-modified virus-specific T lymphocytes to control Epstein-Barr-virus-related lymphoproliferation. *Lancet*, 345:9-13, 1995.

Rooney CM, Smith CA, Ng CY, Loftin SK, Sixbey JW, Gan Y, Srivastava DK, Bowman LC, Krance RA, Brenner MK, Heslop HE. Infusion of cytotoxic T cells for the prevention and treatment of Epstein-Barr virus-induced lymphoma in allogeneic transplant recipients. *Blood*, 92:1549-1555, 1998.

Roskrow MA, Suzuki N, Gan Y, Sixbey JW, Ng CY, Kimbrough S, Hudson M, Brenner MK, Heslop HE, Rooney CM. Epstein-Barr virus (EBV)-specific cytotoxic T lymphocytes for the treatment of patients with EBV-positive relapsed Hodgkin's disease. *Blood*, 91:2925-2934, 1998.

Rowe DT, Rowe M, Evan GI, Wallace L, Farrell P, Rickinson AB. Restricted expression of EBV latent genes and T-lymphocyte-detected membrane antigen in Burkitt's lymphoma cells. *EMBO J*, 5:2599-2607, 1986.

Rowe DT, Webber S, Schauer EM, Reyes J, Green M. Epstein-Barr virus load monitoring: its role in the prevention and management of post-transplant lymphoproliferative disease. *Transpl Infect Dis*, 3:79-87, 2001.

Rowe M, Lear A, Croom-Carter D, Davies A, Rickinson A. Three pathways of Epstein-Barr virus gene activation from EBNA1-positive latency in B lymphocytes. *J Virol*, 66:122-131, 1992.

Rowe M, Khanna R, Jacob CA, Argaet V, Kelly A, Powis S, Belich M, Croom-Carter D, Lee S, Trowsdale J, Moss DJ, Rickinson AP, Burrows SR. Restoration of endogenous antigen processing in Burkitt's lymphoma cells by Epstein-Barr virus latent membrane protein-1: coordinate up-regulation of peptide transporters and HLA-class I antigen expression. *Eur J Immunol*, 25:1374-1384, 1995.

Salghetti SE, Kim SY, Tansey WP. Destruction of Myc by ubiquitin-mediated proteolysis: cancer-associated and transforming mutations stabilize Myc. *EMBO J*, 18:717-726, 1999.

Sallusto F, Lenig D, Forster R, Lipp M, Lanzavecchia A. Two subsets of memory T lymphocytes with distinct homing potentials and effector functions. *Nature*, 401:708-712, 1999.

Sam CK, Prasad U, Pathmanathan R. Serological markers in the diagnosis of histopathological types of nasopharyngeal carcinoma. *Eur J Surg Oncol* 15:357-360, 1989.

Savoie A, Perpete C, Carpentier L, Joncas J, Alfieri C. Direct correlation between the load of Epstein-Barr virus-infected lymphocytes in the peripheral blood of pediatric transplant patients and risk of lymphoproliferative disease. *Blood*, 83:2715-2722, 1994.

Serraino D, Franceschi S, Talamini R, Barra S, Negri E, Carbone A, La Vecchia C. Socioeconomic indicators, infectious diseases and Hodgkin's disease. *Int J Cancer*, 47; 352-357, 1991.

Sharipo A, Imreh M, Leonchiks A, Imreh S, Masucci MG. A minimal glycine-alanine repeat prevents the interaction of ubiquitinated I kappaB alpha with the proteasome: a new mechanism for selective inhibition of proteolysis. *Nat Med*, 4:939-944, 1998.

Sharipo A, Imreh M, Leonchiks A, Brändén CI, Masucci MG. cis-Inhibition of proteasomal degradation by viral repeats: impact of length and amino acid composition. *FEBS Lett*, 499:137-142, 2001.

Shibata D, Weiss LM, Hernandez AM, Nathwani BN, Bernstein L, Levine AM. Epstein-Barr virus-associated non-Hodgkin's lymphoma in patients infected with the human immunodeficiency virus. *Blood*, 81:2102-2109, 1993.

Shiramizu B, Barriga F, Neequaye J, Jafri A, Dalla-Favera R, Neri A, Guttierez M, Levine P, Magrath I. Patterns of chromosomal breakpoint locations in Burkitt's lymphoma: relevance to geography and Epstein-Barr virus association. *Blood*, 77:1516-1526, 1991.

Slobod KS, Taylor GH, Sandlund JT, Furth P, Helton KJ, Sixbey JW. Epstein-Barr virus-targeted therapy for AIDS related primary lymphoma of the central nervous system. *Lancet*, 356:1493-1494, 2000.

Smets F, Latinne D, Bazin H, Reding R, Otte JB, Buts JP, Sokal EM. Ratio between Epstein-Barr viral load and anti-Epstein-Barr virus specific T-cell response as a predictive marker of posttransplant lymphoproliferative disease. *Transplantation*, 73:1603-1610, 2002.

Spence J, Gali RR, Dittmar G, Sherman F, Karin M, Finley D. Cell cycle-regulated modification of the ribosome by a variant multiubiquitin chain. *Cell*, 102:67-76, 2000.

Staege MS, Lee SP, Frisan T, Mautner J, Scholz S, Pajic A, Rickinson AB, Masucci MG, Polack A, Bornkamm GW. MYC overexpression imposes a nonimmunogenic phenotype on Epstein-Barr virus-infected B cells. *Proc Natl Acad Sci USA*, 99:4550-4555, 2002.

Steven NM, Leese AM, Annels NE, Lee SP, Rickinson AB. Epitope focusing in the primary cytotoxic T cell response to Epstein-Barr virus and its relationship to T cell memory. *J Exp Med*, 184:1801-1813, 1996.

Stevens SJ, Verschuuren EA, Pronk I, van Der Bij W, Harmsen MC, The TH, Meijer CJ, van Den Brule AJ, Middeldorp JM. Frequent monitoring of Epstein-Barr virus DNA load in unfractionated whole blood is essential for early detection of posttransplant lymphoproliferative disease in high-risk patients. *Blood*, 97:1165-1171, 2001.

Stevens SJ, Verschuuren EA, Verkuujlen SA, Van Den Brule AJ, Meijer CJ, Middeldorp JM. Role of Epstein-Barr virus DNA load monitoring in prevention and early detection of post-transplant lymphoproliferative disease. *Leuk Lymphoma*, 43:831-840, 2002.

Swinnen LJ, Costanzo-Nordin MR, Fisher SG, O'Sullivan EJ, Johnson MR, Heroux AL, Dizikes GJ, Pifarre R, Fisher RI. Increased incidence of lymphoproliferative disorders after immunosuppression with the monoclonal antibody OKT3 in cardiac transplant recipients. *N Engl J Med*, 323:1723-1728, 1990.

Takanashi M, Li J, Shirakata M, Mori S, Hirai K. Tumorigenicity of mouse BALB/c 3T3 fibroblast cells which express Epstein-Barr virus-encoded LMP1 and show normal growth phenotypes in vitro is correlated with loss of transforming growth factor-beta 1-mediated growth inhibition. *Arch Virol*, 144:241-257, 1999.

Tan R, Xu X, Ogg GS, Hansasuta P, Dong T, Rostron T, Luzzi G, Conlon CP, Screaton GR, McMichael AJ, Rowland-Jones S. Rapid death of adoptively transferred T cells in acquired immunodeficiency syndrome. *Blood*, 93:1506-1510, 1999.

Thomas J, Allday M, Crawford D. Epstein-Barr virus-associated lymphoproliferative disorders in immunocompromised individuals. *Adv Cancer Res*, 57:329-380, 1991.

Thorley-Lawson DA. Epstein-Barr virus: exploiting the immune system. *Nat Rev Immunol*, 1:75-82, 2001.

Tischendorf P, Shramek GJ, Balagtas RC, Deinhardt F, Knospe WH, Noble GR, Maynard JE. Development and persistence of immunity to Epstein-Barr virus in man. *J Inf Dis*, 122:401-409, 1970

Traub NR. Epstein-Barr virus and Nasopharyngeal carcinoma. In: *Infectious causes of cancer: targets for interventions*, Goedert JJ ed., pg: 93-111, Humana Press, Totowa NJ, 2000.

Tussey L, Speller S, Gallimore A, Vessey R. Functionally distinct CD8+ memory T cell subsets in persistent EBV infection are differentiated by migratory receptor expression. *Eur J Immunol*, 30:1823-1829, 2000.

van Baarle D, Hovenkamp E, Callan MF, Wolthers KC, Kostense S, Tan LC, Niesters HG, Osterhaus AD, McMichael AJ, van Oers MH, Miedema F. Dysfunctional Epstein-Barr virus (EBV)-specific CD8(+) T lymphocytes and increased EBV load in HIV-1 infected individuals progressing to AIDS-related non-Hodgkin lymphoma. *Blood*, 98:146-155, 2001.

van Esser JW, Niesters HG, Thijsen SF, Meijer E, Osterhaus AD, Wolthers KC, Boucher CA, Gratama JW, Budel LM, van der Holt B, van Loon AM, Lowenberg B, Verdonck LF, Cornelissen JJ. Molecular quantification of viral load in plasma allows for fast and accurate prediction of response to therapy of Epstein-Barr virus-associated lymphoproliferative disease after allogeneic stem cell transplantation. *Br J Haematol*, 113:814-821, 2001.

van Esser JW, van der Holt B, Meijer E, Niesters HG, Trenschel R, Thijsen SF, van Loon AM, Frassoni F, Bacigalupo A, Schaefer UW, Osterhaus AD, Gratama JW, Lowenberg B, Verdonck LF, Cornelissen JJ. Epstein-Barr virus (EBV) reactivation is a frequent event after allogeneic stem cell transplantation (SCT) and quantitatively predicts EBV-lymphoproliferative disease following T-cell-depleted SCT. *Blood*, 98:972-978, 2001.

Vivanco I, Sawyers CL. The phosphatidylinositol 3-Kinase AKT pathway in human cancer. *Nat Rev Cancer*, 2:489-501, 2002.

Voo KS, Fu T, Wang HY, Tellam J, Heslop HE, Brenner MK, Rooney CM, Wang RF. 2004. Evidence for the presentation of Major Histocompatibility Complex Class I-restricted Epstein-Barr Virus Nuclear Antigen 1 Peptides to CD8+ T Lymphocytes. *J Exp Med*, 199:459-470, 2004.

Wagner HJ, Wessel M, Jabs W, Smets F, Fischer L, Offner G, Bucsky P. Patients at risk for development of posttransplant lymphoproliferative disorder: plasma versus peripheral blood mononuclear cells as material for quantification of Epstein-Barr viral load by using real-time quantitative polymerase chain reaction. *Transplantation*, 72:1012-1019, 2001.

Winberg G, Matskova L, Chen F, Plant P, Rotin D, Gish G, Ingham R, Ernberg I, Pawson T. Latent membrane protein 2A of Epstein-Barr virus binds WW domain E3 protein-ubiquitin ligases that ubiquitinate B-cell tyrosine kinases. *Mol Cell Biol*, 20:8526-8535, 2000.

Wu CJ, Leu CY, Liu ST, Chow KP, Meng CL, Chang YS. Transcriptional activation of NF-kappa B activity by Epstein-Barr virus (EBV) LMP1 as a selective therapeutic strategy for EBV-associated diseases. *Gene Ther*, 5:905-912, 1998.

Wu TC, Mann RB, Epstein JI, MacMahon E, Lee WA, Charache P, Hayward SD, Kurman RJ, Hayward GS, Ambinder RF. Abundant expression of EBER1 small-nuclear RNA in nasopharyngeal carcinoma. A morphologically distinctive target for detection of Epstein-Barr virus in formalin-fixed paraffin-embedded carcinoma specimens. *Am J Pathol*, 138:1461-1469, 1991.

Yamamoto M, Kimura H, Hironaka T, Hirai K, Hasegawa S, Kuzushima K, Shibata M, Morishima T. Detection and quantification of virus DNA in plasma of patients with Epstein-Barr virus-associated diseases. *J Clin Microbiol*, 33:1765-1768, 1995.

Yang J, Tao Q, Flinn IW, Murray PG, Post LE, Ma H, Piantadosi S, Caligiuri MA, Ambinder RF. Characterization of Epstein-Barr virus-infected B cells in patients with posttransplantation lymphoproliferative disease: disappearance after rituximab therapy does not predict clinical response. *Blood*, 96:4055-4063, 2000.

Yang X, Sham JS, Ng MH, Tsao SW, Zhang D, Lowe SW, Cao L. LMP1 of Epstein-Barr virus induces proliferation of primary mouse embryonic fibroblasts and cooperatively transforms the cells with a p16-insensitive CDK4 oncogene. *J. Virol.*, 74:883-891, 2000.

Young LS, Clark D, Sixbey JW, Rickinson AB. Epstein-Barr virus receptors on human pharyngeal epithelia. *Lancet*, i:240-242, 1986.

Ziegler JL, Beckstead JA, Volberding PA, Abrams DI, Levine AM, Lukes RJ, Gill PS, Burkes RL, Meyer PR, Metroka CE. Non-Hodgkin's lymphoma in 90 homosexual men. Relation to generalized lymphadenopathy and the acquired immunodeficiency syndrome. *N Engl J Med*, 311:565-570, 1984.

Ziegler JL, Bluming AZ, Morrow RH, Fass L, Carbone PP. Central nervous system involvement in Burkitt's lymphoma. *Blood*, 36:718-728, 1970.

zur Hausen H, Schulte-Holthausen H. Presence of EB virus nucleic acid homology in a virus-fee line of Burkitt's tumour cells. *Nature*, 227:245-248, 1970.

In: Progress in Virus Research
Editor: Sabine Thebault, pp. 227-266

ISBN 1-59454-438-7

Chapter VIII

Developments in Understanding the Mechanism of Tumorigenesis Mediated by Kaposi's Sarcoma Associated Herpesvirus

Subhash C. Verma, Ke Lan, Zenda L. Woodman and Erle S. Robertson*

201E Johnson Pavilion, 3610 Hamilton Walk
Department of Microbiology and Tumor Virology Program of the Abramson Comprehensive Cancer Center
University of Pennsylvania, School of Medicine
Philadelphia PA 19104

Abstract

Kaposi's sarcoma Associated Herpesvirus (KSHV) is the recently discovered human herpesvirus associated with various human malignancies including Body Cavity Based Lymphomas (BCBLs) or Primary Effusion Lymphomas (PELs) and Multicentric Castleman's Disease (MCDs). The viral DNA was detected in all forms of KS associated tumors and the sequence determination revealed that KSHV belongs to Gammherpesvirus sub-family of Herpesviridae and showed similarity to Herpesvirus saimiri and Rhesus Rhadinovirus. KSHV, like other Herpesviruses, establishes a latent infection and expresses a repertoire of latent genes. These latent genes have been shown to be important for the establishment of latent infection and tumorigenesis. Among the latent genes, Latency Associated Nuclear Antigen (LANA) is detected in all forms of KS-associated malignancies and is critical for episomal maintenance of the viral DNA by anchoring the viral genome to the host chromosome. Besides episomal maintenance, LANA has also been shown to modulate various cellular pathways. It down regulates the

* Corresponding author: Erle S. Robertson, Tel:215-746-0114, Fax: 215-898-9557 E mail address: erle@mail.med.upenn.edu

p53 and pRb tumor suppressor pathways involved in programmed cell death in response to cellular stress which includes viral infection. Additionally, other viral encoded genes are involved in driving cell proliferation and are critical components of the overall mechanism by which KSHV mediates cell transformation. Studies are underway to understand the complex signaling processes of cell cycle regulation and cell proliferation mediated by KSHV.

Kew Words: Tumor Virus, Kaposi's Sarcoma Associated Herpesvirus, Latency Associated Nuclear Antigen, Episomal maintenance, Cell proliferation, Oncogene, Transformation

Introduction

Kaposi Sarcoma lesions were initially documented by Moritz Kaposi, a Hungarian dermatologist working in Vienna. In 1872, he reported a case study of five men with "idiopathic multiple pigmented sarcomas of the skin" including one patient who developed visceral disease in the lungs and gastrointestinal tract (Antman and Chang, 2000). These KS lesions, idiopathic multiple pigmented sarcoma of the skin, were referred to as classic KS lesions and were independent of the immune-suppression stage of the patients. During that time classic KS was of rare concern to clinicians because it did not cause any serious complications in the patients. However, after the onset of the AIDS pandemic, KS was documented as one of the major contributors to morbidity and mortality in AIDS patients. In addition, HIV seronegative homosexual men have also been reported to be at higher risk for developing KS (Ganem, 1997). Post-transplant patients undergoing immunosuppressive therapy to prevent graft rejection were also shown to develop KS neoplasm. Moreover, the KS lesions regressed when immunosuppressive therapies were withdrawn from transplant patients, indicating that the immune status of the patients were important in KS associated malignancies. However, immunosuppression is not the only condition required for the development of KS. There are four distinct clinical variants of KS. i) Classic KS; was originally described by Dr. Moritz Kaposi, predominantly affects elderly men of Mediterranean and Eastern European origin, with a male to female ratio of nearly 15 to 1 (Antman and Chang, 2000). The symptoms are the multiple, firm reddish-brown to purple blue plaques and nodules, usually on the hands and feet progressing towards arm and legs over a period of time. This form of KS is generally the most indolent variant, leading to paranodular skin disease without visceral dissemination (Ganem, 1997). ii) Endemic KS; is a clinically and epidemiologically distinct variant prevalent in equatorial, eastern and southern Africa (Wabinga et al., 1993). Endemic KS; is a clinically more aggressive form presents as localized or indolent nodules or plaques on edematous limbs. iii) Iatrogenic or post-transplant KS; develops in patients on immunosuppressive therapy to prevent graft rejection after organ transplantation. This generally develops within 6 months after starting the immunosuppressive therapy (Regamey et al., 1998). This variant of KS is relatively aggressive, involving lymph nodes mucosa and visceral organs even in the absence of detectable skin lesions. iv) AIDS associated KS; is the most aggressive form and most commonly presented in gay and bisexual males than men with hemophilia or intravenous drug users with similar degrees of immunosuppression, suggesting potential transmission through various high risk sexual routes (Gao et al., 1996b). Clinically this variant is similar to

the body cavities of HIV-positive patients with a high copy number ranging from 40-80 copies per cell (Cesarman et al., 1995a; Cesarman et al., 1995b). PEL's occur predominantly in HIV-seropositive individuals with advanced stage of immunosuppression but are also seen in HIV-seronegative patients (Carbone et al., 1998; Carbone and Gaidano, 1997; Carbone et al., 1996; Nador et al., 1996). Moreover, PEL cells are more frequently found to be co-infected with EBV and KSHV, although EBV negative and KSHV positive PEL has also been described, primarily in HIV-negative patients (Arvanitakis et al., 1996). Most PEL's are thought to originate from post-germinal center B-cells as they are shown to have the hypermutated immunoglobulin genes and markers of late stage B-cell differentiation. PEL's are considered to be clonal due to the presence of clonal immunoglobulin gene rearrangement and monoclonal terminal repeats (TR's) (Fais et al., 1999; Matolcsy et al., 1998).

Multicentric Castleman's Disease (MCD)

Castleman's disease, a polyclonal neoplastic disorder, is an unusual lymphoproliferative disorder thought to be related to immune dysregulation and by interleukin (IL)-6 overexpression (Ablashi et al., 2002; Brandt et al., 1990),. KSHV was detected in almost all cases of MCD occurring in AIDS patients and about 40% of HIV seronegative MCDs. MCD was also found to be frequently associated with KS and is mostly observed in men infected with HIV by sexual contact (Oksenhendler et al., 1996). Castleman's disease is usually described as a polyclonal neoplastic disorder and there are two distinct histopathologic subtypes. The hyaline vascular type; is most common and presents as a solitary mass and can be surgically removed (Bowne et al., 1999; Delmer et al., 1990; Guillem et al., 2000). The other type of MCD, related to plasma cell types and is associated with more generalized lymphoadenopathy and immunological disorders. Unlike PEL cells, co-infection by EBV has not been detected in MCD plasmablasts. The exact role of KSHV in MCD development is not very well understood except that a few open reading frames of KSHV are expressed in MCDs. Germinal centers of hyperblastic lymph nodes of MCD patients produce large quantities of IL-6 which induces the differentiation of B cells. KSHV encodes a viral IL-6 homolog and this could explain the possible reason for the presence of large proportion of plasma cells in lymph nodes. ORF73, encoding latency associated nuclear antigen has also been detected in most of MCDs (Dupin et al., 1999).

Host Range Infectivity of KSHV in Cultured Cells

KSHV has a limited range of efficient infectivity towards varying cell types. Transmission of KSHV DNA from BC-1 cell into EBV-positive Raji cells (EBV non-producing B cell), BJAB (EBV-negative B cell) and owl monkey kidney cells have been shown to occur *in vitro* (Mesri et al., 1996; Moore et al., 1996). Human embryonic kidney cells could also be infected by KSHV isolated from KS lesions but the transmission of the virus as confirmed by the presence of viral DNA could only be detected by PCR amplification (Foreman et al., 1997a). KSHV isolated from BCBL-1 cells could be transmitted to HEK293 cells shown using detection of a spliced KSHV late mRNA as confirmation of transmission (Renne et al., 1998). A large number of other cell lines including B cells, endothelial cells, and fibroblastic origin cells were used as targets for transmission with limited efficiency for transmission (Renne et al., 1998). Primary human monocytes-macrophage from AIDS-associated KS can also be infected by KSHV and treatment with cytokines enhanced viral production. Episomal DNA was also maintained for longer periods of time (Renne et al., 1998). Microvascular endothelial cells transformed with E6 and E7 of human papilomavirus type 16 were shown to be permissive for KSHV transmission and the viral DNA was maintained indefinitely in latent form with expression of limited genes, primarily shown the expression of Latency Associated Nuclear Antigen (LANA) (Moses et al., 1999). Untransformed primary fetal dermal microvasuclar endothelial cells, derived from large blood vessels have successfully been infected with KSHV from PELs (JSC-1, BC-3) (Cannon et al., 2000). Infection with the virus caused the cells to change morphology and become spindle shaped cells, expressing LANA and decrease time for cells to enter mitosis. Lytic cycle proteins were also expressed as demonstrated by immunofluorescence assay.

Seroepidemiology of KSHV Infection

Various proteins encoded by KSHV open reading frames (ORFs) during latent and lytic infection have been detected by patient sera using immunofluorescence assay, ELISA and western blot analysis. LANA encoded by ORF73 has been found to be an important antigen for detection of viral infection (Katano et al., 2000a). ORF65 encoded capsid protein has also been used for the detection of viral infection and been shown to be an authentic antigen for use in serological studies. Antibodies against these two proteins have been used successfully for the detection of KSHV in almost 90% of the patients (Katano et al., 2000a, Miller et al., 1996). Serologic studies, based on immunofluorescence, western blot and ELISA assays, suggest that KSHV is not a ubiquitous virus and does not infect the general population as seen with other herpesvirus including Epstein Barr Virus (EBV). The global prevalence of KSHV shows that its lowest in Asia (0.2% of blood donors in Japan testing positive) (Fujii et al., 1999) but comparatively higher in United States (2-3% blood donors), and approximately 10 fold higher in certain regions of Italy based on blood donors (approximately 28%) (Rezza et al., 1998). Besides the normal human population the seroprevalence of KSHV has been reported to be very high (approximately 50%) in KS endemic areas and also KS is the most common cancers in general population of these areas, including Uganda and some Mediterranean countries (Wabinga et al., 1993).

Genomic Organization of KSHV

The KSHV belongs to the gammaherpesvirus family, Rhadinovirus genera sharing a common genomic organization with the other members of the group (Russo et al., 1996). KSHV from two separate cell lines, PEL and a KS specimen was sequenced separately by two groups (Neipel, Albrecht, and Fleckenstein, 1997; Russo et al., 1996). The sequence data revealed that KSHV has a linear dsDNA genome approximately 165Kbp in size with a central region of low (L-DNA) GC-rich DNA flanked by a highly repetitive GC-rich sequence (H-DNA) of 801 bp reiterated numerous times. The L DNA contains at least 87 ORFs, of which

68 ORFs were conserved with prototype gamma herpes virus HVS and the unique genes were designated as K1 to K15 (Neipel, Albrecht, and Fleckenstein, 1997; Russo et al., 1996). The open reading frames of KSHV L DNA are collinear to the L DNA of *Herpesvirus saimiri*, another primate gammaherpesvirus capable of growth transforming human T-lymphocytes (Neipel, Albrecht, and Fleckenstein, 1997; Russo et al., 1996). A characteristic of gammaherpesviruses is that some of the viral ORFs show similarity to cellular genes suggesting that these genes may have been pirated from the host chromosomes during the process of viral evolution (Moore and Chang, 1998a; Moore and Chang, 1998b; Neipel, Albrecht, and Fleckenstein, 1997). The recent reports have shown that these cellular homologs are important for modulating the immune response of the host and thus helps in tumorigenesis (Russo et al., 1996; Swanton et al., 1997). Some of these genes include the Bcl-2, IL-8R, and MIP-1α, vIL-6, DHFR and the D type viral cyclin, and were shown to be important for progression of KSHV associated tumors (Russo et al., 1996). Comparison to other gammaherpesvirus genomes reveal that the degree of conservation of the cellular homologs varies within members of the family, suggesting that different members may have alternative strategies for bypassing the host immune response. KSHV, like other gammaherpesviruses establish latent infection and persists indefinitely once infected. Various ORFs have been shown to be important for maintaining latent infection. Here we describe the profile of genes transcribed during latent and lytic replication.

Latent Genes Expression Profile

KSHV, like other members of herpesvirus becomes latent in most of the cells it infects. This virus exists as a circular episomal DNA in the nucleus. The KSHV genome encodes over 85 genes, of which less than 10% are expressed during latent infection and the transcripts for the major proteins detected during latent infection are shown in figure 2. All cancer cells in KSHV-associated KS and lymphomas express these latent genes, which drive cell proliferation, block apoptotic pathways and provide evasive tactics from host immune responses. Among these are LANA, a putative viral oncoprotein, viral cyclin (*v*-cyclin) encoded by ORF72 and viral Fas (*v*-FLIP) associated death domain (FADD) interleukin 1B converting enzyme (FLICE) inhibitory protein encoded by ORF71 (Dittmer et al., 1998). These genes lie adjacent to each other in the genome and are co-transcribed on polycistronic mRNAs (Dittmer et al., 1998; Grundhoff and Ganem, 2001). Other viral genes expressed during latency include viral interferon regulatory factor (v-IRF) and kaposin encoded by open reading frame K10 and K12, respectively (Russo et al., 1996). Among all these genes LANA is most consistently detected in all KSHV infected cells in KS, PEL and MCD (Dupin et al., 1999). Various roles have been assigned to these latent proteins in terms of maintenance of latency and cell cycle regulation. LANA has been shown to be the main viral protein, having multifunctional role in episome maintenance, cell cycle regulation and transcription (Ballestas, Chatis, and Kaye, 1999; Cotter and Robertson, 1999; Cotter, Subramanian, and Robertson, 2001; Garber et al., 2001; Grundhoff and Ganem, 2003; Lim et al., 2001; Lim et al., 2003; Lim et al., 2002). The roles of major latent genes are described here.

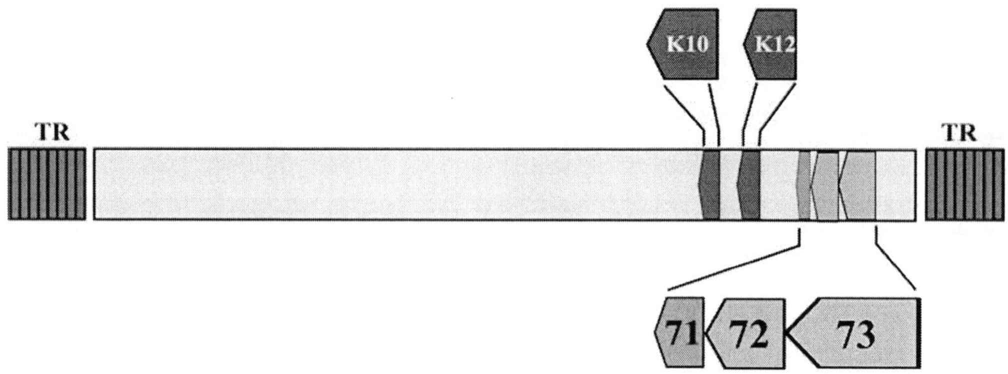

Figure 2. Major transcript expressed during latent infection in all forms of KSHV associated malignancies. K10 encodes for a viral interferon regulatory factor, K12 encodes Kaposin. ORF71, ORF72 and ORF73 encodes for a v-FLIP, v-Cyclin and Latency Associated Nuclear Antigen, respectively.

Role of KSHV Latent Genes in Establishment and Maintenance of Viral Latency: *Latency Associated Nuclear Antigen (LANA)*

Latency associated nuclear antigen (LANA), is a highly immunogenic nuclear protein originally identified by immunofluorescence assay with sera from KS patients (Gao et al., 1996c; Kedes et al., 1996a) and is a major latent protein encoded by ORF 73 of KSHV (Rainbow et al., 1997; Russo et al., 1996). LANA is 1162 amino acids long with an apparent molecular mass of 222-234 kDa on sodium dodecyl sulfate (SDS)-polyacrylamide gel which is larger than the calculated size of the protein. This is most likely due to the central acidic amino acid residues (Rainbow et al., 1997). LANA has three distinct domains (Fig 3A), the N-terminal domain which contains a nuclear localization sequence, a proline rich region, and chromosome binding sequence (5-22aa) shown to be important for tethering to the host chromatin (Piolot et al., 2001a; Rainbow et al., 1997; Shinohara et al., 2002). The central domain, rich in Glutamine (Q)/Glutamic acid (E) residues is reported to vary in size in different isolates, which may account for its slow electrophoretic migration. LANA also contains a leucine zipper motif near the C-terminal domain (Gao et al., 1999). The leucine zipper has been shown to be important for interaction of LANA with various cellular proteins including p53, pRb, Histone H1 and CREB2/ATF (Cotter and Robertson, 1999; Friborg et al., 1999; Lim et al., 2000; Radkov, Kellam, and Boshoff, 2000). The C-terminal domain of LANA comprises a nuclear localization signal sequence, which may be important for its characteristic nuclear speckling (Schwam et al., 2000). The C-terminus has also been shown to bind the terminal repeat DNA of the KSHV genome and is thus crucial for tethering the viral genome to the host chromosome (Cotter, Subramanian, and Robertson, 2001; Garber et al., 2001).

the maintenance of chromosome length and is active in highly degenerative cells including lymphocytes, keratinocytes, hematopoietic progenitor cells and uterine endometrial cells (Harle-Bachor and Boukamp, 1996; Hiyama et al., 1995; Kyo et al., 1997), in tumor derived cell lines and malignant tissues (Kim et al., 1994; Meyerson et al., 1997; Shay and Bacchetti, 1997). Additional evidence suggests that KSHV encoded LANA cooperates with HIV Tat and upregulates the HIV LTR promoter thus suggesting the hypothesis that KSHV co-infection promotes HIV propagation (Hyun et al., 2001; Knight, Cotter, and Robertson, 2001). Furthermore, LANA can also activate the major latent EBV promoter elements (LMP1 and Cp1) in transient studies (Groves et al., 2001). LANA regulates Wnt signaling pathway through stabilizing beta-catenin by binding to the negative regulator GSK-3β, causing a cell cycle-dependent nuclear accumulation of GSK-3β (Fujimuro and Hayward, 2003; Fujimuro et al., 2003). The data available in literature strongly suggests that LANA is a multifunctional protein involved in the activation and repression of transcription, and targets cell cycle regulating proteins. These activities are important in the regulation of the cell cycle and apoptosis in KSHV infected cells.

MHC-I Downregulation by KSHV

Cytotoxic T lymphocytes (CTLs) recognize and neutralize infected cells that display viral peptides in the context of major histocompatibility complex (MHC) class I molecules after binding to the T cell receptor (Berke, 1994). However, KSHV is able to establish latent infection in hosts with an intact immune system suggesting that like other herpes viruses, they affect MHC class I expression (Ahn et al., 1996; Tortorella et al., 2000). It was shown that the KSHV genome encodes two similar proteins K3 and K5 during the lytic cell cycle that is capable of increasing endocytosis of MHC class I molecules to the endolysosomes where they are degraded (Coscoy and Ganem, 2000; Ishido et al., 2000). The N-terminal zinc finger and the central sorting motif of K3 triggers the internalization of MHC class I molecules and its redirection to the trans golgi network. Consequently, the C terminal diacidic cluster region of K3 then targets the MHC chains to the lysosome (Means et al., 2002). In this way, KSHV prevents the cell surface expression of MHC chains and protects itself from CTL cell lysis during the lytic cell cycle. However, as only a subset of the infected cells undergo lytic replication (1 to 5%), KSHV is likely to have a means to protect itself from the immune response during latent infection (Dittmer et al., 1998; Katano et al., 1999; Parravicini et al., 2000; Sarid et al., 1998; Staskus et al., 1997; Talbot et al., 1999; Zhong and Ganem, 1997). Infection of primary dermal microvascular endothelial cells with KSHV resulted in a down regulation of cell surface MHC class I, CD31 (PE-CAM) and CD54 (ICAM-I) immune response modulators. Moreover, this down regulation was correlated with increased levels of LANA, suggesting that during latency LANA plays a role in the down regulation of MHC class I molecules (Tomescu, Law, and Kedes, 2003).

KSHV Encoded Viral Cyclin

Two processes are vital for eukaryotic cell proliferation: 1) the doubling of the DNA genome during S (synthesis) phase and 2) the partitioning of half that genomic information during M (mitosis) phase to daughter cells. Intermediate stages, G1 and G2 phases prepare the cells to undergo these processes and all stages are tightly regulated by a group of cytoplasmic proteins called cyclins and cyclin-dependent kinases (CDK) (Norbury and Nurse, 1992), (Sherr, 1996); (Senderowicz, 2003). Cyclin levels do vary during the cell cycle whereas CDK levels remain stable (Coqueret, 2002). CDK's associate with the appropriate cyclin to be activated. This activation results in the phosphorylation of proteins involved in cell cycle regulation. The cell cycle has a number of checkpoints to prevent the progression of cell division as a result of genetic aberrations such as DNA damage. These checkpoints are vital for the maintenance of genome integrity and genetic stability of daughter cells.

The retinoblastoma protein (pRb) is an essential regulator of the transition from G1 to S phase as it binds to, and represses the transcriptional activity of E2F when hypophosphorylated (G0) (Weintraub, Prater, and Dean, 1992; Flemington, Speck, and Kaelin, 1993). pRb represses E2F-activated genes by modifying nucleosomes when it recruits the histone de-acetylase, HDAC and ATP-dependent nucleosome remodeling complexes, SWI/SNF to promoters comprising E2F sites like cyclin E (Brehm et al., 1998; Zhang et al., 2000). Cyclin D associates with and activates CDK4 and CDK6 to phosphorylate pRb, thus releasing pRb from its E2F repressing activity (Matsushime et al., 1992). This allows E2F to activate the expression of cyclin E, which in turn leads to the expression of cyclin A and thus ensuring mitosis (Ohtani, DeGregori, and Nevins, 1995; Geng et al., 1996; Krek et al., 1994) (Fig 4). Viral invasion of eukaryotic cells normally occur during host cell quiescence (G0) when there is a rate limiting level of nucleotides (Thelander and Reichard, 1979). For successful viral propagation, the virus has to induce its host cells to enter the S phase and this is achieved by disruption of the cellular checkpoints. One of the ways that this is accomplished is by the production of viral homologs of human cyclins.

The KSHV ORF72 encodes v-cyclin, a homolog of the human cyclin D, most similar to cyclin D2. This viral homolog plays an identical role as the human form, interacting with CDK4 and CDK6 and resulting in the phosphorylation of pRb (Chang et al., 1996). The release of E2F repression allows E2F to activate genes needed for the transition from G1 to the S phase. Moreover, pRb phosphorylation by v-cyclin is not susceptible to inhibition by CDK inhibitors (CKIs) $p16^{INK4a}$, $p21^{CIP1}$ and $p27^{KIP1}$, unlike the human form (Swanton et al., 1997). The viral cyclin also enables CDK6 to have broader substrate specificity than cyclin D, as v-cyclin/CDK6 can also phosphorylate histone H1 and the CDK2 substrate, $p27^{KIP1}$. Phosphorylation of the latter results in its ubiquitination and degradation thus, abrogating $p27^{KIP1}$ arrest on the cell cycle (Swanton et al., 1999). Furthermore, v-cyclin has also been implicated in the activation of cyclin A expression in quiescent cells dependent on CDK6 activity and E2F sites comprising the cyclin A promoter (Swanton and Jones, 2001).

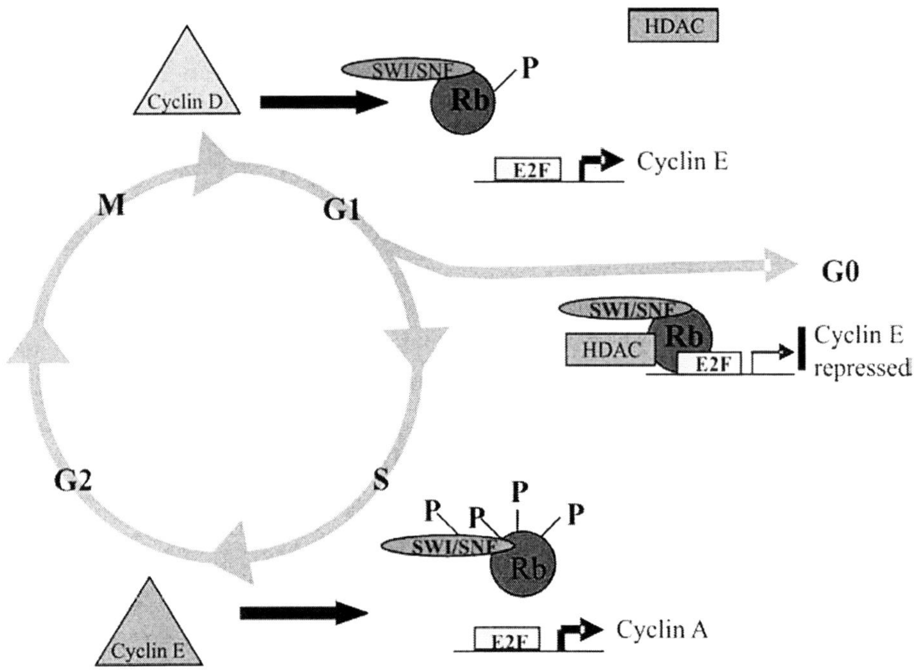

Figure 4. The cell cycle indicating the factors involved in G1/S cell cycle transition.

Understandably, the production of *v*-cyclin during the lytic cycle is highly beneficial for the virus as it ensures an abundant supply of cellular factors needed for viral replication. More complex is the recent discovery that v-cyclin is also expressed during latent KSHV infection (Dittmer, 2003; Davis et al., 1997). In this state, the virus replicates with the human genome and has no need to ensure increased production of cellular factors. It has been suggested that the repression of E2F activity by hypophosphorylated pRb acts as an anti-viral mechanism, locking the host cell in quiescence and thus preventing latent replication (Moore and Chang, 1998a). An alternative explanation for the production of *v*-cyclin during latency is based on the fact that cellular cyclin D, and not cyclins E and A is able to inhibit cellular differentiation of immature myoblasts (Skapek et al., 1995). This suggests that *v*-cyclin has an alternative function during latency which involves the severance of the cellular proliferation pathway from that of cellular differentiation (Swanton and Jones, 2001). Therefore, KSHV v-cyclin seems to have gained additional functions, and might not only serve to mimic the role of its human homolog (Jenner and Boshoff, 2002).

Role of KSHV Encoded vFLIP

Programmed cell death or apoptosis is a mechanism harnessed to protect cells from infectious agents and cancer, and to eradicate unwanted cells. This process is mediated by death receptors which fall part of the tumor necrosis factor (TNF) receptor family and include Fas (CD95), TNF-receptor I (TNF-R1), TNF-related apoptosis-induced ligand (TRAIL) receptor-1 (R-1), and R2 (Gura, 1997). Clustering of the receptors occurs when stimulated by ligand and this recruits the Fas-associated death domain (FADD), an adaptor molecule that leads to the sequestering of caspase-8 (FLICE) to the death-inducing signaling complex (DISC) (Ashkenazi and Dixit, 1998; Chou et al., 1998). Recruitment of FLICE to the DISC results in its activation, and it then dissociates from the complex to initiate the caspase cascade that results in induction of apoptosis (Chinnaiyan et al., 1996; Kischkel et al., 1995). Nuclear factor (NF)-κβ, is a transcription factor involved in the regulation of apoptosis and its activity is induced by stimulation of Fas, TNF-R1 and TRAIL-1 (Ponton, Clement, and Stamenkovic, 1996; Wajant et al., 2000). Over-expression of cellular FLIP was shown to activate NFκβ and inhibition attenuated TNF-α-induced activation of NFκβ. Thus, FLIP not only inhibits apoptosis by interacting with FADD, but it also controls cell survival via NFκβ (Okano et al., 2003).

FLICE inhibitory protein (FLIP), part of DISC modulates the activity of FLICE by preventing procaspase processing and therefore plays an important role in rescuing cells from Fas-induced apoptosis (Wang et al., 2000; Willems et al., 2000). KSHV produces a homolog of cellular FLIP (cFLIP) called vFLIP encoded from ORF71 (Dittmer et al., 1998; Thome et al., 1997). vFLIP comprises two death effector domains which interact with FADD and prevents the activation of FLICE by Fas, thus inhibiting the caspase cascade and induction of apoptosis (Belanger et al., 2001; Thome et al., 1997). Moreover, vFLIP was also shown to be able to induce tumors in mice using Fas-sensitive lymphoma B cell lines transduced with v-FLIP (Djerbi et al., 1999). KSHV-encoded vFLIP, like its cellular counterpart is also able to activate NFκβ unlike any other viral FLIP (Matta et al., 2003). This was demonstrated by vFLIP's ability to activate a promoter construct controlled by NFκβ as well as interacting and activating the kinase involved in the NFκβ pathway, IκB (IKK) by binding to the IKKγ subunit (Chaudhary et al., 1999; Field et al., 2003; Liu et al., 2002). The activation of the NFκβ pathway by this protein is responsible for the transformation phenotype observed in cells expressing vFLIP independently of it ability to inhibit Fas-induced apoptosis (Sun, Zachariah, and Chaudhary, 2003).

KSHV Encoded Kaposin

Kaposin, a predicted Type II membrane protein is encoded by ORF K12 and is highly expressed during latent KSHV infection in PEL and KS (Davis et al., 1997; Staskus et al., 1997). There is controversy surrounding the localization of kaposin, however it was suggested that it may be a multicompartmental protein (Tomkowicz et al., 2002).

Gene expression of ORF K12 gives rise to three alternatively transcribed isoforms, kaposin A, B and C of which kaposin A is preferentially expressed in body cavity based cell lines (Sadler et al., 1999). Kaposin A has been identified as an oncogene with the ability to transform Rat-3 cells and induce formation of tumors in athymic *nu/nu* mice (Muralidhar et al., 2000). In addition it is also able to transform NIH3T3 fibroblasts and promote lymphocyte aggregation and adhesion (Kliche et al., 2001). The transforming ability of kaposin A has been attributed to its interaction with cytohesin-1, a guanine nucleotide exchange factor (GEF) that regulates β2- integrin. Cytohesin-1 catalyses the guanine nucleotide exchange of ADP-ribosylation factors (ARF) GTPases. Phospholipids are required to bind to the myristoylated N-terminus of ARF in order to free the GTP binding site for nucleotide exchange. It has been suggested that kaposin mimics the action of phosphotidinyl inositol by membrane recruitment of cytohesin-1 and thus increases the activity of cytohesin-1 to an unidentified GTPase resulting in actin remodeling and downstream gene activation (Kliche et al., 2001). Therefore, the small hydrophobic (6 kDa) viral protein is able to imitate cellular processes normally carried out by lipid, phosphotidinyl inositol.

Interferon Regulatory Factor (IRF) Encoded By KSHV

Upon viral infection, the host cell initiates a series of anti-inflammatory responses with the aim to eliminate the invading pathogen. This inflammatory response involves the production of inflammatory factors that lead to the activation of T cell cytolytic functions or directly inhibit viral replication. These cytokines act to limit lytic or latent viral persistence and one of the key players in this arena is interferon (IFN). There are two types of IFN, type I IFN (IFNα and IFNβ) and type II IFN (IFNγ) produced by a variety of infected cells, T cells and natural killer cells, respectively (Lengyel, 1982). Type I IFN is encoded by a single *IFNB* gene and 13 *IFNA* genes and the production of IFNα/β is induced upon viral infection which is dependent on the presence of a viral response element (VRE) within the promoter (Kim and Maniatis, 1997; Merika et al., 1998; Wathelet et al., 1998). The VRE of the *IFNA* genes comprises a number of repeats that bind interferon regulatory factors (IRFs), and these are responsible for the differential induction of the various subtypes of IFNα due to two nucleotide substitutions in these repeats and the presence of negative regulatory sequences, 5′ to these genes (Au and Pitha, 2001; Au et al., 1993; Lopez et al., 2000). IRFs mediate the transcriptional response to viral infection and regulate the expression of type I IFN, IFN-activated genes (ISG) and other chemokines and cytokines. Thus far, 9 human IRF's have been identified while KSHV encodes four IRF homologs, three of which have been characterized (Fig 5).

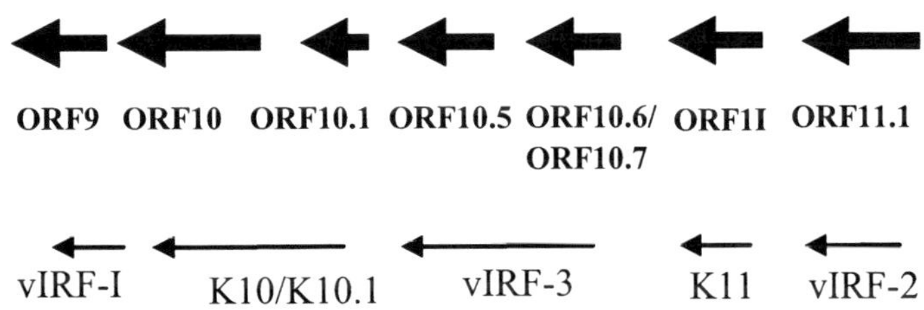

Figure 5. Figure showing the arrangement of vIRF gene cluster on the KSHV genome.

IRF-3 and IRF-7 are phosphorylated at its C-terminus upon viral infection by an unknown kinase leading to activation of *IFNA* and *IFNB* genes. These factors also interact with the histone transacetylases, CBP/p300 and p300-CBP associated factor (PCAF), respectively (Lin et al., 1998; Weaver, Kumar, and Reich, 1998; Yoneyama et al., 1998). IRF-5 is also able to induce IFN and cytokine genes but its expression is limited to specific cell types and is dependent on specific viral infection (Fig 6).

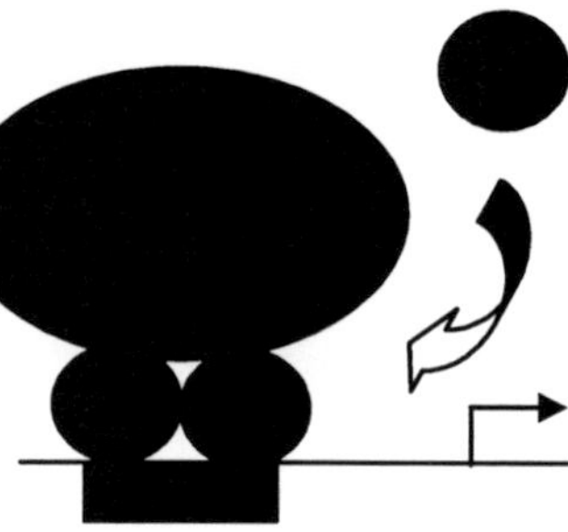

Figure 6. IFN gene activation by cellular IRFs (Barnes, Lubyova, and Pitha, 2002).

KSHV Encoded vIRF-I

This IRF is encoded by ORF K9 of the KSHV genome and functions to repress the activation of ISG by IFN and inhibits the virus-mediated activation of IFN genes (Gao et al., 1997; Li et al., 1998; Zimring, Goodbourn, and Offermann, 1998). Moreover, the expression of *v*IRF-I in B cells prevents the anti-proliferative action of IFN, is able to induce tumors in nude mice, and causes NIH3T3 cells to grow in soft agar (Flowers, Flowers, and Nabel, 1998)). *v*IRF-I, similar to IRF-2 is thus an oncogene capable of transforming cells. *v*IRF is unable to bind with the same level of specificity as cellular IRF and the potency behind its function remains unknown. However, it has been suggested that its inhibition of p53 and CPB/p300, its activation of the c-myc promoter and its ability to induce structural changes in chromatin facilitates its transforming and anti-apoptotic functions (Barnes, Lubyova, and Pitha, 2002).

KSHV Encoded vIRF-2

*v*IRF-2 is a constitutively expressed nuclear protein that unlike cellular IRF is unable to bind to IFN stimulated response elements (ISRE) while able to bind to NFκB responsive oligonucleotides in gel shift assays (Burysek, Yeow & Pitha, 1999). It is able to heterodimerise with transcription factors such as IRF-I, IRF-2, IRF-8, CBP/p300 and Rel/p65 involved in the initial inflammatory response (Burysek, Yeow, and Pitha, 1999). The proapoptotic factor, protein kinase (PKR) plays an important role in the anti-viral response and the association of vIRF-2 with PKR prevents its autophosphorylation and its ability to phosphorylate specific substrates like histone 2A and eukaryote inhibitor factor 2 (eIF-2). This interaction leads to the inhibition of the IFN-induced block on viral protein synthesis (Burysek and Pitha, 2001). vIRF-2 is unable to transform NIH3T3 cells but its heterodimeric protein interactions with cellular targets suggest that it plays an important role in KSHV's anti-apoptotic mechanism (Burysek et al., 1999; Burysek, Yeow, and Pitha, 1999).

KSHV Encoded vIRF-3

Over-expression of vIRF-3 prevented the induction of biologically active type I IFN by viral infection. This could be due to its ability to inhibit p53-induced apoptosis and its association with cellular IRF-3 and CBP/p300 ((Rivas et al., 2001). However, although displaying a similar profile of protein interactions as *v*IRF-I, vIRF-3 was unable to transform cells (Barnes et al, 2002).

Therefore, KSHV has developed a method for producing interferon factors that behave similar to cellular IRF's but interferes with the anti-viral response mechanism triggered by interferon by targeting the cellular IRFs and their regulatory cellular factors, preventing the activation of *IFNA* gene expression (Fig 7) (Barnes, Lubyova, and Pitha, 2002).

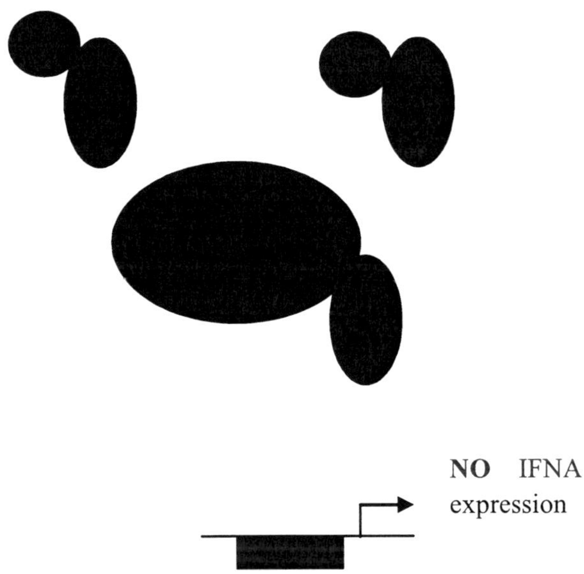

Figure 7. vIRF inhibits the activation of IFN genes (Barnes, Lubyova, and Pitha, 2002).

Lytic Genes Expression Profile

The majority of the KSHV genome remains silent during latent infection most likely due to the methylation of the promoter sequence. Virus replicating latently is able to reactivate to enter lytic replication and production of virus progeny leading to host cell lysis. In KSHV Rta (replication transcription activator), which is encoded by ORF50 is necessary and sufficient for KSHV to reactivate and enter into the lytic replication cycle (Lukac, Kirshner, and Ganem, 1999; Gradoville et al., 2000). Cellular factors like CREB binding protein (CBP) and c-jun have been shown to bind to Rta, upregulating Rta mediated transcription (Gwack et al., 2001). During latent infection most of the promoters including Rta is heavily methylated in PELs and infected PBMCs (Chen et al., 2001). Treatment with TPA leads to reactivation, possibly due to the demethylation of the promoter. The program of gene expression during lytic replication was revealed by DNA array expression profiling of PEL cell lines chronically as well as acutely (Paulose-Murphy et al., 2001; Murphy, 2001). Both studies show that the first genes expressed after the induction of lytic replication are the regulators of downstream gene expression like ORF50 (Rta), K8 (Zta or K-bZIP) and ORF57 (Post transcriptional regulator of gene expression). These are followed by the genes, important for the replication of the viral DNA, Such as DNA polymerase (ORF9) and the processivity factor PF-8 (ORF59) which is generally expressed 10h post-infection. Structural genes important for viral maturation are expressed 24h post-infection. Cellular homologous genes involved in signaling are mostly expressed after ORF50 but before the expression of viral DNA machinery.

Studies have identified an ORF with similarities to the EBV immediate early transactivator, which functions as an immediate early transactivator of KSHV lytic replication. It has been demonstrated that this protein is capable of stimulating KSHV lytic replication *in vitro* similar to the Zta and Rta proteins of EBV (Renne et al., 1996).

Viral Homologs to Cellular Proteins Important For Virus Survival

KSHV encodes a large number of genes with homology to cellular genes and function to help the virus escape the immune surveillance of the host, allowing the virus to propagate efficiently and induce tumor formation. This limits viral dependence on the host cellular machinery and protects the virus against anti-viral mechanisms. Therefore, not only the virus is able to highjack host cellular machinery in order to replicate, it can also produce factors that mimic the function of cellular proteins while thwarting the host cell anti-viral response. These genes are believed to be pirated from the host cells during viral evolution.

***vBcl-2*:** vBcl-2, encoded by ORF16, shows 60% homology with the cellular gene bcl-2 family members such as Bcl-2 and Bax, which has been shown to inhibit virally induced apoptosis (Cheng et al., 1997). *v*bcl-2 transcripts are present predominantly during the lytic phase of viral replication (Sarid et al., 1998; Sarid et al., 1997). VBcl-2 interacts with, and inhibits the proapoptotic function of the Bcl-2 family member, suggesting that KSHV-bcl-2 is capable of enhancing cell survival and thus provides enough time for the assembly of viral progeny. Bcl-2 deregulation has been reported in some human malignancies like follicular lymphomas (Cheng et al., 1997; Friborg et al., 1998), suggesting that ectopic expression of KSHV-bcl-2 may also contribute to tumorigenesis through its anti-apoptotic effect.

vGPCR: The KSHV encoded *v*GPCR (ORF74) is one of the viral encoded potential oncogenes and is most closely related to the IL-8 receptors CXCR1 and CXCR2 (Swanton et al., 1997) that is produced in KS lesions. The protein constitutively signals through protein kinase B and C (PKB/Akt) (Montaner et al., 2001) and activates related adhesion tyrosine kinase (RAFTK), lyn (Munshi et al., 1999), c-jun N-terminal kinase (JNK)/stress-activated protein kinase and p38 mitogen activated protein kinase (MAPK) (Bais et al., 1998). KSHV *v*GPCR has been shown to activate the phosphoinositide pathway (mitogenic signaling pathway) in COS-1 cells and transfection of rat fibroblast with *v*GPCR leads to cell proliferation (Arvanitakis et al., 1997).

vMIP: KSHV encoded Chemokines, vMIP-I, vMIP-II and vMIP-III show maximum similarity to cellular Chemokines i.e. thymus-activation regulated Chemokines (TARC), macrophase derived Chemokine (MDC) and myeloid progenitor inhibitory factor (MPIF)-2 (Ganem, 1997). All these viral macrophage inflammatory proteins have been shown to be angiogenic in a chick chorioallantoic assay, associating them in the pathogenesis of KS (Boshoff et al., 1997a).

vIL-6: The KSHV encoded IL-6 has 24% identity with human IL-6 (Moore et al., 1996). Cultured KS cells have been shown to respond to recombinant hIL-6, suggesting its role in KS pathogenesis. This is also supported by the fact that KS spindle cells express high affinity IL-6 receptor *in vivo* (Moore et al., 1996). PEL cell lines treated with TPA also showed an increase in the expression of vIL-6 from 1% to 50% (Katano et al., 2000b) which supports that vIL6 is a lytic gene. In MCD, vIL-6 might be expressed by some latently infected cells, and thought to be involved in phenotypical maturation of the pre-germinal centre cells.

Understanding the Role of Genes Important for Pathogenesis

Although over 90 genes or open reading frames have been identified in the KSHV genome, only a small subset of these viral genes are typically expressed during the latency. This pattern of viral gene expression allows the virus to escape host immune surveillance and to establish persistent latent infection (Russo et al., 1996; Sarid et al., 1998). Several genes of KSHV have been identified, and demonstrated to interfere with normal cell growth and differentiation, and this data has greatly contributed to our understanding of the mechanism by which KSHV transforms cells. However, several basic but critical questions concerning this virus remain unanswered. Firstly, the target cell of KSHV is still controversial. Secondly, the genes and genomic elements that control the activation and completion of the viral lytic cycle are only partly understood. Finally, nothing is known about the function of KSHV gene products in the context of the whole virus. Except for a single report, all data concerning the transforming activity of KSHV have been derived from experiments with single viral gene products. It is possible that within the context of the whole virus, some of these genes do not contribute to cell transformation.

Genetic Approaches for Assessing KSHV Gene Function

The best way to elucidate the gene function in the context of whole virus is to knock-out specific genes and then determine the effect on the viral life cycle, packaging, infective ability and cell transformation. Understanding the role of specific KSHV gene products in the context of intact virus has long been hampered by the large size of the KSHV genome. This genome has proven difficult to experimentally manipulate *in vitro*, limiting the ability to functionally characterize and create specifically targeted changes in the genetic material. Additionally, because of the lack of an efficient infection system, the generation of KSHV mutants has been difficult and has impeded the delineation of the functions of KSHV genes. With the advent of current molecular cloning technologies, progress in understanding small mammalian DNA tumor viruses advanced rapidly; however, progress in applying such technologies to more complex herpesvirus genomes has been a slower, step-wise process. In EBV studies, several reports have successfully shown that specific genes, including EBNA3A and EBNA3C are essential for cell transformation and EBNA3B is dispensable for cell transformation in the context of the whole virus using traditional homologous recombination strategy. Similar to EBV studies, targeted manipulation of the KSHV genome generally employs one of two broad strategies (Tomkinson, Robertson, and Kieff, 1993; Tomkinson et al., 1993). Recombinant KSHV can be generated by homologous recombination in infected cells with the DNA fragment carrying the alleles of mutations or assembling overlapping inserts of cosmid clones covering the entire viral genome, or more recently, by cloning the viral genome as a bacterial artificial chromosome (BAC). In this chapter we will provide an overview of the use of BAC technologies for KSHV genetic manipulation and the traditional generation of mutant viruses utilizing homologous recombination (Tomkinson, Robertson, and Kieff, 1993; Tomkinson et al., 1993).

BAC Approaches to Herpesvirus Genetics

The use of BAC technology for generating gammaherpesvirus recombinants is still somewhat in its infancy. The stability and successful generation of mutants in the gammaherpesviruses EBV and KSHV will further enhance the capability of this technology in the analysis of the large DNA virus genomes. However, studies utilizing this technology for KSHV have not produced consistent results in targeting specific viral genes. So far, only few papers have shown that recombinant KSHV approaches based on BAC technology to be successful (Vieira et al., 2001; Zhou et al., 2002).

A. Construction of Replacement Plasmids

The cloning limit of BACs exceeds the size of the largest known herpesviruses (hCMV); therefore, BACs can theoretically support and maintain complete herpesvirus genomes. In 1997, the first infectious herpesvirus BAC, the murine cytomegalovirus, was constructed and was subsequently followed by other viruses (Adler, Messerle, and Koszinowski, 2003; Delecluse et al., 1998; Delecluse et al., 2001; Messerle et al., 1997; Saeki et al., 1998; Stavropoulos and Strathdee, 1998; Wagner et al., 1999). The large size and recalcitrance to *in vitro* manipulation of the herpesvirus genome prevented the simple ligation of the viral genome into a BAC by standard molecular cloning procedures. Efficient cloning required a preliminary step in which the BAC plasmid cassette which contain the BAC replicon, GFP gene and drug resistant gene such as hygromycin is introduced into the viral genome essentially through the homologous recombination strategy described above (Fig 8).

B. Cloning of Herpesvirus Genomes into BACs and Purification of BAC Recombinant Virus

The replacement plasmid can be used for transfection of KSHV infected cells. Under continuous selection, the cells which stably express GFP will survive and stable cells can be used for preparation of episomal KSHV DNA. E.coli will then be transformed by the viral DNA. For successful transformation, an intact, circularized form of the recombinant virus must be isolated prior to transformation into *E. coli*. The most common approach takes advantage of the herpesvirus replication cycle, which results in circular replicative intermediates of viral DNA prior to cleavage into unit-length linearized forms for packaging in the virus particle. DNA harvested by Hirt extraction early after transfection gives significant amount of circular replicative intermediates (Hirt, 1967). Once the transformants are obtained, the BAC-KSHV DNA should be reconstituted in mammalian cells. To reconstitute BAC-KSHV, it is transfected into 293 cells followed by a selection and induction procedure. The supernatant from the stable transfected 293 cells can be used for infection of permissive cells, by selection, the BAC-KSHV stable infected cells can be obtained, and then the infective viral stock can be made for further use by induction of these cells (Fig. 9).

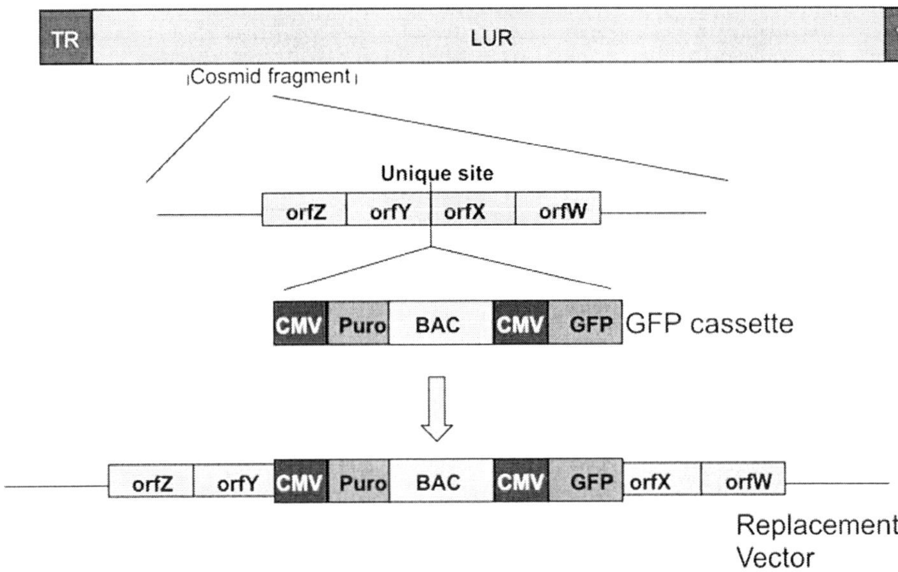

Figure 8. Schematic representation of replacement vector for BAC-KSHV construction.

C. Generation of KSHV Mutants

The main purpose for generating KSHV mutants is to knock-out specific genes. Based on BAC-KSHV backbone, the second site targeting vector can be constructed to introduce selective markers to disrupt KSHV specific genes. The construction of second site targeting vector is similar to the BAC replacement vector. For construction of this vector, a selective marker (like RFP) should be introduced into a specific gene of KSHV, and this vector will be electroporated into E. coli that has been transformed with BAC-KSHV and pGETrec. The second targeting vector will recombine with BAC-KSHV; episomal KSHV DNA can be isolated from bacterial colonies for further infection and characterization.

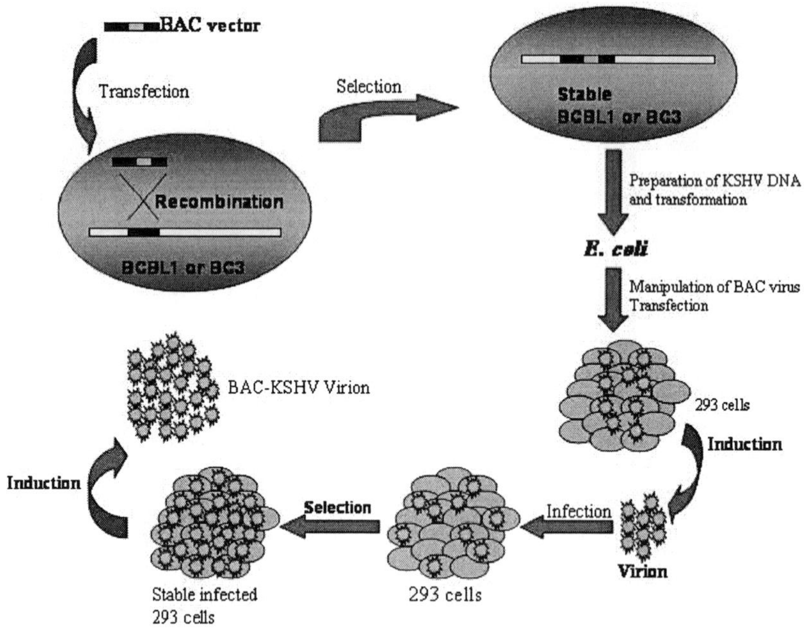

Figure 9. Construction of BAC-herpesvirus by homologous recombination; Efficient cloning of the BAC-herpesvirus requires a preliminary step in which the BAC plasmid cassette is introduced into the viral genome through homologous recombination. Insertion of the BAC sequence is followed by isolation of a BAC recombinant virus. A selection marker (antibiotic resistance) may be employed to facilitate efficient recovery of recombinant virus.

Traditional Homologous Recombination Approaches to KSHV Genetics

Obviously, BAC strategy for genetic analysis of KSHV has some of its own problems and this prevents its application. Manipulation of BAC DNA is not convenient, and more importantly, spontaneous mutation and random deletion usually happen and this would complicate the analysis. Therefore, researchers are always seeking more reliable methods for generating KSHV mutants. There are some successful examples of genetic analysis of EBV by using traditional homologous recombination approaches, and the advantage of this method is that all manipulation of viral DNA is carried out in eukaryotic cells, minimizing spontaneous mutations. Here we will describe the principle behind the preparation of a targeting vector for the production of a general GFP-KSHV that can facilitate the identification of KSHV-infected cells or trace dissemination of virus from a peripheral site of inoculation to different tissues.

A. Construction of Gfp Expression Cassette

This cassette potentially contains several elements including the GFP gene, a eukaryotic promoter to drive GFP gene expression, a drug resistance gene (such as puromycin or neomycin) and a eukaryotic promoter to drive expression of the drug resistant gene. In some cases a polycistronic gene may be constructed using an internal ribosome entry site (such as the IRES of HCV) to facilitate expression of GFP and drug resistance from a common promoter. This cassette should be constructed in a vector background with unique flanking restriction sites such that the cassette can be excised for insertion into the targeting vector (Figure 10A). For viruses such as KSHV which are primarily maintained in tissue culture as latently-infecting genomes, drug selection may allow the establishment of latently-infected producer cell lines that can be used to generate free virus by induction of lytic replication with chemical inducers or viral transactivating genes.

B. Construction of Targeting Vector for Generation of Recombinant Virus

Ideally the GFP cassette should be inserted into a unique restriction site in a region of the genome that does not encode a gene product, eliminating the possibility that the cassette will disrupt a viral open-reading frame and minimizing the potential for disruption of viral gene transcription. Careful consideration of this site is important as the end goal of this experiment is to generate a recombinant GFP-herpesvirus that mimics the functionality of the wild-type virus. Ideally, the aforementioned restriction site should be chosen with careful consideration of available cosmid sequences such that digestion with the appropriate enzyme results in a single cut and linearization of the cosmid. The GFP/drug resistance cassette can then be excised from its plasmid and inserted into the digested cosmid by a standard molecular ligation reaction. A restriction site can be chosen such that the insert is flanked by at least 3 kb of viral DNA sequence on the 5' and 3' termini. Again, flanking sequence should be chosen to minimize the dysregulation of the gene expression pattern of viral genes in the adjacent areas.

An example of such a strategy utilizes the *Pme*I site that lies downstream of ORF 18 and ORF 19 of the KSHV genome (Zhou et al., 2002). Because ORF 18 is transcribed to the right and ORF 19 to the left, this site is downstream of both transcripts minimizing the likelihood that insertion of the GFP cassette will disrupt viral gene expression. Importantly, this is the only *Pme*I restriction site in the available cosmid fragment, permitting unique insertion of the GFP cassette. This site has previously been used in the construction of a GFP KSHV-BAC (Zhou et al., 2002), but the same advantages apply in the construction of a targeting vector for homologous recombination.

C. Mutagenesis Strategies

The desired recombinant virus may include either a manipulated or disrupted viral gene, in contrast to the general introduction of a foreign gene, GFP, as described above. This usually depends on the specific application or genetic question. We will briefly outline approaches for introducing such mutations.

As mentioned above, GFP provides a signal which marks GFP recombinant KSHV-infected cells. However, for addressing specific gene functions in the context of the whole virus, genes can be disrupted and the resultant virus can be assayed for viral replication, packaging, infection, etc. Based on the GFP recombinant virus backbone, another cassette containing a second selectable marker can be introduced into a specific viral ORF by homologous recombination, disrupting the coding region of the targeted gene. The advantage of introduction of another selection marker into the viral genome is to facilitate double selection or selection of cloned infected populations which includes the GFP (initial selection marker) "wild-type" recombinants with green color in infected cells and those that can positively be selected by antibiotics as "mutant" recombinants. This strategy might also employ one of the many fluorescent color proteins available as the second selection marker.

The strategy for construction of this second targeting vector is very similar to that described above for GFP. The primary difference is that the cassette which contains the second selection marker should be introduced into a unique restriction site within the coding region of the viral gene of interest. A cassette encoding RFP under the control of the CMV promoter can be digested from a commercially available construct. For generation of mutant virus, this cassette should be introduced by restriction digest into a cosmid vector which contains a large fragment of viral genome including the target gene of interest. Cells stably infected with GFP recombinant KSHV can then be transfected with the mutant targeting vector. With induction of lytic replication, the targeting vector can recombine with GFP recombinant KSHV. To get a pure population of mutant virus, the supernatant containing GFP "wild-type" virus and GFP "mutant" virus can be used to infect permissive cells. By serial dilution of supernatant, and by employing the selection principle described above, pure mutant virus may be obtained.

D. Transfection and Induction of Lytic Replication

As no robust system for KSHV lytic infection exists *in vitro*, the targeting vector is generally transfected into a virus-infected cell line (typically B cell) by electroporation. Successful electroporation of B cells can be performed with the BioRad Gene Pulser at 260 V and 975 μF. Lytic infection is induced either by co-transfection of a lytic transactivator (Rta for KSHV) or by treatment with chemical agents such as phorbol esters or butyric acid (Nakamura et al., 2003; Tomkinson, Robertson, and Kieff, 1993). While treatment with chemical agents produces a more robust overall lytic response, co-transfection of the transactivator has the advantage of enriching for lytic replication only in the transfected population of cells, potentially reducing the background of wild-type virus with no potential to recombine.

The efficiency of lytic virus induction can be monitored by immunofluorescence of a lytic gene and also by specific PCR analysis of the cell supernatant for viral DNA. PCR analysis may differentiate wild-type from recombinant virus by specific PCR primers that will detect and distinguish the wild-type from recombinant virus DNA. It should, however, be noted that these methods only indicate the production of viral progeny and not the infectious nature of these progeny virions. Returning to the example of the unique PmeI site in the viral cosmid construct, PCR primers for distinguishing wild-type and recombinant virus might be designed as indicated in Figure 8.

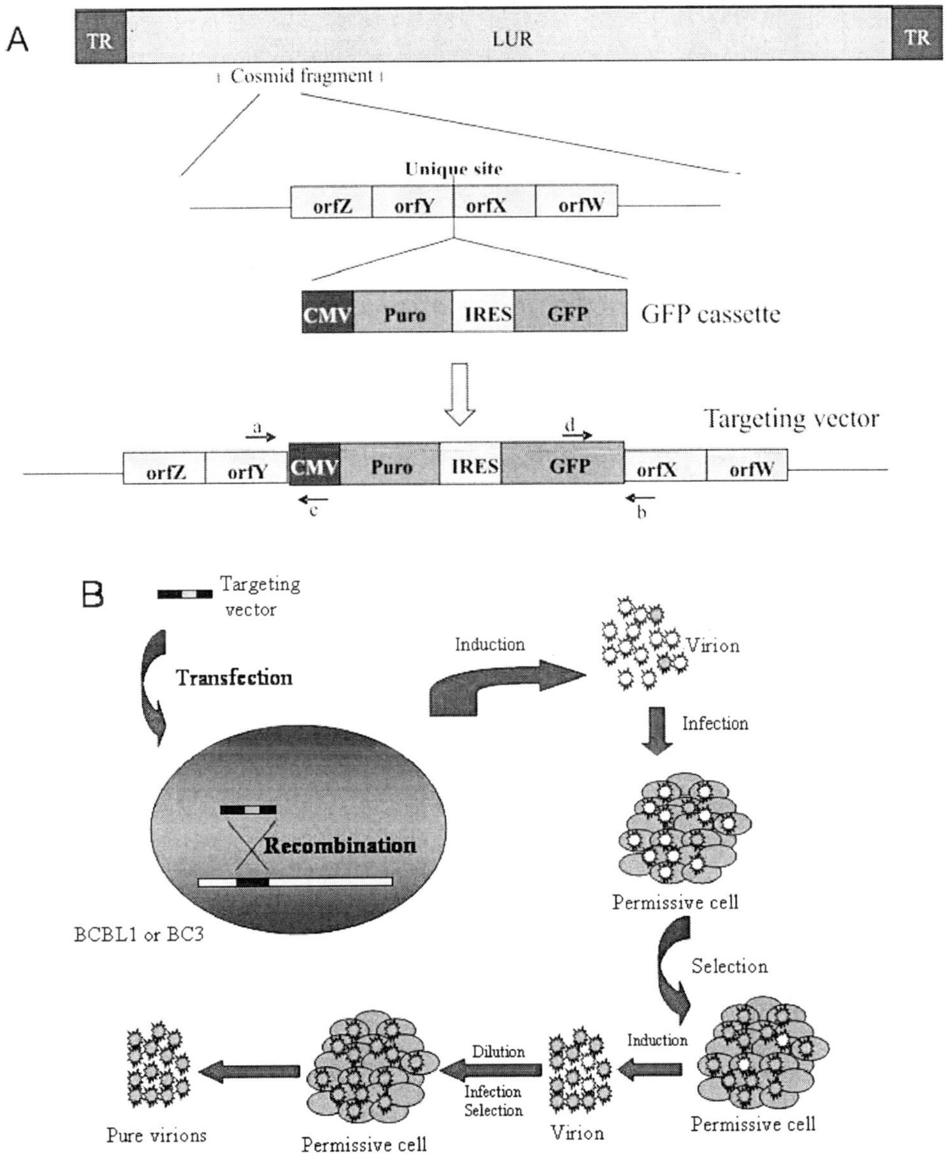

Figure 10. Schematic representation of targeting vector for homologous recombination (A) and strategy for purification of GFP recombinant KSHV (B). A cassette containing the puromycin and GFP genes under the control of the CMV promoter is inserted into a unique restriction site in a commercially available herpesvirus cosmid. a, b, c, and d represent primers for differential determination of wild-type and recombinant virus. Primer a and b amplifies wild-type virus whereas primer pair a:c and d:b amplifies recombinant virus.

E. Selection and Isolation of Recombinant KSHV

For viruses with tissue culture systems that less robustly support lytic infection, induction of latently-infected cells may be necessary as discussed previously. Ideally, upon induction recombinant virus will be produced and egress from the cell into the supernatant where it can be collected for further purification and study. In some cases, recovery of recombinant virus may be enhanced by transfecting latently infected cells with the targeting vector followed by drug selection. Selection will enrich for cells transfected with the targeting vector and should increase the ratio of recombinant to wild-type virus in the cell supernatant. To obtain a pure recombinant virus population, the supernatant can be diluted and used to infect permissive cells at varying dilutions. The infected cells are again drug selected, and under continuous selection, stable recombinant virus-infected cells can be obtained. These latently-infected cell lines can then be induced to generate recombinant virus. After several cycles of limiting supernatant dilution, cell infection, and selection, a pure recombinant virus-producer cell line can be obtained (Fig. 10B).

F. Characterization

To determine if the desired recombinant KSHV was successfully generated, it is necessary to complete genetic and phenotypic characterization of the virus. The first line of characterization involves restriction analysis using enzymes that cut with moderate frequency in KSHV genomes such as BamH1, EcoR1, Not1 and Sal1. It is recommended that analysis of at least three restriction profiles be determined. Any discord between wild-type and recombinant profiles is likely attributable to the desired recombination event. The identity of bands modified by the recombination event should then be confirmed by Southern hybridization with probe/probes specific to sequence immediately flanking the recombination event as well as any genetic material inserted in the site (selectable marker). Once the alteration of the genetic structure of the genome has been confirmed, the integrity of the recombination event should be confirmed by PCR amplification and sequencing of any junctions between the viral and foreign DNA. Phenotypically, the recombinant virus can be characterized by gene expression profiling of genes immediately flanking the recombination event. Such analysis can be done by standard methods such as northern blotting or real-time PCR.

Treatment of Kaposi's Sarcoma

Kaposi's sarcoma has been successfully treated with various methods depending on the severity of the infection. Classic KS, patients with single lesions, simple excision is typically sufficient and can be used to treat recurrence (Brenner et al., 1999), but patients with multiple lesions are often treated with single doses of radiation. In extensive or recurrent disease, patients can undergo a combination of surgery, radiation and chemotherapy. Complete remissions have also been shown to occur by chemotherapy alone with vinblastine, bleomycin, doxorubicin alone or in combination (Antman and Chang, 2000). Endemic KS is usually treated with same drugs used for classic KS patients. In patients with immunosuppresion associated KS, the lesions typically regress with the reduction, modification or cessation of the immunosuppressive chemotheraputic regimen. In endemic KS highly active antiretroviral therapy (HAART) have been successfully used to completely treat the patients (Wit et al., 1998).

Acknowledgement

This work was supported by grants from the Leukemia and Lymphoma Society of America and public health service grants from the NCI CA072510 and CA091792 and from NIDCR DE01436 (to ESR). ESR is a scholar of the Leukemia and Lymphoma Society of America.

References

Ablashi, D. V., Chatlynne, L. G., Whitman, J. E., Jr., and Cesarman, E. (2002). Spectrum of Kaposi's sarcoma-associated herpesvirus, or human herpesvirus 8, diseases. *Clin Microbiol Rev* **15**(3), 439-64.

Adler, H., Messerle, M., and Koszinowski, U. H. (2003). Cloning of herpesviral genomes as bacterial artificial chromosomes. *Rev Med Virol* **13**(2), 111-21.

Ahn, K., Meyer, T. H., Uebel, S., Sempe, P., Djaballah, H., Yang, Y., Peterson, P. A., Fruh, K., and Tampe, R. (1996). Molecular mechanism and species specificity of TAP inhibition by herpes simplex virus ICP47. *Embo J* **15**(13), 3247-55.

Alexander, L., Denekamp, L., Knapp, A., Auerbach, M. R., Damania, B., and Desrosiers, R. C. (2000). The primary sequence of rhesus monkey rhadinovirus isolate 26-95: sequence similarities to Kaposi's sarcoma-associated herpesvirus and rhesus monkey rhadinovirus isolate 17577. *J Virol* **74**(7), 3388-98.

Antman, K., and Chang, Y. (2000). Kaposi's sarcoma. *N Engl J Med* **342**(14), 1027-38.

Arvanitakis, L., Geras-Raaka, E., Varma, A., Gershengorn, M. C., and Cesarman, E. (1997). Human herpesvirus KSHV encodes a constitutively active G-protein-coupled receptor linked to cell proliferation. *Nature* **385**(6614), 347-50.

Arvanitakis, L., Mesri, E. A., Nador, R. G., Said, J. W., Asch, A. S., Knowles, D. M., and Cesarman, E. (1996). Establishment and characterization of a primary effusion (body cavity- based) lymphoma cell line (BC-3) harboring kaposi's sarcoma-associated herpesvirus (KSHV/HHV-8) in the absence of Epstein-Barr virus. *Blood* **88**(7), 2648-54.

Ashkenazi, A., and Dixit, V. M. (1998). Death receptors: signaling and modulation. *Science* **281**(5381), 1305-8.

Au, W. C., and Pitha, P. M. (2001). Recruitment of multiple interferon regulatory factors and histone acetyltransferase to the transcriptionally active interferon a promoters. *J Biol Chem* **276**(45), 41629-37.

Au, W. C., Su, Y., Raj, N. B., and Pitha, P. M. (1993). Virus-mediated induction of interferon A gene requires cooperation between multiple binding factors in the interferon alpha promoter region. *J Biol Chem* **268**(32), 24032-40.

Bais, C., Santomasso, B., Coso, O., Arvanitakis, L., Raaka, E. G., Gutkind, J. S., Asch, A. S., Cesarman, E., Gershengorn, M. C., Mesri, E. A., and Gerhengorn, M. C. (1998). G-protein-coupled receptor of Kaposi's sarcoma-associated herpesvirus is a viral oncogene and angiogenesis activator. *Nature* **391**(6662), 86-9.

Ballestas, M. E., Chatis, P. A., and Kaye, K. M. (1999). Efficient persistence of extrachromosomal KSHV DNA mediated by latency-associated nuclear antigen. *Science* **284**(5414), 641-4.

Ballestas, M. E., and Kaye, K. M. (2001). Kaposi's sarcoma-associated herpesvirus latency-associated nuclear antigen 1 mediates episome persistence through cis-acting terminal repeat (TR) sequence and specifically binds TR DNA. *J Virol* **75**(7), 3250-8.

Barnes, B., Lubyova, B., and Pitha, P. M. (2002). On the role of IRF in host defense. *J Interferon Cytokine Res* **22**(1), 59-71.

Belanger, C., Gravel, A., Tomoiu, A., Janelle, M. E., Gosselin, J., Tremblay, M. J., and Flamand, L. (2001). Human herpesvirus 8 viral FLICE-inhibitory protein inhibits Fas-mediated apoptosis through binding and prevention of procaspase-8 maturation. *J Hum Virol* **4**(2), 62-73.

Berke, G. (1994). The binding and lysis of target cells by cytotoxic lymphocytes: molecular and cellular aspects. *Annu Rev Immunol* **12**, 735-73.

Boshoff, C., Endo, Y., Collins, P. D., Takeuchi, Y., Reeves, J. D., Schweickart, V. L., Siani, M. A., Sasaki, T., Williams, T. J., Gray, P. W., Moore, P. S., Chang, Y., and Weiss, R. A. (1997a). Angiogenic and HIV-inhibitory functions of KSHV-encoded chemokines. *Science* **278**(5336), 290-4.

Boshoff, C., Endo, Y., Collins, P. D., Takeuchi, Y., Reeves, J. D., Schweickart, V. L., Siani, M. A., Sasaki, T., Williams, T. J., Gray, P. W., Moore, P. S., Chang, Y., and Weiss, R. A. (1997b). Angiogenic and HIV-inhibitory functions of KSHV-encoded chemokines [see comments]. *Science* **278**(5336), 290-4.

Bowne, W. B., Lewis, J. J., Filippa, D. A., Niesvizky, R., Brooks, A. D., Burt, M. E., and Brennan, M. F. (1999). The management of unicentric and multicentric Castleman's disease: a report of 16 cases and a review of the literature. *Cancer* **85**(3), 706-17.

Brandt, S. J., Bodine, D. M., Dunbar, C. E., and Nienhuis, A. W. (1990). Dysregulated interleukin 6 expression produces a syndrome resembling Castleman's disease in mice. *J Clin Invest* **86**(2), 592-9.

Brehm, A., Miska, E. A., McCance, D. J., Reid, J. L., Bannister, A. J., and Kouzarides, T. (1998). Retinoblastoma protein recruits histone deacetylase to repress transcription. *Nature* **391**(6667), 597-601.

Brenner, B., Rakowsky, E., Katz, A., Gutman, H., Sulkes, A., Schacter, J., and Fenig, E. (1999). Tailoring treatment for classical Kaposi's sarcoma: comprehensive clinical guidelines. *Int J Oncol* **14**(6), 1097-102.

Burysek, L., and Pitha, P. M. (2001). Latently expressed human herpesvirus 8-encoded interferon regulatory factor 2 inhibits double-stranded RNA-activated protein kinase. *J Virol* **75**(5), 2345-52.

Burysek, L., Yeow, W. S., Lubyova, B., Kellum, M., Schafer, S. L., Huang, Y. Q., and Pitha, P. M. (1999). Functional analysis of human herpesvirus 8-encoded viral interferon regulatory factor 1 and its association with cellular interferon regulatory factors and p300. *J Virol* **73**(9), 7334-42.

Burysek, L., Yeow, W. S., and Pitha, P. M. (1999). Unique properties of a second human herpesvirus 8-encoded interferon regulatory factor (vIRF-2). *J Hum Virol* **2**(1), 19-32.

Cannon, J. S., Ciufo, D., Hawkins, A. L., Griffin, C. A., Borowitz, M. J., Hayward, G. S., and Ambinder, R. F. (2000). A new primary effusion lymphoma-derived cell line yields a highly infectious Kaposi's sarcoma herpesvirus-containing supernatant. *J Virol* **74**(21), 10187-93.

Carbone, A., Cilia, A. M., Gloghini, A., Capello, D., Todesco, M., Quattrone, S., Volpe, R., and Gaidano, G. (1998). Establishment and characterization of EBV-positive and EBV-negative primary effusion lymphoma cell lines harbouring human herpesvirus type-8. *Br J Haematol* **102**(4), 1081-9.

Carbone, A., and Gaidano, G. (1997). HHV-8-positive body-cavity-based lymphoma: a novel lymphoma entity. *Br J Haematol* **97**(3), 515-22.

Carbone, A., Gloghini, A., Vaccher, E., Zagonel, V., Pastore, C., Dalla Palma, P., Branz, F., Saglio, G., Volpe, R., Tirelli, U., and Gaidano, G. (1996). Kaposi's sarcoma-associated herpesvirus DNA sequences in AIDS-related and AIDS-unrelated lymphomatous effusions. *Br J Haematol* **94**(3), 533-43.

Cesarman, E., Chang, Y., Moore, P. S., Said, J. W., and Knowles, D. M. (1995a). Kaposi's sarcoma-associated herpesvirus-like DNA sequences in AIDS-related body-cavity-based lymphomas. *N Engl J Med* **332**(18), 1186-91.

Cesarman, E., Moore, P. S., Rao, P. H., Inghirami, G., Knowles, D. M., and Chang, Y. (1995b). In vitro establishment and characterization of two acquired immunodeficiency syndrome-related lymphoma cell lines (BC-1 and BC-2) containing Kaposi's sarcoma-associated herpesvirus-like (KSHV) DNA sequences. *Blood* **86**(7), 2708-14.

Cesarman, E., Nador, R. G., Aozasa, K., Delsol, G., Said, J. W., and Knowles, D. M. (1996). Kaposi's sarcoma-associated herpesvirus in non-AIDS related lymphomas occurring in body cavities. *Am J Pathol* **149**(1), 53-7.

Chang, Y., Cesarman, E., Pessin, M. S., Lee, F., Culpepper, J., Knowles, D. M., and Moore, P. S. (1994a). Identification of herpesvirus-like DNA sequences in AIDS-associated Kaposi's sarcoma. *Science* **266**(5192), 1865-9.

Chang, Y., Cesarman, E., Pessin, M. S., Lee, F., Culpepper, J., Knowles, D. M., and Moore, P. S. (1994b). Identification of herpesvirus-like DNA sequences in AIDS-associated Kaposi's sarcoma [see comments]. *Science* **266**(5192), 1865-9.

Chang, Y., Moore, P. S., Talbot, S. J., Boshoff, C. H., Zarkowska, T., Godden, K., Paterson, H., Weiss, R. A., and Mittnacht, S. (1996). Cyclin encoded by KS herpesvirus. *Nature* **382**(6590), 410.

Chaudhary, P. M., Jasmin, A., Eby, M. T., and Hood, L. (1999). Modulation of the NF-kappa B pathway by virally encoded death effector domains-containing proteins. *Oncogene* **18**(42), 5738-46.

Chen, J., Ueda, K., Sakakibara, S., Okuno, T., Parravicini, C., Corbellino, M., and Yamanishi, K. (2001). Activation of latent Kaposi's sarcoma-associated herpesvirus by demethylation of the promoter of the lytic transactivator. *Proc Natl Acad Sci U S A* **98**(7), 4119-24.

Cheng, E. H., Nicholas, J., Bellows, D. S., Hayward, G. S., Guo, H. G., Reitz, M. S., and Hardwick, J. M. (1997). A Bcl-2 homolog encoded by Kaposi sarcoma-associated virus, human herpesvirus 8, inhibits apoptosis but does not heterodimerize with Bax or Bak. *Proc Natl Acad Sci U S A* **94**(2), 690-4.

Chinnaiyan, A. M., Tepper, C. G., Seldin, M. F., O'Rourke, K., Kischkel, F. C., Hellbardt, S., Krammer, P. H., Peter, M. E., and Dixit, V. M. (1996). FADD/MORT1 is a common mediator of CD95 (Fas/APO-1) and tumor necrosis factor receptor-induced apoptosis. *J Biol Chem* **271**(9), 4961-5.

Chou, J. J., Matsuo, H., Duan, H., and Wagner, G. (1998). Solution structure of the RAIDD CARD and model for CARD/CARD interaction in caspase-2 and caspase-9 recruitment. *Cell* **94**(2), 171-80.

Coqueret, O. (2002). Linking cyclins to transcriptional control. *Gene* **299**(1-2), 35-55.

Coscoy, L., and Ganem, D. (2000). Kaposi's sarcoma-associated herpesvirus encodes two proteins that block cell surface display of MHC class I chains by enhancing their endocytosis. *Proc Natl Acad Sci U S A* **97**(14), 8051-6.

Cotter, M. A., 2nd, and Robertson, E. S. (1999). The latency-associated nuclear antigen tethers the Kaposi's sarcoma-associated herpesvirus genome to host chromosomes in body cavity-based lymphoma cells. *Virology* **264**(2), 254-64.

Cotter, M. A., 2nd, Subramanian, C., and Robertson, E. S. (2001). The Kaposi's sarcoma-associated herpesvirus latency-associated nuclear antigen binds to specific sequences at the left end of the viral genome through its carboxy-terminus. *Virology* **291**(2), 241-59.

Davis, M. A., Sturzl, M. A., Blasig, C., Schreier, A., Guo, H. G., Reitz, M., Opalenik, S. R., and Browning, P. J. (1997). Expression of human herpesvirus 8-encoded cyclin D in Kaposi's sarcoma spindle cells. *J Natl Cancer Inst* **89**(24), 1868-74.

Delecluse, H. J., Hilsendegen, T., Pich, D., Zeidler, R., and Hammerschmidt, W. (1998). Propagation and recovery of intact, infectious Epstein-Barr virus from prokaryotic to human cells. *Proc Natl Acad Sci U S A* **95**(14), 8245-50.

Delecluse, H. J., Kost, M., Feederle, R., Wilson, L., and Hammerschmidt, W. (2001). Spontaneous activation of the lytic cycle in cells infected with a recombinant Kaposi's sarcoma-associated virus. *J Virol* **75**(6), 2921-8.

Delmer, A., Karmochkine, M., Le Tourneau, A., Bouscary, D., Rio, B., Perez, E., Prudent, J., Bonnet, P., Audouin, J., Diebold, J., and et al. (1990). [Castleman's disease (giant lymph node hyperplasia): clinical, biological and developing polymorphism. Apropos of 4 cases]. *Ann Med Interne (Paris)* **141**(2), 123-8.

Dittmer, D., Lagunoff, M., Renne, R., Staskus, K., Haase, A., and Ganem, D. (1998). A cluster of latently expressed genes in Kaposi's sarcoma-associated herpesvirus. *J Virol* **72**(10), 8309-15.

Dittmer, D., Stoddart, C., Renne, R., Linquist-Stepps, V., Moreno, M. E., Bare, C., McCune, J. M., and Ganem, D. (1999). Experimental transmission of Kaposi's sarcoma-associated herpesvirus (KSHV/HHV-8) to SCID-hu Thy/Liv mice. *J Exp Med* **190**(12), 1857-68.

Dittmer, D. P. (2003). Transcription profile of Kaposi's sarcoma-associated herpesvirus in primary Kaposi's sarcoma lesions as determined by real-time PCR arrays. *Cancer Res* **63**(9), 2010-5.

Djerbi, M., Screpanti, V., Catrina, A. I., Bogen, B., Biberfeld, P., and Grandien, A. (1999). The inhibitor of death receptor signaling, FLICE-inhibitory protein defines a new class of tumor progression factors. *J Exp Med* **190**(7), 1025-32.

Dupin, N., Fisher, C., Kellam, P., Ariad, S., Tulliez, M., Franck, N., van Marck, E., Salmon, D., Gorin, I., Escande, J. P., Weiss, R. A., Alitalo, K., and Boshoff, C. (1999). Distribution of human herpesvirus-8 latently infected cells in Kaposi's sarcoma, multicentric Castleman's disease, and primary effusion lymphoma. *Proc Natl Acad Sci U S A* **96**(8), 4546-51.

Fais, F., Gaidano, G., Capello, D., Gloghini, A., Ghiotto, F., Roncella, S., Carbone, A., Chiorazzi, N., and Ferrarini, M. (1999). Immunoglobulin V region gene use and structure suggest antigen selection in AIDS-related primary effusion lymphomas. *Leukemia* **13**(7), 1093-9.

Field, N., Low, W., Daniels, M., Howell, S., Daviet, L., Boshoff, C., and Collins, M. (2003). KSHV vFLIP binds to IKK-gamma to activate IKK. *J Cell Sci* **116**(Pt 18), 3721-8.

Fiorelli, V., Gendelman, R., Sirianni, M. C., Chang, H. K., Colombini, S., Markham, P. D., Monini, P., Sonnabend, J., Pintus, A., Gallo, R. C., and Ensoli, B. (1998). gamma-Interferon produced by CD8+ T cells infiltrating Kaposi's sarcoma induces spindle cells with angiogenic phenotype and synergy with human immunodeficiency virus-1 Tat protein: an immune response to human herpesvirus-8 infection? *Blood* **91**(3), 956-67.

Flemington, E. K., Speck, S. H., and Kaelin, W. G., Jr. (1993). E2F-1-mediated transactivation is inhibited by complex formation with the retinoblastoma susceptibility gene product. *Proc Natl Acad Sci U S A* **90**(15), 6914-8.

Flowers, C. C., Flowers, S. P., and Nabel, G. J. (1998). Kaposi's sarcoma-associated herpesvirus viral interferon regulatory factor confers resistance to the antiproliferative effect of interferon-alpha. *Mol Med* **4**(6), 402-12.

Foreman, K. E., Friborg, J., Jr., Kong, W. P., Woffendin, C., Polverini, P. J., Nickoloff, B. J., and Nabel, G. J. (1997a). Propagation of a human herpesvirus from AIDS-associated Kaposi's sarcoma. *N Engl J Med* **336**(3), 163-71.

Foreman, K. E., Friborg, J., Jr., Kong, W. P., Woffendin, C., Polverini, P. J., Nickoloff, B. J., and Nabel, G. J. (1997b). Propagation of a human herpesvirus from AIDS-associated Kaposi's sarcoma [see comments]. *N Engl J Med* **336**(3), 163-71.

Friborg, J., Jr., Kong, W., Hottiger, M. O., and Nabel, G. J. (1999). p53 inhibition by the LANA protein of KSHV protects against cell death. *Nature* **402**(6764), 889-94.

Friborg, J., Jr., Kong, W. P., Flowers, C. C., Flowers, S. L., Sun, Y., Foreman, K. E., Nickoloff, B. J., and Nabel, G. J. (1998). Distinct biology of Kaposi's sarcoma-associated herpesvirus from primary lesions and body cavity lymphomas. *J Virol* **72**(12), 10073-82.

Fujii, T., Taguchi, H., Katano, H., Mori, S., Nakamura, T., Nojiri, N., Nakajima, K., Tadokoro, K., Juji, T., and Iwamoto, A. (1999). Seroprevalence of human herpesvirus 8 in human immunodeficiency virus 1-positive and human immunodeficiency virus 1-negative populations in Japan. *J Med Virol* **57**(2), 159-62.

Fujimuro, M., and Hayward, S. D. (2003). The latency-associated nuclear antigen of Kaposi's sarcoma-associated herpesvirus manipulates the activity of glycogen synthase kinase-3beta. *J Virol* **77**(14), 8019-30.

Fujimuro, M., Wu, F. Y., ApRhys, C., Kajumbula, H., Young, D. B., Hayward, G. S., and Hayward, S. D. (2003). A novel viral mechanism for dysregulation of beta-catenin in Kaposi's sarcoma-associated herpesvirus latency. *Nat Med* **9**(3), 300-6.

Ganem, D. (1997). KSHV and Kaposi's sarcoma: the end of the beginning? *Cell* **91**(2), 157-60.

Gao, S. J., Boshoff, C., Jayachandra, S., Weiss, R. A., Chang, Y., and Moore, P. S. (1997). KSHV ORF K9 (vIRF) is an oncogene which inhibits the interferon signaling pathway. *Oncogene* **15**(16), 1979-85.

Gao, S. J., Kingsley, L., Hoover, D. R., Spira, T. J., Rinaldo, C. R., Saah, A., Phair, J., Detels, R., Parry, P., Chang, Y., and Moore, P. S. (1996a). Seroconversion to antibodies against Kaposi's sarcoma-associated herpesvirus-related latent nuclear antigens before the development of Kaposi's sarcoma. *N Engl J Med* **335**(4), 233-41.

Gao, S. J., Kingsley, L., Li, M., Zheng, W., Parravicini, C., Ziegler, J., Newton, R., Rinaldo, C. R., Saah, A., Phair, J., Detels, R., Chang, Y., and Moore, P. S. (1996b). KSHV antibodies among Americans, Italians and Ugandans with and without Kaposi's sarcoma. *Nat Med* **2**(8), 925-8.

Gao, S. J., Kingsley, L., Li, M., Zheng, W., Parravicini, C., Ziegler, J., Newton, R., Rinaldo, C. R., Saah, A., Phair, J., Detels, R., Chang, Y., and Moore, P. S. (1996c). KSHV antibodies among Americans, Italians and Ugandans with and without Kaposi's sarcoma [see comments]. *Nat Med* **2**(8), 925-8.

Gao, S. J., Zhang, Y. J., Deng, J. H., Rabkin, C. S., Flore, O., and Jenson, H. B. (1999). Molecular polymorphism of Kaposi's sarcoma-associated herpesvirus (Human herpesvirus 8) latent nuclear antigen: evidence for a large repertoire of viral genotypes and dual infection with different viral genotypes. *J Infect Dis* **180**(5), 1466-76.

Garber, A. C., Shu, M. A., Hu, J., and Renne, R. (2001). DNA binding and modulation of gene expression by the latency-associated nuclear antigen of Kaposi's sarcoma-associated herpesvirus. *J Virol* **75**(17), 7882-92.

Geng, Y., Eaton, E. N., Picon, M., Roberts, J. M., Lundberg, A. S., Gifford, A., Sardet, C., and Weinberg, R. A. (1996). Regulation of cyclin E transcription by E2Fs and retinoblastoma protein. *Oncogene* **12**(6), 1173-80.

Gradoville, L., Gerlach, J., Grogan, E., Shedd, D., Nikiforow, S., Metroka, C., and Miller, G. (2000). Kaposi's sarcoma-associated herpesvirus open reading frame 50/Rta protein activates the entire viral lytic cycle in the HH-B2 primary effusion lymphoma cell line. *J Virol* **74**(13), 6207-12.

Groves, A. K., Cotter, M. A., Subramanian, C., and Robertson, E. S. (2001). The latency-associated nuclear antigen encoded by Kaposi's sarcoma-associated herpesvirus activates two major essential Epstein-Barr virus latent promoters. *J Virol* **75**(19), 9446-57.

Grundhoff, A., and Ganem, D. (2001). Mechanisms governing expression of the v-FLIP gene of Kaposi's sarcoma-associated herpesvirus. *J Virol* **75**(4), 1857-63.

Grundhoff, A., and Ganem, D. (2003). The latency-associated nuclear antigen of Kaposi's sarcoma-associated herpesvirus permits replication of terminal repeat-containing plasmids. *J Virol* **77**(4), 2779-83.

Guillem, P., Karoui, M., Deroide, G., Herjean, M., Leteurtre, E., and Triboulet, J. P. (2000). [Mesenteric localization of Castleman's disease]. *Gastroenterol Clin Biol* **24**(1), 116-22.

Gura, T. (1997). How TRAIL kills cancer cells, but not normal cells. *Science* **277**(5327), 768.

Gwack, Y., Byun, H., Hwang, S., Lim, C., and Choe, J. (2001). CREB-binding protein and histone deacetylase regulate the transcriptional activity of Kaposi's sarcoma-associated herpesvirus open reading frame 50. *J Virol* **75**(4), 1909-17.

Harle-Bachor, C., and Boukamp, P. (1996). Telomerase activity in the regenerative basal layer of the epidermis inhuman skin and in immortal and carcinoma-derived skin keratinocytes. *Proc Natl Acad Sci U S A* **93**(13), 6476-81.

Hirt, B. (1967). Selective extraction of polyoma DNA from infected mouse cell cultures. *J Mol Biol* **26**(2), 365-9.

Hiyama, K., Hirai, Y., Kyoizumi, S., Akiyama, M., Hiyama, E., Piatyszek, M. A., Shay, J. W., Ishioka, S., and Yamakido, M. (1995). Activation of telomerase in human lymphocytes and hematopoietic progenitor cells. *J Immunol* **155**(8), 3711-5.

Hyun, T. S., Subramanian, C., Cotter, M. A., 2nd, Thomas, R. A., and Robertson, E. S. (2001). Latency-associated nuclear antigen encoded by Kaposi's sarcoma-associated herpesvirus interacts with Tat and activates the long terminal repeat of human immunodeficiency virus type 1 in human cells. *J Virol* **75**(18), 8761-71.

Ishido, S., Wang, C., Lee, B. S., Cohen, G. B., and Jung, J. U. (2000). Downregulation of major histocompatibility complex class I molecules by Kaposi's sarcoma-associated herpesvirus K3 and K5 proteins. *J Virol* **74**(11), 5300-9.

Jenner, R. G., and Boshoff, C. (2002). The molecular pathology of Kaposi's sarcoma-associated herpesvirus. *Biochim Biophys Acta* **1602**(1), 1-22.

Katano, H., Iwasaki, T., Baba, N., Terai, M., Mori, S., Iwamoto, A., Kurata, T., and Sata, T. (2000a). Identification of antigenic proteins encoded by human herpesvirus 8 and seroprevalence in the general population and among patients with and without Kaposi's sarcoma. *J Virol* **74**(8), 3478-85.

Katano, H., Sato, Y., Kurata, T., Mori, S., and Sata, T. (1999). High expression of HHV 8 encoded ORF73 protein in spindle-shaped cells of Kaposi's sarcoma. *Am J Pathol* **155**(1), 47-52.

Katano, H., Sato, Y., Kurata, T., Mori, S., and Sata, T. (2000b). Expression and localization of human herpesvirus 8-encoded proteins in primary effusion lymphoma, Kaposi's sarcoma, and multicentric Castleman's disease. *Virology* **269**(2), 335-44.

Kedes, D. H., Lagunoff, M., Renne, R., and Ganem, D. (1997). Identification of the gene encoding the major latency-associated nuclear antigen of the Kaposi's sarcoma-associated herpesvirus. *J Clin Invest* **100**(10), 2606-10.

Kedes, D. H., Operskalski, E., Busch, M., Kohn, R., Flood, J., and Ganem, D. (1996a). The seroepidemiology of human herpesvirus 8 (Kaposi's sarcoma- associated herpesvirus): distribution of infection in KS risk groups and evidence for sexual transmission [see comments] [published erratum appears in Nat Med 1996 Sep;2(9):1041]. *Nat Med* **2**(8), 918-24.

Kedes, D. H., Operskalski, E., Busch, M., Kohn, R., Flood, J., and Ganem, D. (1996b). The seroepidemiology of human herpesvirus 8 (Kaposi's sarcoma-associated herpesvirus): distribution of infection in KS risk groups and evidence for sexual transmission. *Nat Med* **2**(8), 918-24.

Kim, N. W., Piatyszek, M. A., Prowse, K. R., Harley, C. B., West, M. D., Ho, P. L., Coviello, G. M., Wright, W. E., Weinrich, S. L., and Shay, J. W. (1994). Specific association of human telomerase activity with immortal cells and cancer. *Science* **266**(5193), 2011-5.

Kim, T. K., and Maniatis, T. (1997). The mechanism of transcriptional synergy of an in vitro assembled interferon-beta enhanceosome. *Mol Cell* **1**(1), 119-29.

Kischkel, F. C., Hellbardt, S., Behrmann, I., Germer, M., Pawlita, M., Krammer, P. H., and Peter, M. E. (1995). Cytotoxicity-dependent APO-1 (Fas/CD95)-associated proteins form a death-inducing signaling complex (DISC) with the receptor. *Embo J* **14**(22), 5579-88.

Kliche, S., Nagel, W., Kremmer, E., Atzler, C., Ege, A., Knorr, T., Koszinowski, U., Kolanus, W., and Haas, J. (2001). Signaling by human herpesvirus 8 kaposin A through direct membrane recruitment of cytohesin-1. *Mol Cell* **7**(4), 833-43.

Knight, J. S., Cotter, M. A., 2nd, and Robertson, E. S. (2001). The latency-associated nuclear antigen of Kaposi's sarcoma-associated herpesvirus transactivates the telomerase reverse transcriptase promoter. *J Biol Chem* **276**(25), 22971-8.

Krek, W., Ewen, M. E., Shirodkar, S., Arany, Z., Kaelin, W. G., Jr., and Livingston, D. M. (1994). Negative regulation of the growth-promoting transcription factor E2F-1 by a stably bound cyclin A-dependent protein kinase. *Cell* **78**(1), 161-72.

Krithivas, A., Fujimuro, M., Weidner, M., Young, D. B., and Hayward, S. D. (2002). Protein interactions targeting the latency-associated nuclear antigen of Kaposi's sarcoma-associated herpesvirus to cell chromosomes. *J Virol* **76**(22), 11596-604.

Krithivas, A., Young, D. B., Liao, G., Greene, D., and Hayward, S. D. (2000). Human herpesvirus 8 LANA interacts with proteins of the mSin3 corepressor complex and negatively regulates Epstein-Barr virus gene expression in dually infected PEL cells. *J Virol* **74**(20), 9637-45.

Kyo, S., Takakura, M., Kohama, T., and Inoue, M. (1997). Telomerase activity in human endometrium. *Cancer Res* **57**(4), 610-4.

Lengyel, P. (1982). Biochemistry of interferons and their actions. *Annu Rev Biochem* **51**, 251-82.

Li, M., Lee, H., Guo, J., Neipel, F., Fleckenstein, B., Ozato, K., and Jung, J. U. (1998). Kaposi's sarcoma-associated herpesvirus viral interferon regulatory factor. *J Virol* **72**(7), 5433-40.

Lim, C., Gwack, Y., Hwang, S., Kim, S., and Choe, J. (2001). The transcriptional activity of cAMP response element-binding protein-binding protein is modulated by the latency associated nuclear antigen of Kaposi's sarcoma-associated herpesvirus. *J Biol Chem* **276**(33), 31016-22.

Lim, C., Lee, D., Seo, T., Choi, C., and Choe, J. (2003). Latency-associated nuclear antigen of Kaposi's sarcoma-associated herpesvirus functionally interacts with heterochromatin protein 1. *J Biol Chem* **278**(9), 7397-405.

Lim, C., Sohn, H., Gwack, Y., and Choe, J. (2000). Latency-associated nuclear antigen of Kaposi's sarcoma-associated herpesvirus (human herpesvirus-8) binds ATF4/CREB2 and inhibits its transcriptional activation activity. *J Gen Virol* **81**(Pt 11), 2645-52.

Lim, C., Sohn, H., Lee, D., Gwack, Y., and Choe, J. (2002). Functional dissection of latency-associated nuclear antigen 1 of Kaposi's sarcoma-associated herpesvirus involved in latent DNA replication and transcription of terminal repeats of the viral genome. *J Virol* **76**(20), 10320-31.

Lin, R., Heylbroeck, C., Pitha, P. M., and Hiscott, J. (1998). Virus-dependent phosphorylation of the IRF-3 transcription factor regulates nuclear translocation, transactivation potential, and proteasome-mediated degradation. *Mol Cell Biol* **18**(5), 2986-96.

Liu, L., Eby, M. T., Rathore, N., Sinha, S. K., Kumar, A., and Chaudhary, P. M. (2002). The human herpes virus 8-encoded viral FLICE inhibitory protein physically associates with and persistently activates the Ikappa B kinase complex. *J Biol Chem* **277**(16), 13745-51.

Lopez, S., Island, M. L., Drouin, J., Bandu, M. T., Christeff, N., Darracq, N., Barbey, R., Doly, J., Thomas, D., and Navarro, S. (2000). Repression of virus-induced interferon A promoters by homeodomain transcription factor Ptx1. *Mol Cell Biol* **20**(20), 7527-40.

Lukac, D. M., Kirshner, J. R., and Ganem, D. (1999). Transcriptional activation by the product of open reading frame 50 of Kaposi's sarcoma-associated herpesvirus is required for lytic viral reactivation in B cells. *J Virol* **73**(11), 9348-61.

Matolcsy, A., Nador, R. G., Cesarman, E., and Knowles, D. M. (1998). Immunoglobulin VH gene mutational analysis suggests that primary effusion lymphomas derive from different stages of B cell maturation. *Am J Pathol* **153**(5), 1609-14.

Matsushime, H., Ewen, M. E., Strom, D. K., Kato, J. Y., Hanks, S. K., Roussel, M. F., and Sherr, C. J. (1992). Identification and properties of an atypical catalytic subunit (p34PSK-J3/cdk4) for mammalian D type G1 cyclins. *Cell* **71**(2), 323-34.

Matta, H., Sun, Q., Moses, G., and Chaudhary, P. M. (2003). Molecular genetic analysis of human herpes virus 8-encoded viral FLICE inhibitory protein (vFLIP)-induced NF-kappa B activation. *J Biol Chem*.

Means, R. E., Ishido, S., Alvarez, X., and Jung, J. U. (2002). Multiple endocytic trafficking pathways of MHC class I molecules induced by a Herpesvirus protein. *Embo J* **21**(7), 1638-49.

Merika, M., Williams, A. J., Chen, G., Collins, T., and Thanos, D. (1998). Recruitment of CBP/p300 by the IFN beta enhanceosome is required for synergistic activation of transcription. *Mol Cell* **1**(2), 277-87.

Mesri, E. A., Cesarman, E., Arvanitakis, L., Rafii, S., Moore, M. A., Posnett, D. N., Knowles, D. M., and Asch, A. S. (1996). Human herpesvirus-8/Kaposi's sarcoma-associated herpesvirus is a new transmissible virus that infects B cells. *J Exp Med* **183**(5), 2385-90.

Messerle, M., Crnkovic, I., Hammerschmidt, W., Ziegler, H., and Koszinowski, U. H. (1997). Cloning and mutagenesis of a herpesvirus genome as an infectious bacterial artificial chromosome. *Proc Natl Acad Sci U S A* **94**(26), 14759-63.

Meyerson, M., Counter, C. M., Eaton, E. N., Ellisen, L. W., Steiner, P., Caddle, S. D., Ziaugra, L., Beijersbergen, R. L., Davidoff, M. J., Liu, Q., Bacchetti, S., Haber, D. A., and Weinberg, R. A. (1997). hEST2, the putative human telomerase catalytic subunit gene, is up-regulated in tumor cells and during immortalization. *Cell* **90**(4), 785-95.

Miller, G., Rigsby, M. O., Heston, L., Grogan, E., Sun, R., Metroka, C., Levy, J. A., Gao, S. J., Chang, Y., and Moore, P. (1996). Antibodies to butyrate-inducible antigens of Kaposi's sarcoma-associated herpesvirus in patients with HIV-1 infection. *N Engl J Med* **334**(20), 1292-7.

Monini, P., Carlini, F., Sturzl, M., Rimessi, P., Superti, F., Franco, M., Melucci-Vigo, G., Cafaro, A., Goletti, D., Sgadari, C., Butto, S., Leone, P., Chiozzini, C., Barresi, C., Tinari, A., Bonaccorsi, A., Capobianchi, M. R., Giuliani, M., di Carlo, A., Andrconi, M., Rezza, G., and Ensoli, B. (1999). Alpha interferon inhibits human herpesvirus 8 (HHV-8) reactivation in primary effusion lymphoma cells and reduces HHV-8 load in cultured peripheral blood mononuclear cells. *J Virol* **73**(5), 4029-41.

Montaner, S., Sodhi, A., Pece, S., Mesri, E. A., and Gutkind, J. S. (2001). The Kaposi's sarcoma-associated herpesvirus G protein-coupled receptor promotes endothelial cell survival through the activation of Akt/protein kinase B. *Cancer Res* **61**(6), 2641-8.

Moore, P. S., and Chang, Y. (1998a). Antiviral activity of tumor-suppressor pathways: clues from molecular piracy by KSHV. *Trends Genet* **14**(4), 144-50.

Moore, P. S., and Chang, Y. (1998b). Kaposi's sarcoma-associated herpesvirus-encoded oncogenes and oncogenesis. *J Natl Cancer Inst Monogr*(23), 65-71.

Moore, P. S., Kingsley, L. A., Holmberg, S. D., Spira, T., Gupta, P., Hoover, D. R., Parry, J. P., Conley, L. J., Jaffe, H. W., and Chang, Y. (1996). Kaposi's sarcoma-associated herpesvirus infection prior to onset of Kaposi's sarcoma. *Aids* **10**(2), 175-80.

Moses, A. V., Fish, K. N., Ruhl, R., Smith, P. P., Strussenberg, J. G., Zhu, L., Chandran, B., and Nelson, J. A. (1999). Long-term infection and transformation of dermal microvascular endothelial cells by human herpesvirus 8. *J Virol* **73**(8), 6892-902.

Munshi, N., Ganju, R. K., Avraham, S., Mesri, E. A., and Groopman, J. E. (1999). Kaposi's sarcoma-associated herpesvirus-encoded G protein-coupled receptor activation of c-jun amino-terminal kinase/stress-activated protein kinase and lyn kinase is mediated by related adhesion focal tyrosine kinase/proline-rich tyrosine kinase 2. *J Biol Chem* **274**(45), 31863-7.

Muralidhar, S., Veytsmann, G., Chandran, B., Ablashi, D., Doniger, J., and Rosenthal, L. J. (2000). Characterization of the human herpesvirus 8 (Kaposi's sarcoma-associated herpesvirus) oncogene, kaposin (ORF K12). *J Clin Virol* **16**(3), 203-13.

Murphy, P. M. (2001). Viral exploitation and subversion of the immune system through chemokine mimicry. *Nat Immunol* **2**(2), 116-22.

Nador, R. G., Cesarman, E., Chadburn, A., Dawson, D. B., Ansari, M. Q., Sald, J., and Knowles, D. M. (1996). Primary effusion lymphoma: a distinct clinicopathologic entity associated with the Kaposi's sarcoma-associated herpes virus. *Blood* **88**(2), 645-56.

Nakamura, H., Lu, M., Gwack, Y., Souvlis, J., Zeichner, S. L., and Jung, J. U. (2003). Global changes in Kaposi's sarcoma-associated virus gene expression patterns following expression of a tetracycline-inducible Rta transactivator. *J Virol* **77**(7), 4205-20.

Neipel, F., Albrecht, J. C., and Fleckenstein, B. (1997). Cell-homologous genes in the Kaposi's sarcoma-associated rhadinovirus human herpesvirus 8: determinants of its pathogenicity? *J Virol* **71**(6), 4187-92.

Norbury, C., and Nurse, P. (1992). Animal cell cycles and their control. *Annu Rev Biochem* **61**, 441-70.

Ohtani, K., DeGregori, J., and Nevins, J. R. (1995). Regulation of the cyclin E gene by transcription factor E2F1. *Proc Natl Acad Sci U S A* **92**(26), 12146-50.

Okano, H., Shiraki, K., Inoue, H., Kawakita, T., Yamanaka, T., Deguchi, M., Sugimoto, K., Sakai, T., Ohmori, S., Fujikawa, K., Murata, K., and Nakano, T. (2003). Cellular FLICE/caspase-8-inhibitory protein as a principal regulator of cell death and survival in human hepatocellular carcinoma. *Lab Invest* **83**(7), 1033-43.

Oksenhendler, E., Duarte, M., Soulier, J., Cacoub, P., Welker, Y., Cadranel, J., Cazals-Hatem, D., Autran, B., Clauvel, J. P., and Raphael, M. (1996). Multicentric Castleman's disease in HIV infection: a clinical and pathological study of 20 patients. *Aids* **10**(1), 61-7.

Parravicini, C., Chandran, B., Corbellino, M., Berti, E., Paulli, M., Moore, P. S., and Chang, Y. (2000). Differential viral protein expression in Kaposi's sarcoma-associated herpesvirus-infected diseases: Kaposi's sarcoma, primary effusion lymphoma, and multicentric Castleman's disease. *Am J Pathol* **156**(3), 743-9.

Paulose-Murphy, M., Ha, N. K., Xiang, C., Chen, Y., Gillim, L., Yarchoan, R., Meltzer, P., Bittner, M., Trent, J., and Zeichner, S. (2001). Transcription program of human herpesvirus 8 (kaposi's sarcoma-associated herpesvirus). *J Virol* **75**(10), 4843-53.

Piolot, T., Tramier, M., Coppey, M., Nicolas, J. C., and Marechal, V. (2001a). Close but distinct regions of human herpesvirus 8 latency-associated nuclear antigen 1 are responsible for nuclear targeting and binding to human mitotic chromosomes. *J Virol* **75**(8), 3948-59.

Piolot, T., Tramier, M., Coppey, M., Nicolas, J. C., and Marechal, V. (2001b). Close but distinct regions of human herpesvirus 8 latency-associated nuclear antigen 1 are responsible for nuclear targeting and binding to human mitotic chromosomes. *J Virol* **75**(8), 3948-59.

Platt, G. M., Simpson, G. R., Mittnacht, S., and Schulz, T. F. (1999). Latent nuclear antigen of Kaposi's sarcoma-associated herpesvirus interacts with RING3, a homolog of the Drosophila female sterile homeotic (fsh) gene. *J Virol* **73**(12), 9789-95.

Ponton, A., Clement, M. V., and Stamenkovic, I. (1996). The CD95 (APO-1/Fas) receptor activates NF-kappaB independently of its cytotoxic function. *J Biol Chem* **271**(15), 8991-5.

Radkov, S. A., Kellam, P., and Boshoff, C. (2000). The latent nuclear antigen of Kaposi sarcoma-associated herpesvirus targets the retinoblastoma-E2F pathway and with the oncogene Hras transforms primary rat cells. *Nat Med* **6**(10), 1121-7.

Rainbow, L., Platt, G. M., Simpson, G. R., Sarid, R., Gao, S. J., Stoiber, H., Herrington, C. S., Moore, P. S., and Schulz, T. F. (1997). The 222- to 234-kilodalton latent nuclear protein (LNA) of Kaposi's sarcoma-associated herpesvirus (human herpesvirus 8) is encoded by orf73 and is a component of the latency-associated nuclear antigen. *J Virol* **71**(8), 5915-21.

Regamey, N., Tamm, M., Wernli, M., Witschi, A., Thiel, G., Cathomas, G., and Erb, P. (1998). Transmission of human herpesvirus 8 infection from renal-transplant donors to recipients. *N Engl J Med* **339**(19), 1358-63.

Renne, R., Barry, C., Dittmer, D., Compitello, N., Brown, P. O., and Ganem, D. (2001). Modulation of cellular and viral gene expression by the latency-associated nuclear antigen of Kaposi's sarcoma-associated herpesvirus. *J Virol* **75**(1), 458-68.

Renne, R., Blackbourn, D., Whitby, D., Levy, J., and Ganem, D. (1998). Limited transmission of Kaposi's sarcoma-associated herpesvirus in cultured cells. *J Virol* **72**(6), 5182-8.

Renne, R., Zhong, W., Herndier, B., McGrath, M., Abbey, N., Kedes, D., and Ganem, D. (1996). Lytic growth of Kaposi's sarcoma-associated herpesvirus (human herpesvirus 8) in culture. *Nat Med* **2**(3), 342-6.

Rezza, G., Lennette, E. T., Giuliani, M., Pezzotti, P., Caprilli, F., Monini, P., Butto, S., Lodi, G., Di Carlo, A., Levy, J. A., and Ensoli, B. (1998). Prevalence and determinants of anti-lytic and anti-latent antibodies to human herpesvirus-8 among Italian individuals at risk of sexually and parenterally transmitted infections. *Int J Cancer* **77**(3), 361-5.

Rivas, C., Thlick, A. E., Parravicini, C., Moore, P. S., and Chang, Y. (2001). Kaposi's sarcoma-associated herpesvirus LANA2 is a B-cell-specific latent viral protein that inhibits p53. *J Virol* **75**(1), 429-38.

Russo, J. J., Bohenzky, R. A., Chien, M. C., Chen, J., Yan, M., Maddalena, D., Parry, J. P., Peruzzi, D., Edelman, I. S., Chang, Y., and Moore, P. S. (1996). Nucleotide sequence of the Kaposi sarcoma-associated herpesvirus (HHV8). *Proc Natl Acad Sci U S A* **93**(25), 14862-7.

Sadler, R., Wu, L., Forghani, B., Renne, R., Zhong, W., Herndier, B., and Ganem, D. (1999). A complex translational program generates multiple novel proteins from the latently expressed kaposin (K12) locus of Kaposi's sarcoma-associated herpesvirus. *J Virol* **73**(7), 5722-30.

Saeki, Y., Ichikawa, T., Saeki, A., Chiocca, E. A., Tobler, K., Ackermann, M., Breakefield, X. O., and Fraefel, C. (1998). Herpes simplex virus type 1 DNA amplified as bacterial artificial chromosome in Escherichia coli: rescue of replication-competent virus progeny and packaging of amplicon vectors. *Hum Gene Ther* **9**(18), 2787-94.

Sarid, R., Flore, O., Bohenzky, R. A., Chang, Y., and Moore, P. S. (1998). Transcription mapping of the Kaposi's sarcoma-associated herpesvirus (human herpesvirus 8) genome in a body cavity-based lymphoma cell line (BC-1). *J Virol* **72**(2), 1005-12.

Sarid, R., Sato, T., Bohenzky, R. A., Russo, J. J., and Chang, Y. (1997). Kaposi's sarcoma-associated herpesvirus encodes a functional bcl-2 homologue. *Nat Med* **3**(3), 293-8.

Schwam, D. R., Luciano, R. L., Mahajan, S. S., Wong, L., and Wilson, A. C. (2000). Carboxy terminus of human herpesvirus 8 latency-associated nuclear antigen mediates dimerization, transcriptional repression, and targeting to nuclear bodies. *J Virol* **74**(18), 8532-40.

Senderowicz, A. M. (2003). Small-molecule cyclin-dependent kinase modulators. *Oncogene* **22**(42), 6609-20.

Shay, J. W., and Bacchetti, S. (1997). A survey of telomerase activity in human cancer. *Eur J Cancer* **33**(5), 787-91.

Sherr, C. J. (1996). Cancer cell cycles. *Science* **274**(5293), 1672-7.

Shinohara, H., Fukushi, M., Higuchi, M., Oie, M., Hoshi, O., Ushiki, T., Hayashi, J., and Fujii, M. (2002). Chromosome Binding Site of Latency-Associated Nuclear Antigen of Kaposi's Sarcoma-Associated Herpesvirus Is Essential for Persistent Episome Maintenance and Is Functionally Replaced by Histone H1. *J Virol* **76**(24), 12917-24.

Skapek, S. X., Rhee, J., Spicer, D. B., and Lassar, A. B. (1995). Inhibition of myogenic differentiation in proliferating myoblasts by cyclin D1-dependent kinase. *Science* **267**(5200), 1022-4.

Staskus, K. A., Zhong, W., Gebhard, K., Herndier, B., Wang, H., Renne, R., Beneke, J., Pudney, J., Anderson, D. J., Ganem, D., and Haase, A. T. (1997). Kaposi's sarcoma-associated herpesvirus gene expression in endothelial (spindle) tumor cells. *J Virol* **71**(1), 715-9.

Stavropoulos, T. A., and Strathdee, C. A. (1998). An enhanced packaging system for helper-dependent herpes simplex virus vectors. *J Virol* **72**(9), 7137-43.

Sun, Q., Zachariah, S., and Chaudhary, P. M. (2003). The human herpes virus 8 encoded viral FLICE inhibitory protein (vFLIP) induces cellular transformation via NF-kappa B activation. *J Biol Chem*.

Swanton, C., Card, G. L., Mann, D., McDonald, N., and Jones, N. (1999). Overcoming inhibitions: subversion of CKI function by viral cyclins. *Trends Biochem Sci* **24**(3), 116-20.

Swanton, C., and Jones, N. (2001). Strategies in subversion: de-regulation of the mammalian cell cycle by viral gene products. *Int J Exp Pathol* **82**(1), 3-13.

Swanton, C., Mann, D. J., Fleckenstein, B., Neipel, F., Peters, G., and Jones, N. (1997). Herpes viral cyclin/Cdk6 complexes evade inhibition by CDK inhibitor proteins. *Nature* **390**(6656), 184-7.

Talbot, S. J., Weiss, R. A., Kellam, P., and Boshoff, C. (1999). Transcriptional analysis of human herpesvirus-8 open reading frames 71, 72, 73, K14, and 74 in a primary effusion lymphoma cell line. *Virology* **257**(1), 84-94.

Thelander, L., and Reichard, P. (1979). Reduction of ribonucleotides. *Annu Rev Biochem* **48,** 133-58.

Thome, M., Schneider, P., Hofmann, K., Fickenscher, H., Meinl, E., Neipel, F., Mattmann, C., Burns, K., Bodmer, J. L., Schroter, M., Scaffidi, C., Krammer, P. H., Peter, M. E., and Tschopp, J. (1997). Viral FLICE-inhibitory proteins (FLIPs) prevent apoptosis induced by death receptors. *Nature* **386**(6624)**,** 517-21.

Tomescu, C., Law, W. K., and Kedes, D. H. (2003). Surface downregulation of major histocompatibility complex class I, PE-CAM, and ICAM-1 following de novo infection of endothelial cells with Kaposi's sarcoma-associated herpesvirus. *J Virol* **77**(17)**,** 9669-84.

Tomkinson, B., Robertson, E., and Kieff, E. (1993). Epstein-Barr virus nuclear proteins EBNA-3A and EBNA-3C are essential for B-lymphocyte growth transformation. *J Virol* **67**(4)**,** 2014-25.

Tomkinson, B., Robertson, E., Yalamanchili, R., Longnecker, R., and Kieff, E. (1993). Epstein-Barr virus recombinants from overlapping cosmid fragments. *J Virol* **67**(12)**,** 7298-306.

Tomkowicz, B., Singh, S. P., Cartas, M., and Srinivasan, A. (2002). Human herpesvirus-8 encoded Kaposin: subcellular localization using immunofluorescence and biochemical approaches. *DNA Cell Biol* **21**(3)**,** 151-62.

Tortorella, D., Gewurz, B. E., Furman, M. H., Schust, D. J., and Ploegh, H. L. (2000). Viral subversion of the immune system. *Annu Rev Immunol* **18,** 861-926.

Vieira, J., O'Hearn, P., Kimball, L., Chandran, B., and Corey, L. (2001). Activation of Kaposi's sarcoma-associated herpesvirus (human herpesvirus 8) lytic replication by human cytomegalovirus. *J Virol* **75**(3)**,** 1378-86.

Wabinga, H. R., Parkin, D. M., Wabwire-Mangen, F., and Mugerwa, J. W. (1993). Cancer in Kampala, Uganda, in 1989-91: changes in incidence in the era of AIDS. *Int J Cancer* **54**(1)**,** 26-36.

Wagner, M., Jonjic, S., Koszinowski, U. H., and Messerle, M. (1999). Systematic excision of vector sequences from the BAC-cloned herpesvirus genome during virus reconstitution. *J Virol* **73**(8)**,** 7056-60.

Wajant, H., Haas, E., Schwenzer, R., Muhlenbeck, F., Kreuz, S., Schubert, G., Grell, M., Smith, C., and Scheurich, P. (2000). Inhibition of death receptor-mediated gene induction by a cycloheximide-sensitive factor occurs at the level of or upstream of Fas-associated death domain protein (FADD). *J Biol Chem* **275**(32)**,** 24357-66.

Wang, J., Lobito, A. A., Shen, F., Hornung, F., Winoto, A., and Lenardo, M. J. (2000). Inhibition of Fas-mediated apoptosis by the B cell antigen receptor through c-FLIP. *Eur J Immunol* **30**(1)**,** 155-63.

Wathelet, M. G., Lin, C. H., Parekh, B. S., Ronco, L. V., Howley, P. M., and Maniatis, T. (1998). Virus infection induces the assembly of coordinately activated transcription factors on the IFN-beta enhancer in vivo. *Mol Cell* **1**(4)**,** 507-18.

Weaver, B. K., Kumar, K. P., and Reich, N. C. (1998). Interferon regulatory factor 3 and CREB-binding protein/p300 are subunits of double-stranded RNA-activated transcription factor DRAF1. *Mol Cell Biol* **18**(3)**,** 1359-68.

Weintraub, S. J., Prater, C. A., and Dean, D. C. (1992). Retinoblastoma protein switches the E2F site from positive to negative element. *Nature* **358**(6383)**,** 259-61.

Willems, F., Amraoui, Z., Vanderheyde, N., Verhasselt, V., Aksoy, E., Scaffidi, C., Peter, M. E., Krammer, P. H., and Goldman, M. (2000). Expression of c-FLIP(L) and resistance to CD95-mediated apoptosis of monocyte-derived dendritic cells: inhibition by bisindolylmaleimide. *Blood* **95**(11), 3478-82.

Wit, F. W., Sol, C. J., Renwick, N., Roos, M. T., Pals, S. T., van Leeuwen, R., Goudsmit, J., and Reiss, P. (1998). Regression of AIDS-related Kaposi's sarcoma associated with clearance of human herpesvirus-8 from peripheral blood mononuclear cells following initiation of antiretroviral therapy. *Aids* **12**(2), 218-9.

Yoneyama, M., Suhara, W., Fukuhara, Y., Fukuda, M., Nishida, E., and Fujita, T. (1998). Direct triggering of the type I interferon system by virus infection: activation of a transcription factor complex containing IRF-3 and CBP/p300. *Embo J* **17**(4), 1087-95.

Zhang, H. S., Gavin, M., Dahiya, A., Postigo, A. A., Ma, D., Luo, R. X., Harbour, J. W., and Dean, D. C. (2000). Exit from G1 and S phase of the cell cycle is regulated by repressor complexes containing HDAC-Rb-hSWI/SNF and Rb-hSWI/SNF. *Cell* **101**(1), 79-89.

Zhong, W., and Ganem, D. (1997). Characterization of ribonucleoprotein complexes containing an abundant polyadenylated nuclear RNA encoded by Kaposi's sarcoma-associated herpesvirus (human herpesvirus 8). *J Virol* **71**(2), 1207-12.

Zhou, F. C., Zhang, Y. J., Deng, J. H., Wang, X. P., Pan, H. Y., Hettler, E., and Gao, S. J. (2002). Efficient infection by a recombinant Kaposi's sarcoma-associated herpesvirus cloned in a bacterial artificial chromosome: application for genetic analysis. *J Virol* **76**(12), 6185-96.

Zimring, J. C., Goodbourn, S., and Offermann, M. K. (1998). Human herpesvirus 8 encodes an interferon regulatory factor (IRF) homolog that represses IRF-1-mediated transcription. *J Virol* **72**(1), 701-7.

In: Progress in Virus Research
Editor: Sabine Thebault, pp. 267-299
ISBN 1-59454-438-7

Chapter IX

Retroviral Insertional Mutagenesis and its Application to Genetically Engineered Mice

Linda Wolff[*]
Laboratory of Cellular Oncology, Center for Cancer Research,
National Cancer Institute, National Institutes of Health

Abstract

Retroviruses are used as leukemogenic agents because they can activate genes to be become transforming or inactivate tumor suppressors and, importantly, the cellular genes they alter can be easily identified. During the retrovirus' life cycle, viral DNA integrates into the host DNA by a mechanism that is essentially random. If the mutagenic events contribute to an oncogenic phenotype, than cells carrying such insertions have a selective advantage for growth and/or survival, leading to clonal outgrowth and eventually leukemia. Retrovirus-induced leukemias usually have multiple integrated viruses (proviruses) with more than one contributing to the multistep disease process. Common mechanisms of oncogene activation include promoter insertion, enhancer insertion, and destabilization of mRNA. The retroviral sequences at the site of proviral insertion provide molecular tags to identify genetic loci. Although retroviral tagging has been carried out for nearly two decades, the methodology has recently expanded to include large-scale identification of cancer genes. The recent exploitation of this gene discovery method became possible with several new technical advances in PCR based strategies for cloning virus integration sites, sequencing, databases of sequences from the mouse genome, and informatics. Initial reports on retroviral tagging predominantly focused on loci involved in lymphomagenesis, but new ones include loci associated with myeloid leukemia. Retroviruses are currently being used to mutagenize the genome of mice that are already genetically engineered to be predisposed to cancer, for example transgenic mice and those with targeted deletions. Future experimentation in these cancer

[*] Correspondence: NIH, Building 37, Room 4124, 37 Convent Drive MSC 4263, Bethesda, MD 20892-4263 Telephone 301 496-6763 Fax 301 594-3996 E-mail Lwolff@helix.nih.gov

"sensitized" mice will surely provide a wealth of information regarding genes that cooperate in causing cancer and molecular cancer pathways.

Introduction

Retroviruses were discovered during the early 1900's in association with cancer in chickens and were later shown to be pathogenic in mice and other mammals. An incredible volume of research on retroviruses has contributed to the foundation of our current knowledge of neoplastic disease. Although some pathogenic viruses contain transduced cellular oncogenes that are primarily responsible for the observed pathogenic effects, other cancer-inducing retroviruses do not carry genes beyond those required for replication. It should be noted that most tumors induced by retroviruses without oncogenes have been of mammary tissue derivation or hematopoietic cell origin.

This review will focus on the retroviruses that do not contain a transduced cellular oncogene and their use in the identification of novel cancer genes in murine leukemia. The murine leukemia viruses (MuLVs) cause mutations that can promote oncogenesis and this capacity is linked to the unique molecular events associated with their replication. During the retrovirus' life cycle, viral DNA integrates into the cellular genomic DNA by a mechanism that is essentially random. This can result in insertional mutagenic events that are selected during clonal outgrowth of the neoplasm if the mutagenic events contribute to the oncogenic phenotype. This is often referred to as retroviral tagging, because the integrated virus in the genome can be detected through its specific sequence leading to the discovery of potentially oncogenic neighboring genes. Although retroviruses are capable of infecting a variety of cell types in vivo, it is not uncommon for a single hematopoietic lineage to be affected. This demonstrates the cellular specificity of the oncogenic sequence of events. Analysis of insertional mutagenesis can help unravel unique neoplastic steps for each type of leukemia.

In the last few years there has been a renewed interest in retroviral insertional mutagenesis methodology due to advances in molecular biology that facilitate its application. As a result, powerful large-scale and rapid identification of leukemia disease genes is leading to predictions of cancer pathways. Furthermore, the application of insertional mutagenesis to genetically engineered mice is beginning to flourish, because it can assist in the demonstration that a gene is a tumor suppressor or an oncogene by providing the required additional hits. Although for some disease sensitized mice, leukemia can be demonstrated immediately following introduction or ablation of a gene, there are other examples in which the gene cannot be shown to be a susceptibility gene without generating collaborating genetic events in the mouse. In other examples, the percentage of mice that develop tumors may be small or the latency quite long. Retroviral insertional mutagenesis can, therefore, accelerate disease and help dissect the pathways of cancer initiation and progression.

Insertional mutagenesis can be considered a direct mechanism by which a retrovirus causes cancer and is the central theme of the review. However, one should also be aware that the virus may contribute to cancer through indirect mechanisms as well. These indirect mechanisms have been studied mainly in mouse models and include 1) immune suppression where the virus infection impairs the host immune system causing neoplastic cells, that would otherwise be eliminated, progress to leukemia, 2) stimulation of proliferation through the regeneration of damaged tissue, resulting in release of cytokines and other soluble mediators and 3) generation of recombinant viruses that include endogenous virus sequences and which can alter the cellular tropism of the virus and increase infectivity.

This chapter will provide a description of the first studies of insertional mutagenesis in genetically modified mice. To set the stage for this, I will first describe the mechanism of retroviral replication and review the recent investigations on the specificity of viral integration. I will also describe models of insertional mutagenesis in wild-type mice that are not predisposed to cancer and recent approaches to high throughput screens. This is a powerful technique that will have even more extensive applications in the future and continue to reveal genes that mediate novel and important aspects of cell growth.

The Retrovirus Replication Cycle

Retroviruses contain two identical single-stranded positive sense RNA molecules within a nucleocapsid. Surrounding the capsid is a cell-derived lipid layer containing virus envelope glycosylated surface (SU) protein and envelope transmembrane (TM) protein encoded by the virus *env* gene (Figure 1). Co-packaged with the RNA molecules are two enzymes encoded by the viral gene *pol* that are important for replication, reverse transcriptase (RT) and integrase protein (IN). Internal nonglycosylated structural proteins include the matrix (MA), capsid (CA), and the nucleocapsid (NC) proteins, which are all encoded by the *gag* region of the viral genome (Figure 1). A protease (PR) is also packaged and is encoded by the *pro* gene. For a more detailed description of the virus particle and the replication cycle described below see (Coffin, 1996; Coffin et al., 1997a).

Retroviruses enter the cell through the attachment of their surface glycoproteins to specific plasma membrane receptors. The MuLVs most commonly used for investigations involving insertional mutagenesis are subdivided into two host range groups: ecotropic and amphotropic viruses. The receptors for the ecotropic virus group are found on cells of mouse or rat origin only, whereas the receptors for amphotropic viruses are found on both rodent cells and cells of other mammalian species including man. The ecotropic virus receptor was determined to be a sodium-independent transporter for cationic amino acids and the amphotropic virus receptor is a sodium-dependent phosphate transporter.

After attachment and entry, reverse transcription of viral RNA into DNA occurs in the cytoplasm within the viral particle. This process requires RT and produces a double stranded molecule that is colinear with the RNA template. In addition, it contains terminal duplications called long terminal repeats (LTRs) (see Figure 1). The extra sequences contain cis-acting sequences required for subsequent transcription of the integrated virus and it is believed that they are formed as a result of two template switches. An uncommon feature of retrovirus replication is that it facilitates recombination. This is because template switching takes place between the two copies of RNA present in the virion and is most apparent when two different types of viral genomes become packaged together. For many of these viruses at least one homologous recombination occurs for every replication cycle. This is 100 to 1000-fold higher than for non-homologous recombination.

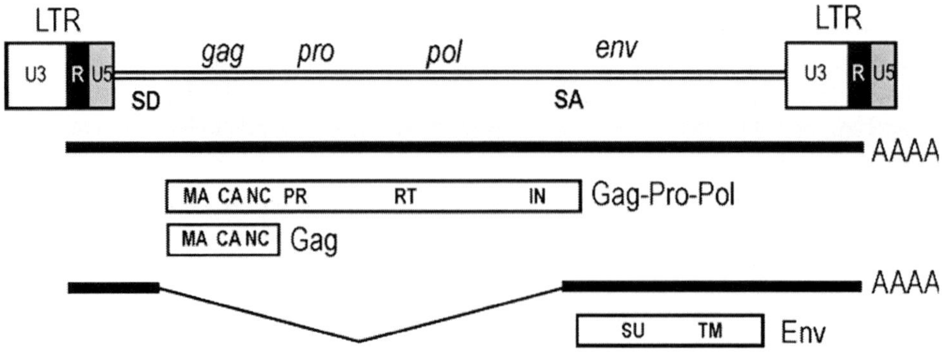

Figure 1. Structure of integrated provirus DNA, expression of messenger RNAs, and proteins. The integrated provirus has two long terminal repeats (LTRs). Transcription begins in the repeat region (R) of the 5' LTR and terminates in the R region of the 3' LTR. Promoter and enhancer elements that drive transcription are found in the U3 region. Two major transcripts are produced, a full size genomic transcript which encodes Gag and the polyprotein Gag-Pro-Pol and a subgenomic spliced message that encodes envelop polyprotein. SD, splice donor; SA, splice acceptor; MA, matrix protein; CA, capsid protein; NC, nucleocapsid protein; PR, protease, RT, reverse transcriptase; SU, envelope surface protein; TM, envelope transmembrane protein.

Integration of viral double stranded DNA into the genomic DNA is the next step in the replication cycle. This feature of the retrovirus is unique among viruses in general and can be clearly distinquished from the aberrant type "integration" observed for some DNA virus infections. Integration facilitates transcription of the viral DNA and provides stable maintenance of the genome because it is replicated along with the cellular genome of dividing cells. Since, for MuLVs, the integration process itself requires cell division, it has been suggested that the viral DNA can only enter the nucleus following nuclear membrane breakdown. The integrated provirus structure has consistent characteristics. The DNA is colinear with the viral RNA and both the viral DNA and cellular genomic DNA have sequence alterations as a result of the integration process. The viral DNA undergoes a loss of 2 bases, usually TT, at its ends and the cellular sequence adjoining the viral DNA (4 to 6 base pairs) is always duplicated. Finally, all integrated viral DNAs have the dinucleotide CA present at the 3' end.

The formation of new retrovirus particles begins after the viral DNA is stably integrated where it acts as a template for transcription of viral RNA. The cell provides the transcriptional and mRNA processing machinery, whereas the LTR contains the required cis-acting elements. There are two RNA transcripts produced, a full genomic message that is packaged into virions and is used to synthesize the Gag protein and the Gag-Pro-Pol fusion protein and subgenomic spliced message used to synthesize envelop proteins (see Figure 1). During translation, a high level of Gag production compared to that of Pro and Pol is accomplished by readthrough suppression at an UAG at the end of the Gag sequences. The Gag and Gag-Pro-Pol proteins mediate budding at the cytoplasmic side of the cellular surface membrane. The RNA is incorporated into the particle and the envelope proteins, encoded by *env*, move into the membrane in the area of budding. Very late or immediately after budding, Gag and Gag-Pro-Pol are processed by the protease PR to generate the individual proteins.

Integration Mechanisms and its Specificity

The integration process begins in the cytoplasm after the DNA is synthesized. The linear viral DNA is bound to what is called a preintegration complex made up of a subset of virion core proteins and most likely cellular proteins as well. While in the cytoplasm, the double stranded DNA of MuLVs is cleaved by viral integrase removing the terminal two bases from each 3' end. Following entry of the viral nucleoprotein complex into the nucleus by an unknown mechanism, the viral DNA binds to host DNA. In a DNA joining step, sometimes called the strand-transfer step, integrase joins the ends of the viral DNA to the target DNA in a direct transesterification reaction. In vitro reconstitution experiments begun in the late 1980's facilitated investigations on the mechanisms of retroviral integration providing much of the information that we have today concerning the these initial stages of integration (Brown et al., 1987;Brown et al., 1989;Bowerman et al., 1989). DNA gaps are repaired by cellular enzymes. It was discovered that the efficiency of stable transduction of retroviral vectors is reduced in cells that are deficient in DNA-dependent protein kinase (DNA-PK) and other components of the nonhomologous end-joining pathway (Daniel et al., 1999), and recently evidence was provided that the retroviral DNA integration process triggers an ATR-dependent DNA damage response similar to that observed during induction of cell cycle checkpoints (Daniel et al., 2003).

It is evident that an enormous number of sites in the genome are susceptible to integration by the virus. However, it seems unlikely that site selection would be totally random and therefore, there have been numerous studies aimed at trying to determine if there is a preference of the virus for specific regions and even for regional hotspots. Many of the early studies have been reviewed by (Rynditch et al., 1998) and the overall initial conclusion was that retroviruses target open chromatin regions or those that are transcribed. This fit well with data that showed that Moloney (Mo)-MuLV preferentially integrated into regions in which there are DNase I-hypersensitivity sites and integration in GC rich areas implicating either the 5'or the 3' end of genes as favored targets of integration (Rohdewohld et al., 1987;Scherdin et al., 1990). Subsequently, it was shown that retroviruses prefer to integrate into nucleosomal rather than nucleosome-free DNA (Pryciak and Varmus, 1992). In fact for HIV it was shown that the integrase favors DNA located in areas of maximum nucleosome distortion (Pruss et al., 1994). On the otherhand integration was shown to be inhibited by site specific DNA binding proteins (Pryciak and Varmus, 1992).

Very recently, a high-throughput method to examine integration sites in MuLV and HIV-expressing cells provided informative results (Wu et al., 2003). For Mo-MuLV, 903 integration sites were mapped and characterized and for HIV, 379 sites. An important finding was that the two viruses had their individual preferences. The MuLV integration sites were evenly distributed both upstream and downstream of transcriptional start sites and tended to integrate in more actively transcribed genes. On the other hand, HIV integration sites were found to be anywhere within the transcriptional unit but not upstream of the transcriptional start. The fact that the viruses had there own preferences suggests that there is interaction of the virus integration complex with specific cellular cofactors. Recent studies on other integrating elements, such as retrotransposons in yeast have shown that each of these elements has very specific targeting and in the case of Ty3 the mechanism of integration involves local tethering to the TFIIIB component of the PolIII transcription factor (Kirchner et al., 1995) . It would not be surprising if it is found that the mechanism of retroviral integration also involves some sort of tethering. In support of this, when retroviral integrase was fused to sequence-specific DNA binding domains, the fused proteins directed integration into the target sites that were the appropriate recognition sequences (Bushman, 2002).

Retroviruses in Mice and Early Investigations on Insertional Mutagenesis

The study of "RNA tumor viruses" and the molecular cloning of the virus components from complex mixtures in tumors emerged in parallel with the discoveries of oncogenes. The components that were responsible for rapid transformation in vivo and in vitro were those viruses that had incorporated sequences from cellular proto-oncogenes into their genomes. The first RNA tumor virus, avian erythroblastosis virus (AEV), was discovered at the turn of the last century. Later, more viruses, both avian and murine, were discovered which caused erytholeukemias, lymphoid and myeloid leukemias, sarcomas and mammary tumors (for a review see Rosenberg and Jolicoeur, 1997). In the mouse, retroviruses carrying oncogenes were always replication defective, because the transduced cellular proto-oncogenes replaced some of the viral sequences required for replication. Examples of these oncogene carrying mouse viruses were Harvey sarcoma virus (Ha-MSV) (Harvey, 1964), Moloney sarcoma virus (MSV)(Moloney, 1966), FBJ-MSV (Finkel et al., 1966) and Abelson murine leukemia virus (Ab-MuLV) (Abelson and Rabstein, 1970). These contained the oncogenes *ras*, *src*, *fos*, and *abl*, respectively, which were transduced by the virus from cellular counterparts.

The complex mixtures of viruses known as "RNA tumor viruses" also contained replication competent retroviruses that did not include an oncogene and provided the necessary replication functions missing in the oncogene-containing viruses. These viruses also induced leukemia on their own. Some of the earliest replication competent retroviruses that were cloned from these mixtures and sequenced were Mo-MuLV (Moloney, 1960;Shinnick et al., 1981) and Friend (Fr)-MuLV (Friend, 1957;Troxler and Scolnick, 1978;Oliff et al., 1980;Perryman et al., 1991). Mo-MuLV induces primarily T-cell lymphomas when inoculated into newborn mice, while Fr-MuLV induces predominantly erythroleukemia. Fr-MuLV can induce myeloid leukemia, as well, in certain strains of mice after a long latency (Shibuya and Mak, 1982;Silver and Fredrickson, 1983;Chesebro et al., 1983;Sola et al., 1985).

It later became apparent that the major mechanism by which the replication competent viruses induce leukemia is through the process of insertional mutagenesis. The pioneering study was carried out by William Hayward and colleagues in 1981(Hayward et al., 1981). They discovered that activation of the cellular proto-oncogene gene c-*myc* by promoter insertion occurred in more than 80% of avian leukosis virus (ALV)-induced bursal lymphomas. The 3' LTR of ALV replaced the function of the cellular promoter in these neoplasms providing constitutive c-*myc* expression. Since this gene activation was observed to be a common event among tumors and it was known that uncontrolled expression of the oncogenic transcription factor c-Myc could provide a proliferative advantage, it was presumed that the rare cells in which this activation occurred expanded preferentially.

Other replication competent viruses were derived from mouse strains such as AKR, C58 and HRS which spontaneously release leukemogenic virus (Kozak C.A. and Ruscetti, 1992). The origin of the first mouse retrovirus, Gross passage A MuLV was actually a spontaneous AKR lymphoma (Gross, 1970;Villemur et al., 1983). Similarly SL3-3, was later isolated from a spontaneous AKR lymphoma (Lenz et al., 1984). Both viruses induce T-cell lymphomas. Gross passage A MuLV and probably SL3-3 virus, are evolutionarily related to the non-leukemogenic endogenous Akv virus. Other strains may have endogenous viruses, that are not expressed and some strains such as BALB/c produce retrovirus only late in life.

Additional leukemogenic viruses were isolated from wild mice in California and include ecotropic Cas-Br-M virus and amphotropic retrovirus 4070A (Gardner et al., 1973) (Hartley and Rowe, 1976). Cas-Br- M, which only replicates in mouse cells, induces a wide spectrum of hematopoietic neoplasms, including T and B cell lymphomas, myelogenous leukemias, and erythroleukemias (Fredrickson et al., 1984;Holmes et al., 1986). Amphotropic retrovirus 4070A, capable of replicating not only mouse cells, but also in rabbit, mink, human guinea pig, and rat cells, is relatively inefficient in inducing disease in newborn mice. It causes T-cell neoplasms and at a very low incidence, myeloid–like tumors and B-cell tumors (Ott et al., 1992). Interestingly, however, in the presence of an acute inflammation induced by pristane, 4070A can induce a high incidence myeloid (monocytic) tumors (Wolff et al., 1991). Replication competent Graffi MuLV was molecularly cloned relatively recently from a virus stock that induces a high incidence of myeloid leukemia (Graffi, 1957;Graffi et al., 1966;Ru et al., 1993).

A few inbred recombinant mouse strains that have been shown to produce high titers of ecotropic virus develop leukemias spontaneously. These are useful leukemia models and have been used extensively to investigate the consequences of insertional mutagenesis. Mice from 21 of 23 AKXD recombinant inbred (RI) mouse strains develop hematopoietic neoplasms at high frequency (Mucenski et al., 1986;Mucenski et al., 1987). These recombinant inbred mice were derived from crosses of AKR/J and DBA/2J mice, strains with a high and low leukemia incidence, respectively, followed by systematic inbreeding. Mice from most of these RI strains succumbed to B-cell lymphomas or T-cell lymphomas or both. However, there is one RI strain which develops primarily myeloid leukemia. (Bedigian et al., 1981;Bedigian et al., 1981;Jenkins et al., 1982;Bedigian et al., 1984). Another RI strain, BXH2, developed from crossing C57BL/6J and C3H/HeJ mice, develops myeloid leukemia at a high frequency. A virus is expressed throughout life in these mice and is presumed to have arisen by recombination between two defective endogenous retroviruses during breeding of the recombinant inbred strain. Myeloblastic or myelomonocytic leukemias that develop in these mice have on average three to four somatically acquired, clonal tumor-specific proviral insertions. A very recent study helps to explain why only 1 of 12 RI BXH strains develop myeloid leukemia (Turcotte et al., 2004). Apparently a mutation in a single locus, *Myls,* was fixed in the BXH2 line and is in part responsible for development of a myeloproliferative syndrome in these mice. This proliferative syndrome is thought to be a predisposing factor for the development of myeloid leukemia. The BXH2 model has been particularly powerful for identification of acute myeloid leukemia genes at retrovirus integration sites and was one of the first in which high-throughput approaches for identification of genes were applied (see below).

Viral and Host Genes that Influence Infection

When choosing the appropriate retroviruses and hosts for tagging oncogenes or tumor suppressors, it is important to consider the influence of viral and cellular genes on the ability of a virus to replicate and be pathogenic in different murine strains and in different tissues.

The murine host has genes that influence the ability of the virus to replicate and may affect spread of virus by altering the immune response to the virus-infected cells (Coffin et al 1997b). The immune response to MuLVs determines the length of productive infection in the mouse. The major histocompatibility complex (MHC) class I and II alleles, as well as a non-MHC gene, Rfv-3 which controls virus specific responses, are involved in the host overcoming infection. MuLVs are most effective in inducing leukemia if they are inoculated into newborn mice because virus replication is maximized in the immunologically immature host (Gisselbrecht et al., 1978;Hasenkrug and Chesebro, 1997).

The *Fv1* gene, which restricts virus replication, is particularly relevant to insertional mutagenesis experiments, because some of the virus types cannot infect all the commonly used mouse strains. The *Fv1* gene affects virus replication at a step following entry into the cell, but before virus integration. The gene was identified following analysis of mouse strains for susceptibility to Fr-MuLV-induced erthroleukemia (Lilly, 1967;Odaka, 1969) and was subsequently found to affect susceptibility to natural variants of MuLV that were described by others as being either N-or B-tropic (Pincus et al., 1971;Pincus et al., 1971). Ecotropic viruses Mo-MuLV and Fr-MuLV are NB tropic, ecotropic Cas-Br-M is N-tropic , and amphotropic retrovirus 4070A is N-tropic (Hartley and Rowe, 1976). The gene has two principal alleles, $Fv1^n$ and $Fv1^b$, which are codominant. Inbred mice with the $Fv1^n$ allele, including NIH, DBA/2N,CBA/J, C3H/HeN, C57L/J, 129/J, NZB/N, and C57BR/J, are permissive for N-tropic viruses, whereas those mice with the $Fv1^b$ allele, including BALB/c, A/J/C57BL/6 and B10.BR/J, are permissive for B-tropic viruses (Pincus et al., 1971). FVB mice commonly used for preparing transgenic mice are $Fv1^{bb}$ (Taketo et al., 1991). Mice heterozygous for the *Fv1* allele are resistant to replication of both N- and B-tropic viruses. The exact mechanism of resistance is not known. Physical mapping of the sequences for *Fv1* tropism on MuLVs localized them to the gene encoding the viral capsid and specifically to two amino acids in the viral capsid protein. It has been thought that the resistance occurs as a result of a direct interaction between the *Fv1* gene product and the viral capsid protein (DesGroseillers and Jolicoeur, 1983;Taylor and Stoye, 2004). The *Fv1* gene encodes a protein that is localized to the *trans*-Golgi network (Best et al., 1996;Yap and Stoye, 2003). It has sequence homology to the endogenous retroviruses of the HERV-L and MuERV-L families (Benit et al., 1997) and an examination of its position within the *gag* gene indicates that it translates into a capsid-like protein. Although the carboxy-terminal third of the protein carries the *Fv1* specificity, both the carboxy-terminus and amino-terminus are required for functioning (Bishop et al., 2001). It has been proposed that the Fv1 protein behaves in a dominant negative manner, interfering with MuLV capsid function (Goff, 1996).

Mechanisms of Retroviral Insertional Mutagenesis that Activate the Transforming Capacity of Cellular Genes

There are several mechanisms by which retroviruses increase the oncogenic potential of cellular oncogenes when they integrate into host DNA. All of these modes of activation require the transcriptional regulatory region of the virus in the long terminal repeat and result in elevated levels of mRNA from the activated host gene. Sometimes a truncated protein product results from the integration, but this is relatively rare (For a recent review see (Mikkers and Berns, 2003).

Promoter insertion can occur in more than one way (See Figure 2A). For example, in the activation of the c-*myc* gene, as observed in ALV induced bursal lymphomas, transcription begins in the 3' LTR generating a hybrid RNA containing both viral and cellular sequences. Often deletions of the 5' end of the virus that include the 5' LTR are observed with this type of activation. It is assumed that the reason for this is that transcription from the 5' end LTR could interfere with transcription from the 3' end (Westaway et al., 1984;Cullen et al., 1984). Several other genes such as *Bmi1*, *Evi1*, *Fli1*, *His1*,*Lck*, and *Ha-ras* have been activated by murine retroviruses using a 3' promoter insertion mechanism (as reviewed in Rosenberg and Jolicoeur, 1997).

Retroviruses can also use the 5' LTR to activate a gene by promoter insertion. Examples in hematopoietic diseases induced by murine retroviruses include *mos*, c-*myc*, c-*myb*, *Gfi1*, and the prolactin receptor (*Prl*r) (as reviewed in (Rosenberg and Jolicoeur, 1997)). Generally read-through transcripts are produced and are spliced using the splice donor (SD) or a cryptic SD from the virus and the next available acceptor site in the host gene. If the virus integrates upstream of the gene or in the first intron of a gene that contains a noncoding first exon, then there will be no protein truncation. However, if the virus integrates in one the first introns, downstream of coding exons, as is common in the activation of *c-myb*, then the amino-terminus of the protein product will be truncated (Shen-Ong and Wolff, 1987;Mukhopadhyaya and Wolff, 1992).

The most common mechanism of activation of cellular oncogenes is thought to be enhancer activation where the gene is under the control of strong enhancer sequences found in the viral LTR (see Figure 2B). The proviruses can be upstream of the gene in the opposite orientation of the gene or downstream of the gene in either orientation. Importantly, enhancer activation can occur over long distances. Examples include integration into the *Mlvi4* and *Mlvi1* loci, which causes transcriptional activation of *c-myc*. These integrations are 30 and 270 kb, respectively, 3' to the *c-myc* promoter (Lazo et al., 1990). Another example is integration into *CB-1*/Fim,3 which activates *Evi1* at a distance of 90 kb (Bartholomew and Ihle, 1991). Integration of a retrovirus in the 3' end of the gene can also result in altered messenger RNA stability. Many RNAs have short half-lives due to destabilizing sequences in their 3' untranslated regions. Therefore, if the virus integrates upstream of these sequences it can result in stabilization of the RNA and, consequently, increased expression. This has been observed for integrations into the transcriptional units of *Pim1* and *c-myb* (Selten et al., 1985;Bies and Wolff, 2004).

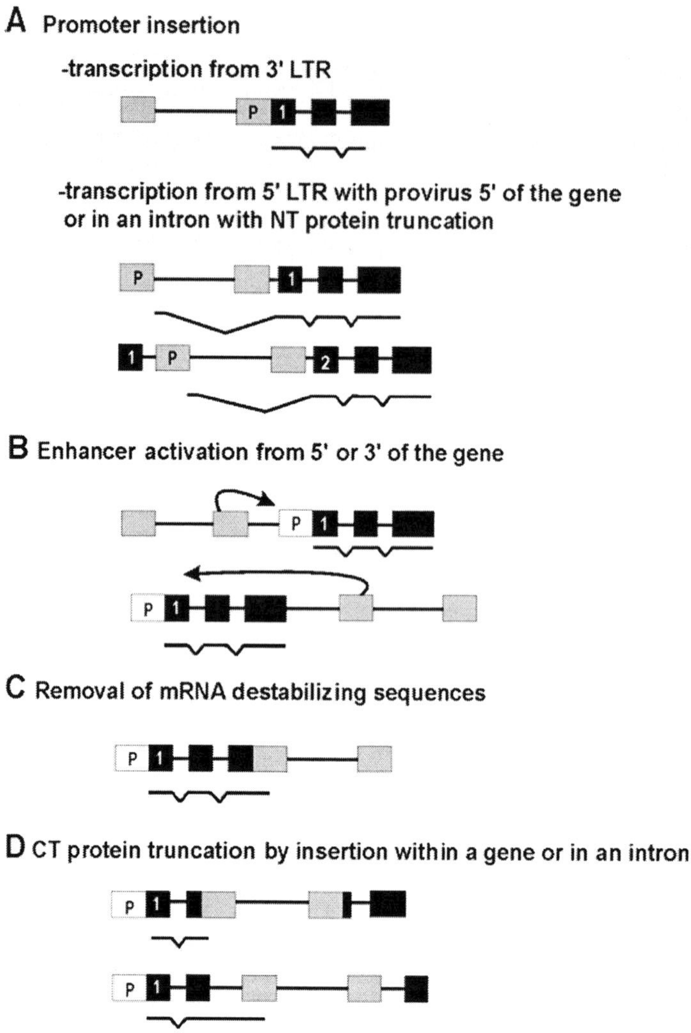

Figure 2. Mechanisms of host gene mutagenesis by retrovirus that result in activation of gene expression or function. Lines underneath exon structures depict mRNAs that are formed. Black boxes, exons; shaded boxes, retroviral LTRs; CT, carboxy-terminal; NT, amino-terminus; P, promoter.

Carboxy-terminal protein truncation is another mechanism utilized by the virus in conjunction with transcriptional enhancement to activate genes. Several examples have been reported in the activation of the *c-myb* gene and result in truncation at different positions (Shen-Ong et al., 1986;Shen-Ong and Wolff, 1987;Mukhopadhyaya and Wolff, 1992;Nazarov and Wolff, 1995). The virus integrates in an exon or intron in the same orientation as *c-myb* and transcription terminates in the viral LTR. Protein translation terminates in the first available termination codon, either in the intron or the LTR. When truncation occurs in exon 9 the encoded protein loses its ability to be targeted to the 26S-proteasome for degradation (Bies and Wolff, 1997). In addition, the protein is missing a region through which its transcriptional activity is regulated. Another example of a carboxy-terminal truncation associated with retroviral integration is the activation of Tpl-2 . The provirus is integrated into intron 7 of the gene giving rise to a constitutively active serine/threonine kinase (Patriotis et al., 1993;Makris et al., 1993).

Mechanisms of Inactivation of Tumor Suppressor Genes

Inactivation of tumor suppressors by retroviral insertion mutagenesis is rare. One of the reasons is that genetic inactivation of many prototype tumor suppressors requires the loss of two alleles (Knudson, Jr., 1971;Knudson, 2001). A case in point is the integration into the Evi2 locus, which was discovered in BXH2 recombinant inbred mice(Largaespada et al., 1995;Largaespada et al., 1996). These mice develop a lifetime viremia due to a horizonally transmitted B-ecotropic retrovirus and the virus causes insertional mutagenic events that invariably lead to myeloid leukemia. It was discovered that the proviruses were positioned in a very large intron of the tumor suppressor Nf1. This gene encodes neurofibromin, a GTPase activating protein (GAP) that negatively regulates Ras. Interestingly, proviral insertions frequently occurred in both alleles (Largaespada et al., 1996). Another example is the tumor suppressor TP53. It is inactivated in some cases of murine erythroleukemia due to independent retroviral integrations on separate alleles of the gene (Selten et al., 1985).

There is the theoretical possibility that a retrovirus could integrate in only one allele of a tumor suppressor gene resulting in loss of expression from that allele and that this mono-allelic loss could produce sensitivity to cancer based on reduced dosage. The proposed likelihood of this type of event comes from recent evidence in man and in mice that, for many described tumor suppressor genes, haplo-insufficiency contributes to neoplastic disease(Quon and Berns, 2001;Santarosa and Ashworth, 2004). When carrying out high-throughput retroviral genetic screens this possibility should be considered.

Mechanisms that Lead to Multiple Retroviral Integrations

MuLVs replicate very efficiently compared to other retrovirus such as the lentiviruses. Consequently, a large number of cells in the mouse are infected and there is a high probability that integration will occur at an oncogenic site. Each activation of an oncogene (or less frequently, inactivation of a tumor suppressor) by the virus results in enrichment of cells in which additional oncogenic events can occur. These secondary events may be spontaneous or could result from additional insults by the virus.

Although there are murine models of retrovirus-induced leukemia where only 1 integration appears to be required to produce disease, this is not usually the case. A single provirus was observed repeatedly in studies of amphotropic retrovirus 4070A-induced myeloid leukemia in mice undergoing an inflammatory response. The single integration was shown to be invariably in the *c-myb* locus or within 100 kb upstream of the c-*myb* locus (Wolff et al., 1991;Koller et al., 1996;Haviernik et al., 2002). This unusual situation may be due to the occurrence of other genetic alterations induced by the inflammatory granuloma (e.g. mutations resulting from the oxidative environment) that could contribution to the stepwise tumor progression and substitute for those caused by insertional mutagenesis in other models.

Normally, multiple proviruses, more than 6, are observed in a single retrovirus-induced tumor of mice, particularly if the mice have not been predisposed to disease by other means such as genetic engineering or chemical carcinogens. Conversion to an oncogenic phenotype is predicted to require, at a minimum, 3 cooperating genetic changes (Hanahan and Weinberg, 2000), and results of recent large scale identification of potential disease loci indicate that many tumors contain multiple common integrations sites (CISs)(Suzuki et al., 2002). Most retrovirus-induced tumors are actually oligoclonal, derived from independently transformed primary cells or from a primary tumor that has undergone a number of different secondary retroviral integrations. Therefore, some of the proviral integrations may be found only in a subset of the primary tumor cells. Transplantation of cells to a recipient mouse can give rise to an increase or decrease in the number of cells that contain a specific integration site, depending upon the ability of the particular provirus to promote progression. It should be noted that not all clonal proviral integration sites are necessarily involved in the multistep process of tumor formation; some proviruses may have integrated early in the clonal selection process, but are bystanders, not involved in the selection towards a transformed phenotype.

Since retrovirus-induced tumors generally have multiple integrations, and it is extremely unlikely that the cells were infected several times simultaneously, it is assumed that integrations occur consecutively (Mikkers and Berns, 2003). If cells become super-infected during tumor progression, normal mechanisms of resistance to re-infection must be overcome. Natural resistance to re-infection occurs when the envelope protein is produced in a cell and binds to its cell surface receptor, where it interferes with subsequent virion attachment. During the evolution of tumors there are two explanations to account for re-infection of cells. One is that during infection recombinant retroviruses are generated which encode an envelope glycoprotein with the capability of binding to a different receptor. An example is production of mink cell focus inducing (MCF) viruses which occurs during Mo-MuLV induced lymphomagenesis, as a consequence of recombination between the infecting virus and endogenous viral sequences (Fan, 1997).

Re-infection of a single cell could also be explained by the presence of integrated proviruses that are mutated or deleted and not capable of producing intact, functional envelope protein that can interfere with virion binding (Wolff and Koller, 1990;Berkowitz and Goff, 1993;Mikkers and Berns, 2003). Such altered proviruses are likely to be the result of recombination events or mistakes made during reverse transcription of viral RNA.

Retrovirus Involvement in Both Early and Late Events in Neoplastic Progression

It is not surprising that investigators have found that some genes are specifically involved in tumor-initiating events, while others promote progression, late in the tumor evolution. Activation of the *c-myb* locus by Mo-MuLV was shown to be an early event in the development of promonocytic leukmemia in young adult BALB/c mice undergoing a pristane-induced chronic inflammation in the peritoneal cavity (Nason-Burchenal and Wolff, 1993). Detection of the activation was possible because virus *gag* RNA sequences are fused to *myb* sequences through predictable splicing events. Using a nested reverse transcriptase (RT) polymerase chain reaction (PCR) assay for *gag-myb* fusion mRNA, the investigators determined the tissue location of the activation and time after virus infection integration into c-*myb* occurred. Incredibly, *gag-myb* mRNA was detected in preleukemic cells as early as 2 weeks following virus infection. By three weeks there was evidence of preleukemic cells expressing *gag-myb* in 83% of the mice receiving both pristane and virus. Therefore, even though replication-competent retroviruses without oncogenes have often been termed "slowly transforming viruses", it is not necessarily because the virus needs to replicate a long time prior to integrating into the appropriate genomic locations. The extra months required for disease development is most likely due to the time required to accumulate the progression events. In the case of the promonocytic leukemia model involving *c-myb*, it was demonstrated that preleukemic cells with activated *c-myb* can remain in a partially transformed state indefinitely. These cells do not accumulate and progress to a leukemia until the mice receive pristane (Nason-Burchenal and Wolff, 1993).

Genes involved in progression can be discovered following transplantation into syngeneic mice of a primary tumor or prolonged culturing of cells lines derived from retrovirus-induced tumors. It is likely that the original tumor already has activated progression genes in a subset of cells in vivo and these facilitate neoplastic development in transplant recipient mice or in culture. The cells with the activated progression genes are selected in the secondary environment. The gene encoding Trp53 was discovered as a target of insertional mutagenesis following transplantation of Fr-MuLV-induced erythroleukemias (Johnson and Benchimol, 1992). Tpl-2/Cot, a serine/threonine protein kinase, (Patriotis et al., 1993;Ceci et al., 1997) and Tiam-1, a guanine nucleotide exchange factor which plays a role in cellular migration, invasion, adhesion and tumor progression, were provirus-associated genes from cells of cultured Mo-MuLV-induced rat T-cell lymphomas (Minard et al., 2004).

Lineage Specificity of Disease Induction by Retroviruses

Whether or not a retrovirus is capable of inducing neoplastic disease within a specific hematopoietic lineage is largely dependent upon its ability to replicate in that lineage. Often this is determined by transcriptional regulatory sequences in the viral LTR, which not only allow the virus to replicate in a tissue, but also determine the ability of the retrovirus to drive transcription of neighboring genes. A classic example of LTR specificity came from an experiment conducted in Nancy Hopkins laboratory, where U3 LTR sequences containing enhancers were exchanged between Fr-MuLV and Mo-MuLV and it was shown that the LTR specified the tissue origin of the leukemia (Chatis et al., 1983;Chatis et al., 1984;Golemis et al., 1989).

Lineage specificity may depend upon the cooperation of other oncogenes already activated through genetic engineering or the use of tumor promoting agents. This is due to the fact that the virus, although it might replicate sufficiently in a tissue, may not cause progressive disease without the provision of cooperating events. For example, Mo-MuLV was first recognized as a virus that caused almost exclusively T-cell lymphomas, however, in the presence of a chronic inflammatory response, it was later shown to be capable of inducing a high incidence of myeloid disease (Wolff et al., 1988). Furthermore, in transgenic mice expressing *c-myc* predominantly in B-lineage cells, Mo-MuLV induces B-cell lymphomas (van Lohuizen et al., 1991). Interestingly, amphotropic retrovirus 4070A was initially considered to be very inefficient for inducing leukemia and when it did cause leukemia it was usually lymphoid in nature. However, this virus has subsequently been found to be a very potent inducer of myeloid leukemia in mice undergoing chronic inflammation (Wolff et al., 1991) or mice predisposed to myeloid disease by expressing the *Cbfb-MYH11* fusion gene (Castilla et al., 2004)(described in more detail below).

High Throughput Identification of Cancer Disease Genes

The tagging of retroviruses in the discovery of new cancer genes has recently been accelerated by new technologies that facilitate a high-throughput approach. The sequences adjacent to integrated proviruses can be rapidly cloned using PCR-based methods instead of the traditional methods of preparing phage libraries of the tumor genomic DNA and screening with virus probes. Furthermore, the sequencing of the mouse and human genome has facilitated the assignment of the virus-tagged sequences to specific genetic locations.

One PCR-based technique for amplifying virus target sites is called inverse PCR (IPCR) (Ochman et al., 1988;Triglia et al., 1988). Restriction endonucleases are used to cut both within the virus and outside the virus in the cellular sequence. The fragment is then ligated to form a circle and used as a template in a PCR reaction using primers homologous to the virus. In another type of PCR-based technique, a double stranded DNA adaptor is ligated to the end of the fragment containing the cellular sequence. An adaptor-specific primer, in addition to a LTR-specific primer, are used to generate sufficient copies of genomic DNA for cloning. In the adaptor-based approach, variations in the adaptor have enhanced the specificity of amplification. One useful adaptor is called a splinkerette (Hengen, 1995;Devon et al., 1995;Mikkers et al., 2002). In this adaptor, the strands are mismatched and the bottom strand is designed to form a strong hairpin. Specificity is enhanced due to the mismatch, which allows amplification only after extension from the viral sequence. In addition, end-repair priming is prevented using this adaptor because the hairpin in the first extension generates a stable double stranded structure that cannot be utilized in subsequent reactions.

A particularly sensitive and specific method of amplifying genomic sequences that flank integrated virus was developed for application to gene therapy. Following retrovirus-mediated transfer of genes to hematopoietic cells, this technique is used to monitor the number of transduced clones in vivo. It combines extension primer tag selection (EPTS) and linear amplification mediated (LAM)-PCR (Schmidt et al., 2001;Schmidt et al., 2001;Kiem et al., 2004). Genomic DNA is digested and subjected to primer extension with a biotinylated specific primer. It is then immobilized onto streptavidin-coated magnetic beads, followed by ligation of an adaptor and nested PCR amplification. Sequencing can be carried out directly without cloning. This technique, as it is used for gene therapy applications, can detect a provirus in 1 out of 100 to 1000 cells and may actually be too sensitive for the identification of cancer-related genes. For retroviral tagging in tumor cell populations, it is desirable to limit amplification to proviruses that are common to a large proportion of the cells. In any case, it is likely that the technique can be easily adapted for this application.

The expansion in the capacity to identify cancer genes through retroviral tagging was due not only to advances in recovering sequences from cells, but also to the availability of published mouse and human genome sequences and related databases. The Ensembl genome database and the Celera Discovery System are extremely valuable for mapping CISs in or near genes (Kerlavage et al., 2002;Hubbard et al., 2002). These databases provide genes, repeats, map markers, as well as other features including single nucleotide polymorphisms (SNPs), phenotypes and disease information, regulatory signals, mRNA expression in the form of Expressed Sequence Tags (ESTs) and tags derived from Serial Analysis of Gene Expression (SAGE).

A retroviral tagged cancer gene database (RTCGD) was very recently set up by Neal Copeland and Nancy Jenkins and their co-workers in the Mouse Cancer Genetics Program, National Cancer Institute (Akagi et al., 2004). It is designed to manage multiple high-throughput insertional mutagenesis projects. Using this database one can compare data derived from a mutagenesis screen with results from multiple tumor models. An advantage of the database is that common insertion sites (CIS) can be identified through the database itself in cases where an insertion site was found once in more than one model system, but not multiple times in a single system. The program includes CIS Lists, with links to other public databases and also to a list of individual clones called the Clone Lists. In the Clone Lists there is information about the tumor model, tumor type, genomic position and orientation of the integrated provirus.

Several high throughput screens have been reported in the last several years. Some of them have been in wild-type (not genetically modified) mice, while others were in mice that were genetically altered and will be discussed further below. The largest of the screens in wild-type mice looked at a total of 194 primary B-and T-cell lymphomas (mostly B-cell) of AKXD and NFS.V$^+$ strain mice from which 884 integration sites were cloned and sequenced (Suzuki et al., 2002). These integration sites were analyzed and compared with additional sites previously cloned from BXH2 myeloid leukemias and from a few AKXD lymphomas. Ninety percent of a total of 1,336 integration sites were mapped to specific chromosomal regions through blast searches of the Celera Discovery System. Forty-one percent of the integration sites were found to be common and these added up to 152 separate CISs. It was general feature of tumors to have multiple CISs, 1-3, and more rarely 4-6. Candidate genes were identified for all but 3 of the 152 CISs; forty were previously published while the others were novel. The CISs were characterized as to the functional types of their protein products and the largest class was noted to be transcription factors or other transcriptional regulators. Other genes found near CISs encode proteins involved in chromatin remodeling, kinases and phosphatases, signaling molecules, cytokines and their receptors, Ras proteins and their regulators, proteins involved in apoptosis, proteins that bind actin, splicing factors and proteins that control the cell cycle. Most of the genes identified fell into pathways known to be involved in cancer including, Ras, Notch, Jak/Stat, and Nfkb. Further analysis of the CIS identified above revealed that 19 CIS were mutated in human leukemia or lymphoma and an additional 9 were improperly expressed, demonstrating the relevance of the mouse studies to human cancer.

Another screen for cancer genes, specifically involved in the development of T-cell lymphomas in wild-type mice identified 19 target genes from a panel of 48 SL3-3-retrovirus neoplasms (Kim et al., 2003). Ras pathway genes, *Rras* and *Rasgrp1*, were the most frequent targets and accounted for 30% of the tumors. Integration with respect to the Rras gene was found to occur from a long distance, 16 to 63 kb 5' to the gene, and correlated with elevated mRNA expression. Since no mutations were detected in *Rras*, an important outcome of this study was the underscoring of the transforming effect of overexpression of this *ras* gene. This contrasts to the findings related to the activation of *Hras1*, *Kras2*, and *Nras* in transformation of mouse and human cells, where mutation of the genes is common. The results of this and other screens can be reviewed in the RTCGD database.

Two other screens were specifically aimed at discovering disease loci for myeloid leukemia. One screen analyzed Cas-Br-M-induced primary leukemias and leukemia derived cells lines (Joosten et al., 2002). The other was a screen of myeloid leukemias induced in FVB/N mice by Graffi1.4 (Erkeland et al., 2004). In both screens, the original cloning of insertion sites only identified a small number of CISs, so the investigators used an additional approach to increase the sensitivity of finding common sites. They prepared a locus-specific primer for each sequence obtained by IPCR. This was used in conjunction with an LTR primer in a PCR amplification of DNA from tumors. These products were examined by Southern blot analysis using radiolabeled LTR- and fragment–specific oligonucleotides as probes. The majority of proviral integration sites were found to be CISs; for example, in the Graffi1.4- induced leukemias, 39 out 41 cloned integration sites were determined to be CISs. Many of the loci that were identified were the same as those previously reported, but many were novel. The finding of so many novel genes that were not found previously in lymphoid or monocytic leukemias suggests that the majority of them are specific for the myeloid lineage. It is clear from these studies that genes will continue to be discovered and the mutagenesis approach to identifying genes has not been saturated.

Insertional Mutagenesis in Genetically Engineered Mice

Retroviruses are extremely valuable in modeling hematopoietic cancers in mice that have been genetically manipulated to increase their susceptibility to this type of neoplastic disease. First, the increased susceptibility to cancer provided by a gene in some cases cannot be demonstrated without provision of secondary virus-induced oncogenic mutations. This is in part due to the fact that spontaneous mutations that activate oncogenes or inactivation of tumor suppressors, and are required for progression to advanced disease, do not always take place in the short murine life span. In the following example, the use of retroviruses helped to verify an important biological role of a cell cycle regulator, previously implicated as a tumor suppressor in human leukemia. Mice deficient in $p15^{Ink4b}$, a cyclin-dependent kinase (CDK) inhibitor, have a low incidence of angiosarcomas, and do not develop leukemia of the myeloid lineage (Latres et al., 2000). However, infection of these mice with retroviruses facilitates the demonstration that these mice have an increased vulnerability to acute myeloid leukemia (AML) (Wolff et al 2003a). More will be discussed about this particular model in the section Modeling of Myeloid Leukemia. Second, the identification of retrovirally tagged genes in combination with genetic abnormalitites can dramatically enhance our understanding of pathways of neoplastic development. Various retroviral targets can be placed in complementary groups that are in the same pathway or similar pathways and can usually be placed in a pathway different from the one in which the altered gene resides. This is well illustrated in studies involved in modeling lymphomagesis in the next section.

Modeling of Lymphomagenesis

Studies employing retroviruses in modeling cancer started with studies on lymphomagenesis. These experiments, carried out by several investigators, clearly demonstrate the power of the insertional mutagenesis approach to gene discovery and the assignment of genes to specific pathways and complementary groups. Table 1 gives examples of animal models and the viruses that were used to infect the mice.

The first in vivo models were mice transgenic for c-*myc* because this gene was demonstrated to be a frequent target of insertional mutagenesis in early studies and was predicted to provide an excellent susceptibility background in which to establish collaborating events (Selten et al., 1984;Corcoran et al., 1984;Adams et al., 1985;O'Donnell et al., 1985;van Lohuizen et al., 1991;Haupt et al., 1991;Scheijen et al., 1997;Mikkers et al., 2002). A series of studies from Anton Berns' laboratory looked at CISs obtained in B-cell lymphomas of Eμ*myc* infected mice and later found that some of these genes (e.g. Pim1 and Bmi1) were themselves oncogenic (van Lohuizen et al., 1989;Alkema et al., 1995). Using Eμc-*myc*, and Eμ*pim1* transgenic mice in combination with retroviral insertional mutagencsis, this group was able to determine a number of complementation groups that collaborated with c-*myc* in the induction of lymphomas. These include the *Bla1/Bmi1* and *Gfi1/Eis1/ Pal1/Evi5* groups that were placed in a complementary groups based on close genetic linkage and the *Pim1/Pim2* group formed on the basis of functional relatedness (van Lohuizen et al., 1989;van Lohuizen et al., 1991;Scheijen et al., 1997). Pim1 and Bmi-1 from two separate complementation groups appear to have quite different roles in transformation, the former being a a cytoplasmic serine-threonie kinase that may be linked to cytokine independence (Domen et al., 1993;Domen et al., 1993;Buckley et al., 1996) and the later functioning as a transcriptional repressor of *Cdkn2a* gene expression affecting both the $p16^{Ink4a}$ and $p14^{Arf}$ pathways (Jacobs et al., 1999). Individual tumors have retroviral activations of only one CIS from each complementation group. Interestingly, studies from James Neil's laboratory on CD2- *Runx2* transgenic mice place *Runx2* in a fourth complementation group (Blyth et al., 2001). This gene encodes a subunit of a transcription factor that is member of a family of transcription factors involved in hematopoiesis (Ogawa et al., 1993;Satake et al., 1995;Meyers et al., 1996).

c-*myc* appears to be a predominant collaborator with both *ras* and v-*abl* in transgenic mice. When EμN*ras* mice, which sporadically develop T cell lymphomas, were infected as newborns with Mo-MuLV, leukemogenesis was significantly accelerated (Haupt et al., 1992). These lymphomas were associated with transcriptional activation of c-*myc* or N-*myc* in 68% of the cases. Eμv-*abl* expressing mice spontaneously develop plasmacytomas, but with Mo-MuLV infection the predominant phenotype was T-cell lymphoma and was associated with only moderate acceleration (Haupt et al., 1993). The primary clonal integrations, representing 42% of the tumors, were found in c-*myc*, *N-myc* and *pim1*.

Table 1. Use of retroviruses to model lymphomagenesis in the mouse

Transgene	Gene targeted for deletion	Tumor type	Mouse Strain	Retrovirus	References
MMTV/Fgf3		T-cell lymphoma	CD1	MMTV	Paquette et al., 1992; Girard et al., 1996
Eμmyc		Pre-B-cell lymphoma	C57BL/6-SJL or CBA-C57BL/6	Mo-MuLV	Adams et al., 1985; Haupt et al., 1991; van Lohuizen et al., 1991; Scheijen et al., 1997; Mikkers et al., 2002a
Eμmyc	Pim1-/-	Pre-B-cell lymphoma	FVB-CBA-C57BL/6-129/Sv-Balb/c	Mo-MuLV	van der Lugt et al., 1995
Eμmyc	Pim1-/- /Pim2-/-	Lymphomas		Mo-MuLV	Mikkers et al., 2002b
EμPim1		T-cell lymphoma	CBA-C57BL/6	Mo-MuLV	van Lohuizen et al., 1989; Scheijen et al.,
Eμmyc + EμPim1		Pre-B-cell lymphoma	CBA-C57BL/6	Mo-MuLV	Verbeek et al., 1991
EμN-ras		T-cell lymphoma	C57BL/6-SJL	Mo-MuLV	Haupt et al., 1992
Eμv-Abl		T-cell lymphoma	C57BL/6-SJL	Mo-MuLV	Haupt et al., 1993
EμBmi1		T-and B- cell lymphoma	FVB	Mo-MuLV	Alkema et al., 1995 Alkema et al., 1997
CD2-myc		T-cell lymphoma	C57BL/6-CBA	Mo-MuLV	Stewart et al., 1993; Baxter et al., 2001
CD2-Runx2		T-cell lymphoma	C57BL/6-CBA	Mo-MuLV	Blyth et al., 2001; Vaillant et al., 1999
E2A-Pbx1		T-cell lymphoma	FVB/N	Mo-MuLV	Dedera et al., 1993; Feldman et al., 1997; Feldman et al., 2000
	Trp53	T- cell lymphoma	129/Sv-NIH	Mo-MuLV	Baxter et al., 1996
CD2-myc	Trp53	T- cell lymphoma	129/Sv-C57BL/6-NIH-CBA/Ca	Mo-MuLV	Baxter et al., 2001
	$p27^{Kip1}$	T-cell lymphoma	C57BL6-129/Sv	Mo-MuLV	Hwang et al., 2002
	Cdkn2a ($p16^{Ink4a}$)	T-,B-cell lymphoma, T-and B-cell mixed, Histiocytic sarcoma	129/Sv-C57BL/6-SJL	Mo-MuLV	Lund et al., 2002; Serrano et al., 1996
	Smad 3	Mostly T-cell lymphoma	C57BL6-129/Sv	MOL4070LTR	J. Letterio, L.Wolff, Personal communication

In recent animal models for leukemogenesis, oncogenes found to play a role in human leukemia have been expressed in mice as transgenes and the mice were subjected to retroviral insertional mutagenesis. One of the first examples came from Michael Cleary's laboratory where mice transgenic for *E2A-PBX1* were made (Dedera et al., 1993). *E2F-PBX1 is* a fusion gene involving the *E2A* and *PBX1* genes and found at a specific chromosomal translocation in pre-B cell acute lymphoblastic leukemias of children (Nourse et al., 1990;Kamps et al., 1991). Both *E2A* and *PBX1* encode transcriptions factors, from the helix-loop-helix and homeodomain families, respectively. The fusion protein, which contains the transcriptional activation domain of E2A and the DNA-binding homeodomain of PBX1, is thought to disrupt Hox regulatory pathways. In contrast to PBX1, which is a repressor acting in a complex with HOX proteins, E2A-PBX1 is a strong activator of transcription. Transgenic mice succumb to T-cell lymphomas by 5 months of age in the absence of retroviral infection. However, infection of newborns with Mo-MuLV results in acceleration of disease with an average latency of 81 days (Feldman et al., 1997). Forty-eight percent of the tumors have clonal retroviral insertions in *Pim1*. It was subsequently determined that *Pim1* and *E2A-PBX1* cooperate in double transgenic mice (Feldman et al., 1997). This laboratory also identified *Notch1* as a gene that was mutated by retrovirus at high frequency and in about half of the cases involving *Notch1*, the virus caused 3' deletions of the gene. The resulting protein was truncated at the carboxy-terminus and missing negative regulatory sequences (Feldman et al., 2000). Mice transgenic for both the *E2A-PBX1* and a carboxy-terminal deletion mutant of *Notch1* were subsequently shown to develop tumors with a shortened latency compared to *E2A-PBX1* single transgenic mice (Feldman et al., 2000).

The power of retroviral tagging has been further exemplified in mice that are deficient for tumor suppressor genes. In the case of the tumor suppressor gene *Trp53*, mice null for the gene developed thymic lymphomas and Mo-MuLV infection of these mice accelerated the disease. However, the combination of virus and loss of the tumor suppressor was only weakly additive. This led James Neil and coworkers to postulate that the retrovirus infection bypasses the need for *Trp53* inactivation because they have overlapping effects (Baxter et al., 1996). The investigators then carried out retroviral mutagenesis experiments in CD2*mycTrp53* +/- mice which further supported this hypothesis. They found that tumors arising at an early age in these mice had a low rate of secondary allele loss, which was in contrast to noninfected CD2*myc Trp53* +/- mice that had a 100% loss at an early stage of leukemogenesis (Baxter et al., 2001).

Cyclin-dependent kinase inhibitors (CDKIs) have been the focus of several recent animal model experiments. One CDKI that has been implicated in human cancers is $p27^{KIP1}$ because its expression is often reduced or absent. Furthermore, mice deficient in $p27^{Kip1}$ develop pituitary adenomas. To study the role of the gene in cancers of blood cells, Hwang and coworkers used Mo-MuLV to induce lymphomas in $p27^{Kip1}$ null mice and found that it accelerated disease in these mice compared to wild-type mice (Hwang et al., 2002). c-*myc* was a clear collaborator, because it was observed to be activated more frequently than other genes in Mo-MuLV-infected null mice. Interestingly, two novel collaborating genes were discovered in these cancer-sensitized mice, the Jun dimerization protein 2 gene (*jundp2*) and an X-linked locus termed *Xpcl1*.

The *CDKN2A* locus is frequently deleted in solid tumors and in lymphomas. It encodes two tumor suppressor genes, $p16^{Ink4a}$ a CDKI of the G1 to S transition of the cell cycle and $p14^{ARF}$ a positive regulator of the p53 pathway. Mo-MuLV infection of *Cdkn2a-/-* mice results in development of both lymphoid (80%; predominantly T-cell) and histiocytic sarcomas (55%) (Lund et al., 2002). A high-throughput approach to retroviral tagging in this model found 37 new CIS. Of these, 19 were new loci specifically associated with the sensitized *Cdkn2a-/-* background and not found in other high-throughput screens.

Furthermore, 14 of the 37 were lymphoid specific. This analysis also showed that there is a preference for MAP kinase signaling in collaboration with loss of *Cdkn2a*. As one would predict, no integrations were found in *Bmi1*, a repressor of $p16^{Ink4a}$ and $p19^{Arf}$ (Jacobs et al., 1999).

Smad3 is involved in signal transduction and transactivation of genes in response to TGFβ, an inducer of growth arrest and apoptosis. Although deletion of the *Smad3* gene has mild affects on the immune system, inoculation of the mice with retrovirus results in a high incidence of both T-cell or myeloid leukemia in a few months (J. Letterio and L. Wolff, personal communication). Future experiments delineating the retrovirus tagged genes should be informative regarding Smad3 collaborators.

Modeling of Myeloid Leukemia

The retrovirus approach to deciphering collaborating events in oncogenesis was at first carried out mainly in studies on lymphomogenesis as described above. A lack of similar studies on myeloid leukemia was due in part to the fact that the retroviruses available for these studies induced primarily lymphoid leukemia. Mo-MuLV was employed in most of the early experiments. Table 2 presents myeloid leukemia modeling experiments performed in recent years, using other viruses.

A landmark study in the modeling of myeloid leukemia and involving the human oncogene, *CBFB-MYHII* was reported recently and amphotropic virus 4070A was the insertional mutagen (Castilla et al., 1999;Castilla et al., 2004). *CBFB-MYHII* fusion gene is an oncogene involved in AML cases involving the chromosomal rearrangement inv(16). The protein encoded by the rearranged chromosome is composed of the first five exons of the β subunit of CBF fused in frame to the carboxy-terminal coiled-coil domain of the smooth muscle myosin heavy chain (SMMHC). The normal CBFβ protein interacts with RUNX1 (AML1) increasing its affinity for DNA and its stability, thereby enhancing its activation or repression of transcription (Ogawa et al., 1993;Huang et al., 2001). This transcriptional complex is involved in promoting definitive hematopoiesis as well as establishing maturation of all lineages of hematopoietic cells (except erythrocytes) (Kundu and Liu, 2003). Mice expressing a Cbfβ-SMMHC from a *Cbfb-MYH11* knockin die at embryonic day 12.5 and have a phenotype which is similar to that of *Cbfb* and *Aml1* null mice (Castilla et al., 1996). There are two theories to explain the inhibition of the normal functions of CBF in human AML by CBFβ-SMMHC: 1) the fusion protein causes sequestration of RUNX1 (Adya et al., 1998) 2) CBFβ-SMMHC acts as a dominant negative, inhibiting the activity of the normal CBF complex. The last theory is supported by a recent study that showed that the carboxy-terminal SMMHC part of the protein interacts with co-repressor mSin3A and histone deacetylase HDAC8 which would convert AML1 from an transactivator to a transcriptional repressor (Durst et al., 2003).

$Cbfb^{+/MYH11}$ chimeric mice develop lymphoid tumors with an incidence of 45% after 1 year (Kundu and Liu, 2001) and following treatment with N-ethyl-N-nitrosourea the mice develop, in 2-6 months, myeloid leukemia with features of human AML (Castilla et al., 1999). To identify genes that collaborate with Cbfb-MYH11 in the pathogenesis of acute myeloid leukemia, a retroviral strategy was used (Castilla et al., 2004). Amphotropic virus

4070A was injected into newborn chimeric mice. The virus is N-tropic and although it would be unable to efficiently infect the C57BL6 blastocyst derived cells of the chimera due to Fv-1 restriction, it could infect the cells from the 129/Sv ES cells. Fifty-four candidate genes were identified at CISs and 6 of these were common. The study identified, for the first time, the C2H2 zinc finger transcription factors Plagl and Plagl2 as genes playing a role in AML. In addition to these and transcription factors Runx2 and c-Myb, genes involved in signaling pathways connected to proliferation and apoptosis were also found as collaborators.

Table 2. Use of retroviruses to model myeloid and erythroid leukemia in the mouse

Transgene	Gene targeted for deletion	Tumor type	Mouse Strain	Retrovirus	References
Cbfb-MYHII		Myeloblastic/ Monoblastic leukemia	C57BL /6-129 sv chimera	Amphotropic retrovirus 4070A	Castilla et al., 2004
SV40 T-Antigen		Histiocytic sarcoma	C57BL /6-C3H	Fr-MuLV	Skalnik et al., 1991 Kone et al., 2002
	Trp53	Erythroid leukemia	75% BALB/c 25% 129sv	Fr-MuLV	Wong et al., 1999
	Nf1	Myeloid leukemia	BXH2	Horizontally transmitted B-tropic virus	Blaydes et al., 2001
	Cdkn2b ($p15^{Ink4b}$)	Myelomonocytic leukemia, Myeloid leukemia with granulocyte maturation	C57BL/ 6-129sv	MOL4070LTR	Wolff et al., 2003a,b
	ICSBP	Myeloid and lymphoid	C57BL/6	Mo-MuLV, 10A1 /Mo-MuL V recombinant	C.Stocking, personal communication

The choice of retroviruses for the application of mutagenesis in myeloid leukemia has not been as straight-forward as for mutagenesis of lymphoid cells. In the latter case, Mo-MuLV has consistently been the virus of choice. To search for cooperating somatic mutations in myeloid leukemia in *Cbfb-MYH11* knock-in mice, as described above, retrovirus 4070A was an appropriate choice. It is capable of infecting 129/Sv cells, and it accelerates leukemia, while at the same time it was not able to induce myeloid leukemia in control wild-type mice. However, the utility of this virus is generally limited due to its *Fv1* tropism. Since it is N-tropic, it cannot infect many mouse strains in which transgenes are expressed or in which transgenics and knockouts have been produced, such as FVB or C57BL/6.

A unique virus was developed recently that is a potent inducer of myeloid leukemia and overcomes the *Fv1* gene limitation of retrovirus 4070A (Wolff et al., 2003). It has already been shown to be a practical and potent mutagenic virus for acceleration of both myeloid and lymphoid leukemia (Wolff et al., 2003; J. Letterio, K. Shannon, personal communication). This virus, called, MOL4070LTR, is a Mo-MuLV based virus in which most of the U3 region of the LTR has been replaced with analogous sequences from 4070A. Like Mo-MuLV, it has a capsid protein that confers NB-tropism and, therefore, it can infect any inbred mouse strain. In contrast to Mo-MuLV, which induces lymphoid leukemia when inoculated into newborn mice, MOL4070LTR induces myeloid and lymphoid leukemia in approximately equal proportions. The myeloid phenotypes are predominantly myelomonocytic leukemia and myeloid leukemia with granulocyte maturation. When used to accelerate leukemia in sensitized mice, it gave a low background of disease up to one year in a C57BL/6-129/Sv mixed background.

Other viruses that are potential insertional mutagens for myeloid leukemia include Fr-MuLV(Kone et al., 2002), Graffi virus(Ru et al., 1993), and a 10A1/Moloney MuLV recombinant virus constructed and tested in the Carol Stocking laboratory (personal communication). Cas-Br-M is also capable of inducing myeloid leukemia, but is limited to infection of $Fv1^{nn}$ mice (Jolicoeur et al., 1983;Holmes et al., 1986). Fr-MuLV induces primarily erythroid leukemia in susceptible strains such as NFS, NIH Swiss, or BALB/c mice in as early as 2 months and induces myeloblastic leukemia or myeloid leukemia with macrophage or granulocyte differentiation in strains that are not highly susceptible to erythroid leukemia (e.g, C57BL/6, DBA/2)(Shibuya and Mak, 1982;Silver and Fredrickson, 1983;Chesebro et al., 1983;Sola et al., 1985). Furthermore, Fr-MuLV has been injection into transgenic mice expressing SV40 T-antigen under a myelomonocytic promoter to facilitate histiocytic sarcomas and to discover collaborating disease genes (Kone et al., 2002). An intriquing outcome from this study was the finding of four proviral insertions in a novel locus *Fim4,* a region syntenic to human 7q32 and commonly deleted in human myelodysplastic syndrome (MDS) and AML.

An alternative approach to inoculating virus into mice is to backcross the genetically modified mouse strain onto the BXH2 genetic background. BXH2 mice have maternally transmitted B-ecotropic endogenous retrovirus, which can act as a mutagen after a few backcross generations. Backcrossing is performed with a genetically modified mouse of a strain that is susceptible to the B-tropic virus infection. This methodology is exemplified in a published search for somatic mutations that cooperate with loss of Nf1, a gene associated with 25% of juvenile myelomonocytic leukemia cases (Blaydes et al., 2001). A single null allele of Nf1 was backcrossed onto the BXH2 genetic background for 3 generations. This resulted in a more rapid development of myeloid leukemia than observed for BXH2 mice and the observation that the *Epi1*locus cooperates with Nf1 in progression to AML.

Insertional mutagenesis has recently been exploited to model myeloid leukemia in mice deficient in a tumor suppressor commonly inactivated in greater than 80% of the cases of human acute myeloid leukemia and in myelodysplastic syndromes. This tumor suppressor *CDKN2B(INK4B)* encodes an INK4 family protein CDKI, $p15^{Ink4b}$ which acts to block the cell cycle at the G1-S transition. Mice deficient in $p15^{Ink4b}$ were inoculated with MOL4070LTR, where it was found to collaborate with the loss of this gene in induction of myeloid leukemia. Although there was no incidence of myeloid leukemia in MOL4070LTR-infected wild-type mice within the first year, there was a 16% incidence of myeloid leukemia in *Ink4b+/-* mice, with hypermethylation of the second allele (Wolff et al., 2003). *Ink4b -/-* mice had a lower incidence suggesting that complete loss of the gene has a negative influence on leukemia induction. Further analysis of proviral integration sites in this model will undoubtedly reveal genes that specifically collaborate with the loss of $p15^{Ink4b}$.

Conclusion

The application of retroviruses for mutagenesis in the mouse has already been demonstrated to be an extremely affective strategy for modeling cancer progression in hematopoietic malignancies. The recent tactic is to infect mice that have been modified to have genetic predispositions to leukemia similar to those found in man (e.g. expression of *E2A-PBX1* or *Cbfb*-MYH11 or deficiency in the tumor suppressors Trp53, $p16^{Ink4a}$ or $p15^{Ink4b}$). In the future this technology will continue to be a powerful way of determining cooperative acitivity of genes in leukemia development. It will also help to distinquish which oncogenes or tumor suppressors are tissue specific and which are common to a wide range of cancers. Multiple investigator input into the RTCGD will significantly expand the potential outcome of retrovirus tagging studies.

Acknowledgement

I thank Sandra Ruscetti for critiquing this review

References

Abelson, H.T., Rabstein, L.S., 1970. Lymphosarcoma: virus-induced thymic-independent disease in mice. *Cancer Res.* 30, 2213-2222.

Adams, J.M., Harris, A.W., Pinkert, C.A., Corcoran, L.M., Alexander, W.S., Cory, S., Palmiter, R.D., Brinster, R.L., 1985. The c-myc oncogene driven by immunoglobulin enhancers induces lymphoid malignancy in transgenic mice. *Nature* 318, 533-538.

Adya, N., Stacy, T., Speck, N.A., Liu, P.P., 1998. The leukemic protein core binding factor beta (CBFbeta)-smooth-muscle myosin heavy chain sequesters CBFalpha2 into cytoskeletal filaments and aggregates. *Mol. Cell Biol.* 18, 7432-7443.

Akagi, K., Suzuki, T., Stephens, R.M., Jenkins, N.A., Copeland, N.G., 2004. RTCGD: retroviral tagged cancer gene database. *Nucleic Acids Res.* 32 Database issue: D523-D527.

Alkema, M.J., van der Lugt, N.M., Bobeldijk, R.C., Berns, A., van Lohuizen, M., 1995. Transformation of axial skeleton due to overexpression of bmi-1 in transgenic mice. *Nature* 374, 724-727.

Alkema, M.J., Jacobs, H., van Lohuizen, M., Berns, A., 1997. Pertubation of B and T cell development and predisposition to lymphomagenesis in Emu Bmi1 transgenic mice require the Bmi1 RING finger. *Oncogene* 15, 899-910.

Bartholomew, C., Ihle, J.N., 1991. Retroviral insertions 90 kilobases proximal to the Evi-1 myeloid transforming gene activate transcription from the normal promoter. *Mol. Cell. Biol.* 11, 1820-1828.

Baxter, E.W., Blyth, K., Donehower, L.A., Cameron, E.R., Onions, D.E., Neil, J.C., 1996. Moloney murine leukemia virus-induced lymphomas in p53-deficient mice: overlapping pathways in tumor development? *J. Virol.* 70, 2095-2100.

Baxter, E.W., Blyth, K., Cameron, E.R., Neil, J.C., 2001. Selection for loss of p53 function in T-cell lymphomagenesis is alleviated by Moloney murine leukemia virus infection in myc transgenic mice. *J. Virol.* 75, 9790-9798.

Bedigian, H.G., Johnson, D.A., Jenkins, N.A., Copeland, N.G., Evans, R., 1984. Spontaneous and induced leukemias of myeloid origin in recombinant inbred BXH mice. *J. Virol.* 51, 586-594.

Bedigian, H.G., Taylor, B.A., Meier, H., 1981. Expression of murine leukemia viruses in the highly lymphomatous BXH-2 recombinant inbred mouse strain. *J. Virol.* 39, 632-640.

Benit, L., De Parseval, N., Casella, J.F., Callebaut, I., Cordonnier, A., Heidmann, T., 1997. Cloning of a new murine endogenous retrovirus, MuERV-L, with strong similarity to the human HERV-L element and with a gag coding sequence closely related to the Fv1 restriction gene. *J. Virol.* 71, 5652-5657.

Berkowitz, R.D., Goff, S.P., 1993. Point mutations in Moloney murine leukemia virus envelope protein: effects on infectivity, virion association, and superinfection resistance. *Virology* 196, 748-757.

Best, S., Le Tissier, P., Towers, G., Stoye, J.P., 1996. Positional cloning of the mouse retrovirus restriction gene Fv1. *Nature* 382, 826-829.

Bies, J., Wolff, L., 1997. Oncogenic activation of c-Myb by carboxyl-terminal truncation leads to decreased proteolysis by the ubiquitin-26S proteasome pathway. *Oncogene* 14, 203-212.

Bies, J., Wolff, L., 2004. C-Myb and Leukemogenesis. In: Jon Frampton (Ed.), Myb Trancription Factors: Their Role in Growth, Differentiation, and Disease. *Kluwer Academic Publishers,* Dordrecht.

Bishop, K.N., Bock, M., Towers, G., Stoye, J.P., 2001. Identification of the regions of Fv1 necessary for murine leukemia virus restriction. *J. Virol.* 75, 5182-5188.

Blaydes, S.M., Kogan, S.C., Truong, B.T., Gilbert, D.J., Jenkins, N.A., Copeland, N.G., Largaespada, D.A., Brannan, C.I., 2001. Retroviral integration at the *Epi1* locus cooperates with *Nf1* gene loss in the progression to acute myeloid leukemia. *J. Virol.* 75, 9427-9434.

Blyth, K., Terry, A., Mackay, N., Vaillant, F., Bell, M., Cameron, E.R., Neil, J.C., Stewart, M., 2001. Runx2: a novel oncogenic effector revealed by in vivo complementation and retroviral tagging. *Oncogene* 20, 295-302.

Bowerman, B., Brown, P.O., Bishop, J.M., Varmus, H.E., 1989. A nucleoprotein complex mediates the integration of retroviral DNA. *Genes Dev.* 3, 469-478.

Brown, P.O., Bowerman, B., Varmus, H.E., Bishop, J.M., 1987. Correct integration of retroviral DNA in vitro. *Cell* 49, 347-356.

Brown, P.O., Bowerman, B., Varmus, H.E., Bishop, J.M., 1989. Retroviral integration: structure of the initial covalent product and its precursor, and a role for the viral IN protein. *Proc. Natl. Acad. Sci.* U. S. A 86, 2525-2529.

Buckley, A.R., Leff, M.A., Buckley, D.J., Magnuson, N.S., de Jong, G., Gout, P.W., 1996. Alterations in pim-1 and c-myc expression associated with sodium butyrate-induced growth factor dependency in autonomous rat Nb2 lymphoma cells. *Cell Growth Differ.* 7, 1713-1721.

Bushman, F., 2002. Targeting retroviral integration? *Mol. Ther.* 6, 570-571.

Castilla, L.H., Wijmenga, C., Wang, Q., Stacy, T., Speck, N.A., Eckhaus, M., Marin-Padilla, M., Collins, F.S., Wynshaw-Boris, A., Liu, P.P., 1996. Failure of embryonic hematopoiesis and lethal hemorrhages in mouse embryos heterozygous for a knocked-in leukemia gene *CBFB-MYH11*. *Cell* 87, 687-696.

Castilla, L.H., Garrett, L., Adya, N., Orlic, D., Dutra, A., Anderson, S., Owens, J., Eckhaus, M., Bodine, D., Liu, P.P., 1999. The fusion gene Cbfb-MYH11 blocks myeloid differentiation and predisposes mice to acute myelomonocytic leukaemia. *Nat. Genet.* 23, 144-146.

Castilla, L.H., Perrat, P., Martinez, N.J., Landrette, S.F., Keys, R., Oikemus, S., Flanegan, J., Heilman, S., Garrett, L., Dutra, A., Anderson, S., Pihan, G.A., Wolff, L., Liu, P.P., 2004. Identification of genes that synergize with Cbfb-MYH11 in the pathogenesis of acute myeloid leukemia. *Proc. Natl. Acad. Sci.* U. S. A 101, 4924-4929.

Ceci, J.D., Patriotis, C.P., Tsatsanis, C., Makris, A.M., Kovatch, R., Swing, D.A., Jenkins, N.A., Tsichlis, P.N., Copeland, N.G., 1997. Tpl-2 is an oncogenic kinase that is activated by carboxy-terminal truncation. *Genes Dev.* 11, 688-700.

Chatis, P.A., Holland, C.A., Hartley, J.W., Rowe, W.P., Hopkins, N., 1983. Role for the 3' end of the genome in determining disease specificity of Friend and Moloney murine leukemia viruses. *Proc. Natl. Acad. Sci.* USA 80, 4408-4411.

Chatis, P.A., Holland, C.A., Silver, J.E., Frederickson, T.N., Hopkins, N., Hartley, J.W., 1984. A 3' end fragment encompassing the transcriptional enhancers of nondefective Friend virus confers erythroleukemogenicity on Moloney leukemia virus. *J. Virol.* 52, 248-254.

Chesebro, B., Portis, J.L., Wehrly, K., Nishio, J., 1983. Effect of murine host genotype on MCF virus expression, latency, and leukemia cell type of leukemias induced by Friend murine leukemia helper virus. Virology 128, 221-233.

Coffin, J.M., 1996. Retroviridae:The Viruses and Their Replication. In: Fields B.N., Knipe D.M., Howley P.M. (Eds.), *Fields Virology* . Vol. 2.

Coffin J.M., Hughes, S.H., Varmus, H.E.(Eds.), 1997a. Retroviruses. *Cold Spring Harbor Laboratory Press.*

Coffin, J.M., Hughes S.H., Varmus H.E., 1997b. The Interactions of retroviruses and their hosts. In: Coffin, J.M., Hughes S.H., Varmus H.E. (Eds.), Retroviruses. *Cold Spring Harbor Press,* New York, pp. 335-342.

Corcoran, L.M., Adams, J.M., Dunn, A.R., Cory, S., 1984. Murine T lymphomas in which the cellular myc oncogene has been activated by retroviral insertion. *Cell* 37, 113-122.

Cullen, B.R., Lomedico, P.T., Ju, G., 1984. Transcriptional interference in avian retroviruses--implications for the promoter insertion model of leukaemogenesis. *Nature* 307, 241-245.

Daniel, R., Kao, G., Taganov, K., Greger, J.G., Favorova, O., Merkel, G., Yen, T.J., Katz, R.A., Skalka, A.M., 2003. Evidence that the retroviral DNA integration process triggers an ATR-dependent DNA damage response. *Proc. Natl. Acad. Sci.* U. S. A 100, 4778-4783.

Daniel, R., Katz, R.A., Skalka, A.M., 1999. A role for DNA-PK in retroviral DNA integration. *Science* 284, 644-647.

Dedera, D.A., Waller, E.K., LeBrun, D.P., Sen-Majumdar, A., Stevens, M.E., Barsh, G.S., Cleary, M.L., 1993. Chimeric homeobox gene E2A-PBX1 induces proliferation, apoptosis, and malignant lymphomas in transgenic mice. *Cell* 74, 833-843.

DesGroseillers, L., Jolicoeur, P., 1983. Physical mapping of the Fv-1 tropism host range determinant of BALB/c murine leukemia viruses. *J. Virol.* 48, 685-696.

Devon, R.S., Porteous, D.J., Brookes, A.J., 1995. Splinkerettes--improved vectorettes for greater efficiency in PCR walking. *Nucleic Acids Res.* 23, 1644-1645.

Domen, J., van der Lugt, N.M., Acton, D., Laird, P.W., Linders, K., Berns, A., 1993a. Pim-1 levels determine the size of early B lymphoid compartments in bone marrow. *J. Exp. Med.* 178, 1665-1673.

Domen, J., van der Lugt, N.M., Laird, P.W., Saris, C.J., Clarke, A.R., Hooper, M.L., Berns, A., 1993b. Impaired interleukin-3 response in Pim-1-deficient bone marrow-derived mast cells. *Blood* 82, 1445-1452.

Erkeland, S.J., Valkhof, M., Heijmans-Antonissen, C., van Hoven-Beijen, A., Delwel, R., Hermans, M.H.A., Touw, I.P., 2004. Large-scale identification of disease genes involved in acute myeloid leukemia. *J.Virol.* 78, 1971-1980.

Durst, K.L., Lutterbach, B., Kummalue, T., Friedman, A.D., Hiebert, S.W., 2003. The inv(16) fusion protein associates with corepressors via a smooth muscle myosin heavy-chain domain. Mol. *Cell Biol.* 23, 607-619.

Fan, H., 1997. Leukemogenesis by Moloney murine leukemia virus: a multistep process. *Trends Microbiol.* 5, 74-82.

Feldman, B.J., Reid, T.R., Cleary, M.L., 1997. Pim1 cooperates with E2a-Pbx1 to facilitate the progression of thymic lymphomas in transgenic mice. *Oncogene* 15, 2735-2742.

Feldman, B.J., Hampton, T., Cleary, M.L., 2000. A carboxy-terminal deletion mutant of Notch1 accelerates lymphoid oncogenesis in E2A-PBX1 transgenic mice. *Blood 96,* 1906-1913.

Finkel, M.P., Biskis, B.O., Jinkins, P.B., 1966. Virus induction of osteosarcomas in mice. *Science* 151, 698-701.

Fredrickson, T.N., Langdon, W.Y., Hoffman, P.M., Hartley, J.W., Morse, H.C., III, 1984. Histologic and cell surface antigen studies of hematopoietic tumors induced by Cas-Br-M murine leukemia virus. *J. Natl. Cancer Inst.* 72, 447-454.

Friend, C., 1957. Leukemia of adult mice caused by a transmissible agent. *Ann. N. Y. Acad. Sci.* 68, 522-532.

Gardner, M.B., Henderson, B.E., Officer, J.E., Rongey, R.W., Parker, J.C., Oliver, C., Estes, J.D., Huebner, R.J., 1973. A spontaneous lower motor neuron disease apparently caused by indigenous type-C RNA virus in wild mice. *J. Natl. Cancer Inst.* 51, 1243-1254.

Girard, L., Hanna, Z., Beaulieu, N., Hoemann, C.D., Simard, C., Kozak, C.A., Jolicoeur, P., 1996. Frequent provirus insertional mutagenesis of Notch1 in thymomas of MMTVD/myc transgenic mice suggests a collaboration of c-myc and Notch1 for oncogenesis. *Genes Dev.* 10, 1930-1944.

Gisselbrecht, S., Pozo, F., Debre, P., Hurot, M.A., Lacombe, M.J., Levy, J.P., 1978. Genetic control of sensitivity to Moloney-virus-induced leukemias in mice. I. Demonstration of multigenic control. *Int. J. Cancer* 21, 626-634.

Goff, S.P., 1996. Operating under a Gag order: a block against incoming virus by the Fv1 gene. *Cell* 86, 691-693.

Golemis, E., Li, Y., Fredrickson, T.N., Hartley, J.W., Hopkins, N., 1989. Distinct segments within the enhancer region collaborate to specify the type of leukemia induced by nondefective Friend and Moloney viruses. *J. Virol.* 63, 328-337.

Graffi, A., 1957. Chloroleukemia in mice. Ann. N. Y. Acad. Sci. 68, 540-558.

Graffi, A., Fey, F., Schramm, T., 1966. Experiments on the hematologic diversification of viral mouse leukemias. *Natl. Cancer Inst. Monogr.* 22, 21-31.

Gross, L., 1970. Oncogenic viruses. *Pergamon Press, Elmsford.*

Hanahan, D., Weinberg, R.A., 2000. The hallmarks of cancer. *Cell* 100, 57-70.

Hartley, J.W., Rowe, W.P., 1976. Naturally occurring murine leukemia viruses in wild mice: characterization of a new "amphotropic" class. *J. Virol.* 19, 19-25.

Harvey, J.J., 1964. An unidentified virus which causes the rapid production of tumours in mice. *Nature* 204, 1104-1105.

Hasenkrug, K.J., Chesebro, B., 1997. Immunity to retroviral infection: the Friend virus model. *Proc. Natl. Acad. Sci.* U. S. A 94, 7811-7816.

Haupt, Y., Alexander, W.S., Barri, G., Klinken, S.P., Adams, J.M., 1991. Novel zinc finger gene implicated as myc collaborator by retrovirally accelerated lymphomagenesis in E mu-myc transgenic mice. *Cell* 65, 753-763.

Haupt, Y., Harris, A.W., Adams, J.M., 1992. Retroviral infection accelerates T lymphomagenesis in E mu-N-ras transgenic mice by activating c-myc or N-myc. *Oncogene* 7, 981-986.

Haupt, Y., Harris, A.W., Adams, J.M., 1993. Moloney virus induction of T-cell lymphomas in a plasmacytomagenic strain of E mu-v-abl transgenic mice. *Int. J. Cancer* 55, 623-629.

Haviernik, P., Festin, S.M., Opavsky, R., Koller, R.P., Barr, N.I., Neil, J.C., Wolff, L., 2002. Linkage on chromosome 10 of several murine retroviral integration loci associated with leukaemia. *J. Gen. Virol.* 83, 819-827.

Hayward, W.S., Neel, B.G., Astrin, S.M., 1981. Activation of a cellular onc gene by promoter insertion in ALV-induced lymphoid leukosis. *Nature* 290, 475-480.

Hengen, P.N., 1995. Vectorette, splinkerette and boomerang DNA amplification. *Trends Biochem. Sci.* 20, 372-373.

Holmes, K., Langdon, W.Y., Fredrickson, T.N., Coffmann, R.L., Hoffman, P.M., Hartley, J., Morse III, H.C., 1986. Analysis of neoplasms induced by Cas-Br-M MuLV tumor extracts. J. *Immunol.* 137, 679-688.

Huang, G., Shigesada, K., Ito, K., Wee, H.J., Yokomizo, T., Ito, Y., 2001. Dimerization with PEBP2beta protects RUNX1/AML1 from ubiquitin-proteasome-mediated degradation. *EMBO J.* 20, 723-733.

Hubbard, T., Barker, D., Birney, E., Cameron, G., Chen, Y., Clark, L., Cox, T., Cuff, J., Curwen, V., Down, T., Durbin, R., Eyras, E., Gilbert, J., Hammond, M., Huminiecki, L., Kasprzyk, A., Lehvaslaiho, H., Lijnzaad, P., Melsopp, C., Mongin, E., Pettett, R., Pocock, M., Potter, S., Rust, A., Schmidt, E., Searle, S., Slater, G., Smith, J., Spooner, W., Stabenau, A., Stalker, J., Stupka, E., Ureta-Vidal, A., Vastrik, I., Clamp, M., 2002. The Ensembl genome database project. *Nucleic Acids Res.* 30, 38-41.

Hwang, H.C., Martins, C.P., Bronkhorst, Y., Randel, E., Berns, A., Fero, M., Clurman, B.E., 2002. Identification of oncogenes collaborating with p27Kip1 loss by insertional mutagenesis and high-throughput insertion site analysis. *Proc. Natl. Acad. Sci.* U. S. A. 99, 11293-11298.

Jacobs, J.J., Kieboom, K., Marino, S., DePinho, R.A., van Lohuizen, M., 1999. The oncogene and Polycomb-group gene bmi-1 regulates cell proliferation and senescence through the ink4a locus. *Nature* 397, 164-168.

Jenkins, N.A., Copeland, N.G., Taylor, B.A., Bedigian, H.G., Lee, B.K., 1982. Ecotropic murine leukemia virus DNA content of normal and lymphomatous tissues of BXH-2 recombinant inbred mice. *J. Virol.* 42, 379-388.

Johnson, P., Benchimol, S., 1992. Friend virus induced murine erythroleukaemia: the p53 locus. *Cancer Surv.* 12, 137-151.

Jolicoeur, P., Nicolaiew, N., DesGroseillers, L., Rassart, E., 1983. Molecular cloning of infectious viral DNA from ecotropic neurotropic wild mouse retrovirus. *J. Virol.* 45, 1159-1163.

Joosten, M., Vankan-Berkhoudt, Y., Tas, M., Lunghi, M., Jenniskens, Y., Parganas, E., Valk, P.J., Lowenberg, B., van den, A.E., Delwel, R., 2002. Large-scale identification of novel potential disease loci in mouse leukemia applying an improved strategy for cloning common virus integration sites. *Oncogene* 21, 7247-7255.

Kamps, M.P., Look, A.T., Baltimore, D., 1991. The human t(1;19) translocation in pre-B ALL produces multiple nuclear E2A-Pbx1 fusion proteins with differing transforming potentials. *Genes Dev.* 5, 358-368.

Kerlavage, A., Bonazzi, V., di Tommaso, M., Lawrence, C., Li, P., Mayberry, F., Mural, R., Nodell, M., Yandell, M., Zhang, J., Thomas, P., 2002. The Celera Discovery System. *Nucleic Acids Res.* 30, 129-136.

Kiem, H.P., Sellers, S., Thomasson, B., Morris, J.C., Tisdale, J.F., Horn, P.A., Hematti, P., Adler, R., Kuramoto, K., Calmels, B., Bonifacino, A., Hu, J., von Kalle, C., Schmidt, M., Sorrentino, B., Nienhuis, A., Blau, C.A., Andrews, R.G., Donahue, R.E., Dunbar, C.E., 2004. Long-term clinical and molecular follow-up of large animals receiving retrovirally transduced stem and progenitor cells: no progression to clonal hematopoiesis or leukemia. *Mol. Ther.* 9, 389-395.

Kim, R., Trubetskoy, A., Suzuki, T., Jenkins, N.A., Copeland, N.G., Lenz, J., 2003. Genome-based identification of cancer genes by proviral tagging in mouse retrovirus-induced T-cell lymphomas. *J. Virol.* 77, 2056-2062.

Kirchner, J., Connolly, C.M., Sandmeyer, S.B., 1995. Requirement of RNA polymerase III transcription factors for in vitro position-specific integration of a retroviruslike element. *Science* 267, 1488-1491.

Knudson, A.G., Jr., 1971. Mutation and cancer: statistical study of retinoblastoma. *Proc. Natl. Acad. Sci.* U. S. A.68, 820-823.

Knudson, A.G., 2001. Two genetic hits (more or less) to cancer. *Nat. Rev. Cancer* 1, 157-162.

Koller, R., Krall, M., Mock, B., Bies, J., Nazarov, V., Wolff, L., 1996. *Mml*1, a new common integration site in murine leukemia virus-induced promonocytic leukemias maps to mouse chromosome 10. *Virology* 224, 224-234.

Kone, J., Arroyo, J., Savinelli, T., Lin, S., Boyd, K., Wu, Y., Nimmakayalu, M., Copeland, N.G., Jenkins, N.A., Qumsiyeh, M., Hu, P., Prescott, A., Wu, H., Yang, L., Roe, B., Perkins, A.S., 2002. F-MuLV acceleration of myelomonocytic tumorigenesis in SV40 large T antigen transgenic mice is accompanied by retroviral insertion at Fli1 and a novel locus, Fim4. *Leukemia* 16, 1827-1834.

Kozak C.A., Ruscetti, S., 1992. Retroviruses in Rodents. In: Levy, J.A. (Ed.), The Retroviridae Volume 1. *Plenum Press,* pp. 405-481.

Kundu, M., Liu, P.P., 2001. Function of the inv(16) fusion gene CBFB-MYH11. *Curr. Opin. Hematol.* 8, 201-205.

Kundu, M., Liu, P.P., 2003. Cbfbeta is involved in maturation of all lineages of hematopoietic cells during embryogenesis except erythroid. *Blood Cells Mol. Dis.* 30, 164-169.

Largaespada, D.A., Shaughnessy, J.D., Jr., Jenkins, N.A., Copeland, N.G., 1995. Retroviral integration at the *Evi-2* locus in BXH-2 myeloid leukemia cell lines disrupts Nfl expression without changes in steady-state Ras-GTP levels. *J. Virol.* 69, 5095-5102.

Largaespada, D.A., Brannan, C.I., Shaughnessy, J.D., Jenkins, N.A., Copeland, N.G., 1996. The neurofibromatosis type 1 (*NF1*) tumor suppressor gene and myeloid leukemia. Curr. Top. Microbiol. *Immunol.* 211, 233-239.

Latres, E., Malumbres, M., Sotillo, R., Martin, J., Ortega, S., Martin-Caballero, J., Flores, J.M., Cordon-Cardo, C., Barbacid, M., 2000. Limited overlapping roles of P15^{INK4b} and P18^{INK4c} cell cycle inhibitors in proliferation and tumorigenesis. *EMBO J.* 19, 3496-3506.

Lazo, P.A., Lee, J.S., Tsichlis, P.N., 1990. Long-distance activation of the Myc protooncogene by provirus insertion in *Mlvi-1* or *Mlvi-4* in rat T-cell lymphomas. *Proc. Natl. Acad. Sci.* USA 87, 170-173.

Lenz, J., Celander, D., Crowther, R.L., Patarca, R., Perkins, D.W., Haseltine, W.A., 1984. Determination of the leukaemogenicity of a murine retrovirus by sequences within the long terminal repeat. *Nature* 308, 467-470.

Lilly, F., 1967. Susceptibility to two strains of Friend leukemia virus in mice. *Science* 155, 461-462.

Lund, A.H., Turner, G., Trubetskoy, A., Verhoeven, E., Wientjens, E., Hulsman, D., Russell, R., DePinho, R.A., Lenz, J., van Lohuizen, M., 2002. Genome-wide retroviral insertional tagging of genes involved in cancer in Cdkn2a-deficient mice. *Nat. Genet.* 32, 160-165.

Makris, A., Patriotis, C., Bear, S.E., Tsichlis, P.N., 1993. Genomic organization and expression of Tpl-2 in normal cells and Moloney murine leukemia virus-induced rat T-cell lymphomas: activation by provirus insertion. *J. Virol.* 67, 4283-4289.

Meyers, S., Lenny, N., Sun, W., Hiebert, S.W., 1996. AML-2 is a potential target for transcriptional regulation by the t(8;21) and t(12;21) fusion proteins in acute leukemia. *Oncogene* 13, 303-312.

Mikkers, H., Allen, J., Berns, A., 2002a. Proviral activation of the tumor suppressor E2a contributes to T cell lymphomagenesis in EmuMyc transgenic mice. *Oncogene* 21, 6559-6566.

Mikkers, H., Allen, J., Knipscheer, P., Romeijn, L., Hart, A., Vink, E., Berns, A., Romeyn, L., 2002b. High-throughput retroviral tagging to identify components of specific signaling pathways in cancer. *Nat. Genet.* 2002. Sep. ;32. (1):153. -9. 32, 153-159.

Mikkers, H., Berns, A., 2003. Retroviral insertional mutagenesis: tagging cancer pathways. *Adv. Cancer Res.* 88, 53-99.

Minard, M.E., Kim, L.S., Price, J.E., Gallick, G.E., 2004. The role of the Guanine nucleotide exchange factor tiaml in cellular migration, invasion, adhesion and tumor progression. *Breast Cancer Res. Treat.* 84, 21-32.

Moloney, J.B., 1960. Biological studies on a lymphoid-leukemia virus extracted from sarcoma 37. I. Origin and introductory investigations. *J. Natl. Cancer Inst.* 24:933-51., 933-951.

Moloney, J.B., 1966. A virus-induced rhabdomyosarcoma of mice. *Natl. Cancer Inst. Monogr* 22:139-42., 139-142.

Mucenski, M.L., Taylor, B.A., Jenkins, N.A., Copeland, N.G., 1986. AKXD recombinant inbred strains: models for studying the molecular genetic basis of murine lymphomas. *Mol. Cell. Biol.* 6, 4236-4243.

Mucenski, M.L., Taylor, B.A., Copeland, N.G., Jenkins, N.A., 1987. Characterization of somatically acquired ecotropic and mink cell focus-forming viruses in lymphomas of AKXD recombinant inbred mice. *J. Virol.* 61, 2929-2933.

Mukhopadhyaya, R., Wolff, L., 1992. New sites of proviral integration associated with murine promonocytic leukemias and evidence for alternate modes of c-*myb* activation. *J. Virol.* 66, 6035-6044.

Nason-Burchenal, K., Wolff, L., 1993. Activation of c-*myb* is an early bone-marrow event in a murine model for acute promonocytic leukemia. *Proc. Natl. Acad. Sci.* USA 90, 1619-1623.

Nazarov, V., Wolff, L., 1995. Novel integration sites at the distal 3' end of the c-*myb* locus in retrovirus-induced promonocytic leukemias. *J. Virol.* 69, 3885-3888.

Nourse, J., Mellentin, J.D., Galili, N., Wilkinson, J., Stanbridge, E., Smith, S.D., Cleary, M.L., 1990. Chromosomal translocation t(1;19) results in synthesis of a homeobox fusion mRNA that codes for a potential chimeric transcription factor. *Cell* 60, 535-545.

O'Donnell, P.V., Fleissner, E., Lonial, H., Koehne, C.F., Reicin, A., 1985. Early clonality and high-frequency proviral integration into the c-myc locus in AKR leukemias. *J. Virol.* 55, 500-503.

Ochman, H., Gerber, A.S., Hartl, D.L., 1988. Genetic applications of an inverse polymerase chain reaction. *Genetics* 120, 621-623.

Odaka, T., 1969. Inheritance of susceptibility to Friend mouse leukemia virus. V. Introduction of a gene responsible for susceptibility in the genetic complement of resistant mice. *J. Virol.* 3, 543-548.

Ogawa, E., Inuzuka, M., Maruyama, M., Satake, M., Naito-Fujimoto, M., Ito, Y., Shigesada, K., 1993a. Molecular cloning and characterization of PEBP2 beta, the heterodimeric partner of a novel Drosophila runt-related DNA binding protein PEBP2 alpha. *Virology* 194, 314-331.

Ogawa, E., Maruyama, M., Kagoshima, H., Inuzuka, M., Lu, J., Satake, M., Shigesada, K., Ito, Y., 1993b. PEBP2/PEA2 represents a family of transcription factors homologous to the products of the Drosophila runt gene and the human AML1 gene. *Proc. Natl. Acad. Sci.* U. S. A 90, 6859-6863.

Oliff, A.I., Hager, G.L., Chang, E.H., Scolnick, E.M., Chan, H.W., Lowy, D.R., 1980. Transfection of molecularly cloned Friend murine leukemia virus DNA yields a highly leukemogenic helper-independent type C virus. *J. Virol.* 33, 475-486.

Ott, D.E., Keller, J., Sill, K., Rein, A., 1992. Phenotypes of murine leukemia virus-induced tumors: influence of 3' viral coding sequences. *J. Virol.* 66, 6107-6116.

Paquette, Y., Doyon, L., Laperriere, A., Hanna, Z., Ball, J., Sekaly, R.P., Jolicoeur, P., 1992. A viral long terminal repeat expressed in CD4+CD8+ precursors is downregulated in mature peripheral CD4-CD8+ or CD4+CD8- T cells. *Mol. Cell Biol.* 12, 3522-3530.

Patriotis, C., Makris, A., Bear, S.E., Tsichlis, P.N., 1993. Tumor progression locus 2 (Tpl-2) encodes a protein kinase involved in the progression of rodent T-cell lymphomas and in T-cell activation. *Proc. Natl. Acad. Sci.* U. S. A 90, 2251-2255.

Perryman, S., Nishio, J., Chesebro, B., 1991. Complete nucleotide sequence of Friend murine leukemia virus, strain FB29. *Nucleic Acids Res.* 19, 6950.

Pincus, T., Hartley, J.W., Rowe, W.P., 1971a. A major genetic locus affecting resistance to infection with murine leukemia viruses. I. Tissue culture studies of naturally occurring viruses. *J. Exp. Med.* 133, 1219-1233.

Pincus, T., Rowe, W.P., Lilly, F., 1971b. A major genetic locus affecting resistance to infection with murine leukemia viruses. II. Apparent identity to a major locus described for resistance to friend murine leukemia virus. *J. Exp. Med.* 133, 1234-1241.

Pruss, D., Reeves, R., Bushman, F.D., Wolffe, A.P., 1994. The influence of DNA and nucleosome structure on integration events directed by HIV integrase. *J. Biol. Chem.* 269, 25031-25041.

Pryciak, P.M., Varmus, H.E., 1992. Nucleosomes, DNA-binding proteins, and DNA sequence modulate retroviral integration target site selection. *Cell* 69, 769-780.

Quon, K.C., Berns, A., 2001. Haplo-insufficiency? Let me count the ways. *Genes Dev.* 15, 2917-2921.

Rohdewohld, H., Weiher, H., Reik, W., Jaenisch, R., Breindl, M., 1987. Retrovirus integration and chromatin structure: Moloney murine leukemia proviral integration sites map near DNase I-hypersensitive sites. *J. Virol.* 61, 336-343.

Rosenberg, N., Jolicoeur, P., 1997. Retroviral Pathogenesis. In: Coffin, J.M., Hughes S.H., Varmus H.E. (Eds.), Retroviruses. *Cold Spring Harbor Press,* New York, pp. 475-586Ru, M., Shustik, C., Rassart, E., 1993.

Graffi murine leukemia virus: molecular cloning and characterization of the myeloid leukemia-inducing agent. *J. Virol.* 67, 4722-4731.

Rynditch, A.V., Zoubak, S., Tsyba, L., Tryapitsina-Guley, N., Bernardi, G., 1998. The regional integration of retroviral sequences into the mosaic genomes of mammals. *Gene* 222, 1-16.

Santarosa, M., Ashworth, A., 2004. Haploinsufficiency for tumour suppressor genes: when you don't need to go all the way. *Biochim. Biophys. Acta.*

Satake, M., Nomura, S., Yamaguchi-Iwai, Y., Takahama, Y., Hashimoto, Y., Niki, M., Kitamura, Y., Ito, Y., 1995. Expression of the Runt domain-encoding PEBP2 alpha genes in T cells during thymic development. *Mol. Cell Biol.* 15, 1662-1670.

Scheijen, B., Jonkers, J., Acton, D., Berns, A., 1997. Characterization of pal-1, a common proviral insertion site in murine leukemia virus-induced lymphomas of c-myc and Pim-1 transgenic mice. *J. Virol.* 71, 9-16.

Scherdin, U., Rhodes, K., Breindl, M., 1990. Transcriptionally active genome regions are preferred targets for retrovirus integration. *J. Virol.* 64, 907-912.

Schmidt, M., Glimm, H., Lemke, N., Muessig, A., Speckmann, C., Haas, S., Zickler, P., Hoffmann, G., von Kalle, C., 2001a. A model for the detection of clonality in marked hematopoietic stem cells. *Ann. N. Y. Acad. Sci.* 938:146-55; discussion 155-6., 146-155.

Schmidt, M., Hoffmann, G., Wissler, M., Lemke, N., Mussig, A., Glimm, H., Williams, D.A., Ragg, S., Hesemann, C.U., von Kalle, C., 2001b. Detection and direct genomic sequencing of multiple rare unknown flanking DNA in highly complex samples. *Hum. Gene Ther.* 12, 743-749.

Selten, G., Cuypers, H.T., Zijlstra, M., Melief, C., Berns, A., 1984. Involvement of c-myc in MuLV-induced T cell lymphomas in mice: frequency and mechanisms of activation. *EMBO J.* 3, 3215-3222.

Selten, G., Cuypers, H.T., Berns, A., 1985. Proviral activation of the putative oncogene Pim-1 in MuLV induced T-cell lymphomas. *EMBO J.* 4, 1793-1798.

Serrano, M., Lee, H., Chin, L., Cordon-Cardo, C., Beach, D., DePinho, R.A., 1996. Role of the INK4a locus in tumor suppression and cell mortality. *Cell* 85, 27-37.

Shen-Ong, G.L., Morse, H.C.3., Potter, M., Mushinski, J.F., 1986. Two modes of c-*myb* activation in virus-induced mouse myeloid tumors. *Mol. Cell. Biol.* 6, 380-392.

Shen-Ong, G.L., Wolff, L., 1987. Moloney murine leukemia virus-induced myeloid tumors in adult BALB/c mice: requirement of c-*myb* activation but lack of v-*abl* involvement. *J. Virol.* 61, 3721-3725.

Shibuya, T., Mak, T.W., 1982. Host control of susceptibility to erythroleukemia and to the types of leukemia induced by Friend murine leukemia virus: initial and late stages. *Cell* 31, 483-493.

Shinnick, T.M., Lerner, R.A., Sutcliffe, J.G., 1981. Nucleotide sequence of Moloney murine leukaemia virus. *Nature* 293, 543-548.

Silver, J.E., Fredrickson, T.N., 1983. A new gene that controls the type of leukemia induced by Friend murine leukemia virus. *J. Exp. Med.* 158, 493-505.

Skalnik, D.G., Dorfman, D.M., Perkins, A.S., Jenkins, N.A., Copeland, N.G., Orkin, S.H., 1991. Targeting of transgene expression to monocyte/macrophages by the gp91-phox promoter and consequent histiocytic malignancies. *Proc. Natl. Acad. Sci.* U. S. A. 88, 8505-8509.

Stewart, M., Cameron, E., Campbell, M., McFarlane, R., Toth, S., Lang, K., Onions, D., Neil, J.C., 1993. Conditional expression and oncogenicity of c-myc linked to a CD2 gene dominant control region. *Int. J. Cancer* 53, 1023-1030.

Sola, B., Heard, J.M., Fichelson, S., Martial, M.A., Pozo, F., Bordereaux, D., Gisselbrecht, S., 1985. Monoclonal proliferation of Friend murine leukemia virus-transformed myeloblastic cells occurs early in the leukemogenic process. *Mol. Cell Biol.* 5, 1009-1014.

Suzuki, T., Shen, H., Akagi, K., Morse, H.C., Malley, J.D., Naiman, D.Q., Jenkins, N.A., Copeland, N.G., 2002. New genes involved in cancer identified by retroviral tagging. *Nat. Genet.* 32, 166-174.

Taketo, M., Schroeder, A.C., Mobraaten, L.E., Gunning, K.B., Hanten, G., Fox, R.R., Roderick, T.H., Stewart, C.L., Lilly, F., Hansen, C.T., ., 1991. FVB/N: an inbred mouse strain preferable for transgenic analyses. *Proc. Natl. Acad. Sci.* U. S. A 88, 2065-2069.

Taylor, W.R., Stoye, J.P., 2004. Consensus structural models for the amino terminal domain of the retrovirus restriction gene Fv1 and the Murine Leukaemia Virus capsid proteins. BMC. *Struct. Biol.* 4, 1.

Triglia, T., Peterson, M.G., Kemp, D.J., 1988. A procedure for in vitro amplification of DNA segments that lie outside the boundaries of known sequences. *Nucleic Acids Res.* 16, 8186.

Troxler, D.H., Scolnick, E.M., 1978. Rapid leukemia induced by cloned friend strain of replicating murine type-C virus. Association with induction of xenotropic-related RNA sequences contained in spleen focus-forming virus. *Virology* 85, 17-27.

Turcotte, K., Gauthier, S., Mitsos, L.M., Shustik, C., Copeland, N.G., Jenkins, N.A., Fournet, J.C., Jolicoeur, P., Gros, P., 2004. Genetic control of myeloproliferation in BXH-2 mice. *Blood* 103, 2343-2350.

Vaillant, F., Blyth, K., Terry, A., Bell, M., Cameron, E.R., Neil, J., Stewart, M., 1999. A full-length Cbfa1 gene product perturbs T-cell development and promotes lymphomagenesis in synergy with myc. *Oncogene* 18, 7124-7134.

van der Lugt, N.M., Domen, J., Verhoeven, E., Linders, K., van der, G.H., Allen, J., Berns, A., 1995. Proviral tagging in E mu-myc transgenic mice lacking the Pim-1 proto-oncogene leads to compensatory activation of Pim-2. *EMBO J.* 14, 2536-2544.

Verbeek, S., van Lohuizen, M., van, d., V, Domen, J., Kraal, G., Berns, A., 1991. Mice bearing the E mu-myc and E mu-pim-1 transgenes develop pre-B-cell leukemia prenatally. *Mol. Cell Biol.* 11, 1176-1179.

van Lohuizen, M., Verbeek, S., Krimpenfort, P., Domen, J., Saris, C., Radaszkiewicz, T., Berns, A., 1989. Predisposition to lymphomagenesis in pim-1 transgenic mice: cooperation with c-myc and N-myc in murine leukemia virus-induced tumors. *Cell* 56, 673-682.

van Lohuizen, M., Verbeek, S., Scheijen, B., Wientjens, E., van der, G.H., Berns, A., 1991. Identification of cooperating oncogenes in E mu-myc transgenic mice by provirus tagging. *Cell* 65, 737-752.

Villemur, R., Rassart, E., DesGroseillers, L., Jolicoeur, P., 1983. Molecular cloning of viral DNA from leukemogenic Gross passage A murine leukemia virus and nucleotide sequence of its long terminal repeat. *J. Virol.* 45, 539-546.

Westaway, D., Payne, G., Varmus, H.E., 1984. Proviral deletions and oncogene base-substitutions in insertionally mutagenized c-myc alleles may contribute to the progression of avian bursal tumors. *Proc. Natl. Acad. Sci.* U. S. A 81, 843-847.

Wolff, L., Koller, R., 1990. Regions of the Moloney murine leukemia virus genome specifically related to induction of promonocytic tumors. *J. Virol.* 64, 155-160.

Wolff, L., Mushinski, J.F., Shen-Ong, G.L., Morse III, H.C., 1988. A chronic inflammatory response. Its role in supporting the development of c-*myb* and c-*myc* related promonocytic and monocytic tumors in BALB/c mice. *J. Immunol.* 141, 681-689.

Wolff, L., Koller, R., Davidson, W., 1991. Acute myeloid leukemia induction by amphotropic murine retrovirus (4070A): clonal integrations involve c-*myb* in some but not all leukemias. *J. Virol.* 65, 3607-3616.

Wolff, L., Garin, M.T., Koller, R., Bies, J., Liao, W., Malumbres, M., Tessarollo, L., Powell, D., Perella, C., 2003a. Hypermethylation of the Ink4b locus in murine myeloid leukemia and increased susceptibility to leukemia in p15(Ink4b)-deficient mice. *Oncogene* 22, 9265-9274.

Wolff, L., Koller, R., Hu, X., Anver, M.R., 2003b. A Moloney murine leukemia virus-based retrovirus with 4070A long terminal repeat sequences induces a high incidence of myeloid as well as lymphoid neoplasms. *J. Virol.* 77, 4965-4971.

Wong, K.S., Li, Y.J., Howard, J., Ben David, Y., 1999. Loss of p53 in F-MuLV induced-erythroleukemias accelerates the acquisition of mutational events that confers immortality and growth factor independence. *Oncogene* 18, 5525-5534.Wu, X., Li, Y., Crise, B., Burgess, S.M., 2003. Transcription start regions in the human genome are favored targets for MLV integration. *Science* 300, 1749-1751.

Yap, M.W., Stoye, J.P., 2003. Intracellular localisation of Fv1. *Virology* 307, 76-89.

In: Progress in Virus Research
Editor: Sabine Thebault, pp. 301-316

ISBN 1-59454-438-7

Chapter X

Insights into Integration Site Selection: Application for Vector Design

Jacqueline Lehmann-Che [*] ***and Ali Saïb***
CNRS UPR9051, Hôpital Saint-Louis
1, Avenue Claude Vellefaux
75475 Paris Cedex 10, France

Abstract

Retroviral vectors are promising tools for gene transfer, in part because retroviral integration precisely links therapeutic sequences to host chromosomes. However, although providing stable transgene expression, vector integration may have deleterious effects, mainly due to insertional mutagenesis. In that sense, the sites of insertion of HIV-1 and MLV have been mainly mapped to RNA pol II highly transcribed genes, strengthening the potential danger of using these viral agents as vectors for gene therapy. Interestingly, other retrovirus-like elements are known to preserve the integrity of the host genome during their replication. This is the case of Ty3, a yeast gypsy-like retroelement, which integrates upstream of RNA pol III dependent promoter start sites without deleterious effects for host survival. Ty 5, another yeast retrotransposon, specifically inserts into regions of silent chromatin. Such site selection is driven by specific interactions occurring between the viral integration machinery and host proteins, allowing a balance between the fitness of the host and the ability of the retrotransposon to propagate and survive in the yeast genome. Understanding the precise interactions between cellular and viral partners involved in the integration process could certainly lead to the elaboration of safer therapeutic vectors, specifically targeting the corrective gene to chosen regions of the host genome allowing a reasonable level of transgene expression without affecting cellular gene transcription. The goal of this review is to summarize our current knowledge on site specificity of integration of retrovirus-like elements, and the recent progress made in integration site selection.

[*] jlehmann@chu-stlouis.fr, tel: 33.1.53.72.40.96 – fax: 33.1.53.72.40.90

Introduction

Gene therapy promises to revolutionize medicine by treating the causes of diseases rather than the symptoms [1, 2]. The principle is very simple, putting corrective genetic material into target deficient cells. This new approach has the important advantage to be applicable to a wide range of human pathologies ranging from monogenic diseases to cardiovascular or infectious illnesses and cancer. The first gene therapy workshop took place in the United States in 1971 dealing with the transfer of therapeutic genes in hereditary diseases (hemoglobinopathies and Lesh-Nyhan syndrome), while the first clinical studies began in 1990 by the use of a retroviral vector to transfer the adenosine deaminase enzyme (ADA) gene in two children suffering from severe combined immunodeficiency (ADA-SCID) [3, 4]. Since then, more than 600 clinical trials took place in the world.

One of the main challenges of gene therapists is to find good vehicles to deliver the corrective gene. Three main types of delivery have been developed using either viruses, non viral vectors (naked DNA, cationic lipid complexed DNA, cationic polymers condensed DNA) or physical ways (injectors and electroporation). Viral vectors still represent the most efficient gene transfer system dealing with 70% of clinical trials (http://www.wiley.co.uk/genetherapy/clinical/). Viruses are highly evolved natural vectors delivering foreign genetic information, their genome, into target cells. The recombinant viral particles generated for gene transfer harbor a modified genome carrying the therapeutic gene in place of parts of the viral genome. This modified virus, which is generally replication defective, will be able to infect the target cell and to initiate the first steps of its biological life cycle, delivering the functional genetic information into the host cell [5]. Many different vectors have been adapted for gene transfer, the most extensively developed being retroviruses [6-8], adenoviruses (Ad) [9, 10] and adeno-associated viruses (AAV) [11], although other viruses such as poxviruses or herpes simplex viruses [12, 13] are also of particular interest [14].

Retroviruses are the most largely used vectors since the beginning of the gene therapy, representing 34.1% of clinical trials in 2002 (http://www.wiley.co.uk/genetherapy/clinical/). Since the first generation of retroviral vectors, approximately 300 clinical trials were conducted and 150 are still active in 2002. Retroviruses are lipid-enveloped particles containing two linear, single strand RNA molecules. After entry in the target cell, the viral RNA genome is reverse transcribed into a double strand DNA, which integrates into host chromosomes, equally transmitted to daughter cells following mitosis like any other cellular genes. This unique property maintains the genetic information life-long in the cell genome and constitutes a major advantage when gene correction has to be continuous. Note that integration into the host genome was accepted to be random and the chance of accidentally disruption or deregulated expression of a host gene extremely low. Distinct families of retroviruses were developed for gene transfer, namely oncoretroviruses (murine and avian C-type retroviruses), lentiviruses and spumaviruses [15-17]. Recently, vectors derived from the murine leukemia oncoretroviruses (MLV) were used in the first definitive cure of a genetic disease by gene therapy. Three young children suffering from SCID-X1 syndrome, characterized by the absence of mature T and natural killer cells, recovered a functional immune system after administration of their own haematopoietic stems cells transduced *ex vivo* with an MLV vector carrying the wild-type γc chain cytokine receptor gene [18, 19].

Since the publication of these data, seven other children were successfully treated. Unfortunately, two of the ten children developed a leukemia-like disorder at least in part because of independent insertions of the murine retroviral vector near the same *LMO2* proto-oncogene, leading to clonal expansion of the corresponding transduced T cells [20, 21]. This represents the first description of insertional mutagenesis following a clinical trial using a murine retroviral vector in humans, raising the old question of the potential danger of such viruses known to cause somatic and germline mutations leading to cancers and inherited disorders in their natural hosts. Indeed, murine leukaemia viruses, which are slow transforming viruses, are successfully used for the identification of essential cellular genes involved in tumorigenesis, a technique called provirus tagging [22, 23].

In that sense, the sites of insertion of HIV-1 and MLV genomes and of derived vectors have been mainly mapped into RNA pol II highly transcribed genes, strengthening the potential danger of their use in gene therapy [24 , 25]. Therefore, understanding the precise mechanisms governing vector integration is an essential prerequisite before going further in the use of retroviruses as tools.

Although the molecular basis of site specificity is unclear and debated in the case of animal retroviruses, much more is known on other retrovirus-like elements known to preserve the integrity of the host genome while actively replicating. This is for example the case of most yeast retrotransposons. Similar to mammalian retroviruses, retrotransposons are flanked by direct repeats (LTRs), contain a similar arrangement of their genes, use similar mechanisms to replicate and share strong reverse transcriptase homologies [26]. However, an extracellular phase is not observed in the case of retrotransposons since they do not encode an envelope glycoprotein. These elements must integrate their DNA into a gene-rich, densely packed and timely haploid genome without suicidal disruption of essential genes [27]. Indeed, *S. pombe* genome is 60% coding whereas that of *S. cerevisiae* reaches 70%. And therefore, integration in such an environment has to be tightly regulated to avoid suicidal insertion events. Initial studies, which were confirmed by the complete genome sequencing of *Saccharomyces* in 1996, revealed that yeast retrotransposons exhibits a strong bias in site integration [28]. For example, Ty3, a yeast gypsy-like element, integrates within a thigh window of 1 to 4 nucleotides upstream of RNA polymerase (Pol) III dependent promoter start sites without deleterious effects either for gene transcription or host survival [29]. Ty 5, another yeast retrotransposons, specifically inserts into regions of silent chromatin [30-32]. Such site selection is driven by specific interactions occurring between the viral integration machinery and host proteins, guiding the integration machinery to proper locations on the genome. This tethering strategy ensures a balance between the fitness of the host and the ability of the retrotransposon to propagate and survive in the yeast genome.

Understanding the precise interactions between cellular and viral partners involved during the integration process could certainly lead to the construction of safer therapeutic vectors, targeting the transgene to specific regions of the host genome without deleterious effects. The goal of this review is to summarize our current knowledge on the site specificity of integration of yeast, insect and animal retroelements, and the recent progress made in the design of site specific integrative vectors.

The Yeast Retrotransposons

The yeast provides an excellent model for studying the molecular basis of integration of retroelements, the influence of cellular proteins, but also of cis-acting sequences on site selection. Yeast LTR-containing retrotransposons display structural as well as mechanistic similarities with animal retroviruses [26]. Encoding for Gag and Pol polyproteins, they are transcribed into a full-length genomic RNA subsequently packaged into virus like particles (VLPs) in which reverse transcription takes place. The newly synthesized viral cDNA, together with certain viral and cellular proteins constituting the pre-integration complex (PIC), is then imported into the nucleus and integrates into the host genome without an extracellular phase [33]. Several families of yeast retrovirus-like elements have been characterized: the copia-like family including Ty1, 2, 4 and 5 elements of *S. cerevisiae*, the gypsy-like family including Ty 3 of *S. cerevisiae* and Tf1 and 2 from *S. pompe* [26].

Early studies have described a strong insertion specificity of Ty3 into regions lying upstream of the tRNA, 5S, and U6 genes in *S. cerevisiae*, three RNA PolIII transcribed genes [34]. Integration occurs specifically within a short window of 1 to 4 nucleotides of the transcription initiation site which do not harbor conserved sequence homologies [29], suggesting that such specificity could be mediated by an interaction between the Ty3 PIC and components of the RNA polIII transcription apparatus. That functional promoter elements are required for site specificity and that mutations affecting binding of RNA PolIII transcription factors at positions distant from the integration site inhibits Ty3 integration, further support this hypothesis [29, 35]. A similar pattern of integration was successfully reproduced *in vitro* using Ty3 VLPs, purified Pol III transcription complexes and target plasmids harboring RNA PolIII dependent promoters [36]. Site specificity precisely required the presence of the Ty3 integrase (IN), TFIIIB and TFIIIC, two high molecular weight cellular complexes recruited on RNA PolIII dependent transcriptional units (for more details see [37]). Recent data demonstrate that TFIIIB is the essential cellular factor tethering the Ty3 PIC to the site of integration since the presence of recombinant TBP and Brf, two components of TFIIIB, are sufficient to trigger site selection. Although the docking specificity is provided by a direct interaction between Brf and the Ty3 IN, Bdp1, another TFIIIB component, contributes to the efficiency of integration by stabilizing the multiprotein complex [38]. In this model, TFIIIC is necessary to specify the orientation of TFIIIB binding on the promoter [38, 39].

Similarly, Ty1 integration occurs preferentially in upstream regions of tRNA genes [28, 40, 41], also under experimental conditions inducing retrotransposition [27, 41]. Although Ty1 integration occurs within a 750 bases window of upstream of RNA PolIII transcribed genes [28], similar to Ty2 and Ty4, the precise molecular events underlying such specificity are not yet known. Thus, both retroelements, Ty1 and Ty3, have developed distinct strategies to finally reach a similar goal, targeting gene-free regions upstream of PolIII-transcribed genes.

Ty5, another copia-like retrotransposon, displays another integration specificity by exclusively integrating near telomeres and adjacent to the silent mating loci (*HMR* and *HML*), sites which tolerate insertion events without deleterious effect for the host [31]. Analysis of *de novo* transposition events into *S. cerevisiae* confirms this targeting preference within or adjacent to transcriptional silencers flanking *HML* and *HMR* or the subtelomeric X repeats [30]. These HM silencers mediate silent chromatin assembly by the recruitment of a large number of cellular proteins including the *Silent Information Regulatory Proteins* (SIR1-4), involved in telomere function and transcriptional inactivation of adjacent genes (for more details see [42]). Although mutations within sequences that disrupt the assembly of silent chromatin at the HM loci abolish Ty5 integration into these particular loci [32], the DNA sequence *per se* is not the critical parameter required for site specificity. In that sense, a natural Ty5 mutant harboring a single amino acid mutation at the boundary of integrase and reverse transcriptase do no longer present such site specificity [43]. This mutation altered probably a domain that mediated Ty5 target choice by interacting with a component of silent chromatin, a hypothesis recently confirmed. Indeed, a specific interaction between a 6 amino acids domain in the C-terminus of the Ty5 integrase (the targeting domain TD) and Sir4p, a structural component of silent chromatin, is sufficient to direct Ty5 integration into silent chromatin [44]. Reciprocally, loss of Sir3p or Sir4p causes a greater than nine-fold decrease in Ty5 targeting into the HM loci [45].

Table 1. Main properties of viral vectors

Vector	Tropism	Packaging capacity	Vector genome forms	Main limitations	Main avantages	References
Retrovirus	broad if pseudotyped Proliferating cells only	8kb	integrated	transduction of proliferating cells only risk of insertional mutagenesis	persistent gene transfert	[6]
Lentivirus	broad if pseudotyped	8 kb	integrated	risk of insertional mutagenesis	persistent gene transfert Transduction of non-proliferating cells	[7-8]
Adenovirus	broad	8 to 30 kb (helper dependent)	episomal	strong inflammatory response transient gene transfert	strong transduction efficiency in proliferating and non-proliferating cells	[9-10]
AAV	broad	< 5kb	episomal >90% integrated < 10%	small packaging capacity	non pathogenic non inflammatory	[11]
HSV-1	particularly neurons	40 to 150 kb	episomal	inflammatory transient gene transfer in cells other than neurons	large packaging capacity strong tropism for neurons	[12-13]

Although belonging to the gypsy-like group of LTR retroelements like Ty3, Tf1 colonizes the genome of the fission yeast *S. pombe* which is largely divergent from *S. cerevisiae*, providing the unique opportunity to compare the molecular basis of site selection in two different host genomes. In contrast to the Ty3 retrotransposon, Tf1 specifically integrates into non-coding regions between RNA PolII transcribed genes [46, 47]. Moreover, Tf1 integration occurs preferentially into the host chromosome III, within insertion sites harboring no apparent sequence similarities [47]. The complete sequence of the *S. pombe* genome reveals that full-length Tf1 insertion occurs in the promoter-proximal regions of RNA PolII-transcribed genes, 100 to 400 bp upstream of coding regions, suggestive for an interaction between RNA PolII specific transcription factors and components of the Tf1 PIC [48]. However, although Tf1 integration occurs near RNA PolII transcriptional units, no significant alteration of gene expression was observed [46]. Recent evidences suggest that Tf1 site specificity occurs through interaction of the chromodomain of IN with histone H3 methylated at lysine 4 [49].

Therefore, many distinct mechanisms have been developed by yeast retrotransposons to achieve a similar goal, replication without host genome disruption. All these strategies are based on a tethering mechanism, guiding the integration machinery to adequate genomic sites [50].

The Drosophila Model

Some drosophila retroelements are also known to integrate into specific regions of the host genome. Zam, a retrovirus-like element present within the genome of *D. melanogaster,* exemplifies a clear-cut specificity of integration. This retroelement, which was initially identified by its insertion at the white locus, encodes for three open reading frames resembling the products of the *gag*, *pol* and *env* of animal retroviruses. The latter encodes an envelope glycoprotein that is typical of retroelements having an extracellular phase [51, 52]. Interestingly, the Zam integration machinery selects the 5' GCGCGCg 3' sequence for integration, reminiscent of the well-characterized property of the sequence specific restriction endonucleases that trigger DNA digestion in bacteria [53]. However, the molecular basis of this specificity is not yet understood.

Another example has also been described in drosophila. As in other eukaryotes, telomeres in *D. melanogaster* are composed of long arrays of repeated DNA sequences but remarkably, in this insect species, these repeats are composed by several copies of two retrotransposons, HeT- A and TART [54]. Both retroelements transpose specifically at the ends of chromosomes in *D. melanogaster*, a phenomenon also described in *D. yakuba* and *D. virilis* [55, 56], maintaining correct chromosome length in these species lacking a functional telomerase. Although the two telomeric retroelements encode very similar Gag proteins, which are both efficiently targeted into the nucleus [57], only HeT-A Gag harbors the determinants targeting it to chromosome ends [58]. Interestingly, TART Gag localizes to telomeres only when HeT-A Gag is expressed, suggestive for an interaction between both molecules [57, 58]. Of note, HeT- A does not encode a reverse transcriptase, a function which is provided *in trans* by TART. Therefore, a symbiotic relationship between these two retroelements is essential to replicate and target chromosome ends. HeT- A and TART are the first transposable elements harboring a bona fide role in cell structure, revealing an unexpected link between telomeres and what is generally considered to be parasitic DNA [59].

Animal Retroviruses

Similarly, the molecular basis of *in vivo* integration of animal retroviruses remains poorly understood. Soon after the discovery that what have been initially called RNA tumor viruses were in fact retroviruses, it has been rapidly shown that tumorigenesis could result from insertional mutagenesis of such animal oncogenic viruses within or near host essential genes, activating or disrupting their expression. Initial studies on retrovirus integration have demonstrated that insertion occurred in a generally sequence non specific fashion but may be influenced by the structure of the chromatin [60], pointing out that retroviral integration occurred randomly along the host genome [61]. However, further studies revealed that integration is finally not a random process but was favoured in open chromatin, near active genes [62, 63]. In that sense, MLV integration was shown to occur within DNAseI-hypersensitive chromatin regions, suggesting that actively transcribed genes are preferred targets for provirus insertion [64-66], while HIV-1 integration was never observed in centromeric alphoid repeats [67]. Conversely, transcriptional active regions are disfavoured sites of integration for avian leukosis viruses (ALV) [68].

Interestingly, a global picture of the integration pattern of a given retrovirus is now possible, thanks to the complete sequencing of the human genome. Indeed, Schröder *et al.* have succeeded in mapping over 500 integration events of HIV-1 and of derived vectors following infection of a human T cell line, revealing that integration preferentially occurred in genes highly transcribed by the RNA PolII [25]. This specificity may therefore favor efficient HIV-1 gene expression, maximizing virus propagation while deleterious to host survival. In the case of MLV, Wu *et al.* have mapped 903 different integration sites in the human genome, revealing that 34% of the integration events landed in transcribed genes, compared to 58% for HIV-1 and 22% if the integration was considered to be at random [24]. MLV integration events distributes evenly upstream and downstream of the transcriptional start site of actively transcribed genes, +/- 1kb from the CpG islands, whereas HIV-1 proviruses are found on the entire length of the transcriptional unit [24].

Such regional preferences along the host genome, in the absence of sequence specificity, suggest that integration may be influenced by specific interaction occurring between host proteins and viral components and/or by a specific chromatin architecture at these sites. Although *in vitro* integration models have identified many cofactors, which seem to be required for efficient integration, the molecular mechanisms involved in animal retrovirus integration still remain unclear.

To get insights into the molecular basis of site specificity, the interactions between host cellular components and incoming viral proteins -those involved in the constitution of the PIC and especially the retroviral integrase- have been studied in great details. Indeed, several studies have implicated the integrase as a key factor determining the site of integration. This protein, a member of the D, D(35)E transposase/IN superfamily of proteins, mediates integration of the viral DNA into the host genome [69].We know for example that the integrase of FIV, HIV and visna virus display distinct preference of integration sites when given an identical DNA target *in vitro* [70, 71]. In the case of HIV-1, the viral model in which most studies were performed, several cellular DNA binding proteins interacting with the integrase were described and may therefore constitute good candidates for directing the PIC to its target site. The integrase interactor 1 (Ini1, also called hSNF5), a subunit of the SWI/SNF chromatin-remodeling complex, was initially isolated by a yeast two hybrid screen for human proteins interacting with the IN [72] and was proposed to stimulate the *in vitro* DNA-joining activity of the IN and to target the viral genome to active genes in a way which remains to be documented [72-75]. Equally, HMG-I(Y), a non-histone chromosomal protein important for transcriptional control and chromosome architecture, and the *barrier-to-*

autointegration factor (BAF), a cellular protein involved in the reorganization of post-mitotic nuclei, have been identified as partners of the HIV-1 IN. Both proteins appear to be required for efficient integration *in vitro*, but their respective role in directing the PIC to precise sites of the host genome was not evaluated [76 , 77-80]

More recently, two other IN-interacting partners were isolated which seem to be critical for directing the PIC to the host chromatin. This is the case of the EED protein, which also interacts with the matrix (MA) protein of HIV-1 [81], the homolog of the mouse embryonic ectoderm development (*eed*) gene product and of the Drosophila esc gene [82, 83]. These genes belong to the family of widely conserved *Polycomb* group of genes, involved in the maintenance of the silent state of chromatin and reduction of DNA accessibility, similar to the Sir proteins already discussed for the yeast model. Intranuclear interaction occurring between the viral proteins MA and IN and EED might not only direct the PIC to the host chromatin but also deregulate silent cellular genes by releasing the binding of EED to factors involved in the transcription machinery, triggering transcriptional activation [81]. Lastly, the lens epithelium-derived growth factor (LEDGF/p75), a protein implicated in the regulation of gene expression and cellular stress response was found to interact with the HIV-1 IN [84]. Interestingly, this interaction is not essential for nuclear accumulation of the HIV-1 IN, but seems to be absolutely required to dock the PIC to the host chromatin ([85] and S. Emiliani, personal communication).

Clearly, more studies are needed to clarify the complex network of interactions occurring during the early steps of infection guiding the retroviral PIC to its site of integration in the host genome.

Targeted Integration by Design

The precise knowledge of the molecular events, the cellular and viral partners guiding the yeast retrotransposon DNA to the host chromatin rapidly led to the development of several technical approaches intending to manipulate such specificity. For example, Ty3 integration could occur in a human tRNA gene inserted into the yeast genome [86], whereas a replicative recombinant MLV harboring a chimeric Ty3/MLV IN protein, in which the C-terminal domain of the MLV integrase was replaced with that of Ty3, was unable to ensure the yeast retrotransposon specificity [87]. Although this assay was unsuccessful, it represented the first step towards the development of engineered IN enzymes in a heterologous organism conferring novel insertion specificities.

More recently, Ty5 integration specificity for heterochromatin was successfully modified by replacing the 6aa TD of the IN with small peptide motifs like a 13aa sequence from Rad9 or a 12aa stretch from NpwBP, two short domains known to interact with protein partners such as FHA-1 or Npw38 respectively [88]. In a tethering assays, these modified Ty5-IN display hot spot insertions within LexA operators of the target plasmid in the presence of the protein partners fused to the LexA binding domain [88]. These results indicate that Ty5 target specificity can be controlled by engineered IN.

Table 2. Integration specificity of main retroelements

Element	Type	Family	Host organism	Integration pattern	Target mechanism	References
Tyl	LTR retrotransposon	copia-like	S. cerevisiae	750bp window upstream of PolIII genes	PolIII transcription apparatus	[28, 41]
Ty3	LTR retrotransposon	gypsy-like	*S. cerevisiae*	1-4 bp window upstream of PolIII genes	Pol III transcription complexes (TFIIIB,TFIIIC)	[28, 35-36]
Ty5	LTR retrotransposon	copia-like	*S. cerevisiae*	heterochromatin at telomeres and silent mating loci	silent chromatin components (Sir4p)	[28, 30, 44]
Tfl	LTR retrotransposon	gypsy-like	S. pombe	100-400bp upstream of PolII genes	unknown	[47-48]
HeT-A and TART	non-LTR retrotransposon		D. melanogaster	telomeres	unknown	[54]
ZAM	non-LTR retrotransposon		*D. melanogaster, D. yakuba, D. virilis*	5'GCGCGCg3'	unknown	[53]
MLV	retrovirus	oncoretrovirus	M. musculus H. sapiens	upstream and downstream of actively Pol II genes	unknown	[24]
HIV	retrovirus	lentivirus	*H. sapiens*	entire length of actively transcribed PolII genes	unknown	[25]

In the case of animal retroviruses, a similar approach, based on chimera between IN and sequence-specific DNA-binding proteins (DBD), was also envisaged. One of the first experiments was performed with a fusion protein between HIV-1 IN and the DBD of the phage lambda repressor (λR) in *in vitro* integration assays. Under these settings, the λR-IN fusion was able to direct integration preferentially in target sequence harbouring λ operators, demonstrating that a tethering mechanism might control integration in the context of animal retroviruses [89]. Similarly, fusion between parts of the *E. coli* LexA protein and the C-terminus of the HIV-1 IN preferentially integrated the viral genome into a DNA domain containing LexA binding sites *in vitro* [90]. However, this approach was unsuccessful in the context of the infectious virus since the fusion protein was hardly incorporated into virus particles [91, 92]. To cope with this limitation, the chimeric IN-LexA protein was delivered *in trans* into IN-defective particles by a secondary fusion with the accessory Vpr protein, leading to the production of infectious viruses harbouring the fusion IN-LexA-Vpr protein [93, 94]. However, the authors did not assess the specificity of site integration of this recombinant virus since there are no naturally occurring LexA-binding sites in mammalian DNA. Note that a similar approach was successfully performed with the DBD of the zing finger protein Zif268 fused to the HIV-1 IN [92] and more recently with the synthetic polydactyl zinc finger protein E2C [109]. Altogether, these data clearly indicate that the IN-fusions do not alter the catalytic function of the retroviral integrase and demonstrate the feasibility of incorporating such fusion proteins into retroviral particles for the development of site-directed retroviral vectors, directing integration to innocuous sites of the host genome, or, alternatively to deleterious DNA sequences such as activated oncogenes or viral genomes which can therefore be inactivated by insertional mutagenesis [92, 94].

In the same idea, efficient site-specific integration could be achieved with enzymes that belong to the site-specific recombinase family [95]. Indeed, certain recombinases could function without any cofactor and therefore in foreign cellular environments. For example, recombinase such as CRE from the *E. coli* phage P1 acts efficiently in mammalian or plant cells, performing not only insertion but also excision of the gene of interest [96]. When integration alone is desired, other specialized recombinases, which require specific cofactors for the excision process, can be used. This is the case of the *Streptomyces* phage ϕC31 integrase which belongs to the resolvase family of recombinases. This integrase catalyses the recombination between attachment *attP* sites on the phage DNA and the *attB* sites on the bacteria genome [97]. The enzyme may be able to mediate integration at naturally occurring pseudo-att sites in eucaryotic genomes having significant similarities with the native att sites [98-100].

Conclusion

Successful proliferation of retrotransposons is due to their ability to identify safe havens in the genome where the integration is not harmful for host survival. This site selection is based on a tethering mechanism of the PIC, especially the integrase, and host DNA-bound proteins. However, in contrast to the yeast model, little is known on the precise mechanics leading to site selection of retroviruses from higher eukaryotes. Studies of retroviral integration at the level of the whole genome, allowed by the richness of the information from the genome project, suggest that a similar tethering strategy was adopted by animal retroviruses such as HIV and MLV to insert their genome into highly transcribed genes. In that sense, it could be interesting to precisely analyze the integration pattern of other retroviruses, such as HTLV-I [101] or foamy viruses. These latter are complex apathogenic animal retroviruses establishing a life long infection in their natural hosts [102, 103]. Interestingly, these viruses present several specific features rendering them suitable for the development of gene therapy vectors [16]. Interestingly, analysis of the few sequenced provirus-cellular DNA junctions available in the literature reveals a preference for non-coding regions of the host genome [104]. If such site specificity is confirmed, this will partly explained the lack of tumorigenesis following long term infection by foamy viruses, contrasting to what has been repeatedly observed for slow transforming retroviruses.

The yeast model has unexpectedly, but significantly, improved our understanding of the integration process regarding animal retroviruses [105]. This provides an excellent system to study the mechanics of retrotransposon integration but also the influence of host genes, which can affect distinct steps of the retrotransposon life [106]. Indeed, functional genomics screen for host factors that influence Ty retrotransposition reveals that several gene products, identified as host defense factors which are able to limit Ty activity, were conserved in other organisms [107], likely providing a starting point for identifying host factors implicated in retroviral restriction of pathogenic viruses [108].

Understanding the stepwise molecular interactions occurring between cell components and viral proteins of the PIC responsible for guiding the viral genome to its integration site will be necessary to fully understand the risk factors and advantages of different retroviral gene-therapy systems. Moreover, this knowledge is clearly indispensable for the development of new generations of engineered safer retroviral vectors harbouring chosen site specificity [105]. For that purpose, comparing the specificity of retroviral integrase, and of other components of the PIC which may influence chromatin docking, and defining the protein domains involved in determining site selection will allow the possibility to engineer this enzyme without loss of *in vivo* function. Moreover, this will certainly open new areas of research for the discovery of antiretroviral therapies targeting early phases of retroviral infection.

References

[1] Pfeifer, A. and I.M. Verma, Gene therapy: promises and problems. *Annu Rev Genomics Hum Genet,* 2001. **2**: p. 177-211.

[2] Mountain, A., Gene therapy: the first decade. *Trends Biotechnol,* 2000. **18**(3): p. 119-28.

[3] Blaese, R.M., et al., T lymphocyte-directed gene therapy for ADA- SCID: initial trial results after 4 years. *Science,* 1995. **270**(5235): p. 475-80.

[4] Blaese, R.M., Steps toward gene therapy: 1. The initial trials. *Hosp Pract* (Off Ed), 1995. **30**(11): p. 33-40.

[5] Kay, M.A., J.C. Glorioso, and L. Naldini, Viral vectors for gene therapy: the art of turning infectious agents into vehicles of therapeutics. *Nat Med,* 2001. **7**(1): p. 33-40.

[6] Kim, S.H., S. Kim, and P.D. Robbins, Retroviral vectors. *Adv Virus Res,* 2000. **55**: p. 545-63.

[7] Jakobsson, J., et al., Lentiviral vectors. *Int Rev Neurobiol,* 2003. **55**: p. 111-22.

[8] Ailles, L.N., L.,). HIV-1-derived lentiviral vectors, in Lentiviral vectors, T. D., Editor. 2000, *Springer.*

[9] Lehtonen, E. and L. Tenenbaum, Adeno-associated viral vectors. *Int Rev Neurobiol,* 2003. **55**: p. 65-98.

[10] St George, J.A., Gene therapy progress and prospects: adenoviral vectors. *Gene Ther,* 2003. **10**(14): p. 1135-41.

[11] Buning, H., et al., AAV-based gene transfer. *Curr Opin Mol Ther*, 2003. **5**(4): p. 367-75.

[12] Burton, E.A., D.J. Fink, and J.C. Glorioso, Gene delivery using herpes simplex virus vectors. *DNA Cell Biol,* 2002. **21**(12): p. 915-36.

[13] Feldman, L.T., Problems in the use of herpes simplex virus as a vector. *Int Rev Neurobiol,* 2003. **55**: p. 99-109.

[14] Thomas, C.E., A. Ehrhardt, and M.A. Kay, Progress and problems with the use of viral vectors for gene therapy. *Nat Rev Genet,* 2003. **4**(5): p. 346-58.

[15] Weber, E., W.F. Anderson, and N. Kasahara, Recent advances in retrovirus vector-mediated gene therapy: teaching an old vector new tricks. *Curr Opin Mol Ther,* 2001. **3**(5): p. 439-53.

[16] Pandya, S., E. Klimatcheva, and V. Planelles, Lentivirus and foamy virus vectors: novel gene therapy tools. *Expert Opin Biol Ther,* 2001. **1**(1): p. 17-40.

[17] Kafri, T., Lentivirus vectors: difficulties and hopes before clinical trials. *Curr Opin Mol Ther,* 2001. **3**(4): p. 316-26.

[18] Cavazzana-Calvo, M., et al., Gene therapy of human severe combined immunodeficiency (SCID)-X1 disease. *Science,* 2000. **288**(5466): p. 669-72.

[19] Hacein-Bey-Abina, S., et al., Sustained correction of X-linked severe combined immunodeficiency by ex vivo gene therapy. *N Engl J Med,* 2002. **346**(16): p. 1185-93.

[20] Hacein-Bey-Abina, S., et al., A serious adverse event after successful gene therapy for X-linked severe combined immunodeficiency. *N Engl J Med,* 2003. **348**(3): p. 255-6.

[21] Hacein-Bey-Abina, S., et al., LMO2-associated clonal T cell proliferation in two patients after gene therapy for SCID-X1. *Science,* 2003. **302**(5644): p. 415-9.

[22] Mikkers, H. and A. Berns, Retroviral insertional mutagenesis: tagging cancer pathways. *Adv Cancer Res,* 2003. **88**: p. 53-99.

[23] Berns, A., Provirus tagging as an instrument to identify oncogenes and to establish synergism between oncogenes. *Arch Virol,* 1988. **102**(1-2): p. 1-18.

[24] Wu, X., et al., Transcription start regions in the human genome are favored targets for MLV integration. *Science,* 2003. **300**(5626): p. 1749-51.

[25] Schroder, A.R., et al., HIV-1 integration in the human genome favors active genes and local hotspots. *Cell,* 2002. **110**(4): p. 521-9.

[26] Sandmeyer, S.B., Yeast retrotransposons. *Curr Opin Genet Dev,* 1992. **2**(5): p. 705-11.

[27] Boeke, J.D. and S.E. Devine, Yeast retrotransposons: finding a nice quiet neighborhood. *Cell,* 1998. **93**(7): p. 1087-9.

[28] Kim, J.M., et al., Transposable elements and genome organization: a comprehensive survey of retrotransposons revealed by the complete Saccharomyces cerevisiae genome sequence. *Genome Res*, 1998. **8**(5): p. 464-78.

[29] Chalker, D.L. and S.B. Sandmeyer, Ty3 integrates within the region of RNA polymerase III transcription initiation. *Genes Dev,* 1992. **6**(1): p. 117-28.

[30] Zou, S., et al., The Saccharomyces retrotransposon Ty5 integrates preferentially into regions of silent chromatin at the telomeres and mating loci. *Genes Dev,* 1996. **10**(5): p. 634-45.

[31] Zou, S., D.A. Wright, and D.F. Voytas, The Saccharomyces Ty5 retrotransposon family is associated with origins of DNA replication at the telomeres and the silent mating locus HMR. *Proc Natl Acad Sci U S A,* 1995. **92**(3): p. 920-4.

[32] Zou, S. and D.F. Voytas, Silent chromatin determines target preference of the Saccharomyces retrotransposon Ty5. *Proc Natl Acad Sci* U S A, 1997. **94**(14): p. 7412-6.

[33] Dang, V.D. and H.L. Levin, Nuclear import of the retrotransposon Tf1 is governed by a nuclear localization signal that possesses a unique requirement for the FXFG nuclear pore factor Nup124p. *Mol Cell Biol,* 2000. **20**(20): p. 7798-812.

[34] Hansen, L.J., D.L. Chalker, and S.B. Sandmeyer, Ty3, a yeast retrotransposon associated with tRNA genes, has homology to animal retroviruses. *Mol Cell Biol,* 1988. **8**(12): p. 5245-56.

[35] Chalker, D.L. and S.B. Sandmeyer, Sites of RNA polymerase III transcription initiation and Ty3 integration at the U6 gene are positioned by the TATA box. *Proc Natl Acad Sci* U S A, 1993. **90**(11): p. 4927-31.

[36] Kirchner, J., C.M. Connolly, and S.B. Sandmeyer, Requirement of RNA polymerase III transcription factors for in vitro position-specific integration of a retroviruslike element. *Science,* 1995. **267**(5203): p. 1488-91.

[37] Geiduschek, E.P. and G.A. Kassavetis, The RNA polymerase III transcription apparatus. *J Mol Biol,* 2001. **310**(1): p. 1-26.

[38] Yieh, L., et al., The Brf and TATA-binding protein subunits of the RNA polymerase III transcription factor IIIB mediate position-specific integration of the gypsy-like element, Ty3. *J Biol Chem,* 2000. **275**(38): p. 29800-7.

[39] Yieh, L., et al., Mutational analysis of the transcription factor IIIB-DNA target of Ty3 retroelement integration. *J Biol Chem,* 2002. **277**(29): p. 25920-8.

[40] Ji, H., et al., Hotspots for unselected Ty1 transposition events on yeast chromosome III are near tRNA genes and LTR sequences. *Cell,* 1993. **73**(5): p. 1007-18.

[41] Devine, S.E. and J.D. Boeke, Integration of the yeast retrotransposon Ty1 is targeted to regions upstream of genes transcribed by RNA polymerase III. *Genes Dev,* 1996. **10**(5): p. 620-33.

[42] Gasser, S.M. and M.M. Cockell, The molecular biology of the SIR proteins. *Gene,* 2001. **279**(1): p. 1-16.

[43] Gai, X. and D.F. Voytas, A single amino acid change in the yeast retrotransposon Ty5 abolishes targeting to silent chromatin. *Mol Cell,* 1998. **1**(7): p. 1051-5.

[44] Xie, W., et al., Targeting of the yeast Ty5 retrotransposon to silent chromatin is mediated by interactions between integrase and Sir4p. *Mol Cell Biol,* 2001. **21**(19): p. 6606-14.

[45] Zhu, Y., et al., Tagging chromatin with retrotransposons: target specificity of the Saccharomyces Ty5 retrotransposon changes with the chromosomal localization of Sir3p and Sir4p. *Genes Dev,* 1999. **13**(20): p. 2738-49.

[46] Behrens, R., J. Hayles, and P. Nurse, Fission yeast retrotransposon Tf1 integration is targeted to 5' ends of open reading frames. *Nucleic Acids Res,* 2000. **28**(23): p. 4709-16.

[47] Singleton, T.L. and H.L. Levin, A long terminal repeat retrotransposon of fission yeast has strong preferences for specific sites of insertion. *Eukaryot Cell,* 2002. **1**(1): p. 44-55.

[48] Bowen, N.J., et al., Retrotransposons and their recognition of pol II promoters: a comprehensive survey of the transposable elements from the complete genome sequence of Schizosaccharomyces pombe. *Genome Res,* 2003. **13**(9): p. 1984-97.

[49] Sandmeyer, S., Integration by design. *Proc Natl Acad Sci* U S A, 2003. **100**(10): p. 5586-8.

[50] Sandmeyer, S., Targeting transposition: at home in the genome. *Genome Res,* 1998. **8**(5): p. 416-8.

[51] Leblanc, P., et al., Invertebrate retroviruses: ZAM a new candidate in D.melanogaster. *Embo J,* 1997. **16**(24): p. 7521-31.

[52] Baldrich, E., et al., Genomic distribution of the retrovirus-like element ZAM in Drosophila. *Genetica,* 1997. **100**(1-3): p. 131-40.

[53] Leblanc, P., B. Dastugue, and C. Vaury, The integration machinery of ZAM, a retroelement from Drosophila melanogaster, acts as a sequence-specific endonuclease. *J Virol,* 1999. **73**(8): p. 7061-4.

[54] Levis, R.W., et al., Transposons in place of telomeric repeats at a Drosophila telomere. *Cell,* 1993. **75**(6): p. 1083-93.

[55] Casacuberta, E. and M.L. Pardue, Coevolution of the telomeric retrotransposons across Drosophila species. *Genetics,* 2002. **161**(3): p. 1113-24.

[56] Casacuberta, E. and M.L. Pardue, Transposon telomeres are widely distributed in the Drosophila genus: TART elements in the virilis group. *Proc Natl Acad Sci* U S A, 2003. **100**(6): p. 3363-8.

[57] Rashkova, S., et al., Gag proteins of the two Drosophila telomeric retrotransposons are targeted to chromosome ends. *J Cell Biol,* 2002. **159**(3): p. 397-402.

[58] Rashkova, S., A. Athanasiadis, and M.L. Pardue, Intracellular targeting of Gag proteins of the Drosophila telomeric retrotransposons. *J Virol,* 2003. **77**(11): p. 6376-84.

[59] Pardue, M.L. and P.G. DeBaryshe, Retrotransposons provide an evolutionarily robust non-telomerase mechanism to maintain telomeres. *Annu Rev Genet,* 2003. **37**: p. 485-511.

[60] Pryciak, P.M. and H.E. Varmus, Nucleosomes, DNA-binding proteins, and DNA sequence modulate retroviral integration target site selection. *Cell,* 1992. **69**(5): p. 769-80.

[61] Withers-Ward, E.S., et al., Distribution of targets for avian retrovirus DNA integration in vivo. *Genes Dev,* 1994. **8**(12): p. 1473-87.

[62] Shih, C.C., J.P. Stoye, and J.M. Coffin, Highly preferred targets for retrovirus integration. *Cell,* 1988. **53**(4): p. 531-7.

[63] Craigie, R., Hotspots and warm spots: integration specificity of retroelements. *Trends Genet,* 1992. **8**(6): p. 187-90.

[64] Scherdin, U., K. Rhodes, and M. Breindl, Transcriptionally active genome regions are preferred targets for retrovirus integration. *J Virol,* 1990. **64**(2): p. 907-12.

[65] Vijaya, S., D.L. Steffen, and H.L. Robinson, Acceptor sites for retroviral integrations map near DNase I-hypersensitive sites in chromatin. *J Virol,* 1986. **60**(2): p. 683-92.

[66] Mooslehner, K., U. Karls, and K. Harbers, Retroviral integration sites in transgenic Mov mice frequently map in the vicinity of transcribed DNA regions. *J Virol,* 1990. **64**(6): p. 3056-8.

[67] Carteau, S., C. Hoffmann, and F. Bushman, Chromosome structure and human immunodeficiency virus type 1 cDNA integration: centromeric alphoid repeats are a disfavored target. *J Virol,* 1998. **72**(5): p. 4005-14.

[68] Weidhaas, J.B., et al., Relationship between retroviral DNA integration and gene expression. *J Virol,* 2000. **74**(18): p. 8382-9.

[69] Asante-Appiah, E. and A.M. Skalka, Molecular mechanisms in retrovirus DNA integration. Antiviral Res, 1997. **36**(3): p. 139-56.

[70] Shibagaki, Y. and S.A. Chow, Central core domain of retroviral integrase is responsible for target site selection. *J Biol Chem,* 1997. **272**(13): p. 8361-9.

[71] Katzman, M. and M. Sudol, Mapping domains of retroviral integrase responsible for viral DNA specificity and target site selection by analysis of chimeras between human immunodeficiency virus type 1 and visna virus integrases. *J Virol,* 1995. **69**(9): p. 5687-96.

[72] Kalpana, G.V., et al., Binding and stimulation of HIV-1 integrase by a human homolog of yeast transcription factor SNF5. *Science,* 1994. **266**(5193): p. 2002-6.

[73] Turelli, P., et al., Cytoplasmic recruitment of INI1 and PML on incoming HIV preintegration complexes: interference with early steps of viral replication. *Mol Cell,* 2001. **7**(6): p. 1245-54.

[74] Miller, M.D. and F.D. Bushman, HIV integration. Ini1 for integration? *Curr Biol,* 1995. **5**(4): p. 368-70.
[75] Morozov, A., E. Yung, and G.V. Kalpana, Structure-function analysis of integrase interactor 1/hSNF5L1 reveals differential properties of two repeat motifs present in the highly conserved region. *Proc Natl Acad Sci* U S A, 1998. **95**(3): p. 1120-5.
[76] Farnet, C.M. and F.D. Bushman, HIV-1 cDNA integration: requirement of HMG I(Y) protein for function of preintegration complexes in vitro. *Cell,* 1997. **88**(4): p. 483-92.
[77] Li, L., et al., Retroviral cDNA integration: stimulation by HMG I family proteins. *J Virol,* 2000. **74**(23): p. 10965-74.
[78] Chen, H. and A. Engelman, The barrier-to-autointegration protein is a host factor for HIV type 1 integration. *Proc Natl Acad Sci* U S A, 1998. **95**(26): p. 15270-4.
[79] Lin, C.W. and A. Engelman, The barrier-to-autointegration factor is a component of functional human immunodeficiency virus type 1 preintegration complexes. *J Virol,* 2003. **77**(8): p. 5030-6.
[80] Mansharamani, M., et al., Barrier-to-autointegration factor BAF binds p55 Gag and matrix and is a host component of human immunodeficiency virus type 1 virions. *J Virol,* 2003. **77**(24): p. 13084-92.
[81] Violot, S., et al., The human polycomb group EED protein interacts with the integrase of human immunodeficiency virus type 1. *J Virol,* 2003. **77**(23): p. 12507-22.
[82] Denisenko, O.N. and K. Bomsztyk, The product of the murine homolog of the Drosophila extra sex combs gene displays transcriptional repressor activity. *Mol Cell Biol,* 1997. **17**(8): p. 4707-17.
[83] Ng, J., et al., Evolutionary conservation and predicted structure of the Drosophila extra sex combs repressor protein. *Mol Cell Biol,* 1997. **17**(11): p. 6663-72.
[84] Cherepanov, P., et al., HIV-1 integrase forms stable tetramers and associates with LEDGF/p75 protein in human cells. *J Biol Chem,* 2003. **278**(1): p. 372-81.
[85] Maertens, G., et al., LEDGF/p75 is essential for nuclear and chromosomal targeting of HIV-1 integrase in human cells. *J Biol Chem,* 2003. **278**(35): p. 33528-39.
[86] Dildine, S.L. and S.B. Sandmeyer, Integration of the yeast retrovirus-like element Ty3 upstream of a human tRNA gene expressed in yeast. *Gene,* 1997. **194**(2): p. 227-33.
[87] Dildine, S.L., et al., A chimeric Ty3/Moloney murine leukemia virus integrase protein is active in vivo. *J Virol,* 1998. **72**(5): p. 4297-307.
[88] Zhu, Y., et al., Controlling integration specificity of a yeast retrotransposon. *Proc Natl Acad Sci* U S A, 2003. **100**(10): p. 5891-5.
[89] Bushman, F.D., Tethering human immunodeficiency virus 1 integrase to a DNA site directs integration to nearby sequences. *Proc Natl Acad Sci* U S A, 1994. **91**(20): p. 9233-7.
[90] Goulaouic, H. and S.A. Chow, Directed integration of viral DNA mediated by fusion proteins consisting of human immunodeficiency virus type 1 integrase and Escherichia coli LexA protein. *J Virol,* 1996. **70**(1): p. 37-46.
[91] Katz, R.A., G. Merkel, and A.M. Skalka, Targeting of retroviral integrase by fusion to a heterologous DNA binding domain: in vitro activities and incorporation of a fusion protein into viral particles. *Virology,* 1996. **217**(1): p. 178-90.
[92] Bushman, F.D. and M.D. Miller, Tethering human immunodeficiency virus type 1 preintegration complexes to target DNA promotes integration at nearby sites. *J Virol,* 1997. **71**(1): p. 458-64.

[93] Holmes-Son, M.L. and S.A. Chow, Integrase-lexA fusion proteins incorporated into human immunodeficiency virus type 1 that contains a catalytically inactive integrase gene are functional to mediate integration. *J Virol,* 2000. **74**(24): p. 11548-56.
[94] Holmes-Son, M.L. and S.A. Chow, Correct integration mediated by integrase-LexA fusion proteins incorporated into HIV-1. *Mol Ther,* 2002. **5**(4): p. 360-70.
[95] Rice, P.A. and T.A. Baker, Comparative architecture of transposase and integrase complexes. *Nat Struct Biol,* 2001. **8**(5): p. 302-7.
[96] Kuhn, R. and R.M. Torres, Cre/loxP recombination system and gene targeting. *Methods Mol Biol,* 2002. **180**: p. 175-204.
[97] Thorpe, H.M. and M.C. Smith, In vitro site-specific integration of bacteriophage DNA catalyzed by a recombinase of the resolvase/invertase family. *Proc Natl Acad Sci* U S A, 1998. **95**(10): p. 5505-10.
[98] Groth, A.C., et al., A phage integrase directs efficient site-specific integration in human cells. *Proc Natl Acad Sci* U S A, 2000. **97**(11): p. 5995-6000.
[99] Thyagarajan, B., et al., Site-specific genomic integration in mammalian cells mediated by phage phiC31 integrase. *Mol Cell Biol,* 2001. **21**(12): p. 3926-34.
[100] Groth, A.C. and M.P. Calos, Phage integrases: biology and applications. *J Mol Biol,* 2004. **335**(3): p. 667-78.
[101] Leclercq, I., et al., Basis of HTLV type 1 target site selection. *AIDS Res Hum Retroviruses,* 2000. **16**(16): p. 1653-9.
[102] Yu, S.F., J. Stone, and M.L. Linial, Productive persistent infection of hematopoietic cells by human foamy virus. *J Virol,* 1996. **70**(2): p. 1250-4.
[103] Lecellier, C.H., et al., Intra- and intercellular trafficking of the foamy virus auxiliary bet protein. *J Virol,* 2002. **76**(7): p. 3388-94.
[104] Neves, M., J. Peries, and A. Saib, Study of human foamy virus proviral integration in chronically infected murine cells. *Res Virol,* 1998. **149**(6): p. 393-401.
[105] Bushman, F.D., Targeting survival: integration site selection by retroviruses and LTR-retrotransposons. *Cell,* 2003. **115**(2): p. 135-8.
[106] Curcio, M.J. and D.J. Garfinkel, New lines of host defense: inhibition of Ty1 retrotransposition by Fus3p and NER/TFIIH. *Trends Genet,* 1999. **15**(2): p. 43-5.
[107] Griffith, J.L., et al., Functional genomics reveals relationships between the retrovirus-like Ty1 element and its host Saccharomyces cerevisiae. *Genetics,* 2003. **164**(3): p. 867-79.
[108] Bieniasz, P.D., Restriction factors: a defense against retroviral infection. *Trends Microbiol,* 2003. **11**(6): p. 286-91.
[109] Wenjie Tan, Kai Zhu, Danid J. Segal, Carlos F. Barbas II, and Samson A. Chow., Fusion proteins consisting of human immunodeficiency virus type 1 integrase and the designed polydactyl zinc finger protein E2C direct integration of viral DNA into specific sites. , *J Virol,* 2003. **78**(3): p. 1301-13.

In: Progress in Virus Research
Editor: Sabine Thebault, pp. 317-324

ISBN 1-59454-438-7

Chapter XI

Entry Inhibitors, the Emerging Family of Anti-HIV Drugs

Christian Callebaut*

Gladstone Institute of Virology and Immunology,
University of California San Francisco, San Francisco, CA

Abstract

Highly active antiretroviral therapy (HAART) based on combinations of drugs that target HIV enzymes has considerably reduced morbidity and mortality from HIV infection. However, despite a large number of antiretroviral drugs approved for the treatment of HIV infection, current drug regimens present challenges, urging for the development of new drugs. Candidate drugs with novel mechanisms of action are under investigation, such as the HIV entry inhibitors. This family of drugs targets the three steps of HIV entry: receptor binding (attachment), co-receptor binding and membrane fusion. Entry inhibitors currently in development are presented in this review. Preliminary results are giving some insight on how these promising family of molecules may be used in future anti-HIV multi-therapies.

Keywords: HIV, receptor, co-receptor, CD4, CXCR4, CCR5, envelope, antiretroviral drugs, entry inhibitors.

Introduction

Among the 20 antiretroviral drugs approved for the treatment of HIV infection (Table 1), many have improved over the last few years, and new drug formulations allow once-daily

* Correspondence should be addressed to Christian Callebaut: Laboratory of Molecular Virology, Gladstone Institute of Virology and Immunology, University of California San Francisco, PO Box 419100, San Francisco, CA 94141-9100. Tel: +1-415-752-2407; Email: ccallebaut@gladstone.ucsf.edu

dosing. Treatment guidelines highly recommend multi-drug combinations of these drugs and their general use has resulted in a significant decrease in HIV mortality. However, current regimens continue to present challenges with issues of tolerability, toxicity, adherence, and drug-resistance, including the emergence of HIV strains resistant to whole drug classes, urging for the development of anti-HIV drugs with new targets. Future progress in treating HIV infection will result from the development of investigational drugs both in existing classes, and in emerging classes, like the entry inhibitors. This new family of entry inhibitors represents almost half of the anti-HIV molecules currently in development.

Table 1. Currently approved anti-HIV drugs (April 2004)

Nucleoside Reverse Transcriptase Inhibitors (NRTI)
zidovudine (ZDV, AZT)
didanosine (ddI)
zalcitabine (ddC)
stavudine (d4T)
lamivudine (3TC)
abacavir (ABC)
emtricitabine (FTC)
tenofovir (TDF) (nucleotide)
Non-Nucleoside Reverse Transcriptase Inhibitors (NNRTI)
nevirapine (NVP)
delavirdine (DLV)
efavirenz (EFV)
Protease Inhibitors (PI)
saquinavir (SQV)
ritonavir (RTV)
indinavir (IDV)
nelfinavir (NFV)
amprenavir (APV)
lopinavir/ritonavir (LPV/r)
atazanavir (ATZ)
fosamprenavir (FPV)
Entry Inhibitors
Enfuvirtide (T-20)

HIV isolates have been initially classified as either macrophage tropic (MT) and non-syncytium-inducing (NSI), or T cell tropic (TT) and syncytium-inducing (SI). It was recently shown that entry of MT/NSI HIV-1 into susceptible cells requires CC-chemokine Receptor 5 (CCR5), a receptor for the CC-chemokines RANTES, MIP-1a, and MIP-1b, whereas entry of TT/SI viruses requires CXC-chemokine Receptor 4 (CXCR4), the receptor for the CXC-chemokine SDF-1. These results have led to the current nomenclature R5 and X4 to denote respective usage of CCR5 or CXCR4 by HIV strains. However, these two HIV types are not equals, as individuals infected with R5 and X4 HIV strains typically transmit only the R5 viruses to recipients. Emergence of X4 viruses is usually very slow, and occurs in only in 50% of the recipients (after 5 to 10 years). Only few mutations in the viral envelope are sufficient to “switch” a virus from R5 to X4. Additionally, HIV strains with dual–tropism (X4R5) may appear during this transition period. These viral strains have the capacity to use either CCR5 or CXCR4 to infect cells. Finally, the emergence of X4 HIV-1 strains correlates with rapidly progressive immunodeficiency and clinical deterioration in patients.

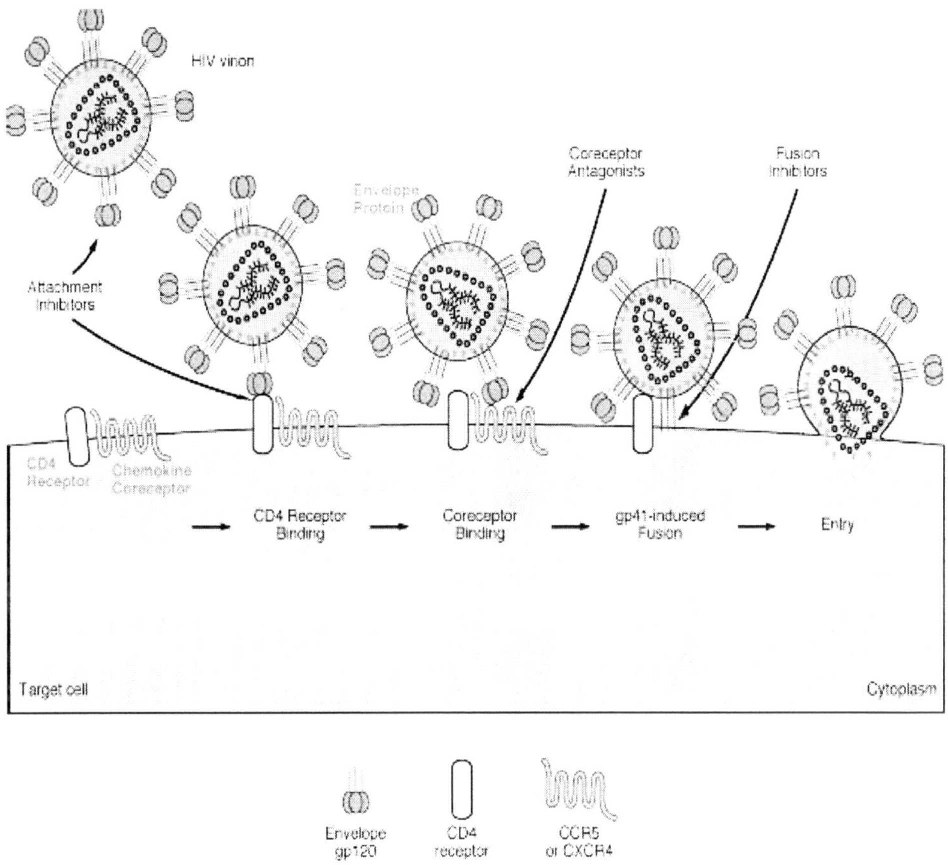

Figure 1.

Although recently characterized, HIV entry became very attractive, as CCR5 and CXCR4 belong to a family of molecules (GPCR) that were successfully targeted for many diseases. HIV entry is mainly composed of three steps: (1) binding of the viral envelope (gp120) to the CD4 receptor; (2) interaction of the gp120 with a co-receptor (chemokine receptor CCR5 or CXCR4); and (3) fusion between the virus particle and the cell membrane (triggered by the viral envelope gp41). HIV entry inhibitors currently in development target each of these steps. Four types of inhibitors are in development to block entry (see Table 2), targeting the binding of the viral envelope gp120 to CD4, or the subsequent binding to one of the chemokine receptor (CCR5 or CXCR4), or at last the fusion between the two membrane induced by the HIV envelope protein (Figure 1).

Table 2 Investigational HIV entry inhibitors (April 2004, non exhaustive)

Name	Step targeted	Type/Class	Dev. Stage	Company	CROI Feb 2004 Abstract #
BMS-488043	CD4 binding	gp120 ligand	Phase II	Bristol-Myers-Squibb	141, 534, 535
TNX-355	CD4 binding	anti-CD4 mAb	Phase I	Tanox/Biogen	536
PRO-542	CD4 binding	anti-CD4 mAb	Phase I	Progenics	537
PRO-140	coreceptor binding	anti-CCR5 mAb	Preclinical	Progenics	537
SCH-D	coreceptor binding	CCR5 antagonist	Phase II	Schering-Plough	140LB
UK 427,857	coreceptor binding	CCR5 antagonist	Phase II	Pfizer	538
GSK-873140	coreceptor binding	CCR5 antagonist	Phase I	GSK	139, 540
AMD-887	coreceptor binding	CCR5 antagonist	Preclinical	AnorMed	539
AMD-070	coreceptor binding	CXCR4 antagonist	Phase I	AnorMed	539
KRH-2731	coreceptor binding	CXCR4 antagonist	Preclinical	Kureha	541
T-1249	membrane fusion	HR peptide (gp41)	stopped	Trimeris/Roche	

CD4 Binding Inhibitors

BMS-488043 is an orally available HIV attachment inhibitor, which binds specifically to the viral envelope (gp120) and blocks the attachment of the gp120 to the CD4 receptor. Several analyses showed that BMS-488043 binds directly to gp120, with a 1:1 stochiometry. The binding was reversible, with a binding constant of 29 nM. In vitro susceptibility assays indicate that BMS-488043 is effective against both X4 and R5 HIV-1 laboratory strains, with potent activity against clade B clinical isolates (EC50 of 37 nM). Interestingly, the resistance profile showed the emergence of resistant viruses with amino acid substitutions predominantly located within the envelope protein gp120. Mutations M426L or M475I, which are located close to the CD4 binding site, conferred higher levels of resistance. A mutation at the CD4 contact site, W427V, which eliminated CD4/gp120 binding, also eliminated the binding of BMS-488043 to gp120. Mutation S375W that stabilizes gp120 binding to CD4 had a reduced BMS-488043 binding. BMS-488043 has a long half-life and a good safety profile in non-infected volunteers. A dose of 800 mg bid is expected to provide adequate plasma concentrations for suppression of subtype-B HIV-1 isolates. This is novel class of attachment inhibitors that target gp120 binding to CD4. (535)

TNX-355 is an IgG4 monoclonal antibody that targets domain 2 of the CD4 receptor. TNX-355 showed potent anti-HIV-1 activity in vitro and in a phase 1a single-dose study in HAART-experienced subjects. Multiple-dose TNX-355 added as a single new agent, showed transient but clinically significant reductions (0.5-1.7 log 10) in HIV RNA in 21 of 22 subjects. Return towards baseline in viral load occurred with reduced susceptibility to TNX-

355. Complete and continuous CD4+ cell coating was observed for a minimum of 14 days after the final dose was administered in each study arm. TNX-355 was overall well tolerated. A phase II study of this agent in combination with optimized background therapy in antiretroviral-experienced patients is planned. (536)

PRO-542 is an antibody-like fusion protein of CD4 and IgG2. The compound binds HIV envelope (gp120) and showed activity against primary viral isolates in vitro, and in animal models. In a single dose phase I study, HIV-infected adults received one intravenous infusion of PRO-542. A dose-dependent HIV RNA reduction was observed. In another study of single and multiple intravenous doses of PRO-542 in 18 HIV-infected subjects, HIV RNA decreases of > 0.7 log10 cpm. Further studies are in progress.

CCR5 Antagonists

PRO-140 is an humanized anti-CCR5 monoclonal antibody. PRO140 recently tested in the hu-PBMC-SCID mouse model of HIV-1 infection. SCID-mice were reconstituted with PBMCs from a CCR5 wild-type donor and infected with a primary HIV-1 R5 isolate. Administration of PRO-140 led to control of viral replication upon repeat dosing. Viral replication rebounded due to the emergence of escape mutants. One variant had switched to CXCR4 usage. Sequencing of the envelope gene revealed two point mutations and one 4 amino acid deletion in the V3 loop of gp120. Analysis of mutations demonstrated that the escape variant had switched to X4 via intermediate dual tropic R5X4 viruses. Additionally, the final X4 variant displayed a remarkably high sensitivity to antibodies, patient sera and viral entry inhibitors in vitro, which was similarly mapped to the V3 loop mutations. The findings underscore the potential of anti-CCR5 therapy but also suggest that coreceptor usage should be closely monitored. Further studies are in progress.

SCH-D is a CCR5 receptor antagonist that has recently replaced SCH-C. SCH-D was proved to be more potent *in vitro* than SCH C with longer half-life and higher bioavailability. Results showed decreased plasma viremia for more than 1 log10 for all doses tested. The drug was well tolerated. One patient with plasma HIV- 1 RNA reduction had evidence of a transient switch to X4 virus after treatment.

UK-427,857 is a new CCR5 antagonist in development. Recent trial showed that when UK-427,857 was administered for 11 days, and plasma viremia was reduced with higher CD4 count, except for one patient. This unique patient was harboring distinct R5, X4, and dual-tropic variants when inadvertently enrolled in a phase IIa, 10-day mono-therapy study. The patient experienced no drop in plasma HIV-1 RNA despite measurable drug exposure and ex vivo UK-427,857 CCR5 receptor occupancy within the anticipated range. Clonal sampling showed a shift from a mixture of viruses using R5 and R5X4 co-receptors at baseline to a mixture of viruses using X4 and R5X4 at day 11. Reversion to R5 predominance was observed after UK-427,857 was stopped.

GSK-873140 is a new orally bioavailable CCR5 receptor antagonist. Interestingly, this antagonist binds CCR5 at a different region than other CCR5 antagonist, and therefore may be synergistic with other CCR5 antagonist. Preliminary data indicated that this compound was well tolerated with no serious adverse events. CCR5 receptor occupancy depended on the drug dose and the sampling time. Receptor-binding analyses 24 hours after a single dose showed median receptor occupancy ranging from 68% to 88%, despite compound concentrations below or near assay limits. At 2 and 12 hours after multiple dosing with GSK-873140, the CCR5 occupancy was greater than 97%. These findings suggest a long half-life for the compound–receptor binding, with once daily or bid dosing schedules. Studies of GSK-873140 in HIV infected individuals are planned.

AMD887, a novel CCR5 inhibitor developed by AnorMED, was evaluated for its in vitro anti-HIV activity. AMD887 was examined for its activity against a wide variety of HIV-1 laboratory strains, primary clinical isolates in different T-cell lines. AMD887 was found to act as a potent CCR5 antagonist that strongly inhibited virus infectivity at a 50% effective concentration (EC 50) of 1 to 10 nM. The compound inhibited different clades of R5 HIV replication in cells expressing CCR5 and PBMC. AMD887 had no activity against X4 or R5/X4 HIV-1 isolates in PBMC. Its anti-HIV potency correlated closely with its potency in inhibiting chemokine binding, chemokine-induced internalization, chemokine-induced Ca 2+ signaling and chemotaxis. AMD887 did not interact with any other chemokine receptor. Further studies are in progress.

CXCR4 Antagonists

AMD070 is a orally bioavailable CXCR4 antagonist that supplanted AMD3100 in its development. Recent results showed the antiviral efficacy of AMD-070 in phase 2 studies. Additionally, the in vitro anti-HIV activity of AMD070 was evaluated. AMD070 was examined for its activity against a wide variety of HIV-1 laboratory strains (R5, X4, R5/X4), primary clinical isolates in different T-cell lines, cells expressing CD4 and CXCR4 or CCR5 and PBMCs. In the presence of AMD070, X4 HIV variants were prevented from infecting PBMCs. When both compounds, AMD887 and AMD070 were combined, no viral replication was measured in PBMCs. Further studies are in progress.

KRH-2731 is a new CXCR4 inhibitor who replaced the former KRH-1636 in development. KRH-2731, is an orally bioavailable. KRH-2731 was orally administered to hu-PBL-SCID mice and tested for its in vivo activity. KRH-2731 efficiently inhibited the replication of all X4 and R5X4 HIV-1 strains tested (EC 50 of 1.0 to 4.2 nM), but did not inhibit that of R5 HIV-1 strains in activated PBMC. KRH-2731 specifically inhibited SDF-1 binding to CXCR4 and SDF-1-induced Ca2+ mobilization. The binding site of KRH-2731 was located in the second and/or third extracellular loops of CXCR4. KRH-2731 successfully suppressed X4 HIV-1 replication in hu-PBL-SCID mice when administered orally at a dose of 10 mg/kg. KRH-2731 has an interesting anti-X4 HIV-1 activity in both in vitro and in vivo. Further studies are in progress.

Fusion Inhibitors

Enfuvirtide (T-20) is the only FDA approved entry inhibitor. It is a peptide inhibitor of HIV fusion that must be given parenterally. The compound binds to a portion of the HIV envelope gp41 molecule and inhibits fusion of the viral and cell membranes. Enfuvirtide is a synthetic 36 residues peptide one of the heptad repeat (HR) of gp41. Accordingly, resistance to T20 is associated with changes in the HR1 domain. Mutations in HR1 were noted in all cases, including the previously unreported changes N42Q/H and N43Q. Four patients demonstrated the loss of both HR1 and HR2 mutations following cessation of Enfuvirtide therapy. The development T-1249, of a second-generation molecule, was recently stopped

Conclusion

Despite dramatic improvements in treatment of HIV-infected patients directly resulting from antiretroviral therapy, better anti-HIV drugs are needed. In addition to innovations that continue to improve the currently available antiretroviral drugs, a number of promising compounds are under investigation. Newer agents in the entry inhibitors family have the potential to enhance activity against drug-resistant virus by targeting the receptor and co-receptor used by the virus to enter the cell. However, a concern in developing CCR5 co-receptor antagonists as a novel class of HIV therapeutics is the possible selection and emergence of CXCR4 using variants during treatment. Since one patient showed a shift from R5 to X4 in a clinical study to test the efficacy of a new CCR5 antagonist, this at least strongly suggests that co-receptor usage should be closely monitored before drug treatment. Another example of this switch was observed in a humanized-animal model, suggesting that this switch might be a general mechanism of virus escape. As X4 HIV has been associated with accelerated progression of HIV disease, therapies directed to CCR5 may need to be directed simultaneously to CXCR4 to avoid such viral escape.

Acknowledgements

I would like to thank all my colleagues at the Gladstone Institutes and UCSF for stimulating discussions. This review was partially based on work presented at the 11th Conference on Retroviruses and Opportunistic Infections Feb 8-11th, 2004, and the UCSF-GIVI CfAR meeting (Biology, Immunology, and Clinical Implications of CCR5 Inhibitors in HIV Infection), hold on Apr 19th, 2004.

References

Abstract numbers from The *11th Conference on Retroviruses and Opportunistic Infections*, Feb 8-11 2004. See Table 2 and text.

139 - Demarest et al. Single and Multiple Dose Escalation Study to Investigate the Safety, Pharmacokinetics, and Receptor Binding of GW873140, a Novel CCR5 Receptor Antagonist, in Healthy Subjects.

140LB - Schurmann et al.

SCH D: Antiviral Activity of a CCR5 Receptor Antagonist.

141 - Hanna et al. Antiviral Activity, Safety, and Tolerability of a Novel, Oral Small-molecule HIV-1 Attachment Inhibitor, BMS-488043, in HIV-1-infected Subjects a

Novel, Oral Small-Molecule HIV-1 Attachment Inhibitor, BMS-488043, in HIV-1-Infected Subjects.

534 - Lin et al. Characterization of a Small Molecule HIV-1 Attachment Inhibitor BMS-488043: Virology, Resistance and Mechanism of Action.

535 - Hanna et al. Safety, Tolerability, and Pharmacokinetics of a Novel, Small-Molecule HIV-1 Attachment Inhibitor, BMS-488043, after Single and Multiple Oral Doses in Healthy Subjects.

536 - Jacobson et al. Phase 1b Study of the Anti-CD4 Monoclonal Antibody TNX-355 in HIV-1-infected Subjects: Safety and Antiretroviral Activity of Multiple Doses.

537 - Franti et al. Control of HIV-1 Replication in the hu-PBL-SCID Mouse Model by an Anti-CCR5 Monoclonal Antibody.

538 - Westby et al. Reversible Predominance of CXCR4 Utilising Variants in a Non-Responsive Dual Tropic Patient Receiving the CCR5 Antagonist UK-427,857.

539 - Schols et al. in vitro Anti-HIV Activity Profile of AMD887, a Novel CCR5 Antagonist, in Combination with the CXCR4 Inhibitor AMD070.

540 - Maeda et al. Determination of Binding Sites of a Unique CCR5 Inhibitor AK602 on Human CCR5.

541 - Murakami et al. KRH-2731: An Orally Bioavailable CXCR4 Antagonist Is a Potent Inhibitor of HIV-1 Infection.

Index

A

B

C

D

E

F

G

H

I

N

O

P

R

S